Infinite dimensional Lie algebras

Third edition

implies $a_{ji} = 0$. The associated *Kac–Moody algebra* $\mathfrak{g}'(A)$ is a complex Lie algebra on $3n$ generators e_i, f_i, h_i ($i = 1, \ldots, n$) and the following defining relations ($i, j, = 1, \ldots, n$):

$$(0.3.1) \qquad \begin{cases} [h_i, h_j] = 0, \quad [e_i, f_i] = h_i, \quad [e_i, f_j] = 0 \text{ if } i \neq j, \\ [h_i, e_j] = a_{ij} e_j, \quad [h_i, f_j] = -a_{ij} f_j, \\ (\operatorname{ad} e_i)^{1-a_{ij}} e_j = 0, \quad (\operatorname{ad} f_i)^{1-a_{ij}} f_j = 0 \text{ if } i \neq j. \end{cases}$$

(The definition given in the main text of the book (see Chapter 1) is different from the above; it is more convenient for a number of reasons. The proof of the fact that the derived algebra of the Lie algebra $\mathfrak{g}(A)$ defined in Chapter 1 coincides with the Lie algebra $\mathfrak{g}'(A)$ defined by relations (0.3.1) has been obtained by Gabber–Kac [1981] under a "symmetrizability" assumption; this proof appears in Chapter 9.)

I came to consider these Lie algebras while trying to understand and generalize the works of Guillemin–Quillen–Singer–Sternberg–Weisfeiler on Cartan's classification. The key idea was to consider arbitrary simple $\mathbb{Z}$-graded Lie algebras $\mathfrak{g} = \bigoplus_j \mathfrak{g}_j$; but since there are too many such Lie algebras, the point was to require the dimension of $\mathfrak{g}_j$ to grow no faster than some polynomial in j. (One can show that Lie algebras of finite depth do satisfy this condition, and that this condition is independent of the gradation.) Such Lie algebras were classified under some technical hypotheses (see Kac [1968 B]). It turned out that in addition to Cartan's four series of Lie algebras of polynomial vector fields, there is another class of infinite-dimensional Lie algebras of polynomial growth, which are called affine Lie algebras (more precisely, they are the quotients of affine Lie algebras by the 1-dimensional center). At the same time, Moody [1968] independently undertook the study of the Lie algebras $\mathfrak{g}'(A)$.

The class of Kac–Moody algebras breaks up into three subclasses. To describe them, it is convenient to assume that the matrix A is *indecomposable* (i.e., there is no partition of the set $\{1, \ldots, n\}$ into two nonempty subsets so that $a_{ij} = 0$ whenever i belongs to the first subset, while j belongs to the second; this is done without loss of generality since the direct sum of matrices corresponds to the direct sum of Kac–Moody algebras). Then there are the following three mutually exclusive possibilities:

a) There is a vector θ of positive integers such that all the coordinates of the vector $A\theta$ are positive. In such case all the principal minors of the matrix A are positive and the Lie algebra $\mathfrak{g}'(A)$ is finite-dimensional.

b) There is a vector δ of positive integers such that $A\delta = 0$. In such case all the principal minors of the matrix A are nonnegative and $\det A = 0$; the algebra $\mathfrak{g}'(A)$ is infinite-dimensional, but is of polynomial growth (moreover, it admits a $\mathbb{Z}$-gradation by subspaces of uniformly bounded dimension). The Lie algebras of this subclass are called *affine Lie algebras*.

c) There is a vector α of positive integers such that all the coordinates of the vector $A\alpha$ are negative. In such case the Lie algebra $\mathfrak{g}'(A)$ is of exponential growth.

The main achievement of the Killing–Cartan theory may be formulated as follows: a simple finite-dimensional complex Lie algebra is isomorphic to one of the Lie algebras of the subclass a). (Note that the classification of matrices of type a) and b) is a rather simple problem.) The existence of the generators satisfying relations (0.3.1) was pointed out by Chevalley [1948] and Harish–Chandra [1951]. (Much later Serre [1966] and Kac [1968 B] showed that these are defining relations.)

It turned out that most of the classical concepts of the Killing–Cartan–Weyl theory can be carried over to the entire class of Kac–Moody algebras, such as the Cartan subalgebra, the root system, the Weyl group, etc. In doing so one discovers a series of new phenomena, which the book treats in detail (see Chapters 1–6). I shall only point out here that $\mathfrak{g}'(A)$ does not always possess a nonzero invariant bilinear form. This is the case if and only if the matrix A is *symmetrizable*, i.e., the matrix DA is symmetric for some invertible diagonal matrix D (see Chapter 2).

§0.4. It is an important property of affine Lie algebras that they possess a simple realization (see Chapters 7 and 8). Here I shall explain this realization for the example of the Kac–Moody algebra associated to the extended Cartan matrix A of a simple finite-dimensional complex Lie algebra $\mathfrak{g}$. (All such matrices are "affine" generalized Cartan matrices; the corresponding algebra $\mathfrak{g}'(A)$ is called a nontwisted affine Lie algebra.) Namely, the affine Lie algebra $\mathfrak{g}'(A)$ is a central extension by the 1-dimensional center of the Lie algebra of polynomial maps of the circle into the simple finite-dimensional complex Lie algebra $\mathfrak{g}$ (so that it is the simplest example of a Lie algebra of the second class mentioned in §0.2).

More precisely, let us consider the Lie algebra $\mathfrak{g}$ in some faithful finite-dimensional representation. Then the Lie algebra $\mathfrak{g}'(A)$ is isomorphic to the Lie algebra on the complex space $(\mathbb{C}[t,t^{-1}]\otimes_{\mathbb{C}}\mathfrak{g})\oplus\mathbb{C}c$ with the bracket

$$[(t^m\otimes a)\oplus\lambda c,(t^n\otimes b)\oplus\mu c]=(t^{m+n}\otimes[a,b])\oplus m\delta_{m,-n}(\operatorname{tr} ab)c,$$

so that $\mathbb{C}c$ is the (1-dimensional) center. This realization allows us to study affine Lie algebras from another point of view. In particular, the algebra of vector fields on the circle (the simplest algebra of the first class) plays an important role in the theory of affine Lie algebras.

Note also that the Lie algebras of the fourth class are closely related to the affine Lie algebras of infinite rank, considered in Chapters 7 and 14.

Unfortunately, no simple realization has been found up to now for any nonaffine infinite-dimensional Kac–Moody algebra. This question appears to be one of the most important open problems of the theory.

§0.5. An important concept missing from the first works in Kac–Moody algebras was the concept of an integrable highest-weight representation (introduced in Kac [1974]). Given a sequence of nonnegative integers $\Lambda = (\lambda_1, \ldots, \lambda_n)$, the ***integrable highest-weight representation*** of a Kac–Moody algebra $\mathfrak{g}'(A)$ is an irreducible representation π_Λ of $\mathfrak{g}'(A)$ on a complex vector space $L(\Lambda)$, which is determined by the property that there is a nonzero vector $v_\Lambda \in L(\Lambda)$ such that

$$\pi_\Lambda(e_i)v_\Lambda = 0 \text{ and } \pi_\Lambda(h_i)v_\Lambda = \lambda_i v_\Lambda \quad (i = 1, \ldots, n).$$

(This terminology is explained by the fact that Λ is called the highest-weight, and the conditions on Λ are necessary and sufficient for being able to integrate π_Λ and obtain a representation of the group.)

Cartan's theorem on the highest-weight asserts that all the representations π_Λ of a complex simple finite-dimensional Lie algebra are finite-dimensional, and that every finite-dimensional irreducible representation is equivalent to one of the π_Λ.

That the representations π_Λ are finite-dimensional (the most nontrivial part of Cartan's theorem) was proved by Cartan by examining the cases, one by one. A purely algebraic proof was found much later by C. Chevalley [1948] and Harish-Chandra [1951] (a "transcendental" proof had been found earlier by H. Weyl). This brief note by Chevalley appears in retrospect as the precursor of the algebraization of the representation theory of Lie groups. This note also contains, in an embryonic form, many of the basic concepts of the theory of Kac–Moody algebras.

The algebraization of the representation theory of Lie groups, which has undergone such an explosive development during the last decade, started with the work Bernstein–Gelfand–Gelfand [1971] on Verma modules (the first nontrivial results about these modules were obtained by Verma [1968]). In particular, using the Verma modules, Bernstein–Gelfand–Gelfand gave a transparent algebraic proof of Weyl's formula for the characters of finite-dimensional irreducible representations of finite-dimensional simple Lie algebras.

At about the same time Macdonald [1972] obtained his remarkable identities. In this work he undertook to generalize the Weyl denominator identity to the case of affine root systems. He remarked that a straightforward generalization is actually false. To salvage the situation he had to add some "mysterious" factors, which he was able to determine as a result of lengthy calculations. The simplest example of Macdonald's identities is the famous Jacobi triple product identity:

$$\prod_{n\geq 1}(1-u^n v^n)(1-u^{n-1}v^n)(1-u^n v^{n-1})$$
$$= \sum_{m\in\mathbb{Z}}(-1)^m u^{\frac{1}{2}m(m+1)} v^{\frac{1}{2}m(m-1)}.$$

The "mysterious" factors which do not correspond to affine roots are the factors $(1-u^n v^n)$.

After the appearance of the two works mentioned above very little remained to be done: one had to place them on the desk next to one another to understand that Macdonald's result is only the tip of the iceberg—the representation theory of Kac–Moody algebras. Namely, it turned out that a simplified version of Bernstein–Gelfand–Gelfand's proof may be applied to the proof of a formula generalizing Weyl's formula, for the formal character of the representation π_Λ of an arbitrary Kac–Moody algebra $\mathfrak{g}'(A)$ corresponding to a symmetrizable matrix A. In the case of the simplest 1-dimensional representation π_0, this formula becomes the generalization of Weyl's denominator identity. In the case of an affine Lie algebra, the generalized Weyl denominator identity turns out to be equivalent to the Macdonald identities. In the process, the "mysterious" factors receive a simple interpretation: they correspond to the so-called imaginary roots (i.e., roots that one should add to the affine roots to obtain all the roots of the affine Lie algebra). Note that the simplest example of the Jacobi triple product identity turns out to be just the generalized denominator identity for the affine Lie algebra corresponding to the matrix $\begin{pmatrix} 2 & -2 \\ -2 & 2 \end{pmatrix}$.

The exposition of these results (obtained by Kac [1974]) may be found in Chapter 10. Chapters 9–14 are devoted to the general theory of highest-weight representations and their applications.

The main tool of the theory of representations with highest-weight is the generalized Casimir operator (see Chapter 2). Unfortunately, the construction of this operator depends on whether the matrix A is symmetrizable. The question whether one can lift the hypothesis of symmetrizability of the matrix A remains open.

Once the integrable highest-weight representations had been introduced, the theory of Kac–Moody algebras got off the ground and has been developing since at an accelerating speed. In the past decade this theory has emerged as a field that has close connections with many areas of mathematics and mathematical physics, such as invariant theory, combinatorics, topology, the theory of modular forms and theta functions, the theory of singularities, finite simple groups, Hamiltonian mechanics, soliton equations, and quantum field theory.

§0.6. This book contains a detailed exposition of the foundations of the theory of Kac–Moody algebras and their integrable representations. Besides the application to the Macdonald identities mentioned above (Chapter 12), the book discusses the application to the classification of finite-order automorphisms of simple finite-dimensional Lie algebras (Chapter 8), and the connection with the theory of modular forms and theta functions (Chapter 13). The last chapter (Chapter 14) discusses the remarkable connection between the representation theory of affine Lie algebras and the Korteweg–de Vries-type equations, discovered by the Kyoto school.

A theory of Lie algebras is usually interesting, insofar as it is related to group theory, and Kac–Moody algebras are no exception. Recently there appeared a series of deep results on groups associated with Kac–Moody algebras. A discussion of these results would require writing another book. I chose to make only a few comments regarding this subject at the end of some chapters.

§0.7. Throughout the book the base field is the field of complex numbers. However, all the results of the book, except, of course, for the ones concerning Hermitian forms and convergence problems, can be extended without difficulty to the case of an arbitrary field of characteristic zero.

§0.8. Motivations are provided at the beginning of each chapter, which ends with related bibliographical comments. The main text of each chapter is followed by exercises (whose total number exceeds 250). Some of them are elementary, others constitute a brief exposition of original works. I hope that these expositions are sufficiently detailed for the diligent reader to reconstruct all the proofs. The square brackets at the end of some exercises contain hints for their solution.

The exposition in the book is practically self-contained. Although I had in mind a reader familiar with the theory of finite-dimensional semisimple Lie algebras, what would suffice for the most part is a knowledge of the elements of Lie algebras, their enveloping algebras and representations. For example, the book of Humphreys [1972] or Varadarajan [1984] is more than sufficient.

One finds a rather extensive bibliography at the end of the book. I hope that the collection of references to mathematical works in the theory of Kac–Moody algebras is at least everywhere dense. This is not at all so in the case of the works in physics. The choice of references in this case was rather arbitrary and often depended on whether I had a copy of the paper or discussed it with the author. The same should be said as regards the references to the works on the other classes of infinite-dimensional Lie algebras.

§0.9. This book is based on lectures given at MIT in 1978, 1980, and 1982, and at the Collège de France in 1981. I would like to thank those who attended for helpful comments and corrections of the notes, in particular F. Arnold, R. Coley, R. Gross, Z. Haddad, M. Haiman, G. Heckman, F. Levstein, A. Rocha, and T. Vongiouklis. I am grateful to M. Duflo, G. Heckman, B. Kupershmidt, and B. Weisfeiler for reading some parts of the manuscript and pointing out errors. I apologize for those errors that remain. My thanks go to F. Rose, B. Katz, and M. Katz without whose help and support this book would never have come out. I also owe thanks to K. Manning and C. Macpherson for help with the language. The book was prepared using D. Knuth's TEX. Finally, I would like on this occasion to express my deep gratitude to D. Peterson, whose collaboration had a great influence not only on this book, but also on most of my mathematical work in the past few years.

The author was supported in part by a Sloan foundation grant and by grants from the National Science Foundation.

July 1983, L'Isle Adam, France.

Preface to the Second Edition

The most important additions reflect recent developments in the theory of infinite-dimensional groups (some key facts, like Proposition 3.8 and Exercise 5.19 are among them) and in the soliton theory (like Exercises 14.37–14.40 which uncover the role of the Virasoro algebra). The most important correction concerns the proof of the complete reducibility Proposition 9.10. The previous proof used Lemma 9.10 b) of the first edition which is false, as Exercise 9.15 shows. A correct version of Lemma 9.10 b) is the new Proposition 10.4 which gives a characterization of integrable highest-weight modules.

In addition to correcting misprints and errors and adding a few dozen of new exercises, I have brought to date the list of references and related bibliographical comments. I want to thank those who have pointed out errors and suggested improvements, in particular: J. Dorfmeister, T. Enright, D. Freed, E. Getzler, E. Gutkin, P. de la Harpe, S. Kumar, B. Kupershmidt, S.-R. Lu, D. Peterson, L.-J. Santharoubane, G. Schwarz, V S. Varadarajan, M. Wakimoto, Z.-X. Wan, X.-D. Wang, Y.-X. Wang, B. Weisfeiler, C.-F. Xie, Y.-C. You, H.-C. Zhang.

April 1985, Cambridge, Massachusetts.

Preface to the Third Edition

This edition differs considerably from the previous ones. Particularly, more emphasis is made on connections to mathematical physics, especially to conformal field theory.

Below is a list of the most important improvements and additions:

Chapter 3. A simplest example of a quantized Kac–Moody algebra, $U_q(sl_2)$, is given, along with its representations (Exercises 3.23 and 3.34).

Chapter 5. The hyperbolic Weyl group theory is applied to the study of the unimodular Lorentzian lattices of rank ≤ 10 (§5.10).

Chapter 6. An explicit construction of all finite type root and coroot lattices is given, along with the associated Weyl group, root systems, etc. (§6.7).

Chapter 7. The field theoretic approach to affine algebras is briefly outlined (§7.7). An explicit construction of all simple finite-dimensional Lie algebras is given in terms of the root lattice and an "asymmetry function" on it (§§7.8–7.10).

Chapter 8. A simple and self-contained proof is given of the basic fact about twisted affine algebras: the equivariant loop algebra $\mathcal{L}(\mathfrak{g}, \sigma, m)$ depends only on the connected component of Aut $\mathfrak{g}$ containing σ (§8.5).

Chapter 9. Elements of the representation theory of the Virasoro algebras are discussed (§9.14). A free field construction of representations of the Virasoro algebra and the affine algebra of type $A_1^{(1)}$ is given (Exercises 9.17–9.20).

Chapter 11. Unitarizability of representations of the Virasoro algebra is discussed (§11.12). A theory of generalized Kac–Moody algebras is outlined (§11.13).

Chapter 12. The Sugawara construction and the coset construction, which are the basic constructions of conformal field theory, are explained. The general branching functions and vacuum pairs are introduced in this context (§§12.8–12.13).

Chapter 13. General branching functions are studied along with string functions. The matrix S of the modular transformations of characters is studied (this was implicit in earlier editions). Explicit estimates of the orders of all poles and of levels of branching functions are given. Asymptotics of characters and branching functions at high temperature limit is studied, along with the related positivity conjecture. The interplay between the modular and conformal invariance constraints is demonstrated (§§13.8–13.14). This is used to study unitarizable representations of the Virasoro algebra, and to calculate the fusion rules (Exercises 13.18–13.26, and 13.34–13.36).

Chapter 14. The homogeneous vertex operator construction is derived via the vertex operator calculus (§14.8). The infinite wedge representation is constructed (§14.9). By making use of the boson-fermion correspondence (§14.10) the whole KP hierarchy is studied (§§14.11 and 14.12). By making use of the principal and homogeneous vertex operator constructions of $A_1^{(1)}$, the whole KdV and NLS hierarchies are described (§14.13). The BKP hierarchy is constructed (Exercises 14.13–14.15). A theory of the infinite Grassmannian and flag manifold is sketched and their connection to the KP and MKP hierarchies is explained (Exercises 14.32, 14.33). A pseudodifferential operator approach to the KP and KdV hierarchies is outlined (Exercises 14.44–14.51). A basis free theory of the Lie algebra and group of type A_∞ is discussed (Exercises 14.55–14.58), and some classical theorems of the theory of algebraic curves are derived from this discussion (Exercises 14.59–14.63).

In addition to correcting misprints and errors and adding some hundred new exercises, I have brought up to date the list of references and related bibliographical comments. The explosion of activity in the field between the second and the third editions, due to a great extent to physicists working in string theory and conformal field theory, made it an impossible task to compile a reasonably complete bibliography. I hope, however, that the collection of references compiled for this edition at least reflects all the major directions of research in the field. Needless to say that every sentence of my bibliographical comments could be prefixed by an "It is my opinion that ...."

I want to thank those who have pointed out errors and suggested improvements, in particular:
R. Borcherds, J. Dixmier, D.Z. Djokovic, E. Frenkel, S. Friedberg, M.-J. Imbens, R. Iyer, R. C. King, M.F.R. Kruelle, J. van de Leur, S.-R. Lu, P. Magyar, G. Rousseau, G. Seligman, M. Wakimoto, Z.-X. Wan, Y.-C. You.

September 1989, Newton, Massachusetts.

Notational Conventions

$\mathbb{Z}$	the set of integers
$\mathbb{Z}_+$	the set of non-negative integers
$\mathbb{N}$	the set of positive integers
$\mathbb{Q}$	the set of rational numbers
$\mathbb{R}$	the set of real numbers
$\mathbb{R}_+$	the set of non-negative real numbers
$\mathbb{C}$	the set of complex numbers
$S^\times$	the set of invertible elements of a ring S
$\operatorname{Re} z$ and $\operatorname{Im} z$	real and imaginary parts of $z \in \mathbb{C}$
$\log z$	for $z \in \mathbb{C}^\times : e^{\log z} = z$ and $-\pi \leq \operatorname{Im} \log z < \pi$
z^α	$= e^{\alpha \log z}$ for $\alpha \in \mathbb{C}$, $z \in \mathbb{C}^\times$
$U \oplus V$ or $\bigoplus_\alpha U_\alpha$	direct sum of vector spaces
$\sum_\alpha U_\alpha$	sum of subspaces of a vector space
$\prod_\alpha U_\alpha$	direct product of vector spaces
kS	the linear k-span of S ($k = \mathbb{Z}, \mathbb{Z}_+, \mathbb{Q}, \mathbb{R}$, or $\mathbb{C}$)
$U \otimes V$	tensor product of vector k-spaces over k ($k = \mathbb{Q}, \mathbb{R}$, or $\mathbb{C}$)
U^*	the dual of a vector k-space over k ($k = \mathbb{Q}, \mathbb{R}$, or $\mathbb{C}$)
k^n	direct sum of n copies of the vector space k ($n \in \mathbb{Z}_+ \cup \{\infty\}$)
I_V or I_n or I	the identity operator on the n-dimensional vector space V
$\langle .,. \rangle$	pairing between a vector space and its dual
$\|u\|^2 = (u\|u)$	square length of a vector u
$\|S\|$	cardinality of a set S
$P \mod Q$	a set of representatives of cosets of an abelian group P with respect to a subgroup Q

$a \equiv b \mod C$	means that $a - b \in C$.
$U(\mathfrak{g})$	universal enveloping algebra of a Lie algebra $\mathfrak{g}$
$U_0(\mathfrak{g}) = U(\mathfrak{g})\mathfrak{g}$	the augmentation ideal of $U(\mathfrak{g})$
$g(v)$ or $g \cdot v$	action of an element g of a Lie algebra or a group on an element v of a module; all modules are assumed to be left modules unless otherwise specified
$G \cdot v$ or $G(v)$	$= \{g \cdot v \mid g \in G\}$ the orbit of v under the action of a group G
$G \cdot V$ or $G(V)$	union of orbits of elements from a set V
$U \cdot M$	linear span of the set $\{u \cdot v \mid u \in U, v \in M\}$, where U is a subspace of an algebra and M is a subset of a module over this algebra.

Chapter 1. Basic Definitions

§1.0. The central object of our study is a certain class of infinite-dimensional Lie algebras alternatively known as contragredient Lie algebras, generalized Cartan matrix Lie algebras or Kac–Moody algebras. Their definition is a rather straightforward "infinite-dimensional" generalization of the definition of semisimple Lie algebras via the Cartan matrix and Chevalley generators. The slight technical difficulty that occurs in the case $\det A = 0$ is handled by introducing the "realization" $(\mathfrak{h}, \Pi, \Pi^\vee)$ of the matrix A. The Lie algebra $\mathfrak{g}(A)$ is then a quotient of the Lie algebra $\tilde{\mathfrak{g}}(A)$ with generators e_i, f_i and $\mathfrak{h}$, and defining relations (1.2.1), by the maximal ideal intersecting $\mathfrak{h}$ trivially. Some of the advantages of this definition as compared to the one given in the introduction, as we will see, are as follows: the definition of roots and weights is natural; the Weyl group acts on a nice convex cone; the characters have a nice region of convergence.

§1.1. We start with a complex $n \times n$ matrix $A = (a_{ij})_{i,j=1}^n$ of rank ℓ and we will associate with it a complex Lie algebra $\mathfrak{g}(A)$.

The matrix A is called a *generalized Cartan matrix* if it satisfies the following conditions:

(C1) $$a_{ii} = 2 \quad \text{for} \quad i = 1, \dots, n;$$

(C2) $$a_{ij} \text{ are nonpositive integers for } i \neq j;$$

(C3) $$a_{ij} = 0 \text{ implies } a_{ji} = 0.$$

Although a deep theory can be developed only for the Lie algebra associated to a generalized Cartan matrix A, it is natural (and convenient) to begin with an arbitrary matrix A.

A *realization* of A is a triple $(\mathfrak{h}, \Pi, \Pi^\vee)$, where $\mathfrak{h}$ is a complex vector space, $\Pi = \{\alpha_1, \dots, \alpha_n\} \subset \mathfrak{h}^*$ and $\Pi^\vee = \{\alpha_1^\vee, \dots, \alpha_n^\vee\} \subset \mathfrak{h}$ are indexed subsets in $\mathfrak{h}^*$ and $\mathfrak{h}$, respectively, satisfying the following three conditions

(1.1.1) both sets Π and $\Pi^\vee$ are linearly independent;

(1.1.2) $$\langle \alpha_i^\vee, \alpha_j \rangle = a_{ij} \quad (i, j = 1, \dots, n);$$

(1.1.3) $$n - \ell = \dim \mathfrak{h} - n.$$

Two realizations $(\mathfrak{h}, \Pi, \Pi^\vee)$ and $(\mathfrak{h}_1, \Pi_1, \Pi_1^\vee)$ are called *isomorphic* if there exists a vector space isomorphism $\phi : \mathfrak{h} \to \mathfrak{h}_1$ such that $\phi(\Pi^\vee) = \Pi_1^\vee$ and $\phi^*(\Pi_1) = \Pi$.

PROPOSITION 1.1. *There exists a unique up to isomorphism realization for every $n \times n$-matrix A. Realizations of matrices A and B are isomorphic if and only if B can be obtained from A by a permutation of the index set.*

Proof. Reordering the indices, if necessary, we can assume that

$$A = \begin{pmatrix} A_1 \\ A_2 \end{pmatrix}$$

where A_1 is an $\ell \times n$ submatrix of rank ℓ. Consider the following $n \times (2n - \ell)$ matrix:

$$C = \begin{pmatrix} A_1 & 0 \\ A_2 & I_{n-\ell} \end{pmatrix}.$$

Taking $\mathfrak{h} = \mathbf{C}^{2n-\ell}$, elements $\alpha_1, \dots, \alpha_n$ to be the first n linear coordinate functions and elements $\alpha_1^\vee, \dots, \alpha_n^\vee$ to be the rows of the matrix C, we obtain a realization of the matrix A.

Conversely, given a realization $(\mathfrak{h}, \Pi, \Pi^\vee)$ we complete Π to a basis by adding elements $\alpha_{n+1}, \dots, \alpha_{2n-\ell} \in \mathfrak{h}^*$ so that we have for some $\ell \times (n - \ell)$ matrix B and invertible $(n - \ell) \times (n - \ell)$-matrix D:

$$(\langle \alpha_i^\vee, \alpha_j \rangle) = \begin{pmatrix} A_1 & B \\ A_2 & D \end{pmatrix}.$$

Adding to $\alpha_{n+1}, \dots$ suitable linear combinations of $\alpha_1, \dots, \alpha_\ell$, we can make $B = 0$. Replacing $\alpha_{n+1}, \dots, \alpha_{2n-\ell}$ by their linear combinations, we can make $D = I$. This proves the uniqueness. The second part of the proposition is now obvious.

□

It is clear that if $(\mathfrak{h}, \Pi, \Pi^\vee)$ is a realization of a matrix A, then $(\mathfrak{h}^*, \Pi^\vee, \Pi)$ is a realization of the transposed matrix ${}^tA$.

Given two matrices A_1 and A_2 and their realizations $(\mathfrak{h}_1, \Pi_1, \Pi_1^\vee)$ and $(\mathfrak{h}_2, \Pi_2, \Pi_2^\vee)$, we obtain a realization of the direct sum $\begin{pmatrix} A_1 & 0 \\ 0 & A_2 \end{pmatrix}$ of the two matrices:

$$(\mathfrak{h}_1 \oplus \mathfrak{h}_2, \Pi_1 \times \{0\} \cup \{0\} \times \Pi_2, \Pi_1^\vee \times \{0\} \cup \{0\} \times \Pi_2^\vee),$$

which is called the *direct sum* of the realizations.

A matrix A (and the corresponding realization) is called *decomposable* if, after reordering the indices (i.e. a permutation of its rows and the same permutation of the columns), A decomposes into a nontrivial direct sum.

It is clear that after reordering the indices, one can decompose A into a direct sum of indecomposable matrices, and the corresponding realization into a direct sum of the corresponding indecomposable realizations.

In analogy with the finite-dimensional theory, we use the following terminology. Π is called the *root basis*, $\Pi^\vee$ the *coroot basis*, elements from Π (resp. $\Pi^\vee$) are called *simple roots* (resp. *simple coroots.*). We also set

$$Q = \sum_{i=1}^{n} \mathbf{Z}\alpha_i, \quad Q_+ = \sum_{i=1}^{n} \mathbf{Z}_+\alpha_i.$$

The lattice Q is called the *root lattice*.

For $\alpha = \sum_i k_i\alpha_i \in Q$ the number $\operatorname{ht}\alpha := \sum_i k_i$ is called the *height* of α.

Introduce a partial ordering $\geq$ on $\mathfrak{h}^*$ by setting $\lambda \geq \mu$ if $\lambda - \mu \in Q_+$.

§1.2. Let $A = (a_{ij})$ be an $n \times n$-matrix over $\mathbf{C}$, and let $(\mathfrak{h}, \Pi, \Pi^\vee)$ be a realization of A. First we introduce an auxiliary Lie algebra $\tilde{\mathfrak{g}}(A)$ with the generators $e_i, f_i (i = 1, \ldots, n)$ and $\mathfrak{h}$, and the following defining relations:

$$(1.2.1) \qquad \begin{cases} [e_i, f_j] = \delta_{ij}\alpha_i^\vee & (i, j = 1, \ldots, n), \\ [h, h'] = 0 & (h, h' \in \mathfrak{h}), \\ [h, e_i] = \langle \alpha_i, h \rangle e_i, & \\ [h, f_i] = -\langle \alpha_i, h \rangle f_i & (i = 1, \ldots, n; h \in \mathfrak{h}). \end{cases}$$

By the uniqueness of the realization of A it is clear that $\tilde{\mathfrak{g}}(A)$ depends only on A.

Denote by $\tilde{\mathfrak{n}}_+$ (resp. $\tilde{\mathfrak{n}}_-$) the subalgebra of $\tilde{\mathfrak{g}}(A)$ generated by $e_1, \ldots, e_n$ (resp. $f_1, \ldots, f_n$). Our first fundamental result is

THEOREM 1.2. a) *$\tilde{\mathfrak{g}}(A) = \tilde{\mathfrak{n}}_- \oplus \mathfrak{h} \oplus \tilde{\mathfrak{n}}_+$ (direct sum of vector spaces).*
b) *$\tilde{\mathfrak{n}}_+$ (resp. $\tilde{\mathfrak{n}}_-$) is freely generated by $e_1, \ldots, e_n$ (resp. $f_1, \ldots, f_n$).*
c) *The map $e_i \mapsto -f_i, f_i \mapsto -e_i (i = 1, \ldots, n)$, $h \mapsto -h (h \in \mathfrak{h})$, can be uniquely extended to an involution $\tilde{\omega}$ of the Lie algebra $\tilde{\mathfrak{g}}(A)$.*
d) *With respect to $\mathfrak{h}$ one has the root space decomposition:*

$$(1.2.2) \qquad \tilde{\mathfrak{g}}(A) = \Big(\bigoplus_{\substack{\alpha \in Q_+ \\ \alpha \neq 0}} \tilde{\mathfrak{g}}_{-\alpha}\Big) \oplus \mathfrak{h} \oplus \Big(\bigoplus_{\substack{\alpha \in Q_+ \\ \alpha \neq 0}} \tilde{\mathfrak{g}}_{\alpha}\Big),$$

where $\tilde{\mathfrak{g}}_\alpha = \{x \in \tilde{\mathfrak{g}}(A) | [h, x] = \alpha(h)x$ for all $h \in \mathfrak{h}\}$. Furthermore, $\dim \tilde{\mathfrak{g}}_\alpha < \infty$, and $\tilde{\mathfrak{g}}_\alpha \subset \tilde{\mathfrak{n}}_\pm$ for $\pm\alpha \in Q_+, \alpha \neq 0$.

e) *Among the ideals of $\tilde{\mathfrak{g}}(A)$ intersecting $\mathfrak{h}$ trivially, there exists a unique maximal ideal $\mathfrak{r}$. Furthermore,*

(1.2.3) $$\mathfrak{r} = (\mathfrak{r} \cap \tilde{\mathfrak{n}}_-) \oplus (\mathfrak{r} \cap \tilde{\mathfrak{n}}_+) \qquad \textit{(direct sum of ideals).}$$

Proof. Let V be the n-dimensional complex vector space with a basis $v_1, \ldots, v_n$ and let λ be a linear function on $\mathfrak{h}$. We define an action of the generators of $\tilde{\mathfrak{g}}(A)$ on the tensor algebra $T(V)$ over V by

α) $f_i(a) = v_i \otimes a$ for $a \in T(V)$;

β) $h(1) = \langle \lambda, h \rangle 1$, and inductively on s,

$$h(v_j \otimes a) = -\langle \alpha_j, h \rangle v_j \otimes a + v_j \otimes h(a) \text{ for } a \in T^{s-1}(V),\ j = 1, \ldots, n;$$

γ) $e_i(1) = 0$, and inductively on s,

$$e_i(v_j \otimes a) = \delta_{ij} \alpha_i^\vee(a) + v_j \otimes e_i(a) \text{ for } a \in T^{s-1}(V),\ j = 1, \ldots, n.$$

This defines a representation of the Lie algebra $\tilde{\mathfrak{g}}(A)$ on the space $T(V)$. To see that, we have to check all of the relations (1.2.1).

The second relation is obvious since $\mathfrak{h}$ operates diagonally. For the first relation we have

$$\begin{aligned}(e_i f_j - f_j e_i)(a) &= e_i(v_j \otimes a) - v_j \otimes e_i(a) \\ &= \delta_{ij} \alpha_i^\vee(a) + v_j \otimes e_i(a) - v_j \otimes e_i(a) = \delta_{ij} \alpha_i^\vee(a),\end{aligned}$$

by α) and γ).

For the fourth relation, we have

$$\begin{aligned}(hf_j - f_j h)(a) &= h(v_j \otimes a) - v_j \otimes h(a) \\ &= -\langle \alpha_j, h \rangle v_j \otimes a + v_j \otimes h(a) - v_j \otimes h(a) \\ &= -\langle \alpha_j, h \rangle f_j(a)\end{aligned}$$

by α) and β).

Finally, the third relation is proved by induction on s. For $s = 0$ it evidently holds. For $s > 0$ take $a = v_k \otimes a_1$, where $a_1 \in T^{s-1}(V)$. We have

$$\begin{aligned}(he_j - e_j h)(v_k \otimes a_1) &= h(\delta_{jk} \alpha_j^\vee(a_1)) + h(v_k \otimes e_j(a_1)) \\ &\qquad - e_j(-\langle \alpha_k, h \rangle (v_k \otimes a_1) + v_k \otimes h(a_1)) \\ &= \delta_{jk} \alpha_j^\vee(h(a_1)) - \langle \alpha_k, h \rangle v_k \otimes e_j(a_1) \\ &\quad + v_k \otimes he_j(a_1) \\ &\quad + \langle \alpha_k, h \rangle \delta_{jk} \alpha_j^\vee(a_1) \\ &\quad + \langle \alpha_k, h \rangle v_k \otimes e_j(a_1) \\ &\quad - \delta_{jk} \alpha_j^\vee h(a_1) - v_k \otimes e_j h(a_1) \\ &= \langle \alpha_j, h \rangle \delta_{jk} \alpha_j^\vee(a_1) + v_k \otimes (he_j - e_j h)(a_1).\end{aligned}$$

To complete the proof, we apply the inductive assumption to the second summand.

Now we can deduce all the statements of the theorem. Using relations (1.2.1), it is easy to show by induction on s that a product of s elements from the set $\{e_i, f_i (i = 1, \dots, n);\ \mathfrak{h}\}$ lies in $\tilde{\mathfrak{n}}_- + \mathfrak{h} + \tilde{\mathfrak{n}}_+$. Let now $u = n_- + h + n_+ = 0$, where $n_\pm \in \tilde{\mathfrak{n}}_\pm$ and $h \in \mathfrak{h}$. Then in the representation $T(V)$ we have $u(1) = n_-(1) + \langle \lambda, h \rangle = 0$. It follows that $\langle \lambda, h \rangle = 0$ for every $\lambda \in \mathfrak{h}^*$ and hence $h = 0$.

Furthermore, using the map $f_i \mapsto v_i$, we see that the tensor algebra $T(V)$ is an associative enveloping algebra of the Lie algebra $\tilde{\mathfrak{n}}_-$. Since $T(V)$ is a free associative algebra, we conclude that $T(V)$ is automatically the universal enveloping algebra $U(\tilde{\mathfrak{n}}_-)$ of $\tilde{\mathfrak{n}}_-$, the map $n_- \mapsto n_-(1)$ being the canonical embedding $\tilde{\mathfrak{n}}_- \to U(\tilde{\mathfrak{n}}_-)$. Hence $n_- = 0$ and a) is proven. Moreover, by the Poincaré–Birkhoff–Witt theorem, $\tilde{\mathfrak{n}}_-$ is freely generated by $f_1, \dots, f_n$. The statement c) is obvious. Now applying $\tilde{\omega}$ we deduce that $\tilde{\mathfrak{n}}_+$ is freely generated by $e_1, \dots, e_n$, proving b).

Using the last two relations of (1.2.1), we have

$$\tilde{\mathfrak{n}}_\pm = \bigoplus_{\substack{\alpha \in Q_+ \\ \alpha \neq 0}} \tilde{\mathfrak{g}}_{\pm\alpha}.$$

We also have the following obvious estimate:

$$\dim \tilde{\mathfrak{g}}_\alpha \leq n^{|\operatorname{ht} \alpha|}. \tag{1.2.4}$$

These together with a) prove d).

To prove e) note that for any ideal $\mathfrak{i}$ of $\tilde{\mathfrak{g}}(A)$ one has (by Proposition 1.5 below):

$$\mathfrak{i} = \bigoplus_\alpha (\tilde{\mathfrak{g}}_\alpha \cap \mathfrak{i}).$$

Hence the sum of ideals which intersect $\mathfrak{h}$ trivially, itself intersects $\mathfrak{h}$ trivially, and the sum of all ideals with this property is the unique maximal ideal $\mathfrak{r}$ which intersects $\mathfrak{h}$ trivially. In particular, we obtain that (1.2.3) is a direct sum of vector spaces. But, clearly, $[f_i, \mathfrak{r} \cap \tilde{\mathfrak{n}}_+] \subset \tilde{\mathfrak{n}}_+$. Hence $[\tilde{\mathfrak{g}}(A), \mathfrak{r} \cap \tilde{\mathfrak{n}}_+] \subset \mathfrak{r} \cap \tilde{\mathfrak{n}}_+$; similarly, $[\tilde{\mathfrak{g}}(A), \mathfrak{r} \cap \tilde{\mathfrak{n}}_-] \subset \mathfrak{r} \cap \tilde{\mathfrak{n}}_-$. This shows that (1.2.3) is a direct sum of ideals.

□

§1.3. Given a complex $n \times n$-matrix A, we can now define the main object of our study: the Lie algebra $\mathfrak{g}(A)$. Let $(\mathfrak{h}, \Pi, \Pi^\vee)$ be a realization of A and let $\tilde{\mathfrak{g}}(A)$ be the Lie algebra on generators e_i, f_i $(i = 1, \ldots, n)$ and $\mathfrak{h}$, and the defining relations (1.2.1). By Theorem 1.2 a) the natural map $\mathfrak{h} \to \tilde{\mathfrak{g}}(A)$ is an imbedding. Let $\mathfrak{r}$ be the maximal ideal in $\tilde{\mathfrak{g}}(A)$, which intersects $\mathfrak{h}$ trivially (see Theorem 1.2 e)). We set:

$$\mathfrak{g}(A) = \tilde{\mathfrak{g}}(A)/\mathfrak{r}.$$

The matrix A is called the *Cartan matrix* of the Lie algebra $\mathfrak{g}(A)$, and n is called the *rank* of $\mathfrak{g}(A)$. The quadruple $(\mathfrak{g}(A), \mathfrak{h}, \Pi, \Pi^\vee)$ is called the *quadruple associated to the matrix A*. Two quadruples $(\mathfrak{g}(A), \mathfrak{h}, \Pi, \Pi^\vee)$ and $(\mathfrak{g}(A_1), \mathfrak{h}_1, \Pi_1, \Pi_1^\vee)$ are called *isomorphic* if there exists a Lie algebra isomorphism $\phi : \mathfrak{g}(A) \to \mathfrak{g}(A_1)$ such that $\phi(\mathfrak{h}) = \mathfrak{h}_1$, $\phi(\Pi^\vee) = \Pi_1^\vee$ and $\phi^*(\Pi_1) = \Pi$.

The Lie algebra $\mathfrak{g}(A)$ whose Cartan matrix is a generalized Cartan matrix is called a *Kac–Moody algebra*.

We keep the same notation for the images of $e_i, f_i, \mathfrak{h}$ in $\mathfrak{g}(A)$. The subalgebra $\mathfrak{h}$ of $\mathfrak{g}(A)$ is called the *Cartan subalgebra*. The elements e_i, f_i $(i = 1, \ldots, n)$ are called the *Chevalley generators*. In fact, they generate the derived subalgebra $\mathfrak{g}'(A) = [\mathfrak{g}(A), \mathfrak{g}(A)]$. Furthermore,

$$\mathfrak{g}(A) = \mathfrak{g}'(A) + \mathfrak{h}$$

($\mathfrak{g}(A) = \mathfrak{g}'(A)$ if and only if $\det A \neq 0$).

We set $\mathfrak{h}' = \sum_{i=1}^n \mathbb{C}\alpha_i^\vee$. Then $\mathfrak{g}'(A) \cap \mathfrak{h} = \mathfrak{h}'$; $\mathfrak{g}'(A) \cap \mathfrak{g}_\alpha = \mathfrak{g}_\alpha$ if $\alpha \neq 0$.

It follows from (1.2.2) that we have the following *root space decomposition* with respect to $\mathfrak{h}$:

$$\mathfrak{g}(A) = \bigoplus_{\alpha \in Q} \mathfrak{g}_\alpha. \tag{1.3.1}$$

Here, $\mathfrak{g}_\alpha = \{x \in \mathfrak{g}(A) | [h, x] = \alpha(h)x \text{ for all } h \in \mathfrak{h}\}$ is the *root space* attached to α. Note that $\mathfrak{g}_0 = \mathfrak{h}$. The number $\operatorname{mult} \alpha := \dim \mathfrak{g}_\alpha$ is called the *multiplicity* of α. Note that

$$\operatorname{mult} \alpha \leq n^{|\operatorname{ht} \alpha|} \tag{1.3.2}$$

by (1.2.4).

An element $\alpha \in Q$ is called a *root* if $\alpha \neq 0$ and $\operatorname{mult} \alpha \neq 0$. A root $\alpha > 0$ (resp. $\alpha < 0$) is called *positive* (resp. *negative*). It follows from (1.2.2) that every root is either positive or negative. Denote by Δ, Δ_+ and Δ_- the sets of all roots, positive and negative roots respectively. Then

$$\Delta = \Delta_+ \cup \Delta_- \quad \text{(a disjoint union)}.$$

Sometimes we will write $\Delta(A), Q(A), \ldots$ in order to emphasize the dependence on A.

Let $\mathfrak{n}_+$ (resp. $\mathfrak{n}_-$) denote the subalgebra of $\mathfrak{g}(A)$ generated by $e_1, \ldots, e_n$ (resp. $f_1, \ldots, f_n$). By Theorem 1.2 a), e) we have the *triangular decomposition*

$$\mathfrak{g}(A) = \mathfrak{n}_- \oplus \mathfrak{h} \oplus \mathfrak{n}_+ \quad \text{(direct sum of vector spaces)}.$$

Note that $\mathfrak{g}_\alpha \subset \mathfrak{n}_+$ if $\alpha > 0$ and $\mathfrak{g}_\alpha \subset \mathfrak{n}_-$ if $\alpha < 0$. In other words, for $\alpha > 0$ (resp. $\alpha < 0$), $\mathfrak{g}_\alpha$ is the linear span of the elements of the form $[\ldots[[e_{i_1}, e_{i_2}], e_{i_3}] \ldots e_{i_s}]$ (resp. $[\ldots[[f_{i_1}, f_{i_2}], f_{i_3}] \ldots f_{i_s}]$), such that $\alpha_{i_1} + \cdots + \alpha_{i_s} = \alpha$ (resp. $= -\alpha$). It follows immediately that

$$\mathfrak{g}_{\alpha_i} = \mathbb{C}e_i, \quad \mathfrak{g}_{-\alpha_i} = \mathbb{C}f_i; \quad \mathfrak{g}_{s\alpha_i} = 0 \quad \text{if } |s| > 1. \tag{1.3.3}$$

Since every root is either positive or negative, (1.3.3) implies the following important fact:

LEMMA 1.3. *If* $\beta \in \Delta_+ \backslash \{\alpha_i\}$, *then* $(\beta + \mathbb{Z}\alpha_i) \cap \Delta \subset \Delta_+$.

It follows from Theorem 1.2 e) that the ideal $\mathfrak{r} \subset \tilde{\mathfrak{g}}(A)$ is $\tilde{\omega}$-invariant (see Theorem 1.2 c)). Hence $\tilde{\omega}$ induces an involutive automorphism ω of the Lie algebra $\mathfrak{g}(A)$, called the *Chevalley involution* of $\mathfrak{g}(A)$. It is determined by

$$\omega(e_i) = -f_i, \ \omega(f_i) = -e_i, \ \omega(h) = -h \text{ if } h \in \mathfrak{h}. \tag{1.3.4}$$

As $\omega(\mathfrak{g}_\alpha) = \mathfrak{g}_{-\alpha}$, we deduce that

$$\operatorname{mult} \alpha = \operatorname{mult}(-\alpha). \tag{1.3.5}$$

In particular,

$$\Delta_- = -\Delta_+.$$

§1.4. The following simple statement is useful.

PROPOSITION 1.4. a) *Let $\mathfrak{g}$ be a Lie algebra, $\mathfrak{h} \subset \mathfrak{g}$ a commutative subalgebra, $e_1, \ldots, e_n$, $f_1, \ldots, f_n$ elements of $\mathfrak{g}$, and let $\Pi^\vee = \{\alpha_1^\vee, \ldots, \alpha_n^\vee\} \subset \mathfrak{h}$, $\Pi = \{\alpha_1, \ldots, \alpha_n\} \subset \mathfrak{h}^*$ be linearly independent sets such that:*

$$(1.4.1) \qquad [e_i, f_j] = \delta_{ij}\alpha_i^\vee \in \mathfrak{h} \quad (i, j = 1, \ldots, n),$$

$$(1.4.2) \quad [h, e_i] = \langle \alpha_i, h \rangle e_i, [h, f_i] = -\langle \alpha_i, h \rangle f_i, \ (h \in \mathfrak{h};\ i = 1, \ldots, n).$$

Suppose that e_i, f_i $(i = 1, \ldots, n)$ and $\mathfrak{h}$ generate $\mathfrak{g}$ as a Lie algebra, and that $\mathfrak{g}$ has no nonzero ideals which intersect $\mathfrak{h}$ trivially. Finally, set $A = (\langle \alpha_i^\vee, \alpha_j \rangle)_{i,j=1}^n$, and suppose that $\dim \mathfrak{h} = 2n - \operatorname{rank} A$. Then $(\mathfrak{g}, \mathfrak{h}, \Pi, \Pi^\vee)$ is the quadruple associated to the matrix A.
b) *Given two $n \times n$-matrices A and A', there exists an isomorphism of the associated quadruples if and only if A' can be obtained from A by reordering the index set.*

Proof follows from Proposition 1.1 and Theorem 1.2 e).

□

COROLLARY 1.4.. *The quadruple associated to a direct sum of matrices A_i, is isomorphic to a direct sum of the quadruples associated to the A_i. The root system of $\mathfrak{g}(A)$ is a union of the root systems of the $\mathfrak{g}(A_i)$.*

Proof. follows from Proposition 1.4 a).

□

§1.5. Now we need a short digression on gradations. Given an abelian group M, a decomposition $V = \bigoplus_{\alpha \in M} V_\alpha$ of the vector space V into a direct sum of its subspaces is called an *M-gradation* of V. A subspace $U \subset V$ is called *graded* if $U = \bigoplus_{\alpha \in M} (U \cap V_\alpha)$. Elements from V_α are called *homogeneous* of degree α. The following fact is widely used in representation theory.

PROPOSITION 1.5. *Let $\mathfrak{h}$ be a commutative Lie algebra, V a diagonalizable $\mathfrak{h}$-module, i.e.*

$$(1.5.1) \quad V = \bigoplus_{\lambda \in \mathfrak{h}^*} V_\lambda, \qquad \text{where } V_\lambda = \{v \in V \mid h(v) = \lambda(h)v \text{ for all } h \in \mathfrak{h}\}.$$

Then any submodule U of V is graded with respect to the gradation (1.5.1).

Proof. Any $v \in V$ can be written in the form $v = \sum_{j=1}^{m} v_j$, where $v_j \in V_{\lambda_j}$, and there exists $h \in \mathfrak{h}$ such that $\lambda_j(h)$ $(j = 1, \ldots, m)$ are distinct. We have for $v \in U$:

$$h^k(v) = \sum_{j=1}^{m} \lambda_j(h)^k v_j \in U \quad (k = 0, 1, \ldots, m-1).$$

This is a system of linear equations with a nondegenerate matrix. Hence all v_j lie in U.

□

One introduces the so-called *formal topology* on a graded vector space $V = \bigoplus_{\alpha \in M} V_\alpha$ as follows. Given a finite subset $F \subset M$, we put $V^F = \bigoplus_{\alpha \in M \setminus F} V_\alpha$, and declare all the subsets V^F of V to be the fundamental system of neighbourhoods of zero. The completion of V in the formal topology is, clearly, $\prod_{\alpha \in M} V_\alpha$. Given a subset C of this (complete) topological vector space, its closure in the formal topology is called the *formal completion* of C.

An *M-gradation* of a Lie algebra $\mathfrak{g}$ is its gradation as a vector space, such that $[\mathfrak{g}_\alpha, \mathfrak{g}_\beta] \subset \mathfrak{g}_{\alpha+\beta}$. For example, (1.3.1) is a Q-gradation of the Lie algebra $\mathfrak{g}(A)$.

In order to introduce an M-gradation in a Lie algebra $\mathfrak{g}$ one chooses a system of generators of $\mathfrak{g}$, say, $a_1, \ldots, a_n$, and elements $\lambda_1, \ldots, \lambda_n \in M$ and prescribes degrees to each a_i : $\deg a_i = \lambda_i$. This defines a (unique) M-gradation of $\mathfrak{g}$ with $\deg a_i = \lambda_i$ if and only if the ideal of relations between the a_i is M-graded. For if $a_1, \ldots, a_n$ is a free system of generators of $\mathfrak{g}$, such a gradation does exist.

Let now $s = (s_1, \ldots, s_n)$ be an n-tuple of integers. Setting

$$\deg e_i = -\deg f_i = s_i \quad (i = 1, \ldots, n), \quad \deg \mathfrak{h} = 0$$

defines a $\mathbb{Z}$-gradation

$$\mathfrak{g}(A) = \bigoplus_{j \in \mathbb{Z}} \mathfrak{g}_j(s),$$

called the *gradation of type s*. Explicitly:

$$\mathfrak{g}_j(s) = \bigoplus_{\alpha} \mathfrak{g}_\alpha,$$

where the sum is taken over $\alpha = \sum_i k_i \alpha_i \in Q$ such that $\sum_i k_i s_i = j$. It is clear that if $s_i > 0$ for all i, then $\mathfrak{g}_0(s) = \mathfrak{h}$ and $\dim \mathfrak{g}_j(s) < \infty$ $(j \in \mathbb{Z})$.

A particularly important gradation is the *principal gradation*. This is the gradation of type $\mathbf{1} = (1, \dots, 1)$. Explicitly:

$$\mathfrak{g}_j(\mathbf{1}) = \bigoplus_{\alpha : \mathrm{ht}\,\alpha = j} \mathfrak{g}_\alpha .$$

Note that

$$\mathfrak{g}_0(\mathbf{1}) = \mathfrak{h}, \quad \mathfrak{g}_{-1}(\mathbf{1}) = \sum_i \mathbf{C} f_i, \quad \mathfrak{g}_1(\mathbf{1}) = \sum_i \mathbf{C} e_i,$$

so that $\mathfrak{n}_\pm = \bigoplus_{j \geq 1} \mathfrak{g}_{\pm j}(\mathbf{1})$.

The following simple lemma is useful for computations in $\mathfrak{g}(A)$.

LEMMA 1.5. *Let $a \in \mathfrak{n}_+$ be such that $[a, f_i] = 0$ for all $i = 1, \dots, n$. Then $a = 0$. Similarly, for $a \in \mathfrak{n}_-$, if $[a, e_i] = 0$ for all $i = 1, \dots, n$, then $a = 0$.*

Proof. Let $a \in \mathfrak{n}_+$ be such that $[a, \mathfrak{g}_{-1}(\mathbf{1})] = 0$. Then it is easy to see that $\sum_{i,j \geq 0} (\mathrm{ad}\, \mathfrak{g}_1(\mathbf{1}))^i (\mathrm{ad}\, \mathfrak{h})^j a$ is a subspace of $\mathfrak{n}_+ \subset \mathfrak{g}(A)$, which is invariant with respect to $\mathrm{ad}\, \mathfrak{g}_1(\mathbf{1})$, $\mathrm{ad}\, \mathfrak{h}$ and $\mathrm{ad}\, \mathfrak{g}_{-1}(\mathbf{1})$ (the condition on a is used only in the last case). Hence if $a \neq 0$, we obtain a nonzero ideal in $\mathfrak{g}(A)$ which intersects $\mathfrak{h}$ trivially. This contradicts the definition of $\mathfrak{g}(A)$.

□

Remark 1.5. Sometimes it is useful to consider the Lie algebra $\mathfrak{g}'(A)$ instead of $\mathfrak{g}(A)$. Let us give a more direct construction of $\mathfrak{g}'(A)$. Denote by $\tilde{\mathfrak{g}}'(A)$ the Lie algebra on generators $e_i, f_i, \alpha_i^\vee$ $(i = 1, \dots, n)$ and defining relations

$$[e_i, f_j] = \delta_{ij} \alpha_i^\vee, [\alpha_i^\vee, \alpha_j^\vee] = 0, [\alpha_i^\vee, e_j] = a_{ij} e_j, [\alpha_i^\vee, f_j] = -a_{ij} f_j .$$

Let Q be a free abelian group on generators $\alpha_1, \dots, \alpha_n$. Introduce a Q-gradation $\tilde{\mathfrak{g}}'(A) = \bigoplus_\alpha \tilde{\mathfrak{g}}'_\alpha$ setting

$$\deg e_i = \alpha_i = -\deg f_i, \quad \deg \alpha_i^\vee = 0.$$

There exists a unique maximal Q-graded ideal $\mathfrak{r} \subset \tilde{\mathfrak{g}}'(A)$ intersecting $\tilde{\mathfrak{g}}'_0$ $(= \sum_i \mathbf{C} \alpha_i^\vee)$ trivially. Then

$$\mathfrak{g}'(A) = \tilde{\mathfrak{g}}'(A)/\mathfrak{r}.$$

Note that this definition works for an infinite n as well.

§**1.6.** The following statement is an application of Lemma 1.5.

PROPOSITION 1.6. *The center of the Lie algebra* $\mathfrak{g}(A)$ *or* $\mathfrak{g}'(A)$ *is equal to*

$$\mathfrak{c} := \{h \in \mathfrak{h} | \langle \alpha_i, h\rangle = 0 \text{ for all } i = 1, \ldots, n\}.$$

Furthermore, $\dim \mathfrak{c} = n - l$.

Proof. Let c lie in the center; write $c = \sum_i c_i$ with respect to the principal gradation. Then $[c, \mathfrak{g}_{-1}(\mathbf{1})] = 0$ implies $[c_i, \mathfrak{g}_{-1}(\mathbf{1})] = 0$ and hence, by Lemma 1.5, $c_i = 0$ for $i > 0$. Similarly, $c_i = 0$ for $i < 0$. Hence $c \in \mathfrak{h}$ and $[c, e_i] = \langle \alpha_i, c\rangle e_i = 0$ implies that $\langle \alpha_i, c\rangle = 0$ $(i = 1, \ldots, n)$. Conversely, if $c \in \mathfrak{h}$ and the latter condition holds, c commutes with all Chevalley generators and, therefore, lies in the center. Finally, $\mathfrak{c} \subset \mathfrak{h}'$ since in the contrary case, $\dim \mathfrak{c} > n - \ell$ and Π would not be a linearly independent set.

□

Another application of Lemma 1.5 is

LEMMA 1.6. *Let* $I_1, I_2 \subset \{1, \ldots, n\}$ *be disjoint subsets such that* $a_{ij} = a_{ji} = 0$ *whenever* $i \in I_1$, $j \in I_2$. *Let* $\beta_s = \sum_{i \in I_s} k_i^{(s)} \alpha_i$ $(s = 1, 2)$. *Suppose that* $\alpha = \beta_1 + \beta_2$ *is a root of the Lie algebra* $\mathfrak{g}(A)$. *Then either* β_1 *or* β_2 *is zero.*

Proof. Let $i \in I_1, j \in I_2$. Then $[\alpha_i^\vee, e_j] = 0, [\alpha_j^\vee, e_i] = 0, [e_i, f_j] = 0, [e_j, f_i] = 0$. Using Lemma 1.5, one checks immediately that $[e_i, e_j] = 0, [f_i, f_j] = 0$. Denote by $\mathfrak{g}^{(s)}$ the subalgebra of $\mathfrak{g}(A)$ generated by e_i, f_i with $i \in I_s$. We have proved that $\mathfrak{g}^{(1)}$ and $\mathfrak{g}^{(2)}$ commute. Since $\mathfrak{g}_\alpha$ lies in the subalgebra generated by $\mathfrak{g}^{(1)}$ and $\mathfrak{g}^{(2)}$, we deduce that $\mathfrak{g}_\alpha$ lies either in $\mathfrak{g}^{(1)}$ or in $\mathfrak{g}^{(2)}$.

□

§**1.7.** We conclude the chapter with a description of the structure of ideals of $\mathfrak{g}(A)$.

PROPOSITION 1.7. a) $\mathfrak{g}(A)$ *is simple if and only if* $\det A \neq 0$ *and for each pair of indices* i *and* j *the following condition holds*

(1.7.1) *there exist indices* $i_1, i_2, \ldots, i_s$ *such that* $a_{ii_1} a_{i_1 i_2} \ldots a_{i_s j} \neq 0$.

b) Provided that (1.7.1) *holds, every ideal of* $\mathfrak{g}(A)$ *either contains* $\mathfrak{g}'(A)$ *or is contained in the center.*

Proof. The conditions in a) are obviously necessary. Now suppose that the conditions are satisfied and let $\mathfrak{i} \subset \mathfrak{g}(A)$ be a nonzero ideal. Then $\mathfrak{i}$ contains a nonzero element $h \in \mathfrak{h}$. Since $\det A \neq 0$, we have $\mathfrak{c} = 0$ by Proposition 1.6 and hence $[h, e_j] = ae_j \neq 0$ for some j. Hence $e_j \in \mathfrak{i}$ and $\alpha_j^\vee = [e_j, f_j] \in \mathfrak{i}$. Now it follows from (1.7.1) that $e_j, f_j, \alpha_j^\vee \in \mathfrak{i}$ for all j. Since $\det A \neq 0$, $\mathfrak{h}$ is a linear span of $\alpha_j^\vee$'s and we obtain that $\mathfrak{i} = \mathfrak{g}(A)$, proving a).

The proof of b) is similar.

□

§1.8. Exercises.

1.1. Show that if $a_{ij} = 0$ implies $a_{ji} = 0$, then (1.7.1) is equivalent to the indecomposability of A.

1.2. Let $A' = (a_{ij})_{i,j=1}^{n'}$ be a submatrix of A of rank ℓ'. One can choose a subspace $\mathfrak{h}'$ of $\mathfrak{h}$ containing $\Pi'^\vee = \{\alpha_1^\vee, \ldots, \alpha_{n'}^\vee\}$ of dimension $2n' - \ell'$ such that $\Pi' = \{\alpha_1, \ldots, \alpha_{n'}\}|_{\mathfrak{h}'}$ is a linearly independent set. Then $(\mathfrak{h}, \Pi', \Pi'^\vee)$ is a realization of A'. Set $Q' = \sum_{i=1}^{n'} \mathbf{Z}\alpha_i$, and let $\mathfrak{g}(A) = \oplus_{\alpha \in Q} \mathfrak{g}_\alpha$ be the Lie algebra associated to A. Then

$$\mathfrak{g}(A') \simeq \mathfrak{h}' \oplus \Big(\bigoplus_{\alpha \in Q' \setminus \{0\}} \mathfrak{g}_\alpha \Big).$$

1.3. Show that if $(\mathfrak{h}, \Pi, \Pi^\vee)$ satisfy (1.1.1 and 2), then $\dim \mathfrak{h} \geq 2n - \ell$.

1.4. Suppose that A satisfies condition (1.7.1). Then, provided that there is no root α such that $\alpha|_{\mathfrak{h}'} = 0$, the Lie algebra $\mathfrak{g}'(A)/\mathfrak{c}$ is simple.

1.5. Show that $\mathrm{mult}(\alpha_i + s\alpha_j) \leq 1$ and $\mathrm{mult}\, 2(\alpha_i + \alpha_j) \leq 1$ in every $\mathfrak{g}(A)$.

1.6. Let $A = \begin{pmatrix} 2 & -3 \\ -3 & 2 \end{pmatrix}$. Show that $\mathrm{mult}(2\alpha_1 + 3\alpha_2) = 2$ in $\mathfrak{g}(A)$. Show that for an arbitrary 2×2-matrix A, $\mathrm{mult}(2\alpha_1 + 3\alpha_2) \leq 2$; find the conditions when it is $= 2$.

1.7. Define the Lie algebra $\mathfrak{g}'(A)$ in characteristic $p > 0$ as in Remark 1.5. Prove that the Lie algebra $\mathfrak{g}'\left(\begin{pmatrix} 2 & \lambda \\ -1 & 2 \end{pmatrix}\right)$ of characteristic 3 is a simple Lie algebra of dimension 10 for any $\lambda \neq 0, -1$, and that two such algebras corresponding to $\lambda = \lambda_1$ and $\lambda = \lambda_2$ are nonisomorphic unless $\lambda_1 = \lambda_2$ or $\lambda_1 = -\lambda_2 - 1$.

1.8. A direct sum of vector spaces $\mathfrak{g}_{-1} \oplus \mathfrak{g}_0 \oplus \mathfrak{g}_1$ is called a local Lie algebra if one has bilinear maps: $\mathfrak{g}_i \times \mathfrak{g}_j \to \mathfrak{g}_{i+j}$ for $|i|, |j|, |i+j| \leq 1$, such that anticommutativity and Jacobi identity hold whenever they make sense. Prove that there exists a unique $\mathbb{Z}$-graded Lie algebra $\tilde{\mathfrak{g}} = \bigoplus_i \mathfrak{g}_i$ such that $\mathfrak{g}_{-1} \oplus \mathfrak{g}_0 \oplus \mathfrak{g}_1$ is a given local Lie algebra and $\bigoplus_{i>0} \mathfrak{g}_{\pm i}$ are free Lie algebras on $\mathfrak{g}_{\pm 1}$.

1.9. Let $\mathfrak{g}$ be a Lie algebra, $\mathfrak{h} \subset \mathfrak{g}$ a finite dimensional diagonalizable subalgebra, $\mathfrak{g} = \bigoplus_{\alpha \in Q} \mathfrak{g}_\alpha$ the root space decomposition such that $\mathfrak{g}_0 = \mathfrak{h}$. Show that $\operatorname{der} \mathfrak{g} = (\operatorname{der} \mathfrak{g})_0 + \operatorname{ad} \mathfrak{g}$, where $(\operatorname{der} \mathfrak{g})_0$ consists of endomorphisms preserving the root space decomposition.
[Choose $h \in \mathfrak{h}$ such that $\langle \alpha, h \rangle \neq 0$ for all $\alpha \in Q, \alpha \neq 0$. Adding to $d \in \operatorname{der} \mathfrak{g}$ an inner derivation, one can assume that $d(h) \in \mathfrak{h}$. Deduce that $d(\mathfrak{h}) \subset \mathfrak{h}$ and $d(\mathfrak{g}_\alpha) \subset \mathfrak{g}_\alpha$].

1.10. Deduce from Exercise 1.9 that all derivations of the Lie algebra $\mathfrak{g}(A)/\mathfrak{c}$ are inner, provided that A has no zero rows.

1.11. Show that ad induces an isomorphism

$$\mathfrak{g}(A)/\mathfrak{c} \to \operatorname{der} \mathfrak{g}'(A),$$

if A is a generalized Cartan matrix and $\alpha|_{\mathfrak{h}'} \neq 0$ for every $\alpha \in \Delta$.
[Show that no root is equal to a simple root when restricted to $\mathfrak{h}'$].

1.12. Let $\mathfrak{g}$ be a complex semisimple finite-dimensional Lie algebra with the Cartan matrix A. Then a choice of a Cartan subalgebra $\mathfrak{h} \subset \mathfrak{g}$ and a root basis $\Pi \subset \mathfrak{h}^*$ provides $\mathfrak{g}$ with a structure of a quadruple associated to the matrix A.

1.13. (This is for a less advanced reader.) Prove that the Lie algebra $\mathfrak{g} = sl_{\ell+1}$ of traceless $(\ell+1) \times (\ell+1)$ matrices with the usual bracket $[A, B] = AB - BA$ is a Kac–Moody algebra. In more detail, let E_{ij} $(i, j = 1, \ldots, \ell+1)$ denote the standard basis of the space of all $(\ell+1) \times (\ell+1)$-matrices. Let $\mathfrak{h}$ be the space of all traceless diagonal matrices. Then

$$\alpha_i^\vee = E_{ii} - E_{i+1,i+1} \ (i = 1, \ldots, \ell)$$

form a basis of $\mathfrak{h}$. Define $\epsilon_i \in \mathfrak{h}^*$ $(i = 1, \ldots, \ell+1)$ by

$$\epsilon_i(\operatorname{diag}(a_1, \ldots, a_{\ell+1})) = a_i.$$

Then

$$\alpha_i = \epsilon_i - \epsilon_{i+1} \ (i = 1, \ldots, \ell)$$

form a basis of $\mathfrak{h}^*$. Set

$$\Pi = \{\alpha_1, \ldots, \alpha_\ell\},\ \Pi^\vee = \{\alpha_1^\vee, \ldots, \alpha_\ell^\vee\}.$$

Then $(\mathfrak{h}, \Pi, \Pi^\vee)$ is a realization of the matrix

$$A = \begin{pmatrix} 2 & -1 & 0 & \ldots & 0 & 0 \\ -1 & 2 & -1 & \ldots & 0 & 0 \\ \cdots & \cdots & \cdots & \cdots & \cdots & \cdots \\ 0 & 0 & 0 & \ldots & -1 & 2 \end{pmatrix}$$

The root space decomposition with respect to $\mathfrak{h}$ is

$$\mathfrak{g} = \mathfrak{h} \oplus (\bigoplus_{i \neq j} \mathbf{C} E_{ij}).$$

Set $e_i = E_{i,i+1}$, $f_i = E_{i+1,i}$ $(i = 1, \ldots, \ell)$. Show that $\mathfrak{g} = \mathfrak{g}(A)$, with the Chevalley generators e_i, f_i. Show that $\{\epsilon_i - \epsilon_j\ (i \neq j)\}$ is the set of all roots, $\{\epsilon_i - \epsilon_j\ (i < j)\}$ being the set of positive roots, and that $\mathfrak{n}_\pm$ are the subalgebras of strictly upper and strictly lower triangular matrices. Show that the Chevalley involution is $a \mapsto -{}^t a$.

1.14. Let $\mathfrak{g} = \oplus_j \mathfrak{g}_j$ be a $\mathbf{Z}$-graded Lie algebra such that every graded ideal is trivial, $\mathfrak{g}_{-1} + \mathfrak{g}_0 + \mathfrak{g}_1$ is finite-dimensional and generates $\mathfrak{g}$, and the $\mathfrak{g}_0$-modules $\mathfrak{g}_{-1}$ and $\mathfrak{g}_1$ are irreducible and contragredient. Show that $\mathfrak{g}$ is isomorphic to a Lie algebra $\mathfrak{g}(A)$.

1.15. Let $\mathfrak{h}'$ and V be two n-dimensional complex vector spaces with bases $\Pi^\vee = \{\alpha_1^\vee, \ldots, \alpha_n^\vee\}$ and $\{v_1, \ldots, v_n\}$ respectively. Define a map $\varphi : \mathfrak{h}' \to V^*$ by $\langle \varphi(\alpha_i^\vee), v_j \rangle = a_{ij}$. Choose a subspace $\mathfrak{h}_0$ of V^* complementary to $\varphi(\mathfrak{h}')$, and let $\mathfrak{h} = \mathfrak{h}_0 \oplus \mathfrak{h}'$. Define $\alpha_i \in \mathfrak{h}^*$ by $\langle \alpha_i, h_0 + \sum c_j \alpha_j^\vee \rangle = \langle \alpha_i, h_0 \rangle + \sum_j a_{ji} c_j$, and let $\Pi = \{\alpha_1, \ldots, \alpha_n\}$. Show that $(\mathfrak{h}, \Pi, \Pi^\vee)$ is a realization of $A = (a_{ij})$.

1.16. Let $\mathrm{Aut}(A)$ be the group of permutations σ of the set $\{1, \ldots, n\}$ such that $a_{ij} = a_{\sigma(i),\sigma(j)}$. This group acts by automorphism of $\mathfrak{g}'(A)$ by $\sigma(e_i) = e_{\sigma(i)}$, $\sigma(f_i) = f_{\sigma(i)}$, $\sigma(\alpha_i^\vee) = \alpha_{\sigma(i)}^\vee$. Using Exercise 1.15, show that this extends (non-canonically) to the whole Kac–Moody algebra $\mathfrak{g}(A)$.

§1.9. Bibliographical notes and comments.

The study of Kac–Moody algebras was started independently by Kac [1967], [1968 A, B], and Moody [1967], [1968], [1969].

The proof of a statement much more general than Proposition 1.1 can be found in Vinberg [1971].

Theorem 1.2 should be probably attributed to Chevalley [1948]. In this 2-page note (presented by E. Cartan) Chevalley introduces a general algebraic approach to the construction of finite-dimensional simple Lie algebras and their finite-dimensional representations. A detailed exposition of this was given by Harish-Chandra [1951] and Jacobson [1962]. The proof in Kac [1968 B] and Moody [1968] is a simple adaptation of these.

The material of §§1.5 and 1.6 is taken from Kac [1968 B].

Exercises 1.4, 1.8 and 1.14 are taken from Kac [1968 B]. A somewhat different approach to the construction of graded Lie algebras was developed by Kantor [1968], [1970]. Exercise 1.7 is taken from Weisfeiler–Kac [1971]. Exercise 1.11 is taken from Berman [1976]. Exercises 1.15 and 1.16 are taken from Kac–Peterson [1987].

The rest of the material of Chapter 1 is fairly standard.

The problem of isomorphism of Kac–Moody algebras has been settled quite recently. Namely, as shown by Peterson–Kac [1983], any two maximal ad-diagonalizable subalgebras of a Kac–Moody algebra are conjugate, and hence two Kac–Moody algebras are isomorphic if and only if their Cartan matrices can be obtained from each other by reordering the index set. The question for arbitrary $\mathfrak{g}(A)$ remains open.

Chapter 2. The Invariant Bilinear Form and the Generalized Casimir Operator

§2.0. In this chapter we introduce two important tools of our theory, the invariant bilinear form and the generalized Casimir operator Ω. The operator Ω is a "second order" operator which, in contrast to the finite-dimensional theory, does not lie in the universal enveloping algebra of $\mathfrak{g}(A)$ and is not defined for all representations of $\mathfrak{g}(A)$. However, Ω is defined in the so-called restricted representations, and commutes with the action of $\mathfrak{g}(A)$. Remarkably, one can manage to prove a number of results (including some classical ones) using only Ω.

§2.1. Note that rescaling the Chevalley generators: $e_i \mapsto e_i$, $f_i \mapsto \epsilon_i f_i$ $(i = 1, \ldots, n)$ where ϵ_i are nonzero numbers, we get $\alpha_i^\vee \mapsto \epsilon_i \alpha_i^\vee$, which extends to an isomorphism $\mathfrak{h} \mapsto \mathfrak{h}$ (nonunique, if $\det A = 0$). This extends to an isomorphism: $\mathfrak{g}(A) \to \mathfrak{g}(DA)$, where $D = \operatorname{diag}(\epsilon_1, \ldots, \epsilon_n)$.

An $n \times n$ matrix $A = (a_{ij})$ is called *symmetrizable* if there exists an invertible diagonal matrix $D = \operatorname{diag}(\epsilon_1, \ldots, \epsilon_n)$ and a symmetric matrix $B = (b_{ij})$, such that

$$A = DB. \tag{2.1.1}$$

The matrix B is then called a *symmetrization* of A and $\mathfrak{g}(A)$ is called a *symmetrizable* Lie algebra.

Let A be a symmetrizable matrix with a fixed decomposition (2.1.1) and let $(\mathfrak{h}, \Pi, \Pi^\vee)$ be a realization of A. Fix a complementary subspace $\mathfrak{h}''$ to $\mathfrak{h}' = \sum \mathbf{C}\alpha_i^\vee$ in $\mathfrak{h}$, and define a symmetric bilinear $\mathbf{C}$-valued form $(.|.)$ on $\mathfrak{h}$ by the following two equations:

$$(\alpha_i^\vee | h) = \langle \alpha_i, h \rangle \epsilon_i \text{ for } h \in \mathfrak{h}, i = 1, \ldots, n; \tag{2.1.2}$$

$$(h' | h'') = 0 \text{ for } h', h'' \in \mathfrak{h}''. \tag{2.1.3}$$

Since $\alpha_1^\vee, \ldots, \alpha_n^\vee$ are linearly independent and since (by (2.1.1 and 2)) we have

$$(\alpha_i^\vee | \alpha_j^\vee) = b_{ij}\epsilon_i\epsilon_j \ (i, j = 1, \ldots, n), \tag{2.1.4}$$

there is no ambiguity in the definition of $(.|.)$.

LEMMA 2.1. a) *The kernel of the restriction of the bilinear form* $(.|.)$ *to* $\mathfrak{h}'$ *coincides with* $\mathfrak{c}$.
b) *The bilinear form* $(.|.)$ *is nondegenerate on* $\mathfrak{h}$.

Proof. a) follows from Proposition 1.6. If now for all $h \in \mathfrak{h}$, we have $0 = (\sum_{i=1}^{n} c_i\alpha_i^\vee|h) = \langle\sum_{i=1}^{n} c_i\epsilon_i\alpha_i, h\rangle$, then $\sum_{i=1}^{n} c_i\epsilon_i\alpha_i = 0$ and hence $c_i = 0$, $i = 1, \ldots, n$, proving b).

□

Since the bilinear form $(.|.)$ is nondegenerate, we have an isomorphism $\nu : \mathfrak{h} \to \mathfrak{h}^*$ defined by

$$\langle\nu(h), h_1\rangle = (h|h_1), \quad h, h_1 \in \mathfrak{h},$$

and the induced bilinear form $(.|.)$ on $\mathfrak{h}^*$.

It is clear from (2.1.2) that

$$\nu(\alpha_i^\vee) = \epsilon_i\alpha_i, \quad i = 1, \ldots, n. \tag{2.1.5}$$

Hence from (2.1.4) we deduce:

$$(\alpha_i|\alpha_j) = b_{ij} = a_{ij}/\epsilon_i, \quad i, j = 1, \ldots, n. \tag{2.1.6}$$

§2.2. Our next basic result is the following theorem.

THEOREM 2.2. *Let* $\mathfrak{g}(A)$ *be a symmetrizable Lie algebra. Fix a decomposition (2.1.1) of* A. *Then there exists a nondegenerate symmetric bilinear* $\mathbb{C}$*-valued form* $(\,.\,|\,.\,)$ *on* $\mathfrak{g}(A)$ *such that:*

a) $(\,.\,|\,.\,)$ *is invariant, i.e.* $([x, y]|z) = (x|[y, z])$ *for all* $x, y, z \in \mathfrak{g}(A)$.

b) $(\,.\,|\,.\,)|_{\mathfrak{h}}$ *is defined by (2.1.2 and 3) and is nondegenerate.*

c) $(\mathfrak{g}_\alpha|\mathfrak{g}_\beta) = 0$ *if* $\alpha + \beta \neq 0$.

d) $(\,.\,|\,.\,)|_{\mathfrak{g}_\alpha+\mathfrak{g}_{-\alpha}}$ *is nondegenerate for* $\alpha \neq 0$, *and hence* $\mathfrak{g}_\alpha$ *and* $\mathfrak{g}_{-\alpha}$ *are nondegenerately paired by* $(\,.\,|\,.\,)$.

e) $[x, y] = (x|y)\nu^{-1}(\alpha)$ *for* $x \in \mathfrak{g}_\alpha$, $y \in \mathfrak{g}_{-\alpha}$, $\alpha \in \Delta$.

Proof. Consider the principal $\mathbb{Z}$-gradation (see §1.5)

$$\mathfrak{g}(A) = \bigoplus_{j \in \mathbb{Z}} \mathfrak{g}_j$$

and set $\mathfrak{g}(N) = \bigoplus_{j=-N}^{N} \mathfrak{g}_j$ for $N = 0, 1, \ldots$. Define a bilinear symmetric form $(\,.\,|\,.\,)$ on $\mathfrak{g}(0) = \mathfrak{h}$ by (2.1.2 and 3) and extend it to $\mathfrak{g}(1)$ by

$$(2.2.1) \qquad \begin{aligned} &(e_i|f_j) = \delta_{ij}\epsilon_i \ (i,j = 1,\ldots,n); \\ &(\mathfrak{g}_0|\mathfrak{g}_{\pm 1}) = 0; \quad (\mathfrak{g}_{\pm 1}|\mathfrak{g}_{\pm 1}) = 0. \end{aligned}$$

Then the form $(\,.\,|\,.\,)$ on $\mathfrak{g}(1)$ satisfies condition a) as long as both $[x, y]$ and $[y, z]$ lie in $\mathfrak{g}(1)$. Indeed, it is sufficient to check that

$$([e_i, f_j]|h) = (e_i|[f_j, h]) \text{ for } h \in \mathfrak{h},$$

or, equivalently,

$$\delta_{ij}(\alpha_i^\vee|h) = \delta_{ij}\epsilon_i\langle\alpha_j, h\rangle$$

which is true, due to (2.1.2).

Now we extend $(\,.\,|\,.\,)$ to a bilinear form on the space $\mathfrak{g}(N)$ by induction on $N \geq 1$ so that $(\mathfrak{g}_i|\mathfrak{g}_j) = 0$ if $|i|, |j| \leq N$ and $i + j \neq 0$, and also condition a) is satisfied as long as both $[x, y]$ and $[y, z]$ lie in $\mathfrak{g}(N)$. Suppose that this is already defined on $\mathfrak{g}(N-1)$; then we have only to define $(x|y)$ for $x \in \mathfrak{g}_{\pm N}, y \in \mathfrak{g}_{\mp N}$. We can write $y = \sum_i [u_i, v_i]$, where u_i and v_i are homogeneous elements of nonzero degree which lie in $\mathfrak{g}(N-1)$. Then $[x, u_i] \in \mathfrak{g}(N-1)$ and we set

$$(2.2.2) \qquad (x|y) = \sum_i ([x, u_i]|v_i).$$

In order to show that this is well defined, we prove that if $i, j, s, t \in \mathbb{Z}$ are such that $|i + j| = |s + t| = N$, $i + j + s + t = 0$, $|i|, |j|, |s|, |t| < N$ and $x_i \in \mathfrak{g}_i$, $x_j \in \mathfrak{g}_j$, $x_s \in \mathfrak{g}_s$, $x_t \in \mathfrak{g}_t$, then we have (on $\mathfrak{g}(N-1)$)

$$(2.2.3) \qquad ([[x_i, x_j], x_s]|x_t) = (x_i|[x_j, [x_s, x_t]]).$$

Indeed, using the invariance of $(\,.\,|\,.\,)$ on $\mathfrak{g}(N-1)$ and the Lie algebra axioms, we have

$$\begin{aligned} ([[x_i, x_j], x_s]|x_t) &= ([[x_i, x_s], x_j]|x_t) - ([[x_j, x_s], x_i]|x_t) \\ &= ([x_i, x_s]|[x_j, x_t]) + (x_i|[[x_j, x_s], x_t]) \\ &= (x_i|[x_s, [x_j, x_t]] + [[x_j, x_s], x_t]) \\ &= (x_i|[x_j, [x_s, x_t]]). \end{aligned}$$

If now $x = \sum_i [u'_i, v'_i]$, then by the definition (2.2.2) and by (2.2.3) we have

$$(x|y) = \sum_i ([x, u_i]|v_i) = \sum_i (u'_i|[v'_i, y]).$$

Hence this is independent of the choice of the expressions for x and y.

It is clear from the definition that a) holds on $\mathfrak{g}(N)$ whenever $[x, y]$ and $[y, z]$ lie in $\mathfrak{g}(N)$. Hence we have constructed a bilinear form $(.|.)$ on $\mathfrak{g}$ such that a) and b) hold. Its restriction to $\mathfrak{h}$ is nondegenerate by Lemma 2.1 b). The form $(.|.)$ satisfies c) since for $h \in \mathfrak{h}$, $x \in \mathfrak{g}_\alpha$ and $y \in \mathfrak{g}_\beta$ we have, by the invariance property:

$$0 = ([h, x]|y) + (x|[h, y]) = (\langle\alpha, h\rangle + \langle\beta, h\rangle)(x|y).$$

The verification of e) is standard. For $x \in \mathfrak{g}_\alpha$, $y \in \mathfrak{g}_{-\alpha}$, where $\alpha \in \Delta$, and $h \in \mathfrak{h}$, we have

$$([x, y] - (x|y)\nu^{-1}(\alpha)|h) = (x|[y, h]) - (x|y)\langle\alpha, h\rangle = 0.$$

Now e) follows from b).

It follows from b), c) and e) that the bilinear form $(.|.)$ is symmetric.

If d) fails, then, by c), the form $(.|.)$ is degenerate. Let $\mathfrak{i} = \mathrm{Ker}(.|.)$. It is an ideal and by b), we have $\mathfrak{i} \cap \mathfrak{h} = 0$, which contradicts the definition of $\mathfrak{g}(A)$.

□

§2.3. Suppose that $A = (a_{ij})$ is a symmetrizable generalized Cartan matrix. Fix a decomposition

$$A = \mathrm{diag}(\epsilon_1, \ldots, \epsilon_n)\,(b_{ij})_{i,j=1}^n \tag{2.3.1}$$

where ϵ_i are positive rational numbers and (b_{ij}) is a symmetric rational matrix. Such a decomposition always exists. Indeed, (2.3.1) is equivalent to a system of homogeneous linear equations and inequalities over $\mathbb{Q}$ with unknowns ϵ_i^{-1} and b_{ij}:

$$\epsilon_i^{-1} \neq 0; \mathrm{diag}(\epsilon_1^{-1}, \ldots, \epsilon_n^{-1})A = (b_{ij});\; b_{ij} = b_{ji}.$$

By definition, it has a solution over $\mathbb{C}$. Hence, it has a solution over $\mathbb{Q}$. We can assume that A is indecomposable. Then for any $1 < j \leq n$ there exists a sequence $1 = i_1 < i_2 < \cdots < i_{k-1} < i_k = j$ such that $a_{i_s, i_{s+1}} < 0$. We have:

$$a_{i_s,i_{s+1}}\epsilon_{i_{s+1}} = a_{i_{s+1},i_s}\epsilon_{i_s} (s = 1, \ldots, k-1) \tag{2.3.2}$$

Hence $\epsilon_j\epsilon_1 > 0$ for all j, completing the proof.

From (2.3.2) we also deduce

Remark 2.3. If A is indecomposable, then the matrix diag $(\epsilon_1, \ldots, \epsilon_n)$ is uniquely determined by (2.3.1) up to a constant factor.

Fix a nondegenerate bilinear symmetric form $(.|.)$ associated to the decomposition (2.3.1) as defined in §2.1 From (2.1.6) we deduce:

$$(\alpha_i|\alpha_i) > 0 \text{ for } i = 1, \ldots, n; \tag{2.3.3}$$

$$(\alpha_i|\alpha_j) \leq 0 \text{ for } i \neq j; \tag{2.3.4}$$

$$\alpha_i^\vee = \frac{2}{(\alpha_i|\alpha_i)}\nu^{-1}(\alpha_i). \tag{2.3.5}$$

Hence we obtain the usual expression for the generalized Cartan matrix:

$$A = \left(\frac{2(\alpha_i|\alpha_j)}{(\alpha_i|\alpha_i)}\right)_{i,j=1}^n$$

We extend the bilinear form $(.|.)$ from $\mathfrak{h}$ to an invariant symmetric bilinear form $(.|.)$ on the entire Kac–Moody algebra $\mathfrak{g}(A)$. By Theorem 2.2 such a form exists and satisfies all the properties stated there. It is easy to show that such a form is also unique (see Exercise 2.2). The bilinear form $(.|.)$ on the Kac–Moody algebra $\mathfrak{g}(A)$ provided by Theorem 2.2 and satisfying (2.3.3) is called a *standard* invariant form.

§2.4. Let A be a symmetrizable matrix, let $\mathfrak{g}(A)$ be the associated Lie algebra and let $(.|.)$ be a bilinear form on $\mathfrak{g}(A)$ provided by Theorem 2.2. Given a root α, by Theorem 2.2 d) we can choose dual bases $\{e_\alpha^{(i)}\}$ and $\{e_{-\alpha}^{(i)}\}$ of $\mathfrak{g}_\alpha$ and $\mathfrak{g}_{-\alpha}$, i.e., such bases that $(e_\alpha^{(i)}|e_{-\alpha}^{(j)}) = \delta_{ij}$ $(i, j = 1, \ldots, \text{mult}\,\alpha)$. Then for $x \in \mathfrak{g}_\alpha$ and $y \in \mathfrak{g}_{-\alpha}$ we have

$$(x|y) = \sum_i (x|e_{-\alpha}^{(i)})(y|e_\alpha^{(i)}). \tag{2.4.1}$$

The following lemma is crucial for many computations:

LEMMA 2.4. *If α, $\beta \in \Delta$ and $z \in \mathfrak{g}_{\beta-\alpha}$, then we have in $\mathfrak{g}(A) \otimes \mathfrak{g}(A)$*

$$\sum_s e_{-\alpha}^{(s)} \otimes [z, e_\alpha^{(s)}] = \sum_s [e_{-\beta}^{(s)}, z] \otimes e_\beta^{(s)}. \tag{2.4.2}$$

Proof. We define the bilinear form $(.|.)$ on $\mathfrak{g}(A) \otimes \mathfrak{g}(A)$ by: $(x \otimes y | x_1 \otimes y_1) = (x|x_1)(y|y_1)$. Pick $e \in \mathfrak{g}_\alpha$ and $f \in \mathfrak{g}_{-\beta}$. It suffices to check that pairing both sides of (2.4.2) with $e \otimes f$ gives the same result. We have:

$$\begin{aligned}\sum_s (e_{-\alpha}^{(s)} \otimes [z, e_\alpha^{(s)}] | e \otimes f) &= \sum_s (e_{-\alpha}^{(s)} | e)([z, e_\alpha^{(s)}] | f) \\ &= \sum_s (e_{-\alpha}^{(s)} | e)(e_\alpha^{(s)} | [f, z]) = (e | [f, z])\end{aligned}$$

by Theorem 2.2 a) and (2.4.1). Similarly,

$$\sum_s ([e_{-\beta}^{(s)}, z] \otimes e_\beta^{(s)} | e \otimes f) = \sum_s (e_{-\beta}^{(s)} | [z, e])(e_\beta^{(s)} | f) = ([z, e] | f).$$

Applying again Theorem 2.2 a) gives the result.

□

COROLLARY 2.4. *In the notation of Lemma 2.4, we have*

$$\sum_s [e_{-\alpha}^{(s)}, [z, e_\alpha^{(s)}]] = - \sum_s [[z, e_{-\beta}^{(s)}], e_\beta^{(s)}] \text{ in } \mathfrak{g}(A), \tag{2.4.3}$$

$$\sum_s e_{-\alpha}^{(s)} [z, e_\alpha^{(s)}] = - \sum_s [z, e_{-\beta}^{(s)}] e_\beta^{(s)} \text{ in } U(\mathfrak{g}(A)). \tag{2.4.4}$$

Proof. Apply to (2.4.2) the linear maps from $\mathfrak{g}(A) \otimes \mathfrak{g}(A)$ to $\mathfrak{g}(A)$ and to $U(\mathfrak{g}(A))$, defined by $x \otimes y \mapsto [x, y]$ and $x \otimes y \mapsto xy$, respectively.

□

§2.5. Let $\mathfrak{g}(A)$ be a Lie algebra associated to a matrix A, $\mathfrak{h}$ the Cartan subalgebra, $\mathfrak{g} = \bigoplus_\alpha \mathfrak{g}_\alpha$ the root space decomposition with respect to $\mathfrak{h}$. A $\mathfrak{g}(A)$-module (resp. $\mathfrak{g}'(A)$-module) V is called *restricted* if for every $v \in V$, we have $\mathfrak{g}_\alpha(v) = 0$ for all but a finite number of positive roots α.

It is clear that every submodule or quotient of a restricted module is restricted, and that the direct sum or tensor product of a finite number of restricted modules is also restricted. Examples of restricted modules will be constructed later (see Exercise 2.9 and Chapter 9).

Assume now that A is symmetrizable and that $(.|.)$ is a bilinear form provided by Theorem 2.2.

Given a restricted $\mathfrak{g}(A)$-module V, we introduce a linear operator Ω on the vector space V, called the (generalized) *Casimir operator*, as follows.

First, introduce a linear function $\rho \in \mathfrak{h}^*$ by equations

$$\langle \rho, \alpha_i^\vee \rangle = \frac{1}{2} a_{ii} \ (i = 1, \ldots, n).$$

If $\det A = 0$, this does not define ρ uniquely, and we pick any solution. It follows from (2.1.5 and 6) that

$$(\rho|\alpha_i) = \frac{1}{2}(\alpha_i|\alpha_i) \quad (i = 1, \ldots, n). \tag{2.5.1}$$

Further, for each positive root α we choose a basis $\{e_\alpha^{(i)}\}$ of the space $\mathfrak{g}_\alpha$, and let $\{e_{-\alpha}^{(i)}\}$ be the dual basis of $\mathfrak{g}_{-\alpha}$. We define an operator Ω_0 on V by

$$\Omega_0 = 2 \sum_{\alpha \in \Delta_+} \sum_i e_{-\alpha}^{(i)} e_\alpha^{(i)}.$$

One could easily check that this is independent of the choice of bases (see Exercise 2.7). Since for each $v \in V$, only a finite number of summands $e_{-\alpha}^{(i)} e_\alpha^{(i)}(v)$ are nonzero, Ω_0 is well defined on V. Let $u_1, u_2, \ldots$ and $u^1, u^2, \ldots$ be dual bases of $\mathfrak{h}$. The generalized Casimir operator is defined by

$$\Omega = 2\nu^{-1}(\rho) + \sum_i u^i u_i + \Omega_0.$$

We record the following simple formula:

$$\sum_i \langle \lambda, u^i \rangle \langle \mu, u_i \rangle = (\lambda|\mu), \tag{2.5.2}$$

which is clear from

$$\lambda = \sum_i \langle \lambda, u^i \rangle \nu(u_i) = \sum_i \langle \lambda, u_i \rangle \nu(u^i). \tag{2.5.3}$$

We make one more simple computation. For $x \in \mathfrak{g}_\alpha$ one has

$$\left[\sum_i u^i u_i, x\right] = \sum_i \langle \alpha, u^i \rangle x u_i + \sum_i u^i \langle \alpha, u_i \rangle x$$
$$= \sum_i \langle \alpha, u^i \rangle \langle \alpha, u_i \rangle x + x\left(\sum_i u^i \langle \alpha, u_i \rangle + u_i \langle \alpha, u^i \rangle\right).$$

Hence, we have

$$\left[\sum_i u^i u_i, x\right] = x((\alpha|\alpha) + 2\nu^{-1}(\alpha)) \text{ for } x \in \mathfrak{g}_\alpha. \tag{2.5.4}$$

§2.6. Consider the root space decomposition of $U(\mathfrak{g}(A))$ with respect to $\mathfrak{h}$:

$$U(\mathfrak{g}(A)) = \bigoplus_{\beta \in Q} U_\beta, \text{ where}$$
$$U_\beta = \{x \in U(\mathfrak{g}(A)) | [h, x] = \langle \beta, h \rangle x \text{ for all } h \in \mathfrak{h}\}.$$

Put $U'_\beta = U(\mathfrak{g}'(A)) \cap U_\beta$, so that $U(\mathfrak{g}'(A)) = \bigoplus_\beta U'_\beta$. Now we prove the following important theorem:

THEOREM 2.6. *Let $\mathfrak{g}(A)$ be a symmetrizable Lie algebra.*
a) *If V is a restricted $\mathfrak{g}'(A)$-module and $u \in U'_\alpha$, then:*

$$[\Omega_0, u] = -u\left(2(\rho|\alpha) + (\alpha|\alpha) + 2\nu^{-1}(\alpha)\right). \tag{2.6.1}$$

b) *If V is a restricted $\mathfrak{g}(A)$-module, then Ω commutes with the action of $\mathfrak{g}(A)$ on V.*

Proof. b) follows immediately from a) and (2.5.4). If a) holds for $u \in U'_\alpha$ and $u_1 \in U'_\beta$, then it holds for $uu_1 \in U'_{\alpha+\beta}$. Indeed

$$\begin{aligned}
[\Omega_0, uu_1] &= [\Omega_0, u]u_1 + u[\Omega_0, u_1] \\
&= -u(2(\rho|\alpha) + (\alpha|\alpha) + 2\nu^{-1}(\alpha))u_1 \\
&\qquad - uu_1(2(\rho|\beta) + (\beta|\beta) + 2\nu^{-1}(\beta)) \\
&= -uu_1(2(\rho|\alpha) + (\alpha|\alpha) + 2\nu^{-1}(\alpha) \\
&\qquad + 2(\alpha|\beta) + 2(\rho|\beta) + (\beta|\beta) + 2\nu^{-1}(\beta)) \\
&= -uu_1(2(\rho|\alpha+\beta) + (\alpha+\beta|\alpha+\beta) + 2\nu^{-1}(\alpha+\beta)).
\end{aligned}$$

Hence, since e_{α_i}, $e_{-\alpha_i}$ $(i = 1, \ldots, n)$ generate $\mathfrak{g}'(A)$, it suffices to check (2.6.1) for $u = e_{\alpha_i}$ or $e_{-\alpha_i}$. Applying (2.4.4) to $z = e_{\alpha_i}$ and using Lemma 1.3, we have:

$$\begin{aligned}
[\Omega_0, e_{\alpha_i}] &= 2 \sum_{\alpha \in \Delta_+} \sum_s ([e_{-\alpha}^{(s)}, e_{\alpha_i}] e_\alpha^{(s)} + e_{-\alpha}^{(s)} [e_\alpha^{(s)}, e_{\alpha_i}]) \\
&= 2[e_{-\alpha_i}, e_{\alpha_i}] e_{\alpha_i} + 2 \sum_{\alpha \in \Delta_+ \setminus \{\alpha_i\}} \Big(\sum_s [e_{-\alpha}^{(s)}, e_{\alpha_i}] e_\alpha^{(s)} \\
&\quad + \sum_s e_{-\alpha+\alpha_i}^{(s)} [e_{\alpha-\alpha_i}^{(s)}, e_{\alpha_i}] \Big) \\
&= -2\nu^{-1}(\alpha_i) e_{\alpha_i} = -2(\alpha_i|\alpha_i) e_{\alpha_i} - 2 e_{\alpha_i} \nu^{-1}(\alpha_i).
\end{aligned}$$

Thanks to (2.5.1), this is (2.6.1) for $u = e_{\alpha_i}$. Similarly, $[\Omega_0, e_{-\alpha_i}] = 2e_{-\alpha_i}[e_{\alpha_i}, \ e_{-\alpha_i}] = 2e_{-\alpha_i}\nu^{-1}(\alpha_i)$, which, by (2.5.1), is (2.6.1) for $u = e_{-\alpha_i}$.

□

COROLLARY 2.6. *If, under the hypotheses of Theorem 2.6 b), there exists $v \in V$ such that $e_i(v) = 0$ for all $i = 1, \ldots, n$, and $h(v) = \langle \Lambda, h \rangle v$ for some $\Lambda \in \mathfrak{h}^*$ and all $h \in \mathfrak{h}$, then*

$$\Omega(v) = (\Lambda + 2\rho|\Lambda)v. \tag{2.6.2}$$

If, furthermore, $U(\mathfrak{g}(A))v = V$, then

$$\Omega = (\Lambda + 2\rho|\Lambda)I_V. \tag{2.6.3}$$

Proof. Formula (2.6.2) follows from the definition of Ω and formula (2.5.2). Formula (2.6.3) follows from (2.6.2) and Theorem 2.6.

□

Remark 2.6. One can define the Q-gradation $U(\mathfrak{g}'(A)) = \bigoplus_{\beta \in Q} U'_\beta$ and the map $\nu^{-1} : Q \to \mathfrak{h}'$ without using the Lie algebra $\mathfrak{g}(A)$. (Here, as in Remark 1.5, the symbol Q denotes the free abelian group on generators $\alpha_1, \ldots, \alpha_n$.) Indeed, the Q-gradation of $U(\mathfrak{g}'(A))$ is induced by the Q-gradation of $\mathfrak{g}'(A)$ defined in Remark 1.5. If A is symmetrizable, we fix a decomposition (2.1.1) and define ν^{-1} as a homomorphism of abelian groups such that $\nu^{-1}(\alpha_i) = \epsilon_i^{-1}\alpha_i^\vee$ $(i = 1, \ldots, n)$. These definitions work for infinite n as well.

§2.7. Let A be an $n \times n$-matrix over $\mathbf{R}$. Let $(\mathfrak{h}_{\mathbf{R}}, \Pi, \Pi^\vee)$ be a realization of the matrix A over $\mathbf{R}$, i.e., $\mathfrak{h}_{\mathbf{R}}$ is a vector space of dimension $2n - \ell$ over $\mathbf{R}$, so that $(\mathfrak{h} := \mathbf{C} \otimes_{\mathbf{R}} \mathfrak{h}_{\mathbf{R}}, \Pi, \Pi^\vee)$ is the realization of A over $\mathbf{C}$.

We define the *compact form* $\mathfrak{k}(A)$ of $\mathfrak{g}(A)$ as follows. Denote by ω_0 the antilinear automorphism of $\mathfrak{g}(A)$ determined by:

$$\begin{aligned} \omega_0(e_i) &= -f_i, \\ \omega_0(f_i) &= -e_i \quad (i = 1, \dots, n), \\ \omega_0(h) &= -h \text{ for } h \in \mathfrak{h}_{\mathbf{R}}. \end{aligned}$$

We call ω_0 the ***compact involution*** of $\mathfrak{g}(A)$. The existence of ω_0 is proved by the same argument as that of ω in §1.3. Then $\mathfrak{k}(A)$ is defined as the fixed point set of ω_0; this is a real Lie algebra whose complexification is $\mathfrak{g}(A)$. Note that this definition of the compact form coincides with the usual one in the finite-dimensional case.

Now let A be a symmetrizable matrix over $\mathbf{R}$ and let $(. | .)$ be a standard form on $\mathfrak{g}(A)$. Define a Hermitian form on $\mathfrak{g}(A)$ by:

$$(x|y)_0 := -(\omega_0(x)|y).$$

Theorem 2.2 implies the following properties of this Hermitian form. The restriction of $(. | .)_0$ to $\mathfrak{g}_\alpha$ is nondegenerate for all $\alpha \in \Delta \cup \{0\}$; $(\mathfrak{g}_\alpha | \mathfrak{g}_\beta)_0 = 0$ if $\alpha \neq \beta$; the operators $\operatorname{ad} u$ and $-\operatorname{ad} \omega_0(u)$ for $u \in \mathfrak{g}(A)$ are adjoint to each other, i.e., $([u, x]|y)_0 = -(x|[\omega_0(u), y])_0$ for all $x, y \in \mathfrak{g}(A)$; in particular, the restriction of $(. | .)_0$ to $\mathfrak{k}(A)$ is a nondegenerate invariant $\mathbf{R}$-bilinear form. We will return to the study of the Hermitian form $(. | .)_0$ after developing some representation theory.

§2.8. The following variation of the above results is very useful for applications.

PROPOSITION 2.8. *Let $\mathfrak{g}$ be a Lie algebra with an invariant non-degenerate bilinear form $(. | .)$, let $\{x_i\}$ and $\{y_i\}$ be dual bases (i.e. $(x_i \mid y_j) = \delta_{ij}$), and let V be a $\mathfrak{g}$-module such that for every pair of elements $u, v \in V$, $x_i(u) = 0$ or $y_i(v) = 0$ for all but a finite number of i. Then the operator*

$$\Omega_2 := \sum_i x_i \otimes y_i$$

is defined on $V \otimes V$ and commutes with the action of $\mathfrak{g}$.

Proof. We have to show that for every $z \in \mathfrak{g}$ one has:

$$\sum_i ([z, x_i] \otimes y_i + x_i \otimes [z, y_i]) = 0. \tag{2.8.1}$$

Write: $[z, x_i] = \sum_j \alpha_{ij} x_j$, $[z, y_i] = \sum_j \beta_{ji} y_j$; taking inner product of the first equation with y_j and the second with x_j, we obtain:

$$\alpha_{ij} = ([z, x_i] | y_j), \quad \beta_{ji} = ([z, y_i] | x_j).$$

Using invariance of $(. | .)$, we deduce:

$$\alpha_{ij} = (z \,|\, [x_i, y_j]), \quad \beta_{ji} = (z \,|\, [y_i, x_j]).$$

It follows that $\alpha_{ij} = -\beta_{ij}$, proving (2.8.1).

□

§2.9. Here we consider the most degenerate example, the Lie algebra $\mathfrak{g}(0)$ associated to the $n \times n$ zero matrix (including $n = \infty$). In this case $[e_i, e_j] = 0$, $[f_i, f_j] = 0$, $[e_i, f_j] = \delta_{ij} \alpha_i^\vee$, $(i, j = 1, \ldots, n)$, so that

$$\mathfrak{g}(0) = \mathfrak{h} \oplus \sum_i \mathbb{C} e_i \oplus \sum_i \mathbb{C} f_i.$$

The center of $\mathfrak{g}(0)$ is $\mathfrak{c} = \sum_{i=1}^n \mathbb{C} \alpha_i^\vee$. Furthermore, $\dim \mathfrak{h} = 2n$ and one can choose elements $d_1, \ldots, d_n \in \mathfrak{h}$ such that

$$\mathfrak{h} = \mathfrak{c} + \sum_{i=1}^n \mathbb{C} d_i,$$

and

$$[d_i, e_j] = \delta_{ij} e_j, \; [d_i, f_j] = -\delta_{ij} f_j \; (i, j = 1, \ldots, n).$$

One defines a nondegenerate symmetric invariant bilinear form on the basis of $\mathfrak{g}(0)$ by:

$$(e_i | f_i) = 1, \; (\alpha_i^\vee | d_i) = 1, \; \text{all the others} = 0.$$

Note that $\rho = 0$ and the Casimir operator is

$$\Omega = 2 \sum_i \alpha_i^\vee d_i + 2 \sum_i f_i e_i.$$

Set $\mathfrak{c}_1 = \sum \mathbb{C}(\alpha_i^\vee - \alpha_j^\vee) \subset \mathfrak{c}$. Then the Lie algebra $\mathfrak{g}'(0)/\mathfrak{c}_1$ is a *Heisenberg Lie algebra of order n*, i.e., it has a basis e_i, f_i $(i = 1, \ldots, n)$, z, such that $[e_i, f_j] = \delta_{ij} z$ $(i, j = 1, \ldots, n)$, and all the other brackets are zero.

§2.10. Exercises.

2.1. The matrix $A = (a_{ij})$ is symmetrizable if and only if

$$a_{ij} = 0 \text{ implies } a_{ji} = 0 \text{ and}$$
$$a_{i_1 i_2} a_{i_2 i_3} \dots a_{i_k i_1} = a_{i_2 i_1} a_{i_3 i_2} \dots a_{i_1 i_k} \text{ for all } i_1, \dots, i_k.$$

2.2. Show that the bilinear form $(\,.\,|\,.\,)$ is uniquely defined by properties a) and b) of Theorem 2.2.

2.3. Let $(\,.\,|\,.\,)$ be a nondegenerate symmetric invariant bilinear form on $\mathfrak{g}(A)$. Show that the matrix A is symmetrizable, that $(\,.\,|\,.\,)|_{\mathfrak{h}}$ is defined by (2.1.2 and 3) for some choice of $\mathfrak{h}''$, and that $(\,.\,|\,.\,)$ satisfies all the properties of Theorem 2.2.
[Set $\epsilon_i = (e_i | f_i)$].

2.4. Let $\mathfrak{g} = \bigoplus_i \mathfrak{g}_i$ be a $\mathbb{Z}$-graded Lie algebra, which is generated by $\mathfrak{g}_{-1} + \mathfrak{g}_0 + \mathfrak{g}_1$. Show that an invariant symmetric bilinear form on the subspace $\mathfrak{g}_{-1} + \mathfrak{g}_0 + \mathfrak{g}_1$ such that $(\mathfrak{g}_i | \mathfrak{g}_j) = 0$ whenever $i + j \neq 0$, can be (uniquely) extended to such a form on the whole $\mathfrak{g}$. (Here the property of invariance is understood to hold whenever it makes sense.)

2.5. Let $(\,.\,|\,.\,)_1$ and $(\,.\,|\,.\,)_2$ be two nondegenerate symmetric invariant bilinear forms on $\mathfrak{g}(A)$ and assume that A is indecomposable. Show that there exists an automorphism ϕ of $\mathfrak{g}(A)$ which leaves $\mathfrak{g}'(A)$ pointwise fixed and preserves $\mathfrak{h}$, such that the bilinear forms $(x|y)_1$ and $(\phi(x)|\phi(y))_2$ are proportional. Show that any two invariant bilinear forms on $\mathfrak{g}'(A)$ are proportional.

2.6. Show that the adjoint representation of $\mathfrak{g}(A)$ is restricted if and only if $\dim \mathfrak{g}(A) < \infty$.

2.7. Let $\{x_i\}$ and $\{y_i\}$ be bases of $\mathfrak{n}_+$ and $\mathfrak{n}_-$, dual with respect to an invariant bilinear form on $\mathfrak{g}$. Show that $\Omega_0 = 2\sum_i y_i x_i$ is independent of the choice of these bases. Prove an analogous statement for Ω_2.

2.8. Let $\mathfrak{g} = \mathfrak{g}(A)$ be a simple finite-dimensional Lie algebra. Choose a basis $v_1, \dots, v_d$ of $\mathfrak{g}$ and the dual basis $v^1, \dots, v^d$ with respect to the form $(\,.\,|\,.\,)$. Set $\Omega = \sum_i v_i v^i$. Show that Ω coincides with the Casimir operator defined in §2.5. Show that ρ is the half-sum of positive roots of $\mathfrak{g}$.

2.9. Show that the $\tilde{\mathfrak{g}}(A)$-module $T(V)$ constructed in the proof of Theorem 1.2 has a unique maximal submodule $J(V)$. Show that $T(V)/J(V)$ is an irreducible restricted $\mathfrak{g}(A)$-module.

2.10. Let $\mathfrak{g}_0$ be a finite-dimensional Lie algebra with the bracket $[\, ,\,]_0$ and a nondegenerate invariant symmetric bilinear form $(.|.)$. Let $\bar{d}$ be a derivation of the Lie algebra $\mathfrak{g}_0$ such that $(\bar{d}(x)|y) = -(x|\bar{d}(y))$ for $x, y \in \mathfrak{g}_0$. Set $\mathfrak{g} = \mathfrak{g}_0 \oplus \mathbb{C}c \oplus \mathbb{C}d$, where c and d are some symbols, and define a bracket $[\, ,\,]$ on $\mathfrak{g}$ by:

$$[x, y] = [x, y]_0 + (\bar{d}(x)|y)c \text{ for } x, y \in \mathfrak{g}_0,$$
$$[c, x] = 0 \text{ for } x \in \mathfrak{g};\ [d, x] = \bar{d}(x).$$

Check that this is a Lie algebra operation. Extend the bilinear form $(.|.)$ from $\mathfrak{g}_0$ to $\mathfrak{g}$ by setting

$$(d|d) = (c|c) = 0,\ (c|d) = 1,\ (c|\mathfrak{g}_0) = 0,\ (d|\mathfrak{g}_0) = 0.$$

Show that this is a nondegenerate symmetric invariant bilinear form on $\mathfrak{g}$.

2.11. Let $\mathfrak{g}$ be a solvable n-dimensional Lie algebra with a nondegenerate invariant symmetric bilinear form. Show that either $\mathfrak{g}$ is an orthogonal direct sum of $(n-1)$-dimensional and 1-dimensional Lie algebras, or $\mathfrak{g}$ can be constructed as in Exercise 2.10 from an $(n-2)$-dimensional Lie algebra $\mathfrak{g}_0$.

[Let $\mathfrak{i}$ be a 1-dimensional ideal in $\mathfrak{g}$. Show that $\mathfrak{i}$ lies in the center (using the fact that coadjoint orbits are even-dimensional). If $\mathfrak{i}$ is isotropic, consider $\mathfrak{i}^{\perp}$ and the Lie algebra $\mathfrak{g}_0 = \mathfrak{i}^{\perp}/\mathfrak{i}$.]

§2.11. Bibliographical notes and comments.

Theorem 2.2 is due to Kac [1968 B] and Moody [1968]. For finite-dimensional semi-simple Lie algebras this result is known due to the existence of the Killing–Cartan form, which cannot be defined in the infinite-dimensional case.

The generalized Casimir operator Ω was introduced by Kac [1974]. The idea of its definition is borrowed from physics. We take the usual definition of the Casimir operator (see Exercise 2.8):

$$\Omega = \sum_{\alpha>0}\sum_i \left(e_{-\alpha}^{(i)}\, e_{\alpha}^{(i)} + e_{\alpha}^{(i)}\, e_{-\alpha}^{(i)}\right) + \sum_i u_i u^i$$

we rewrite it by using commutation relations:

$$\Omega = \sum_{\alpha>0} \nu^{-1}(\alpha) + 2\sum_{\alpha>0}\sum_i e_{-\alpha}^{(i)}\, e_{\alpha}^{(i)} + \sum_i u_i u^i,$$

and then replace the first summand, which makes no sense, by a "finite" quantity $2\nu^{-1}(\rho)$.

The proof of Theorem 2.6 is along the lines of Kac–Peterson [1984 B].

Some of the exercises, such as Exercises 2.10 and 2.11, seem to be new. (The simplification of these exercises in the present edition was pointed out to me by G. Favre and L.J. Santharoubane.) The rest of the material of Chapter 2 is fairly standard.

A complete system of "higher order" Casimir operators was constructed by Kac [1984], using ideas of Feigin–Fuchs [1983A].

Chapter 3. Integrable Representations of Kac–Moody Algebras and the Weyl Group

§3.0. In this chapter we begin a systematic study of the Kac–Moody algebras. Recall that this is the Lie algebra $\mathfrak{g}(A)$ associated to a generalized Cartan matrix A. The main object of the chapter is the Weyl group W of a Kac–Moody algebra, which is a generalization of the classical Weyl group in the finite-dimensional theory. However, in contrast to the finite-dimensional case, W is infinite and the union of the W-translates of the fundamental chamber is a convex cone, which does not coincide with the whole "real" Cartan subalgebra $\mathfrak{h}_{\mathbb{R}}$.

§3.1. Let us first make some remarks on the duality of Kac–Moody algebras. Let A be a generalized Cartan matrix; then its transpose ${}^tA$ is again a generalized Cartan matrix. Let $(\mathfrak{h}, \Pi, \Pi^\vee)$ be a realization of A; then $(\mathfrak{h}^*, \Pi^\vee, \Pi)$ is clearly a realization of ${}^tA$. So, if $(\mathfrak{g}(A), \mathfrak{h}, \Pi, \Pi^\vee)$ is the quadruple associated to A, then $(\mathfrak{g}({}^tA), \mathfrak{h}^*, \Pi^\vee, \Pi)$ is the quadruple associated to ${}^tA$. The Kac–Moody algebras $\mathfrak{g}(A)$ and $\mathfrak{g}({}^tA)$ are called *dual* to each other.

Note that the *dual root lattice* of $\mathfrak{g}(A)$:

$$Q^\vee := \sum_{i=1}^{n} \mathbb{Z}\alpha_i^\vee$$

is the root lattice of $\mathfrak{g}({}^tA)$. Furthermore, denote by $\Delta^\vee \subset Q^\vee$ $(\subset \mathfrak{h}' \subset \mathfrak{h})$ the root system $\Delta({}^tA)$ of $\mathfrak{g}({}^tA)$. This is called the *dual root system* of $\mathfrak{g}(A)$. In contrast to the finite-dimensional case, there is no natural bijection between Δ and $\Delta^\vee$.

§3.2. Recall some well-known results about representations of the Lie algebra $s\ell_2(\mathbf{C})$. Let

$$e = \begin{pmatrix} 0 & 1 \\ 0 & 0 \end{pmatrix}, h = \begin{pmatrix} 1 & 0 \\ 0 & -1 \end{pmatrix}, f = \begin{pmatrix} 0 & 0 \\ 1 & 0 \end{pmatrix}$$

be the standard basis of $s\ell_2(\mathbf{C})$. Then

$$[e, f] = h, \ [h, e] = 2e, \ [h, f] = -2f.$$

By an easy induction on k we deduce the following relations in the universal enveloping algebra of $s\ell_2(\mathbf{C})$:

$$[h, f^k] = -2kf^k, \ [h, e^k] = 2ke^k, \tag{3.2.1}$$

$$[e, f^k] = -k(k-1)f^{k-1} + kf^{k-1}h. \tag{3.2.2}$$

LEMMA 3.2. a) *Let V be an $s\ell_2(\mathbf{C})$-module and let $v \in V$ be such that*

$$h(v) = \lambda v \quad \textit{for some } \lambda \in \mathbf{C}.$$

Set $v_j = (j!)^{-1} f^j(v)$. Then:

$$h(v_j) = (\lambda - 2j)v_j. \tag{3.2.3}$$

If, in addition, $e(v) = 0$, then:

$$e(v_j) = (\lambda - j + 1)v_{j-1}. \tag{3.2.4}$$

b) *For each integer $k \geq 0$ there exists a unique, up to isomorphism, irreducible $(k+1)$-dimensional $s\ell_2(\mathbf{C})$-module. In some $\mathbf{C}$-basis $\{v_j\}_{j=0}^k$ of the space of this module the action of $s\ell_2(\mathbf{C})$ looks as follows:*

$$h(v_j) = (k - 2j)v_j; \ f(v_j) = (j+1)v_{j+1}; \quad e(v_j) = (k + 1 - j)v_{j-1}.$$

Here $j = 0, \ldots, k$ and we assume that $v_{k+1} = 0 = v_{-1}$.

Proof. Formulas (3.2.3 and 4) follow from (3.2.1 and 2). Let now V be an irreducible $(k+1)$-dimensional $s\ell_2(\mathbf{C})$-module. Let $u \in V$ be an eigenvector of h with eigenvalue μ. It follows from (3.2.1) that if $e^s(u) \neq 0$, then it is again an eigenvector for h with eigenvalue $\mu + 2s$. Since V is finite dimensional, there exists $v = e^s(u) \neq 0$, such that $e(v) = 0$ and $h(v) = \lambda v$. We set $v_j = (j!)^{-1} f^j(v)$. As V is finite dimensional, we deduce from (3.2.4) that λ is a nonnegative integer, say m, that $\{v_j\}$ are linearly independent for $j = 0, \ldots, m$ and that $v_{m+1} = 0$. Hence $m = k$ and b) follows.

□

§3.3. Let $\mathfrak{g}(A)$ be a Kac–Moody algebra and let e_i, f_i $(i = 1, \dots, n)$ be its Chevalley generators. Set $\mathfrak{g}_{(i)} = \mathbb{C}e_i + \mathbb{C}\alpha_i^\vee + \mathbb{C}f_i$; then $\mathfrak{g}_{(i)}$ is isomorphic to $s\ell_2(\mathbb{C})$, with standard basis $\{e_i, \alpha_i^\vee, f_i\}$.

We can deduce now the following relations between the Chevalley generators:

$$(3.3.1) \qquad (\operatorname{ad} e_i)^{1-a_{ij}} e_j = 0; \quad (\operatorname{ad} f_i)^{1-a_{ij}} f_j = 0, \text{ if } i \neq j.$$

We prove the second relation; the first one follows by making use of the Chevalley involution ω (see (1.3.4)).

Denote $v = f_j$, $\theta_{ij} = (\operatorname{ad} f_i)^{1-a_{ij}} f_j$. Consider $\mathfrak{g}(A)$ as a $\mathfrak{g}_{(i)}$-module by restricting the adjoint representation. We have:

$$\alpha_i^\vee(v) = -a_{ij} v; \; e_i(v) = 0 \text{ if } i \neq j.$$

Hence Lemma 3.2 a) together with the properties (C1) and (C2) of A imply

$$[e_i, \theta_{ij}] = (1 - a_{ij})(-a_{ij} - (1 - a_{ij}) + 1)(\operatorname{ad} f_i)^{-a_{ij}} f_j = 0 \text{ if } i \neq j.$$

It is also clear that e_k commutes with θ_{ij} if $k \neq i$, $k \neq j$ (by relations (1.2.1)), and also if $k = j$ but $a_{ij} \neq 0$. Finally, if $k = j$ and $a_{ij} = 0$ we have:

$$[e_j, \theta_{ij}] = [e_j, [f_i, f_j]] = a_{ji} f_i = 0 \text{ (by (C3))}.$$

So, $[e_k, \theta_{ij}] = 0$ for all k and we apply Lemma 1.5.

□

§3.4. Now we need a general fact about a module V over a Lie algebra $\mathfrak{g}$. One says that an element $x \in \mathfrak{g}$ is *locally nilpotent* on V if for any $v \in V$ there exists a positive integer N such that $x^N(v) = 0$.

LEMMA 3.4. a) *Let $y_1, y_2, \dots$ be a system of generators of a Lie algebra $\mathfrak{g}$ and let $x \in \mathfrak{g}$ be such that $(\operatorname{ad} x)^{N_i} y_i = 0$ for some positive integers N_i, $i = 1, 2, \dots$. Then $\operatorname{ad} x$ is locally nilpotent on $\mathfrak{g}$.*

b) *Let $v_1, v_2, \dots$ be a system of generators of a $\mathfrak{g}$-module V, and let $x \in \mathfrak{g}$ be such that $\operatorname{ad} x$ is locally nilpotent on $\mathfrak{g}$ and $x^{N_i}(v_i) = 0$ for some positive integers N_i, $i = 1, 2, \dots$. Then x is locally nilpotent on V.*

Proof. Since $\operatorname{ad} x$ is a derivation of $\mathfrak{g}$, one has the Leibnitz formula:

$$(\operatorname{ad} x)^k [y, z] = \sum_{i=0}^{k} \binom{k}{i} [(\operatorname{ad} x)^i y, (\operatorname{ad} x)^{k-i} z],$$

proving a) by induction on the length of commutators in the y_i. b) follows from the following formula (for $\lambda = \mu = 0$):

$$(3.4.1)\quad (x-\lambda-\mu)^k a = \sum_{s=0}^{k} \binom{k}{s} ((\operatorname{ad} x - \lambda)^s a)(x-\mu)^{k-s}, \quad k \geq 0, \ \lambda, \mu \in \mathbb{C},$$

which holds in any associative algebra. In order to prove (3.4.1), note that $\operatorname{ad} x = L_x - R_x$, where L_x and R_x are the operators of left and right multiplication by x, and that L_x and R_x commute (by associativity). Now we apply the binomial formula to $L_x - \lambda - \mu = (\operatorname{ad} x - \lambda) + (R_x - \mu)$.

□

Applying the binomial formula to $\operatorname{ad} x = L_x - R_x$, we obtain another useful formula (in any associative algebra):

$$(3.4.2)\qquad (\operatorname{ad} x)^k a = \sum_{s=0}^{k} (-1)^s \binom{k}{s} x^{k-s} a x^s .$$

§**3.5.** Lemma 3.5. $\operatorname{ad} e_i$ *and* $\operatorname{ad} f_i$ *are locally nilpotent on* $\mathfrak{g}(A)$.

Proof. By relations (1.2.1) and (3.3.1), we have: $(\operatorname{ad} e_i)^{|a_{ij}|+1} x = 0 = (\operatorname{ad} f_i)^{|a_{ij}|+1} x$, if x is e_j or f_j. Also, $(\operatorname{ad} e_i)^2 h = 0 = (\operatorname{ad} f_i)^2 h$ if $h \in \mathfrak{h}$. Now we can apply Lemma 3.4a).

□

§**3.6.** A $\mathfrak{g}(A)$-module V is called $\mathfrak{h}$-*diagonalizable* if $V = \bigoplus_{\lambda \in \mathfrak{h}^*} V_\lambda$, where $V_\lambda = \{v \in V \,|\, h(v) = \langle \lambda, h \rangle v \text{ for } h \in \mathfrak{h}\}$. As usual, V_λ is called a *weight space*, $\lambda \in \mathfrak{h}^*$ is called a *weight* if $V_\lambda \neq 0$, and $\dim V_\lambda$ is called the *multiplicity* of λ and is denoted by $\operatorname{mult}_V \lambda$. Similarly, one defines an $\mathfrak{h}'$-diagonalizable $\mathfrak{g}'(A)$-module, its weights, etc.

An $\mathfrak{h}$-(resp. $\mathfrak{h}'$-)diagonalizable module over a Kac–Moody algebra $\mathfrak{g}(A)$ (resp. $\mathfrak{g}'(A)$) is called *integrable* if all e_i and f_i ($i = 1, \ldots, n$) are locally nilpotent on V.

Note that the underlying module of the adjoint representation of a Kac–Moody algebra is an integrable module by Lemma 3.5. The following important proposition justifies the term "integrable."

PROPOSITION 3.6. *Let V be an integrable $\mathfrak{g}(A)$-module.*

a) *As a $\mathfrak{g}_{(i)}$-module, V decomposes into a direct sum of finite dimensional irreducible $\mathfrak{h}$-invariant modules (hence the action of $\mathfrak{g}_{(i)}$ on V can be "integrated" to the action of the group $SL_2(\mathbb{C})$).*

b) *Let $\lambda \in \mathfrak{h}^*$ be a weight of V and let α_i be a simple root of $\mathfrak{g}(A)$. Denote by M the set of all $t \in \mathbb{Z}$ such that $\lambda + t\alpha_i$ is a weight of V, and let $m_t = \text{mult}_V(\lambda + t\alpha_i)$. Then*

(i) *M is the closed interval of integers $[-p, q]$, where p and q are both either nonnegative integers or ∞ and $p - q = \langle \lambda, \alpha_i^\vee \rangle$ when both p and q are finite; if $\text{mult}_V \lambda < \infty$, then p and q are finite;*

(ii) *$e_i : V_{\lambda+t\alpha_i} \to V_{\lambda+(t+1)\alpha_i}$ is an injection for $t \in [-p, -\frac{1}{2}\langle \lambda, \alpha_i^\vee \rangle)$; in particular, the function $t \mapsto m_t$ increases on this interval;*

(iii) *the function $t \mapsto m_t$ is symmetric with respect to $t = -\frac{1}{2}\langle \lambda, \alpha_i^\vee \rangle$;*

(iv) *if both λ and $\lambda + \alpha_i$ are weights, then $e_i(V_\lambda) \neq 0$.*

Proof. By (3.2.2) we have:

$$e_i f_i^k(v) = k(1 - k + \langle \lambda, \alpha_i^\vee \rangle) f_i^{k-1}(v) + f_i^k e_i(v) \text{ for } v \in V_\lambda. \tag{3.6.1}$$

We deduce that the subspace

$$U = \sum_{k,m \geq 0} f_i^k e_i^m(v)$$

is $(\mathfrak{g}_{(i)} + \mathfrak{h})$-invariant. As e_i and f_i are locally nilpotent on V, $\dim U < \infty$. By the Weyl complete reducibility theorem applied to the $\mathfrak{g}_{(i)}$-module U, the latter decomposes into a direct sum of finite dimensional $\mathfrak{h}$-invariant irreducible $\mathfrak{g}_{(i)}$-modules (cf. Exercise 3.11). So, each $v \in V$ lies in a direct sum of finite dimensional $\mathfrak{h}$-invariant irreducible $\mathfrak{g}_{(i)}$-modules, and a) follows.

For the proof of b) we use a) and Lemma 3.2 b). Set $U = \sum_{k \in \mathbb{Z}} V_{\lambda + k\alpha_i}$; this is a $(\mathfrak{g}_{(i)} + \mathfrak{h})$-module, which is a direct sum of finite dimensional irreducible $(\mathfrak{g}_{(i)} + \mathfrak{h})$-modules. Let $p = -\inf M$, $q = \sup M$. Both p and q are nonnegative as $0 \in M$. Now all the statements of b) follow from Lemma 3.2 b), as $\langle \lambda + t\alpha_i, \alpha_i^\vee \rangle = 0$ for $t = -\frac{1}{2}\langle \lambda, \alpha_i^\vee \rangle$. □

The following corollary of Proposition 3.6 b) (i) and (iii) is very useful.

COROLLARY 3.6. a) *If λ is a weight of an integrable $\mathfrak{g}(A)$-module V and $\lambda + \alpha_i$ (resp. $\lambda - \alpha_i$) is not a weight, then $\langle \lambda, \alpha_i^\vee \rangle \geq 0$ (resp. $\langle \lambda, \alpha_i^\vee \rangle \leq 0$).*

b) *If λ is a weight of V, then $\lambda - \langle \lambda, \alpha_i^\vee \rangle \alpha_i$ is also a weight of the same multiplicity.*

Remark 3.6. Let V be an integrable $\mathfrak{g}'(A)$-module. Then, clearly, Proposition 3.6 and Corollary 3.6, with $\mathfrak{h}$ replaced by $\mathfrak{h}'$, still hold. Furthermore, the local nilpotency of e_i and f_i on V guarantees that V is h_i-diagonalizable, and hence $\mathfrak{h}'$-diagonalizable provided that $n < \infty$. This follows from a general fact which will be proved in §3.8.

§3.7. Now we introduce the important notion of the Weyl group of a Kac–Moody algebra $\mathfrak{g}(A)$. For each $i = 1, \dots, n$ we define the *fundamental reflection* r_i of the space $\mathfrak{h}^*$ by

$$r_i(\lambda) = \lambda - \langle \lambda, \alpha_i^\vee \rangle \alpha_i, \ \lambda \in \mathfrak{h}^*.$$

It is clear that r_i is a reflection since its fixed point set is $T_i = \{\lambda \in \mathfrak{h}^* | \langle \lambda, \alpha_i^\vee \rangle = 0\}$, and $r_i(\alpha_i) = -\alpha_i$.

The subgroup W of $GL(\mathfrak{h}^*)$ generated by all fundamental reflections is called the *Weyl group* of $\mathfrak{g}(A)$. We will write $W(A)$ when necessary to emphasize the dependence on A.

The action of r_i on $\mathfrak{h}^*$ induces the dual fundamental reflection $r_i^\vee$ on $\mathfrak{h}$ (for the dual algebra $\mathfrak{g}({}^tA)$). Hence the Weyl groups of dual Kac–Moody algebras are contragredient linear groups; this allows us to identify these groups.

The following proposition is an immediate consequence of Corollary 3.6 b) and Lemma 3.5.

PROPOSITION 3.7. a) *Let V be an integrable module over a Kac–Moody algebra $\mathfrak{g}(A)$. Then* $\operatorname{mult}_V \lambda = \operatorname{mult}_V w(\lambda)$ *for every $\lambda \in \mathfrak{h}^*$ and $w \in W$. In particular, the set of weights of V is W-invariant.*

b) *The root system Δ of $\mathfrak{g}(A)$ is W-invariant, and* $\operatorname{mult} \alpha = \operatorname{mult} w(\alpha)$ *for every $\alpha \in \Delta, w \in W$.*

The following fact will be needed later.

LEMMA 3.7. *If $\alpha \in \Delta_+$ and $r_i(\alpha) < 0$, then $\alpha = \alpha_i$. In other words, $\Delta_+ \backslash \{\alpha_i\}$ is r_i-invariant.*

Proof follows from Lemma 1.3.

□

§3.8. In this section we outline a somewhat different approach to the Weyl group. Let a be a locally nilpotent operator on a vector space V. Then we can define the exponential

$$\exp a := I_V + \frac{1}{1!}a + \frac{1}{2!}a^2 + \dots,$$

which has the usual properties, in particular,

$$\exp ka = (\exp a)^k \quad (k \in \mathbb{Z}).$$

Let b be another operator on V, such that $(\operatorname{ad} a)^N b = 0$ for some N. Then one knows the following formula:

$$(\exp a)b(\exp -a) = (\exp(\operatorname{ad} a))(b). \tag{3.8.1}$$

This easily follows by using formula (3.4.2).

LEMMA 3.8. *Let π be an integrable representation of $\mathfrak{g}(A)$ on a vector space V. For $i = 1, \dots, n$ set*

$$r_i^\pi = (\exp f_i)(\exp -e_i)(\exp f_i).$$

Then

a) $r_i^\pi(V_\lambda) = V_{r_i(\lambda)}$

b) $r_i^{\text{ad}} \in \operatorname{Aut} \mathfrak{g}(A)$

c) $r_i^{\text{ad}}|_{\mathfrak{h}} = r_i$.

Proof. Let $v \in V_\lambda$. Then $h(r_i^\pi(v)) = \langle \lambda, h \rangle r_i^\pi(v)$ if $\langle \alpha_i, h \rangle = 0$. Hence for a) and c) we have only to check that $\alpha_i^\vee(r_i^\pi(v)) = -\langle \lambda, \alpha_i^\vee \rangle r_i^\pi(v)$. This follows from

$$(r_i^\pi)^{-1} \alpha_i^\vee r_i^\pi = -\alpha_i^\vee. \tag{3.8.2}$$

By (3.8.1) it is sufficient to check (3.8.2) only in the adjoint representation of $s\ell_2(\mathbb{C})$; using (3.8.1) again, one has to check (3.8.2) in the 2-dimensional natural representation of $s\ell_2(\mathbb{C})$. But in this representation we have

$$\exp f_i = \begin{pmatrix} 1 & 0 \\ 1 & 1 \end{pmatrix}, \exp(-e_i) = \begin{pmatrix} 1 & -1 \\ 0 & 1 \end{pmatrix}, r_i^\pi = \begin{pmatrix} 0 & -1 \\ 1 & 0 \end{pmatrix},$$

and (3.8.2) is clear.

b) follows from (3.8.1) applied to the adjoint representation. □

Remark 3.8. Let (V, π) be an integrable $\mathfrak{g}(A)$-module whose kernel lies in $\mathfrak{h}$. By Proposition 3.6, the action of the subalgebra $\mathfrak{g}_{(i)}$ $(i = 1, \ldots, n)$ on V can be integrated to a representation $\pi_i : SL_2(\mathbf{C}) \to GL(V)$. The groups $\pi_i(SL_2(\mathbf{C}))$ generate a subgroup G^π in $GL(V)$. The group G^π can be viewed as an "infinite dimensional" group associated to the Lie algebra $\mathfrak{g}(A)$. The elements $r_i^\pi = \pi_i\left(\begin{smallmatrix} 0 & -1 \\ 1 & 0 \end{smallmatrix}\right)$ $(i = 1, \ldots, n)$ generate a subgroup $\widetilde{W}^\pi \subset G^\pi$. The group $\widetilde{W}^\pi$ contains an abelian normal subgroup D^π generated by $(r_i^\pi)^2$ $(i = 1, \ldots, n)$ such that $W(A) \simeq \widetilde{W}^\pi / D^\pi$.

We conclude this subsection by the proof of a general fact a special case of which was mentioned in §3.6. One says that an element x of a Lie algebra $\mathfrak{g}$ is *locally finite* on a $\mathfrak{g}$-module V if any $v \in V$ lies in a x-invariant finite-dimensional subspace. Note that locally nilpotent and diagonizable elements are locally finite and that (3.8.1) still holds if a is locally finite on V and the linear span of the $(\operatorname{ad} a)^j b$ is finite-dimensional.

PROPOSITION 3.8. *Let $\mathfrak{g}$ be a Lie algebra and let V be a $\mathfrak{g}$-module such that $\mathfrak{g}$ is generated by the set F_V of all ad-locally finite elements which act locally finitely on V. Then*

a) *$\mathfrak{g}$ is spanned over $\mathbf{C}$ by F_V. In particular, if $\mathfrak{g}$ is generated by the set F of all ad-locally finite elements, then $\mathfrak{g}$ is spanned by F.*

b) *If $\dim \mathfrak{g} < \infty$, then V is $\mathfrak{g}$-locally finite (i.e. any element of V lies in a finite-dimensional $\mathfrak{g}$-submodule of V).*

Proof. Let $\mathfrak{g}_V$ denote the $\mathbf{C}$-span of F_V and let $a \in F_V$. Using (3.8.1) we deduce that $\mathfrak{g}_V$ is invariant with respect to automorphisms $\exp t(\operatorname{ad} a)$. Since $\lim_{t \to 0}((\exp t \operatorname{ad} a)b - b)/t = [a, b]$, it follows that $[a, \mathfrak{g}_V] \subset \mathfrak{g}_V$, proving a). b) follows from a) and the Poincaré–Birkhoff–Witt theorem.

□

§3.9. Let A be a symmetrizable generalized Cartan matrix and let $(.|.)$ be a standard invariant bilinear form on $\mathfrak{g}(A)$.

PROPOSITION 3.9. *The restriction of the bilinear form $(.|.)$ to $\mathfrak{h}^*$ is W-invariant.*

Proof. As $|r_i(\alpha_i)|^2 = |-\alpha_i|^2 = |\alpha_i|^2 \neq 0$, it suffices to check that $(\lambda|\alpha_i) = 0$ implies $(r_i(\lambda)|\alpha_i) = 0$. But $r_i(\lambda) = \lambda$ by (2.3.5).

□

For a converse statement see Exercise 3.3.

§3.10. We return to the study of the Weyl group W of a Kac–Moody algebra $\mathfrak{g}(A)$. Let us start with the following technical lemma.

LEMMA 3.10. *If α_i is a simple root and $r_{i_1}\dots r_{i_t}(\alpha_i) < 0$ then there exists s $(1 \leq s \leq t)$ such that*

$$r_{i_1}\dots r_{i_s}\dots r_{i_t}r_i = r_{i_1}\dots r_{i_{s-1}}r_{i_{s+1}}\dots r_{i_t}. \tag{3.10.1}$$

Proof. Set $\beta_k = r_{i_{k+1}}\dots r_{i_t}(\alpha_i)$ for $k < t$ and $\beta_t = \alpha_i$. Then, by the hypothesis, $\beta_0 < 0$; on the other hand, $\beta_t > 0$. Hence, for some s we have: $\beta_{s-1} < 0$, $\beta_s > 0$. But $\beta_{s-1} = r_{i_s}(\beta_s)$; hence by Lemma 3.7, $\beta_s = \alpha_{i_s}$, and we obtain:

$$\alpha_{i_s} = w(\alpha_i), \text{ where } w = r_{i_{s+1}}\dots r_{i_t}. \tag{3.10.2}$$

But

$$w(\alpha_i) = \alpha_j \ (w \in W) \Rightarrow w(\alpha_i^\vee) = \alpha_j^\vee. \tag{3.10.3}$$

Indeed, $w = \tilde{w}|_{\mathfrak{h}}$ for some $\tilde{w}$ from the subgroup in Aut $\mathfrak{g}$ generated by r_i^{ad} $(i = 1, \dots, n)$ (see §3.8). Applying $\tilde{w}$ to both sides of the equality $[\mathfrak{g}_{\alpha_i}, \mathfrak{g}_{-\alpha_i}] = \mathbb{C}\alpha_i^\vee$, we obtain (by Lemma 3.8 b)) $\mathbb{C}w(\alpha_i^\vee) = \mathbb{C}\alpha_j^\vee$. Since $w(\alpha_i)(w(\alpha_i^\vee)) = \langle\alpha_i, \alpha_i^\vee\rangle = 2$, we get $w(\alpha_i^\vee) = \alpha_j^\vee$.

Now we can conclude from (3.10.2 and 3) that

$$r_{i_s} = wr_iw^{-1}.$$

Multiplying both sides of this by $r_{i_1}\dots r_{i_{s-1}}$ on the left and by $r_{i_{s+1}}\dots r_{i_t}r_i$ on the right completes the proof. □

§3.11. The expression $w = r_{i_1}\dots r_{i_s} \in W$ is called *reduced* if s is minimal possible among all representations of $w \in W$ as a product of the r_i. Then s is called the *length* of w and is denoted by $\ell(w)$. Note that $\det_{\mathfrak{h}^*} r_i = -1$ and hence

$$\det_{\mathfrak{h}^*} w = (-1)^{\ell(w)} \text{ for } w \in W. \tag{3.11.1}$$

The following lemma is an important corollary of Lemma 3.10.

LEMMA 3.11. *Let $w = r_{i_1} \dots r_{i_t} \in W$ be a reduced expression and let α_i be a simple root. Then we have*

a) $\ell(wr_i) < \ell(w)$ *if and only if* $w(\alpha_i) < 0$.

b) $w(\alpha_{i_t}) < 0$.

c) *(Exchange condition) If $\ell(wr_i) < \ell(w)$, then there exists s, such that* $1 \le s \le t$ *and*

$$r_{i_s} r_{i_{s+1}} \dots r_{i_t} = r_{i_{s+1}} \dots r_{i_t} r_i.$$

Proof. By Lemma 3.10 (applied to w), $w(\alpha_i) < 0$ implies that $\ell(wr_i) < \ell(w)$. If now $w(\alpha_i) > 0$, then $wr_i(\alpha_i) < 0$ and hence $\ell(w) = \ell(wr_i^2) < \ell(wr_i)$, proving a). b) follows immediately from a). Finally, if $\ell(wr_i) < \ell(w)$, then a) implies $w(\alpha_i) < 0$ and applying Lemma 3.10 to w we deduce the exchange condition from (3.10.1), multiplying it by $(r_{i_1} \dots r_{i_{s-1}})^{-1}$ on the left and by r_i on the right.

□

§3.12. Now we are in a position to study the geometric properties of the action of the Weyl group. Recall the definition of $\mathfrak{h}_\mathbb{R} \subset \mathfrak{h}$ from §2.7. Note that $\mathfrak{h}_\mathbb{R}$ is stable under W since $Q^\vee \subset \mathfrak{h}_\mathbb{R}$. The set

$$C = \{h \in \mathfrak{h}_\mathbb{R} | \langle \alpha_i, h \rangle \ge 0 \text{ for } i = 1, \dots, n\}$$

is called the *fundamental chamber*. The sets $w(C)$, $w \in W$, re called *chambers*, and their union

$$X = \bigcup_{w \in W} w(C)$$

is called the *Tits cone*. We clearly have the corresponding dual notions of $C^\vee$ and $X^\vee$ in $\mathfrak{h}_\mathbb{R}^*$.

PROPOSITION 3.12. a) *For $h \in C$, the group $W_h = \{w \in W | w(h) = h\}$ is generated by the fundamental reflections which it contains.*

b) *The fundamental chamber C is a fundamental domain for the action of W on X, i.e., any orbit $W \cdot h$ of $h \in X$ intersects C in exactly one point. In particular, W operates simply transitively on chambers.*

c) $X = \{h \in \mathfrak{h}_\mathbb{R} | \langle \alpha, h \rangle < 0$ *only for a finite number of* $\alpha \in \Delta_+\}$. *In particular, X is a convex cone.*

d) $C = \{h \in \mathfrak{h}_\mathbb{R} |$ *for every* $w \in W$, $h - w(h) = \sum_i c_i \alpha_i^\vee$ *where* $c_i \ge 0\}$.

e) *The following conditions are equivalent:*

(i) $|W| < \infty$;

(ii) $X = \mathfrak{h}_{\mathbb{R}}$;

(iii) $|\Delta| < \infty$;

(iv) $|\Delta^\vee| < \infty$.

f) *If $h \in X$, then $|W_h| < \infty$ if and only if h lies in the interior of X.*

Proof. Let $w \in W$ and let $w = r_{i_1} \dots r_{i_s}$ be a reduced expression of w. Take $h \in C$ and suppose that $h' = w(h) \in C$. We have $\langle \alpha_{i_s}, h\rangle \geq 0$ and therefore $\langle w(\alpha_{i_s}), h'\rangle \geq 0$. But by Lemma 3.11 b), $w(\alpha_{i_s}) < 0$, and hence $\langle w(\alpha_{i_s}), h'\rangle \leq 0$. So, $\langle w(\alpha_{i_s}), h'\rangle = 0$ and $\langle \alpha_{i_s}, h\rangle = 0$. Hence $r_{i_s}(h) = h$ and both a) and b) follow by induction on $\ell(w)$.

Set $X' = \{h \in \mathfrak{h}_{\mathbb{R}} | \langle \alpha, h\rangle < 0$ only for a finite number of $\alpha \in \Delta_+\}$. It is clear that $C \subset X'$ and it follows from §3.7 that X' is r_i-invariant. Hence $X' \supset X$. In order to prove the reverse inclusion, let $h \in X'$ and set $M_h = \{\alpha \in \Delta_+ | \langle \alpha, h\rangle < 0\}$. By definition, $|M_h|$ is finite. If $M_h \neq \emptyset$, then $\alpha_i \in M_h$ for some i. But then it follows from Lemma 3.7, that $|M_{r_i(h)}| < |M_h|$. Induction on $|M_h|$ completes the proof of c).

The inclusion $\supset$ of d) is obvious. We prove the reverse inclusion by induction on $s = \ell(w)$. For $\ell(w) = 1$, d) is the definition of C. If $\ell(w) > 1$, et $w = r_{i_1} \dots r_{i_s}$. We have $h - w(h) = (h - r_{i_1} \dots r_{i_{s-1}}(h)) + r_{i_1} \dots r_{i_{s-1}}(h - r_{i_s}(h))$ and we apply the inductive assumption to the first summand and Lemma 3.11b) for $\Delta^\vee$ to the second summand, completing the proof.

Now we prove e). (i) $\Rightarrow$ (ii) since an element h' of $W \cdot h$ with maximal height of $(h' - h)$ lies in C. In order to show (ii) $\Rightarrow$ (iii) take h in the interior of C. Then $\langle \alpha, -h\rangle < 0$ for all $\alpha \in \Delta_+$ and hence $|\Delta| < \infty$ by c). (iii) $\Rightarrow$ (i) because of

$$\{w(\alpha) = \alpha \text{ for } w \in W \text{ and all } \alpha \in \Delta\} \Rightarrow w = 1. \tag{3.12.1}$$

To prove (3.12.1), note that if a reduced expression $w = r_{i_1} \dots r_{i_t}$ is non-trivial, then Lemma 3.11 b) implies that $w(\alpha_{i_t}) < 0$, contradiction. The fact that (iv) is equivalent to (i) follows by using the dual root system.

Finally, to prove f) we may assume that $h \in C$. Then f) follows from a) by applying the equivalence of e)i and e)ii to W_h operating on $\mathfrak{h}/\mathbb{C}h$.

□

Note that, by Proposition 3.12 b), the Weyl group W operates simply transitively on chambers, the stabilizer of a point from the interior of a chamber being trivial.

§3.13. Recall that the group with generators $r_1, \ldots, r_n$ and defining relations

(3.13.1) $$r_i^2 = 1 \ (i = 1, \ldots, n); \ (r_i r_j)^{m_{ij}} = 1 \ (i, j = 1, \ldots, n)$$

is called a *Coxeter* group. Here m_{ij} are positive integers or ∞ (we use the convention $x^\infty = 1$ for any x).

PROPOSITION 3.13. *The group W is a Coxeter group, where the m_{ij} are given by the following table:*

$a_{ij}a_{ji}$	0	1	2	3	≥ 4
m_{ij}	2	3	4	6	∞

Proof. First we check relations (3.13.1). The relation $r_i^2 = 1$ is obvious. Furthermore, the subspace $\mathfrak{t} := \mathbf{R}\alpha_i + \mathbf{R}\alpha_j$ is invariant with respect to both r_i and r_j, and the matrices of r_i and r_j in the basis α_i, α_j of $\mathfrak{t}$ are $\begin{pmatrix} -1 & -a_{ij} \\ 0 & 1 \end{pmatrix}$ and $\begin{pmatrix} 1 & 0 \\ -a_{ji} & -1 \end{pmatrix}$. Hence the matrix of $r_i r_j$ in this basis is

(3.13.2) $$\begin{pmatrix} -1 + a_{ij}a_{ji} & a_{ij} \\ -a_{ji} & -1 \end{pmatrix}$$

and we obtain

(3.13.3) $$\det{}_{\mathfrak{t}}(r_i r_j - \lambda I) = \lambda^2 + (2 - a_{ij}a_{ji})\lambda + 1.$$

Hence $r_i r_j$ has an infinite order if $a_{ij}a_{ji} \geq 4$ and the order is given by the table in the rest of the cases ($\mathfrak{h}^* = \mathfrak{t} \oplus \{\lambda \in \mathfrak{h}^* | \langle \lambda, \alpha_i^\vee \rangle = \langle \lambda, \alpha_j^\vee \rangle = 0\}$ if $a_{ij}a_{ji} \neq 4$).

Now we can refer to an abstract fact that relations (3.13.1) and the exchange condition (see Lemma 3.11 c)) imply that the group in question is a Coxeter group (see Bourbaki [1968], Ch. IV, no. 1.6). A more transparent geometric approach is outlined in Exercise 3.10. □

§3.14. Exercises.

3.1. Let A be a symmetrizable matrix and let $A = DB$ be a decomposition of the form (2.3.1). Show that ${}^tA$ is also symmetrizable; more explicitly ${}^tA = D^\vee B^\vee$, where $D^\vee = D^{-1}$, $B^\vee = DBD$. Show that the corresponding standard forms on $\mathfrak{h}$ and $\mathfrak{h}^*$ (defined in §2.1) induce each other.

3.2. Let e, f, h be the standard basis of $sl_2(\mathbf{C})$. Show that one has in the enveloping algebra:

$$\left[\frac{e^m}{m!}, \frac{f^n}{n!}\right] = \sum_{j=1}^{\min(m,n)} \frac{f^{n-j}}{(n-j)!} \binom{h-m-n-2j}{j} \frac{e^{m-j}}{(m-j)!}.$$

Deduce that if f is locally nilpotent on a $sl_2(\mathbf{C})$-module V, then V^e is h-diagonalizable.

[Prove this for $m = 1$ by induction on n, then use induction on m].

3.3. There exists a nondegenerate symmetric W-invariant bilinear form on $\mathfrak{h}$ if and only if A is symmetrizable. Any such form can be extended from $\mathfrak{h}$ to an invariant nondegenerate symmetric bilinear form on the entire Lie algebra $\mathfrak{g}(A)$. This form satisfies all the conclusions of Theorem 2.2.

3.4. Let V be an integrable $\mathfrak{g}(A)$-module. Show that

$$(r_i^\pi)^2(v) = (-1)^{\langle\lambda,\alpha_i^\vee\rangle} v \text{ for } v \in V_\lambda.$$

3.5. The set Δ_+ is uniquely defined by the following properties:

(i) $\Pi \subset \Delta_+ \subset Q_+$; $2\alpha_i \notin \Delta_+$ for $i = 1, \dots, n$;

(ii) if $\alpha \in \Delta_+, \alpha \neq \alpha_i$, then the set $\{\alpha + k\alpha_i; k \in \mathbf{Z}\} \cap \Delta_+$ is a "string" $\{\alpha - p\alpha_i, \dots, \alpha + q\alpha_i\}$, where $p, q \in \mathbf{Z}_+$ and $p - q = \langle\alpha, \alpha_i^\vee\rangle$;

(iii) if $\alpha \in \Delta_+ \backslash \Pi$ then $\alpha - \alpha_i \in \Delta_+$ for some i.

3.6. Prove that $\ell(w) = |\{\alpha \in \Delta_+ | w(\alpha) < 0\}|$.

3.7. Let $A = \begin{pmatrix} 2 & -a \\ -b & 2 \end{pmatrix}$ be a generalized Cartan matrix of rank 2, and assume that $ab \geq 4$. Show that the Weyl group $W(A)$ is an infinite dihedral group. Let $ab > 4$; then $(x\alpha_1 + y\alpha_2 | x\alpha_1 + y\alpha_2) = bx^2 - abxy + ay^2$ is a W-invariant quadratic form on $\mathfrak{h}^*_{\mathbf{R}}$, and $X^\vee \cup -X^\vee = \{\lambda \in \mathfrak{h}^*_{\mathbf{R}} | (\lambda|\lambda) < 0\} \cup \{0\}$.

3.8. Let $A = \begin{pmatrix} 2 & -2 & 0 \\ -2 & 2 & -1 \\ 0 & -1 & 2 \end{pmatrix}$, and let $W = W(A)$ be the associated Weyl group. Show that the map $r_1 \mapsto \begin{pmatrix} 1 & 0 \\ 0 & -1 \end{pmatrix}, r_2 \mapsto \begin{pmatrix} -1 & 1 \\ 0 & 1 \end{pmatrix}, r_3 \mapsto \begin{pmatrix} 0 & 1 \\ 1 & 0 \end{pmatrix}$ induces an isomorphism $\pi : W \to PGL_2(\mathbf{Z})$.

[Use the fact that $PSL_2(\mathbf{Z})$ is generated by elementary matrices $\pi(r_2)\pi(r_1)$ and $\pi(r_1)\pi(r_3)$].

3.9. Let $A = (a_{ij})$ be the matrix of Exercise 3.8, $\mathfrak{h}$ the Cartan subalgebra of $\mathfrak{g}(A), \alpha_1, \alpha_2, \alpha_3$ simple roots. Define a standard bilinear form on $\mathfrak{h}$ by $(\alpha_i|\alpha_j) = a_{ij}$. Set $\gamma_1 = -\alpha_1 - \alpha_2$, $\gamma_2 = \frac{1}{2}\alpha_1$, $\gamma_3 = -\alpha_1 - \alpha_2 - \alpha_3$. Define a map $\mu : \mathfrak{h}^* \to S_2(\mathbf{C})$ (symmetric 2×2 matrices) by $\mu(a\gamma_1 + b\gamma_2 + c\gamma_3) = \begin{pmatrix} a & b/2 \\ b/2 & c \end{pmatrix}$, and define the action of $PGL_2(\mathbf{Z})$ on $S_2(\mathbf{C})$ by $g(S) = gS({}^tg)$. Check that μ is a W-equivariant map, and that $(\alpha|\alpha) = -2\det\mu(\alpha)$ for $\alpha \in \mathfrak{h}^*$. Using Proposition 3.12 b), (5.10.2) and Exercise 3.8, deduce that a quadratic form $ax^2 + bxy + cy^2$ such that $a, b, c \in \mathbf{Z}$, $4ac \geq b^2$ and $a \geq 0$ can be transformed by $GL_2(\mathbf{Z})$ to a unique quadratic form such that $a \geq c \geq b \geq 0$.

3.10. In this exercise we outline a geometric proof of Proposition 3.13. Let W' be the Coxeter group on generators r_i ($i = 1, \ldots, n$) and relations (3.13.1), and let $\pi : W' \to W$ be the canonical homomorphism. We construct a topological space $U = W' \times C/(\sim)$, where W' is equipped with the discrete topology, the fundamental chamber C with the metric topology, and $\sim$ is the following equivalence relation: $(w_1, x) \sim (w_2, y) \Leftrightarrow \{x = y$ and $w_1^{-1}w_2$ lies in the subgroup of W' generated by those r_i which fix $x\}$. Define an action of W' on U by: $w(w_1, x) = (ww_1, x)$. This is obviously well defined. Show that there exists a unique continuous W'-equivariant map $\phi : U \to X$ such that $\phi(1, x) = x$ for $x \in C$ (W' operates on X via π). Let $Y = \{x | x$ is fixed by at least three reflections from $W\}$ and set $X' = X \backslash Y, U' = U \backslash \phi^{-1}(Y)$. Show that $\phi' : U' \to X'$ is a covering map. Deduce that ϕ' is a homeomorphism and hence π is an isomorphism.

3.11. Show that for an irreducible $\mathfrak{h}$-diagonalizable module over $s\ell_2(\mathbf{C})$ all weight spaces are 1-dimensional. Classify these modules. Classify the ones which are integrable. Prove that every finite-dimensional $sl_2(\mathbf{C})$-module is completely reducible.

[Use the Casimir operator and Exercise 3.2].

3.12. Let $w = r_{i_1} \ldots r_{i_t}$ be a reduced expression of $w \in W$. Let $\beta \in \Delta_+$ be such that $w^{-1}(\beta) < 0$; show that the sequence β, $r_{i_1}(\beta)$, $r_{i_2}r_{i_1}(\beta), \ldots$ contains a unique simple root, say $\alpha_{j(\beta)}$. Let $\lambda \in \mathfrak{h}^*$; show that

$$\lambda - w(\lambda) = \sum_{\beta : w^{-1}(\beta) < 0} \langle \lambda, \alpha^\vee_{j(\beta)} \rangle \beta.$$

[Use the identity $\lambda - w_1 r_i(\lambda) = (\lambda - w_1(\lambda)) + w_1(\lambda - r_i(\lambda))$ and induction on t.]

3.13. Show that the action of r_i^{ad} on $\mathfrak{g}(A)$ preserves every invariant bilinear form and commutes with the Chevalley involution.

3.14. **Show that $0 \to \mathfrak{c} \to \mathfrak{g}'(A) \to \mathfrak{g}'(A)/\mathfrak{c} \to 0$ is the universal central extension of $\mathfrak{g}'(A)/\mathfrak{c}$, where A is a generalized Cartan matrix.**

[Let $\varphi : \mathfrak{g} \to \mathfrak{g}'(A)/\mathfrak{c}$ be an epimorphism with a finite-dimensional kernel. Then $\varphi^{-1}(e_i + f_i + (\mathfrak{h}'/\mathfrak{c}))$ is isomorphic to $\tilde{\mathfrak{g}}_i \oplus \tilde{\mathfrak{h}}$, where $\tilde{\mathfrak{g}}_i \simeq s\ell_2(\mathbb{C})$ and $\tilde{\mathfrak{h}}$ is commutative, and is completely reducible on $\mathfrak{g}$. Let $\tilde{e}_i, \tilde{\alpha}_i^\vee, \tilde{f}_i \in \tilde{\mathfrak{g}}_i$ be the preimages of $e_i, \alpha_i^\vee, f_i$; show that they satisfy relations (1.2.1) and hence generate a quotient $\mathfrak{g}'$ of $\mathfrak{g}'(A)$ by a central ideal. Show that $\mathfrak{g}$ is a direct sum of $\mathfrak{g}'$ and a central ideal from $\tilde{\mathfrak{h}}$].

3.15. **Show that Interior $X = \{h \in \mathfrak{h}_\mathbb{R} | \langle \alpha, h \rangle \leq 0$ only for a finite number of $\alpha \in \Delta_+\}$.**

3.16. **An element x of a Lie algebra $\mathfrak{g}$ is called *locally finite* if ad x is locally finite on $\mathfrak{g}$. Put $\mathfrak{g}_{\text{fin}}$ = linear span of $\{x \in \mathfrak{g} | x$ is locally finite $\}$. For a $\mathfrak{g}$-module V, put $V_{\text{fin}} = \{v \in V | \dim \sum_{k \geq 0} \mathbb{C}x^k(v) < \infty$ for every locally finite element x of $\mathfrak{g}\}$. Show that $\mathfrak{g}_{\text{fin}}$ is a subalgebra of $\mathfrak{g}$ and V_{fin} is a $\mathfrak{g}$-submodule of V.**

3.17. **Let $\mathfrak{g}$ be a Lie algebra such that $\mathfrak{g} = \mathfrak{g}_{\text{fin}}$; then $\mathfrak{g}$ is called *integrable*. Let V be a $\mathfrak{g}$-module such that $V = V_{\text{fin}}$; then V is called *integrable*. Let G^* be a free group on generators $S := \{x \in \mathfrak{g} | x$ is locally finite $\}$. Given an integrable representation π of $\mathfrak{g}$ on V, define a representation $I(\pi)$ of G^* on V by $I(\pi)(x) = \sum_{n \geq 0} \frac{1}{n!}\pi(x)^n$, $x \in S$. Let N^* be the intersection of kernels of all $I(\pi)$, where π runs over all integrable $\mathfrak{g}$-modules. Put $G = G^*/N^*$; denote by $\exp x$ the image of $x \in S$ under the canonical homomorphism $G^* \to G$. The group G is called the *group associated to the integrable Lie algebra* $\mathfrak{g}$. Show that if $\mathfrak{g}$ is a simple finite-dimensional Lie algebra, then G is the associated connected simply connected Lie group, and $x \mapsto \exp x$ is the exponential map.** (It is a deep fact established in Peterson–Kac [1983] that the above definition of an integrable module over a Kac–Moody algebra coincides with that of §3.6.)

3.18. Show that any Kac–Moody algebra $\mathfrak{g}(A)$ is integrable. (As shown in Peterson–Kac [1983], the associated group $G(A)$ constructed in Exercise 3.17 is a central extension of the group constructed in Remark 3.8.)

3.19. Let $\mathfrak{g}$ be an integrable Lie algebra and G the associated group. Then G acts on the space of every integrable representation V of $\mathfrak{g}$, so that $\exp x$ is locally finite on V for every $x \in S$ (defined in Exercise 3.17). Open problem: let V be a G-module such that every $\exp tx$ is locally finite on V and is a differentiable function in t on every finite-dimensional x-invariant subspace. Show that V can be "differentiated" to an integrable $\mathfrak{g}$-module.

3.20. Let A be the extended Cartan matrix of a complex connected simply connected algebraic group $\overset{\circ}{G}$. Show that the group G associated to the Kac–Moody algebra $\mathfrak{g}(A)$ is a central extension of the group $\overset{\circ}{G}(\mathbb{C}[t,t^{-1}])$.

3.21. Let $\mathfrak{g}$ be the Lie algebra of polynomial vector fields $\sum_{i=1}^{n} P_i \frac{\partial}{\partial x_i}$ such that $\sum_i \frac{\partial P_i}{\partial x_i} = \text{const}$. Show that $\mathfrak{g}$ is an integrable Lie algebra and that the associated group G is a central extension of a group of biregular automorphisms of $\mathbb{C}^n$. Show that $\mathfrak{g}$ is generated by all locally finite elements of the Lie algebra of all polynomial vector fields.

3.22. Show that the group G^{π} described in Remark 3.8 depends, up to isomorphism, only on the $\mathbb{Z}$-span of the set of weights of π. Show that if the $\mathfrak{g}(A)$-module V is irreducible then the kernel of the adjoint action of $\widetilde{W}^{\pi}$ on $\mathfrak{g}(A)$ lies in $\{\pm I_V\}$.

3.23. Let q be a complex number $\neq 0, \pm 1$. Let $U_q(s\ell_2)$ denote an associative algebra on generators e, f, a and a^{-1} and the following defining relations:

$$aa^{-1} = a^{-1}a = 1, \quad aea^{-1} = q^2 e, \quad afa^{-1} = q^{-2} f,$$
$$ef - fe = (a - a^{-1})/(q - q^{-1}).$$

Let $k \in \mathbb{Z}_+$ and let V be the $(k+1)$-dimensional vector space with a basis $v_0, \ldots, v_k$. For $j \in \mathbb{N}$ let $[j] = (q^j - q^{-j})/(q - q^{-1})$. Show that formulas (we assume $v_{-1} = 0, v_{k+1} = 0$) : $e(v_i) = \pm[k - i + 1]v_{i-1}$, $f(v_i) = [i+1]v_{i+1}$, $a(v_i) = \pm q^{k-2i} v_i$ define a representation of the algebra $U_q(s\ell_2)$ on V. Show that this representation is irreducible if and only if q is not an n-th root of unity, where $n \leq k$, and that, under this assumption, these are the only two irreducible $(k+1)$-dimensional $U_q(s\ell_2)$-modules. Show that as $q \to 1$, we get the $(k+1)$-dimensional irreducible representation of $s\ell_2(\mathbb{C})$ (or rather of $U(s\ell_2(\mathbb{C}))$) described by Lemma 3.2b).

3.24. Let for $s \in \mathbb{Z}$ and $j \in \mathbb{N}$: $[a; s] = (aq^s - a^{-1}q^{-s})/(q - q^{-1})$ and

$$\begin{bmatrix} a;s \\ j \end{bmatrix} = [a;s][a;s-1]\ldots[a;s-j+1]/[j]!.$$

Show that we have the following identity in $U_q(s\ell_2)$ (which is a "q-analogue" of that of Exercise 3.2):

$$\left[\frac{e^m}{[m]!}, \frac{f^n}{[n]!}\right] = \sum_{j=1}^{\min(m,n)} \frac{f^{n-j}}{[n-j]!} \begin{bmatrix} a;-m-n+2j \\ j \end{bmatrix} \frac{e^{m-j}}{[m-j]!}.$$

3.25. (Open problem.) Classify all simple $\mathbb{Z}$-graded (by finite-dimensional subspaces) integrable Lie algebras.

§3.15. Bibliographical notes and comments.

The Weyl group of a Kac–Moody algebra was introduced by Kac [1968 B] and Moody [1968]. The exposition of §§3.6–3.9 follows mainly Kac [1968 B]. The importance of the category of integrable modules was pointed out by Frenkel-Kac [1980] and Tits [1981].

The Tits cone was introduced by Tits in the "symmetrizable" case. In the framework of general groups generated by reflections it was introduced by Vinberg [1971] (see also Looijenga [1980]). The exposition of §§3.10–3.13 follows mainly Kac–Peterson [1984 A].

Exercise 3.2 is due to Steinberg [1967] and Benoist [1987]. Exercise 3.8 is due to Vinberg (see Piatetsky-Shapiro–Shafarevich [1971]). Exercise 3.9 is taken from Feingold–Frenkel [1983]. Exercise 3.10 is taken from Vinberg [1971]. Exercise 3.14 was independently found by Tits [1982]. Exercises 3.16–3.20 are based on Peterson–Kac [1983], Kac–Peterson [1983] and Kac [1985 B]. The notion of an integrable Lie algebra seems to be new.

Hopefully, one can develop a general theory of integrable Lie algebras, associated groups and their integrable representations (cf. Kac [1985 B], Abe–Takeuchi [1989]). As shown in Peterson–Kac [1983] and Kac–Peterson [1983], [1984 B], [1985 A,B], [1987] one can go quite far in this direction in the case of arbitrary Kac–Moody algebras (some important previous work was done by Kac [1969 B], Moody–Teo [1972], Marcuson [1975], and Tits [1981], [1982]). Various aspects of the theory of Kac–Moody groups may be found in Arabia [1986], Kumar [1985], [1987A], [1988], Kostant-Kumar [1986], [1987], Mathieu [1986C], [1988], [1989A], Slodowy [1985B], Tits [1985], [1987], Bausch-Rousseau [1989], and others.

The algebra $U_q(s\ell_2)$ constructed in Exercise 3.23 is the simplest example, first considered by Kulish–Reshetikhin [1983], of a "quantized Kac–Moody algebra" introduced independently by Drinfeld [1985], [1986] and Jimbo [1985]. This new field, known under the name of quantum groups, is in a process of rapid development. Exercise 3.24 is due to Kac (cf. Lusztig [1988]).

Chapter 4. A Classification of Generalized Cartan Matrices

§4.0. In order to develop the theory of root systems of Kac–Moody algebras we need to know some properties of generalized Cartan matrices. It is convenient to work in a slightly more general situation. Unless otherwise stated, we shall deal with a real $n \times n$ matrix $A = (a_{ij})$ which satisfies the following three properties:

(m1) A is indecomposable;

(m2) $a_{ij} \leq 0$ for $i \neq j$;

(m3) $a_{ij} = 0$ implies $a_{ji} = 0$.

Note that a generalized Cartan matrix satisfies (m2) and (m3) and we can assume (m1) without loss of generality.

We adopt the following notation: for a real column vector $u = {}^t(u_1, u_2, \ldots)$ we write $u > 0$ if all $u_i > 0$, and $u \geq 0$ if all $u_i \geq 0$.

§4.1. Recall the following fundamental fact from the theory of linear inequalities:

> *A system of real homogeneous linear inequalities $\lambda_i > 0$, $i = 1, \ldots, m$, has a solution if and only if there is no nontrivial linear dependence with nonnegative coefficients among the λ_i.*

We shall use a slightly different form of this statement.

LEMMA 4.1. *If $A = (a_{ij})$ is an arbitrary real $m \times s$ matrix for which there is no $u \geq 0$, $u \neq 0$, such that $({}^tA)u \geq 0$, then there exists $v > 0$ such that $Av < 0$.*

Proof. Set $\lambda_i = \sum_j a_{ij} x_j$, where the x_j form the standard basis of the dual of $\mathbf{R}^s$. Then the lemma is a consequence of the "fundamental fact" for the system of inequalities:

$$\begin{cases} -\lambda_i > 0 & (i = 1, \ldots, m) \\ x_i > 0 & (i = 1, \ldots, s). \end{cases}$$

□

§4.2. We need one more lemma.

LEMMA 4.2. *If A satisfies* (m1)–(m3), *then $Au \geq 0$, $u \geq 0$ imply that either $u > 0$ or $u = 0$.*

Proof. Let $Au \geq 0$, $u \geq 0$, $u \neq 0$. We reorder the indices so that $u_i = 0$ for $i \leq s$ and $u_i > 0$ for $i > s$. Then by (m2) and (m3), $Au \geq 0$ implies that $a_{ij} = a_{ji} = 0$ for $i \leq s$ and $j > s$, in contradiction with (m1).

□

§4.3. Now we can prove the central result of the chapter.

THEOREM 4.3. *Let A be a real $n \times n$ matrix satisfying* (m1), (m2), and (m3). *Then one and only one of the following three possibilities holds for both A and ${}^tA$:*

(Fin) $\det A \neq 0$; *there exists $u > 0$ such that $Au > 0$; $Av \geq 0$ implies $v > 0$ or $v = 0$;*

(Aff) $\operatorname{corank} A = 1$; *there exists $u > 0$ such that $Au = 0$; $Av \geq 0$ implies $Av = 0$;*

(Ind) *there exists $u > 0$ such that $Au < 0$; $Av \geq 0$, $v \geq 0$ imply $v = 0$.*

Proof. Replacing v by $-v$ we obtain that in cases (Fin) and (Aff) there is no $v \geq 0$ such that $Av \leq 0$ and $Av \neq 0$. Therefore each of (Fin) and (Aff) is not compatible with (Ind). Also (Fin) and (Aff) exclude each other because rank A differs. Now we shall show that each A together with ${}^tA$ is of type (Fin), (Aff), or (Ind).

Consider the convex cone

$$K_A = \{u \mid Au \geq 0\}.$$

By Lemma 4.2, the cone K_A can cross the boundary of the cone $\{u \mid u \geq 0\}$ only at the origin; hence we have

$$K_A \cap \{u \mid u \geq 0\} \subset \{u \mid u > 0\} \cup \{0\}.$$

Therefore, the property

$$K_A \cap \{u \mid u \geq 0\} \neq \{0\} \tag{4.3.1}$$

is possible only in the following two cases:

1) $K_A \subset \{u \mid u > 0\} \cup \{0\}$, or

2) $K_A = \{u \mid Au = 0\}$ is a 1-dimensional subspace.

Now, 1) is equivalent to (Fin); indeed, det $A \neq 0$ because K_A does not contain a 1-dimensional subspace. Clearly, 2) is equivalent to (Aff). We also proved that (4.3.1) implies that there is no $u \geq 0$ such that $Au \leq 0$, $Au \neq 0$. By Lemma 4.1 it follows that if (4.3.1) holds, then both A and ${}^tA$ are of type (Fin) or (Aff). If (4.3.1) does not hold, then both A and ${}^tA$ are of type (Ind), again by Lemma 4.1.

□

Referring to cases (Fin), (Aff), or (Ind), we shall say that A is of *finite*, *affine*, or *indefinite type*, respectively.

COROLLARY 4.3. *Let A be a matrix satisfying* (m1)–(m3). *Then A is of finite (resp. affine or indefinite) type if and only if there exists $\alpha > 0$ such that $A\alpha > 0$ (resp. $= 0$ or < 0).*

§4.4. We proceed to investigate the properties of the matrices of finite and affine types. Recall that a matrix of the form $(a_{ij})_{i,j \in S}$, where $S \subset \{1, \ldots, n\}$, is called a *principal submatrix* of $A = (a_{ij})$; we shall denote it by A_S. The determinant of a principal submatrix is called a *principal minor*.

LEMMA 4.4. *If A is of finite or affine type, then any proper principal submatrix of A decomposes into a direct sum of matrices of finite type.*

Proof. Let $S \subset \{1, \ldots, n\}$ and let A_S be the corresponding principal submatrix. For a vector u, define u_S similarly. Now, if there exists $u > 0$ such that $Au \geq 0$, then $A_S u_S \geq 0$ and $= 0$ only if $a_{ij} = 0$ for $i \in S$, $j \notin S$. The latter case is impossible since A is indecomposable. Now the lemma follows from Theorem 4.3.

□

§4.5. LEMMA 4.5. *A symmetric matrix A satisfying* (m1)–(m3) *is of finite (resp. affine) type if and only if A is positive-definite (resp. positive-semidefinite of rank $n-1$).*

Proof. If A is positive-semidefinite, then it is of finite or affine type, since otherwise there is $u > 0$ such that $Au < 0$ and therefore ${}^tuAu < 0$. The cases (Fin) and (Aff) are distinguished by the rank.

Now let A be of finite or affine type. Then there exists $u > 0$ such that $Au \geq 0$. Therefore, for $\lambda > 0$ one has: $(A + \lambda I)u > 0$, hence $A + \lambda I$ is of finite type by Theorem 4.3. Hence $\det(A + \lambda I) \neq 0$ for all $\lambda > 0$ and all the eigenvalues of A are nonnegative.

□

§4.6. LEMMA 4.6. *Let $A = (a_{ij})$ be a matrix of finite or affine type such that $a_{ii} = 2$ and $a_{ij}a_{ji} = 0$ or ≥ 1. Then A is symmetrizable. Moreover, if*

(4.6.1) $\quad a_{i_1 i_2} a_{i_2 i_3} \dots a_{i_s i_1} \neq 0$ *for some distinct* $i_1, \dots, i_s$ *with* $s \geq 3$,

then A is of the form

$$\begin{pmatrix} 2 & -u_1 & & 0 & -u_n^{-1} \\ -u_1^{-1} & 2 & & 0 & 0 \\ & & \ddots & & \\ 0 & 0 & & 2 & -u_{n-1} \\ -u_n & 0 & & -u_{n-1}^{-1} & 2 \end{pmatrix},$$

where $u_1, \dots, u_n$ are some positive numbers such that $u_1 \dots u_n = 1$.

Proof. It is clear that the second statement implies the first one (cf. Exercise 2.1). Suppose now that (4.6.1) holds. Taking the smallest possible s in (4.6.1), we see that there exists a principal submatrix B of A of the form:

$$\begin{pmatrix} 2 & -b_1 & & 0 & -b'_s \\ -b'_1 & 2 & & 0 & 0 \\ & & \ddots & & \\ 0 & 0 & & 2 & -b_{s-1} \\ -b_s & 0 & & -b'_{s-1} & 2 \end{pmatrix}.$$

By Lemma 4.4, B is of finite or affine type and therefore by Theorem 4.3 there exists $u > 0$ such that $Bu \geq 0$. Replacing B by the matrix $(\operatorname{diag} u)^{-1} B (\operatorname{diag} u)$, we may assume that ${}^t u = (1, \dots, 1)$. But then $Bu \geq 0$ implies that the sum of the entries of B is nonnegative:

(4.6.2) $$2s - \sum_{i=1}^{s} (b_i + b'_i) \geq 0.$$

Since $b_i b'_i \geq 1$, we have $b_i + b'_i \geq 2$; hence, by (4.6.2) we obtain that $b_i + b'_i = 2$ and therefore $b_i = b'_i = 1$ for all i. As in this case $\det B = 0$, Lemmas 4.4 and 4.5 imply that $B = A$.

□

§4.7. We proceed to classify all generalized Cartan matrices of finite and affine type. For this it is convenient to introduce the so-called Dynkin diagrams. Let $A = (a_{ij})_{i,j=1}^{n}$ be a generalized Cartan matrix. We associate with A a graph $S(A)$, called the ***Dynkin diagram*** of A as follows. If $a_{ij}a_{ji} \le 4$ and $|a_{ij}| \ge |a_{ji}|$, the vertices i and j are connected by $|a_{ij}|$ lines, and these lines are equipped with an arrow pointing toward i if $|a_{ij}| > 1$. If $a_{ij}a_{ji} > 4$, the vertices i and j are connected by a bold-faced line equipped with an ordered pair of integers $|a_{ij}|$, $|a_{ji}|$.

It is clear that A is indecomposable if and only if $S(A)$ is a connected graph. Note also that A is determined by the Dynkin diagram $S(A)$ and an enumeration of its vertices. We say that $S(A)$ is of finite, affine, or indefinite type if A is of that type.

Now we summarize the results obtained above for generalized Cartan matrices.

PROPOSITION 4.7. *Let A be an indecomposable generalized Cartan matrix.*

a) *A is of finite type if and only if all its principal minors are positive.*

b) *A is of affine type if and only if all its proper principal minors are positive and* $\det A = 0$.

c) *If A is of finite or affine type, then any proper subdiagram of $S(A)$ is a union of (connected) Dynkin diagrams of finite type.*

d) *If A is of finite type, then $S(A)$ contains no cycles. If A is of affine type and $S(A)$ contains a cycle, then $S(A)$ is the cycle $A_\ell^{(1)}$ from Table* Aff 1.

e) *A is of affine type if and only if there exists $\delta > 0$ such that $A\delta = 0$; such a δ is unique up to a constant factor.*

Proof. To prove a) and b) note that by Lemma 4.6, if A is of finite or affine type, it is symmetrizable, i.e., there exists a diagonal matrix D with positive entries on the diagonal such that DA is symmetric (see §2.3). Now a) and b) follow from Lemma 4.5. c) follows from Lemma 4.4, d) follows from Lemma 4.6, e) follows from Theorem 4.3.

□

§4.8. Now we can list all generalized Cartan matrices of finite and affine type.

THEOREM 4.8. a) *The Dynkin diagrams of all generalized Cartan matrices of finite type are listed in Table* Fin.

b) *The Dynkin diagrams of all generalized Cartan matrices of affine type are listed in Tables* Aff 1–3 *(all of them have $\ell + 1$ vertices).*

c) *The numerical labels in Tables* Aff 1–3 *are the coordinates of the unique vector $\delta = {}^t(a_0, a_1, \ldots, a_\ell)$ such that $A\delta = 0$ and the a_i are positive relatively prime integers.*

Proof. First, we prove c). Note that $A\delta = 0$ means that $2a_i = \sum_j m_j a_j$ for all i, where the summation is taken over the j's which are connected with i, and $m_j = 1$ unless the number of lines connecting i and j is $s > 1$ and the arrow points toward i; then $m_j = s$. Now c) is easily checked case by case.

It follows from c) and Proposition 4.7 e) that all diagrams in Tables Aff 1–3 are of affine type. Since all diagrams from Table Fin appear as subdiagrams of diagrams in Tables Aff 1–3, we deduce, by Proposition 4.7 c), that all diagrams in Table Fin are of finite type.

It remains to show that if A is of finite (resp. affine) type, then $S(A)$ appears in Table Fin (resp. Aff). This is an easy exercise. We do it by induction on n. First, det $A \geq 0$ immediately gives

(4.8.1) A_2, C_2 and G_2 are the only finite type diagrams of rank 2; $A_1^{(1)}$ and $A_2^{(2)}$ are the only affine type diagrams of rank 2.

(4.8.2) A_3, B_3, C_3 are the only finite type diagrams of rank 3; $A_2^{(1)}$, $C_2^{(1)}$, $G_2^{(1)}$, $D_3^{(2)}$ $A_4^{(2)}$, $D_4^{(3)}$ are the only affine type diagrams of rank 3.

Furthermore, by Proposition 4.7 d) we have

(4.8.3) $S(A)$ has a cycle, then $S(A) = A_\ell^{(1)}$.

By Proposition 4.7 and inductive assumption we have

(4.8.4) Any proper connected subdiagram of $S(A)$ appears in Table Fin.

Now let $S(A)$ be of finite type. Then $S(A)$ does not appear in Tables Aff 1–3, has no cycles by (4.8.3), each of its branch vertices is of type D_4 by (4.8.2 and 4), and it has at most one branch vertex by (4.8.4). By (4.8.4), for $S(A)$ with a branch vertex, the only possibilities are D_ℓ, E_6, E_7, E_8. Similarly we show that if $S(A)$ is not simply-laced (i.e., has multiple edges), then it is B_ℓ, C_ℓ, F_4, or G_2. A simply-laced diagram with no cycles and branch vertices is A_ℓ.

Now let $S(A)$ be of affine type. By (4.8.3) we can assume that $S(A)$ has no cycles. By (4.8.4), $S(A)$ is obtained from a diagram of Table Fin by adding one vertex in such a way that any subdiagram is from Table Fin. Using (4.8.1 and 2) it is not difficult to see that only the diagrams from Tables Aff 1–3 may be obtained in this way.

□

We fix once and for all an enumeration of the vertices of the diagrams of generalized Cartan matrices A in tables below as follows. In Table Fin vertices are enumerated by symbols $\alpha_1, \ldots, \alpha_\ell$. Each diagram $X_\ell^{(1)}$ of Table Aff 1 is obtained from the diagram X_ℓ by adding one vertex, which we enumerate by the symbol α_0, keeping the enumeration of the rest of

TABLE Fin

A_ℓ	o — o — $\cdots$ — o — o α_1 α_2 $\alpha_{\ell-1}$ α_ℓ	$(\ell+1)$
B_ℓ	o — o — $\cdots$ — o $\Rightarrow$ o α_1 α_2 $\alpha_{\ell-1}$ α_ℓ	(2)
C_ℓ	o — o — $\cdots$ — o $\Leftarrow$ o α_1 α_2 $\alpha_{\ell-1}$ α_ℓ	(2)
D_ℓ	o — o — $\cdots$ — o — o (with o α_ℓ attached to $\alpha_{\ell-2}$) α_1 α_2 $\alpha_{\ell-2}$ $\alpha_{\ell-1}$	(4)
E_6	o — o — o — o — o (with o α_6 attached to α_3) α_1 α_2 α_3 α_4 α_5	(3)
E_7	o — o — o — o — o — o (with o α_7 attached to α_3) α_1 α_2 α_3 α_4 α_5 α_6	(2)
E_8	o — o — o — o — o — o — o (with o α_8 attached to α_5) α_1 α_2 α_3 α_4 α_5 α_6 α_7	(1)
F_4	o — o $\Rightarrow$ o — o α_1 α_2 α_3 α_4	(1)
G_2	o $\Rrightarrow$ o α_1 α_2	(1)

TABLE Aff 1

Type	Diagram (labels)
$A_1^{(1)}$	 o⟺o 1 1
$A_\ell^{(1)} (\ell \geq 2)$	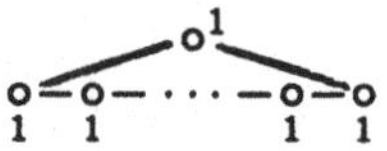 o–o– ··· –o–o, with an extra vertex o (label 1) joined to both end vertices 1 1 ··· 1 1
$B_\ell^{(1)} (\ell \geq 3)$	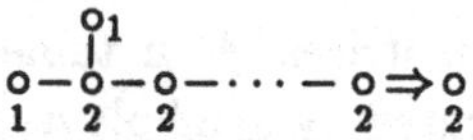 o–o–o– ··· –o⇒o, with a vertex o (label 1) attached to the second vertex 1 2 2 ··· 2 2
$C_\ell^{(1)} (\ell \geq 2)$	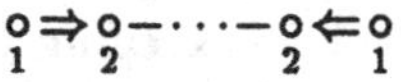 o⇒o– ··· –o⇐o 1 2 ··· 2 1
$D_\ell^{(1)} (\ell \geq 4)$	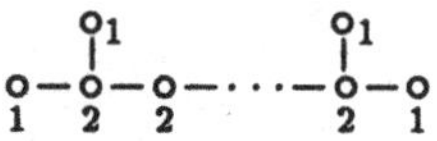 o–o–o– ··· –o–o, with a vertex o (label 1) attached to the second vertex and one (label 1) to the second-to-last vertex 1 2 2 ··· 2 1
$G_2^{(1)}$	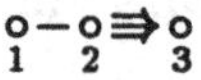 o–o⇛o 1 2 3
$F_4^{(1)}$	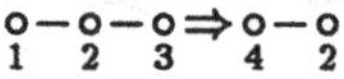 o–o–o⇒o–o 1 2 3 4 2
$E_6^{(1)}$	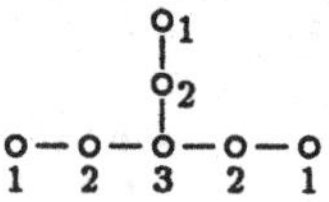 o–o–o–o–o, with a vertical branch o (label 2)–o (label 1) attached to the middle vertex 1 2 3 2 1
$E_7^{(1)}$	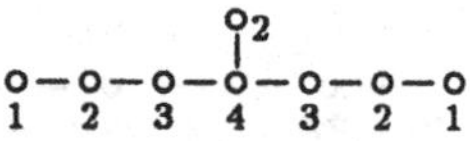 o–o–o–o–o–o–o, with a vertex o (label 2) attached to the middle vertex 1 2 3 4 3 2 1
$E_8^{(1)}$	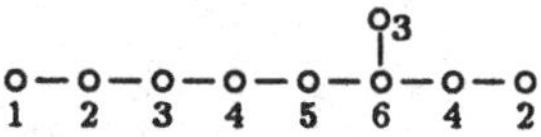 o–o–o–o–o–o–o–o, with a vertex o (label 3) attached to the sixth vertex 1 2 3 4 5 6 4 2

the vertices as in Table Fin. Vertices of the diagrams of Tables Aff 2 and Aff 3 are enumerated by the symbols $\alpha_0, \ldots, \alpha_\ell$. The numbers in parentheses in Table Fin are $\det A$. The numerical labels in Tables Aff are coefficients of a linear dependence between the columns of A.

TABLE Aff 2

$A_2^{(2)}$ $$\underset{\alpha_0}{\overset{2}{\circ}} \Lleftarrow \underset{\alpha_1}{\overset{1}{\circ}}$$

$A_{2\ell}^{(2)} (\ell \geq 2)$ $$\underset{\alpha_0}{\overset{2}{\circ}} \Leftarrow \underset{\alpha_1}{\overset{2}{\circ}} - \cdots - \underset{\alpha_{\ell-1}}{\overset{2}{\circ}} \Leftarrow \underset{\alpha_\ell}{\overset{1}{\circ}}$$

$A_{2\ell-1}^{(2)} (\ell \geq 3)$ $$\underset{\alpha_1}{\overset{1}{\circ}} - \underset{\alpha_2}{\overset{\overset{\alpha_0 \overset{1}{\circ}}{|}}{\circ}}{}^{2} - \underset{\alpha_3}{\overset{2}{\circ}} - \cdots - \underset{\alpha_{\ell-1}}{\overset{2}{\circ}} \Leftarrow \underset{\alpha_\ell}{\overset{1}{\circ}}$$

$D_{\ell+1}^{(2)} (\ell \geq 2)$ $$\underset{\alpha_0}{\overset{1}{\circ}} \Leftarrow \underset{\alpha_1}{\overset{1}{\circ}} - \cdots - \underset{\alpha_{\ell-1}}{\overset{1}{\circ}} \Rightarrow \underset{\alpha_\ell}{\overset{1}{\circ}}$$

$E_6^{(2)}$ $$\underset{\alpha_0}{\overset{1}{\circ}} - \underset{\alpha_1}{\overset{2}{\circ}} - \underset{\alpha_2}{\overset{3}{\circ}} \Leftarrow \underset{\alpha_3}{\overset{2}{\circ}} - \underset{\alpha_4}{\overset{1}{\circ}}$$

TABLE Aff 3

$D_4^{(3)}$ $$\underset{\alpha_0}{\overset{1}{\circ}} - \underset{\alpha_1}{\overset{2}{\circ}} \Lleftarrow \underset{\alpha_2}{\overset{1}{\circ}}$$

§4.9. We conclude the chapter with the following characterization of Kac–Moody algebras associated with generalized Cartan matrices of finite type (cf. Proposition 3.12 e)).

PROPOSITION 4.9. *Let A be an indecomposable generalized Cartan matrix. Then the following conditions are equivalent:*

(i) *A is a generalized Cartan matrix of finite type;*

(ii) *A is symmetrizable and the bilinear form $(\,.\,|\,.\,)_{\mathfrak{h}_{\mathbb{R}}}$ is positive-definite;*

(iii) $|W| < \infty$;

(iv) $|\Delta| < \infty$;

(v) *$\mathfrak{g}(A)$ is a simple finite-dimensional Lie algebra;*

(vi) *there exists $\alpha \in \Delta_+$ such that $\alpha + \alpha_i \notin \Delta$ for all $i = 1, \ldots, n$.*

Proof. By Lemma 4.6 and Proposition 4.7, the bilinear form $(\,.\,|\,.\,)_{\mathfrak{h}_{\mathbb{R}}}$ is positive-definite if A is of finite type; thus, (i) $\Rightarrow$ (ii). If (ii) holds, then by Proposition 3.9, the group W lies in the orthogonal group $O((\,.\,|\,.\,))$ and hence is compact. Since W preserves the lattice Q, it is discrete. Hence $|W| < \infty$, which proves (ii) $\Rightarrow$ (iii). The implication (iii) $\Rightarrow$ (iv) follows from Proposition 3.12 e); (iv) $\Rightarrow$ (v) is clear by Proposition 1.7 a), and for (v) $\Rightarrow$ (vi) we can take α to be a root of maximal height. Finally, let $\alpha + \alpha_i \notin \Delta$ for all i. Corollary 3.6 a) implies that $\langle \alpha, \alpha_i^\vee \rangle \geq 0$ for all i. But then it follows by Theorem 4.3 that A is of finite or affine type, and in the latter case $\langle \alpha, \alpha_i^\vee \rangle = 0$ for all i. But in this case, by Lemma 1.5, $\alpha - \alpha_i \in \Delta_+$ for some i. Hence, if A is of affine type, Proposition 3.6 b) implies that $\alpha + \alpha_i \in \Delta_+$, a contradiction. This proves the last implication (vi) $\Rightarrow$ (i).

□

Remark 4.9. A root of a finite root system Δ which satisfies condition (vi) of Proposition 4.9 is called a ***highest*** root. It is easy to see that there exists a unique such root (cf. Proposition 6.4 in Chapter 6); it is given by the formula

$$\theta = \sum_{i=1}^{\ell} a_i \alpha_i,$$

where a_i are the labels of the extended Dynkin diagram from Table Aff 1.

§4.10. Exercises.

4.1. An indecomposable generalized Cartan matrix A is said to be of strictly hyperbolic type (resp. hyperbolic type) if it is of indefinite type and any connected proper subdiagram of $S(A)$ is of finite (resp. finite or affine) type. Show that a matrix $\begin{pmatrix} 2 & -a \\ -b & 2 \end{pmatrix}$, such that a and b are positive integers, is of finite (resp. affine or strictly hyperbolic) type if and only if $ab \leq 3$ (resp. $ab = 4$ or $ab > 4$). Show that there is only a finite number of hyperbolic matrices of order ≥ 3 and that the order of a strictly hyperbolic (resp. hyperbolic) matrix is ≤ 5 (resp. ≤ 10). (Note a discrepancy with Chapter 5 where we assume a hyperbolic matrix to be symmetrizable.)

4.2. Let A be of type $T_{p,q,r}$ ($p \geq q \geq r$), i.e., let its Dynkin diagram be of the form

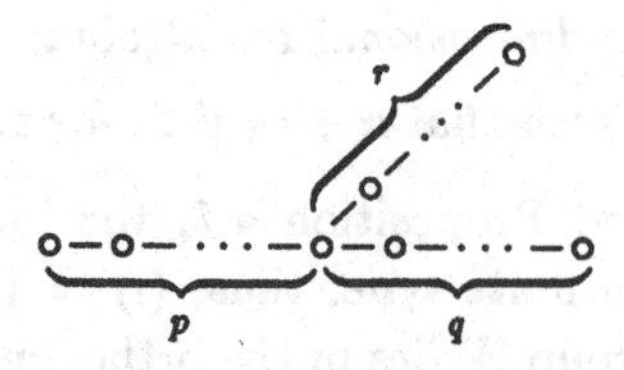

Set $c = \frac{1}{p} + \frac{1}{q} + \frac{1}{r}$. Then A is of finite (resp. affine or indefinite) type if and only if $c > 1$ (resp. $c = 1$, $c < 1$). Show that for $c < 1$, the signature of A

is $(++\cdots+-)$. Show that A is hyperbolic if and only if $(p,q,r)=(4,3,3)$ or $(5,4,2)$ or $(7,3,2)$. Show that $\det A = pq+pr+qr-pqr$.

[To prove the statement about the signature, delete the branch point.]

4.3. Introduce the following diagrams with n vertices ($n \geq 5$):

$E_n(=T_{n-3,3,2})$ o—o— ⋯ —o—o—o—o (branch node attached to the third vertex from the right)

AE_n o—o— ⋯ —o—o—o (with an arc joining the first vertex to the second vertex from the right)

BE_n o⇐o— ⋯ —o—o—o—o (branch node attached to the third vertex from the right)

CE_n o⇒o— ⋯ —o—o—o—o (branch node attached to the third vertex from the right)

DE_n o—o— ⋯ —o—o—o—o (branch nodes attached to the second vertex and to the third vertex from the right)

Show that E_{10}, BE_{10}, CE_{10}, and DE_{10} are all hyperbolic matrices of rank 10. Show that a hyperbolic matrix of rank 7, 8, or 9 is one of the following list: $T_{4,3,3}$, $T_{5,4,2}$, AE_n, BE_n, CE_n, DE_n ($n=7,8$, or 9).

4.4. Classify all (strictly) hyperbolic matrices.

4.5. Show that E_{10} is the only symmetric hyperbolic matrix with determinant -1. Show that $\operatorname{diag}(1/2,1,\ldots,1)(BE_{10})$ is an integral symmetric matrix with determinant -1.

4.6. Show that if A is symmetrizable and hyperbolic, then the corresponding "symmetrized" matrix has signature $(++\cdots+-)$.

4.7. Let A be a strictly hyperbolic matrix. Show that the group $\operatorname{Aut} X$ of all linear automorphisms of the Tits cone X acts with a compact fundamental domain.

4.8. Let A be a hyperbolic matrix. Show that if A is symmetrizable, then X is linearly isomorphic to an upper cone over a unit ball. Show that, conversely, if A is a hyperbolic matrix of order 3, such that $\operatorname{Aut} X$ operates transitively on the interior of X, then A is symmetrizable.

4.9. Show that the real Tits cone X for the matrix $\begin{pmatrix} 2 & -1 & -1 \\ -2 & 2 & -1 \\ -1 & -1 & 2 \end{pmatrix}$ provides an example of a cone in $\mathbf{R}^3$ such that $\operatorname{Aut} X$ has a compact fundamental domain, but does not act transitively on the interior of X.

4.10. Using Exercise 1.4, show that if A is an indecomposable generalized Cartan matrix and A is not from Table Aff, then the Lie algebra $\mathfrak{g}'(A)/\mathfrak{c}$ is simple.

4.11. Let $A = (a_{ij})$ be a symmetric matrix. Set $t_k = \sum_i a_{ik}c_i$, where c_k are nonzero numbers. Check that

$$\sum_{i,k} a_{ik}x_ix_k = \sum_k \frac{t_kx_k^2}{c_k} - \frac{1}{2}\sum_{i,k} c_ic_ka_{ik}\left(\frac{x_i}{c_i} - \frac{x_k}{c_k}\right)^2.$$

Assume that $a_{ij} \leq 0$ for $i \neq j$. Deduce that if there exist positive numbers $c_1, \ldots, c_n$ such that $t_k = 0$ $(k = 1, \ldots, n)$, then the quadratic form $\sum_{i,k} a_{ik}\xi_i\xi_k$ is positive-semidefinite. Use this to prove Proposition 4.7 e). Deduce also that this quadratic form with an indecomposable A is positive-definite if and only if there is no $c_1 > 0, \ldots, c_n > 0$ such that $t_k \leq 0$ $(k = 1, \ldots, n)$.

4.12. Let S' and S'' be affine Dynkin diagrams. Let A be a generalized Cartan matrix with the Dynkin diagram

$$S(A) = \boxed{\;S'\;} - \circ - \boxed{\;S''\;}.$$

Show that $\det A = 0$.

4.13. Show that an indecomposable generalized Cartan matrix is affine if and only if it is degenerate and all its principal minors are nonnegative.

4.14. Show that the following is a complete list of connected Dynkin diagrams of generalized Cartan matrices of infinite order such that any principal minor of finite order is positive:

A_∞ ⋯ —o—o—o— ⋯

$A_{+\infty}$ o—o—o— ⋯

B_∞ o⇐o—o—o— ⋯

C_∞ o⇒o—o—o— ⋯

D_∞ o—o—o—o— ⋯ (with an extra node attached above the second node)

§4.11. Bibliographical notes and comments.

Theorem 4.3 is due to Vinberg [1971], as well as most of the results of §§4.4–4.7. The list of affine diagrams appears in Kac [1967] and Moody [1967]. The notation appears naturally in the context of the theory of finite order automorphisms (Kac [1969 A]), as we shall see in Chapter 8.

Exercises 4.8 and 4.9 are taken from Vinberg–Kac [1967]. Exercise 4.11 is taken from Bourbaki [1968]. The rest of the exercises are fairly standard.

Chapter 5. Real and Imaginary Roots

§5.0. In this chapter we give an explicit description of the root system Δ of a Kac–Moody algebra $\mathfrak{g}(A)$. Our main instrument is the notion of an imaginary root, which has no counterpart in the finite-dimensional theory. As an application, we derive the structure of the automorphism groups of integral unimodular Lorentzian lattices of dimension ≤ 10.

§5.1. A root $\alpha \in \Delta$ is called *real* if there exists $w \in W$ such that $w(\alpha)$ is a simple root. Denote by Δ^{re} and Δ_+^{re} the sets of all real and positive real roots respectively.

Let $\alpha \in \Delta^{re}$; then $\alpha = w(\alpha_i)$ for some $\alpha_i \in \Pi$, $w \in W$. Define the *dual* (real) *root* $\alpha^\vee \in \Delta^{\vee re}$ by $\alpha^\vee = w(\alpha_i^\vee)$. This is independent of the choice of the presentation $\alpha = w(\alpha_i)$. Indeed, we have to show that the equality $u(\alpha_i) = \alpha_j$ implies $u(\alpha_i^\vee) = \alpha_j^\vee$ for $u \in W$. But this is (3.10.3). Thus we have a canonical W-equivariant bijection $\Delta^{re} \to \Delta^{\vee re}$. By an easy induction on ht α, one shows, using Proposition 5.1 e) below, that $\alpha > 0$ if and only if $\alpha^\vee > 0$.

We define a reflection r_α with respect to $\alpha \in \Delta^{re}$ by

$$r_\alpha(\lambda) = \lambda - \langle \lambda, \alpha^\vee \rangle \alpha, \quad \lambda \in \mathfrak{h}^*.$$

Since $\langle \alpha, \alpha^\vee \rangle = 2$, this is a reflection, and since $r_\alpha = w r_i w^{-1}$ if $\alpha = w(\alpha_i)$, it lies in W. Note that r_{α_i} is the fundamental reflection r_i.

The following proposition shows that real roots have all the "classical" properties.

PROPOSITION 5.1. *Let α be a real root of a Kac–Moody algebra $\mathfrak{g}(A)$. Then*

a) mult $\alpha = 1$.

b) *$k\alpha$ is a root if and only if $k = \pm 1$.*

c) *If $\beta \in \Delta$ then there exist nonnegative integers p and q related by the equation*

$$p - q = \langle \beta, \alpha^\vee \rangle,$$

such that $\beta + k\alpha \in \Delta \cup \{0\}$ if and only if $-p \leq k \leq q$, $k \in \mathbb{Z}$.

d) *Suppose that A is symmetrizable and let $(.|.)$ be a standard invariant bilinear form on $\mathfrak{g}(A)$. Then*

(i) $(\alpha|\alpha) > 0$;
(ii) $\alpha^\vee = 2\nu^{-1}(\alpha)/(\alpha|\alpha)$.
(iii) if $\alpha = \sum_i k_i\alpha_i$, *then* $k_i(\alpha_i|\alpha_i) \in (\alpha|\alpha)\mathbf{Z}$.

e) *Provided that $\pm\alpha \notin \Pi$, there exists i such that*

$$|\operatorname{ht} r_i(\alpha)| < |\operatorname{ht}\alpha|.$$

Proof. All the statements a)–d) are clear if α is a simple root: a) holds by definition, b) holds due to (1.3.3), c) follows from Proposition 3.6b), and d) is (2.3.3 and 5). Now a), b), and c) follow from Proposition 3.7b), while d) (i) and (ii) follow from Proposition 3.9. Statement d) (iii) follows from the fact that $\alpha^\vee \in \sum \mathbf{Z}\alpha_i^\vee$ and the following formula:

$$\alpha^\vee = \sum_i \frac{(\alpha_i|\alpha_i)}{(\alpha|\alpha)} k_i\alpha_i^\vee. \tag{5.1.1}$$

Finally, suppose the contrary to e); we may assume that $\alpha > 0$. But then $-\alpha \in C^\vee$ and, by Proposition 3.12 d) for the dual root system, $-\alpha + w(\alpha) \geq 0$ for any $w \in W$. Taking w such that $w(\alpha) \in \Pi$, we arrive at a contradiction.

□

The following lemma will be needed in the sequel.

LEMMA 5.1. *Suppose that A is symmetrizable. Then the set of all $\alpha = \sum_i k_i\alpha_i \in Q$ such that*

$$k_i(\alpha_i|\alpha_i) \in (\alpha|\alpha)\mathbf{Z} \quad \textit{for all} \quad i \tag{5.1.2}$$

is W-invariant.

Proof. It suffices to check that $r_i(\alpha)$ again satisfies (5.1.2), i.e., that $(k_i - (\alpha|\alpha_i^\vee))(\alpha_i|\alpha_i) \in (\alpha|\alpha)\mathbf{Z}$. This is equivalent to

$$2(\alpha|\alpha_i) \in (\alpha|\alpha)\mathbf{Z}, \tag{5.1.3}$$

which follows from (5.1.2):

$$2(\alpha|\alpha_i) = \sum_j \frac{2(\alpha_j|\alpha_i)}{(\alpha_j|\alpha_j)} k_j(\alpha_j|\alpha_j) = \sum_j a_{ji}k_j(\alpha_j|\alpha_j) \in (\alpha|\alpha)\mathbf{Z}.$$

□

Let A be a symmetrizable generalized Cartan matrix, and let $(.|.)$ be a standard invariant bilinear form (see §2.3). Then, given a real root α we have $(\alpha|\alpha) = (\alpha_i|\alpha_i)$ for some simple root α_i. We call α a *short* (resp. *long*) real root if $(\alpha|\alpha) = \min_i(\alpha_i|\alpha_i)$ (resp. $= \max_i(\alpha_i|\alpha_i)$). These are independent of the choice of a standard form. Note that if A is symmetric, then all simple roots and hence all real roots have the same square length. If A is not symmetric and $S(A)$ is equipped with m arrows pointing in the same direction, then there are simple roots of exactly $m+1$ different square lengths since an arrow in $S(A)$ points to a shorter simple root. It follows that if A is a nonsymmetric matrix from Table Fin, then every root is either short or long. Furthermore, if A is a nonsymmetric matrix from Table Aff, and A is not of type $A_{2\ell}^{(2)}$ with $l > 1$, then every real root is either short or long; for the type $A_{2\ell}^{(2)}$ with $l > 1$ there are real roots of three different lengths.

In this chapter, we shall normalize $(.|.)$ such that the $(\alpha_i|\alpha_i)$ are relatively prime positive integers for each connected component of $S(A)$. For example, if A is symmetric, then $(\alpha_i|\alpha_i) = 1$ for all i, and all real roots are short (= long).

§5.2. A root α which is not real is called an *imaginary root*. Denote by Δ^{im} and Δ_+^{im} the sets of imaginary and positive imaginary roots, respectively. By definition,

$$\Delta = \Delta^{re} \cup \Delta^{im} \quad \text{(disjoint union)}.$$

It is also clear that $\Delta^{im} = \Delta_+^{im} \cup (-\Delta_+^{im})$.

The following properties of imaginary roots are useful.

PROPOSITION 5.2. a) *The set Δ_+^{im} is W-invariant.*

b) *For $\alpha \in \Delta_+^{im}$ there exists a unique root $\beta \in -C^\vee$ (i.e., $\langle\beta, \alpha_i^\vee\rangle \leq 0$ for all i) which is W-equivalent to α.*

c) *If A is symmetrizable and $(.|.)$ is a standard invariant bilinear form, then a root α is imaginary if and only if $(\alpha|\alpha) \leq 0$.*

Proof. As $\Delta_+^{im} \subset \Delta_+ \backslash \Pi$ and the set $\Delta_+ \backslash \{\alpha_i\}$ is r_i-invariant (by Lemma 1.3), it follows that Δ_+^{im} is r_i-invariant for all i and hence W-invariant, proving a). Let $\alpha \in \Delta_+^{im}$ and let β be an element of minimal height in $W \cdot \alpha \subset \Delta_+$. Then $-\beta \in C^\vee$. Such a β is unique in the orbit $W \cdot \alpha$ by Proposition 3.12 b), proving b). Let α be an imaginary root; we may assume, by b), that $-\alpha \in C^\vee$ (since $(.|.)$ is W-invariant). Let $\alpha = \sum_i k_i\alpha_i$,

$k_i \geq 0$; then $(\alpha|\alpha) = \sum_i k_i(\alpha|\alpha_i) = \sum_i \frac{1}{2}|\alpha_i|^2 k_i \langle\alpha, \alpha_i^\vee\rangle \leq 0$ by (2.3.3 and 5). The converse holds by Proposition 5.1 d).

□

§5.3. For $\alpha = \sum_i k_i\alpha_i \in Q$ we define the *support* of α (written $\operatorname{supp}\alpha$) to be the subdiagram of $S(A)$ which consists of the vertices i such that $k_i \neq 0$, and of all the edges joining these vertices. By Lemma 1.6, $\operatorname{supp}\alpha$ is connected for every root α. Set:

$$K = \{\alpha \in Q_+\backslash\{0\} \mid \langle\alpha, \alpha_i^\vee\rangle \leq 0 \text{ for all } i \text{ and } \operatorname{supp}\alpha \text{ is connected}\}.$$

Lemma 5.3. *In the above notation,* $K \subset \Delta_+^{im}$.

Proof. Let $\alpha = \sum_i k_i\alpha_i \in K$. Set

$$\Omega_\alpha = \{\gamma \in \Delta_+ \mid \gamma \leq \alpha\}.$$

The set Ω_α is finite, and it is nonempty because the simple roots, which appear in the decomposition of α, lie in Ω_α. Let $\beta = \sum_i m_i\alpha_i$ be an element of maximal height in Ω_α. It follows from Corollary 3.6 a) that

$$\operatorname{supp}\beta = \operatorname{supp}\alpha. \tag{5.3.1}$$

First, we prove that $\alpha \in \Delta_+$. Suppose the contrary; then $\alpha \neq \beta$. By definition:

$$\beta + \alpha_i \notin \Delta_+ \text{ if } k_i > m_i. \tag{5.3.2}$$

Let A_1 be the principal submatrix of A corresponding to $\operatorname{supp}\alpha$. If A_1 is of finite type, then $\{\alpha \in Q_+ \mid \langle\alpha, \alpha_i^\vee\rangle \leq 0 \text{ for all } i\} = \{0\}$ and there is nothing to prove. If A_1 is not of finite type, then, by Proposition 4.9, we have

$$P := \{j \in \operatorname{supp}\alpha \mid k_j = m_j\} \neq \emptyset. \tag{5.3.3}$$

Let R be a connected component of the subdiagram $(\operatorname{supp}\alpha)\backslash P$.

From (5.3.2) and Corollary 3.6 a) we deduce that

$$\langle\beta, \alpha_i^\vee\rangle \geq 0 \text{ if } i \in R. \tag{5.3.4}$$

Set $\beta' = \sum_{i\in R} m_i\alpha_i$. Since $\operatorname{supp}\alpha$ is connected, (5.3.1) and (5.3.4) imply

$$\langle\beta', \alpha_i^\vee\rangle \geq 0 \text{ if } i \in R; \qquad \langle\beta', \alpha_j^\vee\rangle > 0 \text{ for some } j \in R. \tag{5.3.5}$$

Therefore, by Theorem 4.3, the diagram R is of finite type.

On the other hand, set

$$\alpha' = \sum_{i \in R} (k_i - m_i)\alpha_i \ .$$

Since supp α' is a connected component of supp$(\alpha - \beta)$, we obtain that

$$(5.3.6) \qquad \langle \alpha', \alpha_i^\vee \rangle = \langle \alpha - \beta, \alpha_i^\vee \rangle \text{ for } i \in R.$$

But $\langle \alpha, \alpha_i^\vee \rangle \leq 0$ since $\alpha \in K$, hence by (5.3.4 and 6):

$$\langle \alpha', \alpha_i^\vee \rangle \leq 0 \text{ for } i \in R.$$

This contradicts the fact that R is of finite type. Thus, we have proved that $\alpha \in \Delta_+$.

But 2α also satisfies all the hypotheses of the lemma; hence $2\alpha \in \Delta_+$ and, by Proposition 5.1 b), $\alpha \in \Delta_+^{im}$.

□

§5.4. Lemma 5.3 yields the following description of the set of imaginary roots.

THEOREM 5.4. $\Delta_+^{im} = \bigcup_{w \in W} w(K)$.

Proof. Lemma 5.3 and Proposition 5.2 a) prove the inclusion $\supset$. The reverse inclusion holds by Proposition 5.2 b) and by the fact that supp α is connected for every root α (by Lemma 1.6).

□

§5.5. The following proposition shows that the properties of imaginary roots differ drastically from those of real roots.

PROPOSITION 5.5. *If $\alpha \in \Delta_+^{im}$ and r is a nonzero (rational) number such that $r\alpha \in Q$, then $r\alpha \in \Delta^{im}$. In particular, $n\alpha \in \Delta^{im}$ if $n \in \mathbb{Z}\backslash\{0\}$.*

Proof. By Proposition 5.2 b) we can assume that $\alpha \in -C^\vee \cap Q_+$; since $\alpha \in \Delta$, it follows that supp α is connected and hence $\alpha \in K$. Hence $r\alpha \in K$ for $r > 0$ and therefore, by Lemma 5.3, $r\alpha \in \Delta^{im}$.

□

§5.6. Now we prove the existence theorem for imaginary roots.

THEOREM 5.6. *Let A be an indecomposable generalized Cartan matrix.*
a) If A is of finite type, then the set Δ^{im} is empty.
b) If A is of affine type, then

$$\Delta_+^{im} = \{n\delta \ (n = 1, 2, \dots)\},$$

where $\delta = \sum_{i=0}^{\ell} a_i \alpha_i$, and the a_i are the labels of $S(A)$ in Table Aff.

c) If A is of indefinite type, then there exists a positive imaginary root $\alpha = \sum_i k_i \alpha_i$ such that $k_i > 0$ and $\langle \alpha, \alpha_i^\vee \rangle < 0$ for all $i = 1, \dots, n$.

Proof. Recall (see Chapter 4) that the set $\{\alpha \in Q_+ \mid \langle \alpha, \alpha_i^\vee \rangle \leq 0,\ i = 1, \dots, n\}$ is zero for A of finite type, is equal to $\mathbb{Z}\delta$ for A of affine type, and there exists $\alpha = \sum_i k_i \alpha_i$ such that $k_i > 0$ and $\langle \alpha, \alpha_i^\vee \rangle < 0$ for all i if A is of indefinite type. The theorem now follows from Theorem 5.4.

□

§5.7. It follows from Proposition 4.7 that if α is a *null-root*, i.e., $\alpha \in \Delta(A)$ is such that $\alpha|_{\mathfrak{h}'} = 0$, then $\operatorname{supp} \alpha$ is a diagram of affine type which is a connected component of $S(A)$ and $\alpha = k\delta$, $k \in \mathbb{Z}$. We proceed to describe the isotropic roots.

PROPOSITION 5.7. *Let A be symmetrizable. A root α is isotropic (i.e., $(\alpha|\alpha) = 0$) if and only if it is W-equivalent to an imaginary root β such that* $\operatorname{supp} \beta$ *is a subdiagram of affine type of $S(A)$ (then $\beta = k\delta$).*

Proof. Let α be an isotropic root and let $(.|.)$ be a standard bilinear form. We can assume that $\alpha > 0$. Then $\alpha \in \Delta_+^{im}$ by Proposition 5.1 d), and α is W-equivalent to an imaginary root $\beta \in K$ such that $\langle \beta, \alpha_i^\vee \rangle \leq 0$ for all i, by Proposition 5.2 b). Let $\beta = \sum_{i \in P} k_i \alpha_i$ and $P = \operatorname{supp} \beta$. Then $(\beta|\beta) = \sum_{i \in P} k_i (\beta|\alpha_i) = 0$, where $k_i > 0$ and $(\beta|\alpha_i) = \frac{1}{2}|\alpha_i|^2 \langle \beta, \alpha_i^\vee \rangle \leq 0$ for $i \in P$. Therefore $\langle \beta, \alpha_i^\vee \rangle = 0$ for all $i \in P$, and P is a diagram of affine type. Conversely, if $\beta = k\delta$ is an imaginary root for a diagram of affine type, then $(\beta|\beta) = k^2(\delta|\delta) = k^2 \sum_i a_i (\delta|\alpha_i) = 0$, since $\langle \delta, \alpha_i^\vee \rangle = 0$ for all i.

□

§5.8. Now we give a description of the Tits cone X in terms of imaginary roots.

PROPOSITION 5.8. a) *If A is of finite type, then $X = \mathfrak{h}_{\mathbb{R}}$.*
b) *If A is of affine type, then*

$$X = \{h \in \mathfrak{h}_{\mathbb{R}} \mid \langle \delta, h \rangle > 0\} \cup \mathbb{R}\nu^{-1}(\delta).$$

c) *If A is of indefinite type, then*

$$\overline{X} = \{h \in \mathfrak{h}_{\mathbb{R}} \mid \langle \alpha, h \rangle \geq 0 \text{ for all } \alpha \in \Delta_+^{im}\}, \tag{5.8.1}$$

where $\overline{X}$ denotes the closure of X in the metric topology of $\mathfrak{h}_{\mathbb{R}}$.

Proof. a) holds by Proposition 3.12 e). If A is of affine type, then $\Delta^{re}+k\delta = \Delta^{re}$ for some k (see Proposition 6.3 d) from Chapter 6). Using Proposition 3.12 c), we deduce immediately that

$$\{h \in \mathfrak{h}_{\mathbb{R}} \mid \langle \delta, h \rangle > 0\} \subset X \text{ and } \{h \in \mathfrak{h}_{\mathbb{R}} \mid \langle \delta, h \rangle < 0\} \cap X = \emptyset.$$

If $\langle \delta, h \rangle = 0$ and $h \notin \mathbb{R}\nu^{-1}(\delta)$, then $\langle \alpha_i, h \rangle < 0$ for some i, completing the proof of b).

In order to prove c), denote by X' the right-hand side of (5.8.1). By Proposition 5.2 a), X' is W-invariant; also it is obvious that $X' \supset C$. Hence $X' \supset X$.

To prove the converse inclusion it is sufficient to show that for $h \in X'$ such that $\langle \alpha_i, h \rangle \in \mathbb{Z}$ $(i = 1, \ldots, n)$ there exists only a finite number of positive real roots γ such that $\langle \gamma, h \rangle < 0$ (see Proposition 3.12 c)). Recall that by Theorem 5.6 c) there exists $\beta \in \Delta_+^{im}$ such that $\langle \beta, \alpha_i^\vee \rangle < 0$ for all i. If $\gamma \in \Delta_+^{re}$, then $r_\gamma(\beta) = \beta + s\gamma \in \Delta_+^{im}$, where $s = -\langle \beta, \gamma^\vee \rangle \geq \operatorname{ht} \gamma^\vee$. As $h \in X'$, we have $\langle \beta + s\gamma, h \rangle \geq 0$. Hence there is only a finite number of real roots γ such that $\langle \gamma, h \rangle \leq -1$, which is the same as $\langle \gamma, h \rangle < 0$.

□

Proposition 5.8 has a nice geometric interpretation. Define the *imaginary cone* Z as the convex hull in $\mathfrak{h}_{\mathbb{R}}^*$ of the set $\Delta_+^{im} \cup \{0\}$. Then the cones $\overline{Z}$ and $\overline{X}$ are dual to each other:

$$\overline{X} = \{h \in \mathfrak{h}_{\mathbb{R}} \mid \langle \alpha, h \rangle \geq 0 \text{ for all } \alpha \in \overline{Z}\}.$$

In particular, Z is a convex cone (cf. Proposition 3.12 c)). Note also that $\overline{Z} \subset -\overline{X}^\vee$. Exercises 5.10 e) and 5.12 give another description of the cone $\overline{Z}$.

In the next subsection we will need

LEMMA 5.8. *The limit rays (in metric topology) for the set of rays* $\{\mathbb{R}_+\alpha \mid \alpha \in \Delta_+^{re}\}$ *lie in* $\overline{Z}$.

Proof. We can assume that the Cartan matrix A is indecomposable. In the finite type case there is nothing to prove since $|\Delta| < \infty$ by Proposition 4.9. In the affine case the result follows from the description of Δ in Chapter 6. In the indefinite case we choose $\beta \in \Delta_+^{im}$ such that $\langle \beta, \alpha_i^\vee \rangle < 0$ for all i (see Theorem 5.6 c)). Then $\langle \beta, \alpha^\vee \rangle \leq -\operatorname{ht} \alpha^\vee$ for $\alpha \in \Delta_+^{re}$ and $r_\alpha(\beta) = \beta + k\alpha \in \Delta_+^{im}$, where $k \geq 1$, proving the lemma in this case also.

□

§5.9. A linearly independent set of roots $\Pi' = \{\alpha_1', \alpha_2', \dots\}$ is called a ***root basis*** of Δ if each root α can be written in the form $\alpha = \pm \sum_i k_i \alpha_i'$, where $k_i \in \mathbb{Z}_+$.

PROPOSITION 5.9. *Let A be an indecomposable generalized Cartan matrix. Then any root basis Π' of Δ is W-conjugate to Π or to $-\Pi$.*

Proof. Set $Q_+' = \sum_i \mathbb{Z}_+ \alpha_i'$. By Theorem 5.4, the set of rays through $\alpha \in \Delta_+^{im}$ is dense in Z, which is convex. It follows that Δ_+^{im} lies in Q_+' or $-Q_+'$, and changing the sign if necessary we can assume that $\Delta_+^{im} \subset Q_+'$. It follows by Lemma 5.8 that the set $\Delta_+ \cap (-Q_+')$ is finite. If this set is nonempty, it contains a simple root α_i. But then $|\Delta_+ \cap (-r_i(Q_+'))| < |\Delta_+ \cap (-Q_+')|$. After a finite number of such steps we get $\Delta_+ \subset w(Q_+')$ for some $w \in W$ and hence $\Pi = w(\Pi')$.

□

Remark 5.9. Π is W-conjugate to $-\Pi$ if and only if A is of finite type.

§5.10. A generalized Cartan matrix A is called a matrix of ***hyperbolic*** type if it is indecomposable symmetrizable of indefinite type, and if every proper connected subdiagram of $S(A)$ is of finite or affine type.

Note that if A is symmetrizable, then a standard invariant bilinear form $(.|.)$ can be normalized such that $(\alpha_i|\alpha_j)$ are integers. Hence $a = \min_{\alpha \in Q : |\alpha|^2 > 0} |\alpha|^2$ exists and is a positive number.

LEMMA 5.10. *Let A be a generalized Cartan matrix of finite, affine or hyperbolic type, and let* $\alpha = \sum_j k_j \alpha_j \in Q$.

a) *If* $|\alpha|^2 \le a$, *then either* α *or* $-\alpha$ *lies in* Q_+.
b) *If* α *satisfies* (5.1.2), *then either* α *or* $-\alpha$ *lies in* Q_+.

Proof. Suppose the contrary to a); then $\alpha = \beta - \gamma$, where $\beta, \gamma \in Q_+ \backslash \{0\}$, and the supports P_1 and P_2 of β and γ have no common vertices. Then

$$a \ge |\alpha|^2 = |\beta|^2 + |\gamma|^2 + 2(-\beta|\gamma). \tag{5.10.1}$$

There are two possibilities: (i) both P_1 and P_2 are of finite type; (ii) P_1 is of finite type, P_2 is of affine type, and they are joined by an edge in $S(A)$. In case (i) we have $|\beta|^2 \ge a$, $|\gamma|^2 \ge a$, and $(-\beta|\gamma) \ge 0$, which contradicts (5.10.1). In case (ii) we have $|\beta|^2 \ge a$, $|\gamma|^2 \ge 0$, and $(-\beta|\gamma) > 0$, which again contradicts (5.10.1). The proof of b) is exactly the same using the observation that $|\beta|^2, |\gamma|^2 \in \mathbb{Z}|\alpha|^2$, since:

$$\begin{aligned} |\beta|^2 &= \sum_i k_i^2 |\alpha_i|^2 + \sum_{i<j} k_i k_j 2(\alpha_i|\alpha_j) \\ &= \sum_i k_i(k_i|\alpha_i|^2) + \sum_{i<j} a_{ij} k_j (k_i|\alpha_i|^2) \in \mathbb{Z}|\alpha|^2. \end{aligned}$$

□

PROPOSITION 5.10. *Let A be a generalized Cartan matrix of finite, affine, or hyperbolic type. Then*

a) *The set of all short real roots is*

$$\{\alpha \in Q \mid |\alpha|^2 = a = \min_i |\alpha_i|^2\}.$$

b) *The set of all real roots is*

$$\{\alpha = \sum_j k_j \alpha_j \in Q \mid |\alpha|^2 > 0 \text{ and } k_j |\alpha_j|^2 / |\alpha|^2 \in \mathbb{Z} \text{ for all } j\}.$$

c) *The set of all imaginary roots is*

$$\{\alpha \in Q \backslash \{0\} \mid |\alpha|^2 \le 0\}.$$

d) *If A is affine, then there exist roots of intermediate squared length m if and only if $A = A_{2\ell}^{(2)}$, $\ell \ge 2$. The set of such roots coincides with $\{\alpha \in Q \mid |\alpha|^2 = m\}$.*

Proof. Let $\alpha \in Q$ be such that $|\alpha|^2 = a$. Then $|w(\alpha)|^2 = a$ for $w \in W$ and hence, by Lemma 5.10 a), $w(\alpha) \in \pm Q_+$ for every $w \in W$. Without loss of generality we may assume that $\alpha \in Q_+$. Let β be an element of minimal height among $(W\cdot\alpha)\cap Q_+$. As $(\beta|\beta) = a > 0$, we have $(\beta|\alpha_i) > 0$ for some i. If now $\beta \neq \alpha_i$, then $r_i(\beta) \in Q_+$ and $\mathrm{ht}(r_i(\beta)) < \mathrm{ht}(\beta)$, a contradiction with the choice of β. This shows that α is a real root, proving a). The proof of d) is similar.

By Proposition 5.1 b) (i) and (iii), Δ^{re} lies in the set given by b). The proof of the reverse inclusion is the same as that of a), using Lemmas 5.1 and 5.10 b).

Let now $\alpha \in Q\backslash\{0\}$ and $|\alpha|^2 \leq 0$. By Lemma 5.10 we may assume that $\alpha \in Q_+$. The same argument as above shows that $W\cdot\alpha \subset Q_+$. As above, we choose an element $\beta \in W\cdot\alpha$ of minimal height. Then $\langle\beta,\alpha_i^\vee\rangle \leq 0$ for all i. Furthermore, $\mathrm{supp}\,\beta$ is connected, since otherwise $\beta = \gamma_1 + \gamma_2$, where $\mathrm{supp}\,\gamma_1$ and $\mathrm{supp}\,\gamma_2$ are disjoint unions of finite type diagrams which are disjoint, and, moreover, are not connected by an edge in $S(A)$; then $|\beta|^2 = |\gamma_1|^2 + |\gamma_2|^2 > 0$, a contradiction. So, $\beta \in K$ and therefore $\beta \in \Delta^{im}$ by Lemma 5.3; hence $\alpha \in \Delta^{im}$. Now c) follows by Proposition 5.2 c).

□

Note that Propositions 5.10 c) and 5.8 c) give the following explicit description of the closure of the Tits cone in the hyperbolic case:

$$\overline{X} \cup -\overline{X} = \{h \in \mathfrak{h}_{\mathbb{R}} \mid (h|h) \leq 0\}. \tag{5.10.2}$$

We obtain also the following corollary of Propositions 5.9 and 5.10. Note that an automorphism σ of the Dynkin diagram $S(A)$, induces an automorphism of the root lattice Q by $\sigma(\alpha_i) = \alpha_{\sigma(i)}$; denote the group of all such automorphisms by $\mathrm{Aut}(A)$. Another subgroup of $\mathrm{Aut}\,Q$ is the Weyl group W. Note that $\sigma r_i \sigma^{-1} = r_{\sigma(i)}$ and that $W \cap \mathrm{Aut}(A) = 1$ by Lemma 3.11 b). Thus $\mathrm{Aut}\,Q \supset \mathrm{Aut}(A) \ltimes W$.

COROLLARY 5.10. a) *If A is indecomposable, then the group of all automorphisms of Q preserving Δ is $\pm\,\mathrm{Aut}(A) \ltimes W$.*

b) *If A is of a symmetric matrix of finite, affine, or hyperbolic type, then the group of all automorphisms of Q preserving $(.|.)$ is $\pm\,\mathrm{Aut}(A) \ltimes W$.*

Proof. If $\sigma \in \mathrm{Aut}\,Q$ and $\sigma(\Delta) \subset \Delta$, then $\Pi' := \sigma(\Pi)$ is a root basis of Δ. By Proposition 5.9, $\pm w(\Pi') = \Pi$ for some $w \in W$, proving a). b) follows from a) and Proposition 5.10 a) and c).

□

Remark 5.10. a) Suppose that in Corollary 5.10, A is not symmetric but the square length of one of the simple roots, say α_1, is 1 and that of all $\alpha_i \neq \alpha_1$ is 2. Then

$$\Delta^{re} = \{\alpha \in Q \mid (\alpha|\alpha) = 1 \text{ or } 2\}.$$

Indeed, by Proposition 5.10 b) the only additional condition occurs if $(\alpha|\alpha) = 2$, and in this case $k_1 \in 2\mathbb{Z}$; but $(\alpha|\alpha) \equiv k_1^2 \bmod 2$, hence $(\alpha|\alpha) = 2$ implies that $k_1 \in 2\mathbb{Z}$. Therefore, by Corollary 5.10 a), the conclusion of Corollary 5.10 b) still holds.

b) By Remark 5.9, if A is of finite type, one can drop $\pm$ from the conclusion of Corollary 5.10 b).

§5.11. In this section we apply the results of §5.10 to the study of the standard Lorentzian lattice Λ_n ($n \geq 3$) with the bilinear form $(.|.)$ which in some $\mathbb{Z}$-basis $v_0, v_1, \ldots, v_{n-1}$ is given by

$$|\sum_{i=0}^{n-1} x_i v_i|^2 = -x_0^2 + x_1^2 + \cdots + x_{n-1}^2. \tag{5.11.1}$$

(This is the only odd unimodular integral n-dimensional Lorentzian lattice.) First, we choose another basis of Λ_n:

$$\alpha_1 = v_1 - v_2, \ldots, \ \alpha_{n-2} = v_{n-2} - v_{n-1}, \ \alpha_{n-1} = v_{n-1},$$
$$\alpha_n = -v_0 - v_1 - v_2 - v_3 \ \ (\text{resp. } = -v_0 - v_1 - v_2)$$
$$\text{if } n \geq 4 \ \ (\text{resp. } n = 3).$$

This shows that Λ_n is the root lattice corresponding to the generalized Cartan matrix $(2(\alpha_i|\alpha_j)/(\alpha_i|\alpha_i))$ with the following Dynkin diagram BE_n:

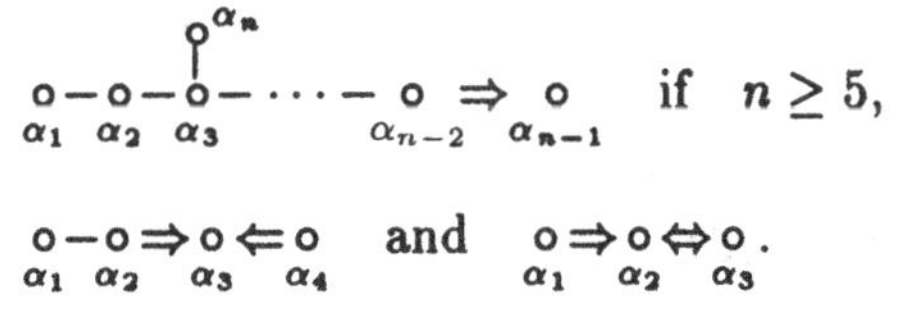

```
          o α_n
          |
o — o — o — ··· — o ⇒ o      if  n ≥ 5,
α_1 α_2 α_3      α_{n-2} α_{n-1}

o — o ⇒ o ⇐ o     and     o ⇒ o ⇔ o .
α_1 α_2 α_3 α_4           α_1  α_2  α_3
```

Note that in the basis $\{v_i\}$, the fundamental reflection r_i for $i \leq n-2$ is the transposition of v_i and v_{i+1}, r_{n-1} is the sign change of v_{n-1}, and r_n is represented by the following matrix:

$$R_n = \begin{pmatrix} 2 & 1 & 1 & 1 & \\ -1 & 0 & -1 & -1 & \\ -1 & -1 & 0 & -1 & \\ -1 & -1 & -1 & 0 & \\ & & & & I_{n-4} \end{pmatrix} \text{ if } n \geq 4, \quad \text{and} \quad R_3 = \begin{pmatrix} 3 & 2 & 2 \\ -2 & -1 & -2 \\ -2 & -2 & -1 \end{pmatrix}.$$

Denoting by $\overset{\circ}{W}_n$ the group of all permutations of $v_1, \ldots, v_{n-1}$ with simultaneous arbitrary sign changes, we have thus obtained that the Weyl group W_n of type BE_n is generated by $\overset{\circ}{W}_n$ and R_n. The fundamental chamber $C_n^\vee \subset \mathbf{R}^n = \mathbf{R} \otimes_{\mathbf{Z}} \Lambda_n$ of the group W_n is given by the following inequalities:

$$k_1 \geq k_2 \geq \cdots \geq k_{n-1} \geq 0, \qquad k_0 \geq k_1 + k_2 + k_3 \qquad (k_3 = 0 \text{ if } n = 3).$$

Note that $C_n^\vee \subset L_n^+ := \{\alpha = \sum_{i=0}^{n-1} x_i v_i \mid x_0 \geq 0, -x_0^2 + x_1^2 + \cdots + x_{n-1}^2 \geq 0\}$, the upper half of the light cone. Thus, the dual Tits cone $X_n^\vee$ lies in L_n^+. Note that $W_n \subset O_{\Lambda_n}^+ := \{g \in GL_n(\mathbf{R}) \mid g \cdot L_n^+ = L_n^+, g \cdot \Lambda_n = \Lambda_n\}$.

Now, BE_n is hyperbolic if and only if $n \leq 10$. It follows from §5.10 that, for $3 \leq n \leq 10$ we have:

$$\Delta^{im} = \{\alpha \in \Lambda_n \mid (\alpha|\alpha) \leq 0\} \backslash \{0\}, \tag{5.11.2}$$

$$\Delta^{re} = \{\alpha \in \Lambda_n \mid (\alpha|\alpha) = 1 \text{ or } 2\}. \tag{5.11.3}$$

(If $n \geq 4$, (5.11.3) follows from Remark 5.10. For $n = 3$, due to Proposition 5.10 b), all $\alpha \in \Lambda_3$ with $(\alpha|\alpha) = 1$ are roots, but the $\alpha = k_1\alpha_1 + k_2\alpha_2 + k_3\alpha_3 \in \Lambda_3$ with $(\alpha|\alpha) = 2$ should satisfy the additional condition $k_2, k_3 \in 2\mathbf{Z}$; however, this easily follows from $(\alpha|\alpha) = 2k_1(k_1 - k_2) + (k_2 - k_3)^2 = 2$.) Thus:

$$\Delta = \{\alpha \in \Lambda_n \mid (\alpha|\alpha) \leq 2\} \backslash \{0\}. \tag{5.11.4}$$

It follows from Corollary 5.10 a) that

$$W_n = O_{\Lambda_n}^+ \quad \text{for} \quad 3 \leq n \leq 10. \tag{5.11.5}$$

COROLLARY 5.11. *Let $3 \leq n \leq 10$. Then*

a) *All integral solutions of the equation*

$$-x_0^2 + x_1^2 + \cdots + x_n^2 = 1 \ (\textit{resp.} = 2)$$

form a W_n-orbit of the vector v_1 if $n > 3$ and a union of W_n-orbits of v_1 and $v_0 + v_1 + v_2$ if $n = 3$ (resp. form a union of W_n-orbits of $v_1 + v_2$ and $v_0 + v_1 + v_2 + v_3$ if $n > 3$ and a W_n-orbit of $v_1 + v_2$ if $n = 3$).

b) *All integral solutions of the equation*

$$-x_0^2 + x_1^2 + \cdots + x_n^2 = 0$$

form a $\pm W_n$-orbit of $v_0 + v_1$ if $n \leq 9$ and a union of two $\pm W_n$-orbits of $v_0 + v_1$ and $3v_0 + v_1 + v_2 + \cdots + v_9$ if $n = 10$.

□

We conclude this section with the following general construction of hyperbolic lattices. Let $\overset{\circ}{Q}$ be the root lattice corresponding to a generalized Cartan matrix of finite type X_ℓ. Let $\alpha_1, \ldots, \alpha_\ell$ be simple roots and θ be the highest root (see Remark 4.9). We normalize the standard invariant form on $\overset{\circ}{Q}$ by the condition $(\theta|\theta) = 2$. Let H_2 be the 2-dimensional hyperbolic lattice with basis u_1, u_2 and bilinear form $(u_1|u_2) = 1$, $(u_1|u_1) = (u_2|u_2) = 0$. Let Q be the orthogonal direct sum of lattices $\overset{\circ}{Q}$ and H_2. Then Q is the root lattice corresponding to the Dynkin diagram X_ℓ^H obtained from $X_\ell^{(1)}$ by adding an addition vertex α_{-1} joined with the vertex α_0 by a simple edge. This follows from the following choice of basis of Q:

$$\alpha_{-1} = u_1 + u_2, \qquad \alpha_0 = -u_1 - \theta, \qquad \alpha_1, \ldots, \alpha_\ell.$$

The lattice Q (resp. diagram X_ℓ^H) is called the *canonical hyperbolic extension* of the lattice $\overset{\circ}{Q}$ (resp. diagram X_ℓ). Note that

$$\det X_\ell^H = -\det X_\ell. \tag{5.11.6}$$

Note that if $X = A, B, C$, or D, then $X_\ell^H = XE_{\ell+2}$ (cf. Exercise 4.3), and that $E_6^H = T_{4,3,3}$, $E_7^H = T_{5,4,2}$, $E_8^H = E_{10} = T_{7,3,2}$ (cf. Exercise 4.2).

The most interesting example is $E_8^H = E_{10}$. The corresponding root lattice Q is the (only) even integral unimodular Lorentzian 10-dimensional lattice. Since E_{10} is of hyperbolic type, we can apply the results of §5.10. Thus, the Weyl group of E_{10} coincides with the group of automorphisms of the root lattice Q preserving the upper half of the light cone, $\Delta = \{\alpha \in Q \mid (\alpha|\alpha) \leq 2\} \backslash \{0\}$, etc.

§5.12. In conclusion, let us make one useful observation. Recall that $\mathfrak{g}(A) = \tilde{\mathfrak{g}}(A)/\mathfrak{r}$, where $\mathfrak{r}$ is the maximal ideal intersecting $\mathfrak{h}$ trivially. However all the proofs in Chapters 3, 4, and 5 used only the fact that

$$(\operatorname{ad} e_i)^{N_{ij}} e_j = 0 = (\operatorname{ad} f_i)^{N_{ij}} f_j \text{ for all } i \neq j \text{ and some } N_{ij}. \tag{5.12.1}$$

In other words, we have the following:

PROPOSITION 5.12. *Let $\mathfrak{g}$ be a quotient algebra of the Lie algebra $\tilde{\mathfrak{g}}(A)$ by a nontrivial Q-graded ideal such that (5.12.1) holds. Then all the statements of Chapters 3, 4, and 5 for $\mathfrak{g}(A)$ hold for the Lie algebra $\mathfrak{g}$ as well.*

We deduce the following:

COROLLARY 5.12. *Let $\mathfrak{g}$ be as in Proposition 5.12. Then*

a) *The root system of $\mathfrak{g}$ is the same as that of $\mathfrak{g}(A)$, the multiplicities of real roots being equal to 1.*

b) *If A is of finite type, then $\mathfrak{g} = \mathfrak{g}(A)$.*

Proof. a) follows from the proofs of Proposition 5.1 a) and Theorem 5.4, while b) follows from a).

□

Remark 5.12. We shall see in Chapter 9 that Corollary 5.12 b) holds for an arbitrary symmetrizable generalized Cartan matrix.

§5.13. Exercises.

5.1. Show that for $\alpha \in \Delta^{re}(A)$ one has:

$$[\mathfrak{g}_\alpha, \mathfrak{g}_{-\alpha}] = \mathbb{C}\alpha^\vee.$$

Show that if A is a nonsymmetrizable 3×3 matrix and $\alpha = \alpha_1 + \alpha_2 + \alpha_3$, then $\alpha \in \Delta^{im}(A)$ and $\dim[\mathfrak{g}_\alpha, \mathfrak{g}_{-\alpha}] > 1$ (cf. Theorem 2.2 e)).

5.2. Show that $\dim[\mathfrak{g}_\alpha, \mathfrak{g}_{-\alpha}] = 1$ for all $\alpha \in \Delta(A)$ if and only if A is symmetrizable.

5.3. If $\dim \mathfrak{g}(A) = \infty$, then $|\Delta^{re}| = \infty$.

5.4. The set $\Delta_+(A)$ is uniquely defined by the properties (i) and (ii) of Exercise 3.5 and the following property

(iii)′ if $\alpha \in \Delta_+$, then $\operatorname{supp} \alpha$ is connected.

[Let Δ'_+ satisfy (i), (ii), and (iii)′. Then $\Delta'_+ \backslash \{\alpha_i\}$ is r_i-invariant, hence $\Delta_+^{re} \subset \Delta'_+$. If now $\alpha \in \Delta'_+ \backslash \Delta_+^{re}$, then $W(\alpha) \subset \Delta'_+$ and β of minimal height from $W(\alpha)$ lies in K.]

5.5. Let $A = (a_{ij})$ be a symmetric generalized Cartan matrix. Show that

$$\{\alpha \in Q \backslash \{0\} \mid (\alpha|\alpha) \le 1\} \supset \Delta.$$

Show that the converse inclusion holds if and only if A is of finite, affine, or hyperbolic type.

5.6. Let $A = (a_{ij})$ be a finite, affine, or hyperbolic matrix, let $B = (b_{ij})$ be the corresponding "symmetrized" matrix, and let $B(x) = \sum b_{ij}x_i x_j$ be the associated quadratic form. Show that all the integral solutions of the equation $B(x) = 0$ are of the form $sw(\delta)$, where $s \in \mathbb{Z}$, $w \in W(A)$, and δ is the indivisible imaginary root of $\Delta(A')$, where A' is a principal affine submatrix of A. (In particular, any solution is 0 if A is strictly hyperbolic.) Show that if A is symmetric, then all the integral solutions of the equation $\sum a_{ij}x_i x_j = 2$ are of the form $w(\alpha_i)$, where $w \in W$, $\alpha_i \in \Pi$.

5.7. Show that the sublattice of the lattice Λ_n consisting of vectors with even sum of coordinates is isomorphic to the root lattice of type D^H_{n-2}.

5.8. Show that the Weyl group W is a subgroup of index 3 in the group of automorphisms preserving $(.|.)$ of the root lattice of type C_4.

5.9. Let $\alpha \in \Delta_+^{im}$ be such that $-\alpha \in C^\vee$ and $\langle \alpha, \alpha_i^\vee \rangle \neq 0$ for some $i \in \operatorname{supp} \alpha$. Then the subdiagram $\{i \in \operatorname{supp} \alpha \mid \langle \alpha, \alpha_i^\vee \rangle = 0\} \subset S(A)$ is a union of diagrams of finite type.

[Denote by T a connected component of this subdiagram and let $\alpha = \sum\limits_i k_i \alpha_i$; set $\beta = \sum\limits_{i \in T} k_i \alpha_i$. Then $\langle \beta, \alpha_i^\vee \rangle \geq 0$ for all i and > 0 for some $i \in T$.]

In Exercises 5.10–5.15 we assume A to be indecomposable.

5.10. An imaginary root α is called *strictly imaginary* if for every $\gamma \in \Delta^{re}$ either $\alpha + \gamma$ or $\alpha - \gamma$ is a root. Denote by Δ^{sim} the set of strictly imaginary roots.

a) If $\alpha \in \Delta_+$ and $\langle \alpha, \alpha_i^\vee \rangle < 0$ $(i = 1, \dots, n)$, then $\alpha \in \Delta^{sim}$. Deduce that if $\alpha \in \Delta^{im}$, $r_\gamma(\alpha) \neq \alpha$ for all $\gamma \in \Delta^{re}$, then $\alpha \in \Delta^{sim}$.

b) If $\alpha \in \Delta_+^{sim}$ and $\langle \alpha, \alpha_i^\vee \rangle \leq 0$ for $i = 1, \dots, n$, then $\alpha + \beta \in \Delta_+$ for every $\beta \in \Delta_+$.

c) If $\alpha \in \Delta_+^{sim}$, $\beta \in \Delta_+^{im}$, then $\alpha + \beta \in \Delta_+^{im}$.

d) Δ_+^{sim} is a semigroup.

e) $\overline{Z} = \overline{\mathbf{R}_+ \Delta_+^{sim}}$.

5.11. Let $\Lambda_i^\vee$ denote the linear function on $Q_\mathbb{R} := \mathbb{R} \otimes_\mathbb{Z} Q$ defined by $\langle \Lambda_i^\vee, \alpha_j \rangle = \delta_{ij}$ $(j = 1, \dots, n)$. Then $\overline{Z} = \{\alpha \in Q_\mathbb{R} \cap -\overline{X}_\mathbb{R}^\vee \mid \langle w(\Lambda_s^\vee), \alpha \rangle \geq 0$ for all $w \in W$ and those $s = 1, \dots, n$ for which the principal submatrix $(a_{ij})_{i,j \neq s}$ is of indefinite type$\}$.

5.12. $\overline{Z}$ is the convex hull of the set of limit points for $\mathbf{R}_+ \Delta_+^{re}$.

5.13. If A is a matrix of finite or affine type, $\beta \in \Delta$, $\alpha \in \Delta^{re}$, then the string $\{\beta + k\alpha, k \in \mathbb{Z}\}$ contains at most five roots. Show that if A is of indefinite type, then the number of roots in a string can be arbitrarily large.

5.14. Given $\beta \in \Delta$ and $\alpha \in \Delta^{re}$, the string $\{\beta + k\alpha\}$ contains at most four real roots.

5.15. Let A be symmetrizable of indefinite type. Then the following conditions are equivalent:

(i) A is of hyperbolic type;

(ii) $\overline{X} \cup -\overline{X} = \{h \in \mathfrak{h}_{\mathbb{R}} \mid (h|h) \leq 0\}$;

(iii) $\overline{Z} = -\overline{X}^{\vee}$;

(iv) $\Delta^{im} = \{\alpha \in Q\backslash\{0\} \mid (\alpha|\alpha) \leq 0\}$.

5.16. Let A be symmetrizable and let $(.|.)$ be a standard form. Let α, $\beta \in \Delta_+^{im}$. Then $(\alpha|\beta) \leq 0$.

[One can assume that $-\alpha \in C^{\vee}$.]

5.17. Under the hypotheses of Exercise 5.16 assume that $(\alpha|\beta) < 0$. Then $\alpha + \beta \in \Delta_+^{im}$.

[Since the cone $X^{\vee}$ is convex, we can assume that $-(\alpha+\beta) \in C^{\vee}$. But $\operatorname{supp}\alpha$ and $\operatorname{supp}\beta$ are connected, and since $(\alpha|\beta) < 0$, $\operatorname{supp}(\alpha+\beta)$ is connected and we can apply Lemma 5.3.]

5.18. Under the hypotheses of Exercise 5.16, assume that $\alpha+\beta \in \Delta_+^{im}$ and that α and β are not proportional isotropic roots. Then $(\alpha|\beta) < 0$.

[Use Exercise 5.9.]

5.19. Let α, $\beta \in \Delta_+^{re}$ be such that $\langle\alpha, \beta^{\vee}\rangle \geq 0$. Show that $\langle\beta, \alpha^{\vee}\rangle \geq 0$. Furthermore, show that $(\mathbb{Z}_+\alpha + \mathbb{Z}_+\beta) \cap \Delta = \{\alpha, \beta, \alpha+\beta\} \cap \Delta_+^{re}$.

[To show that $\langle\beta, \alpha^{\vee}\rangle \geq 0$, one can asssume that α is a simple root; then $\langle\alpha, \beta^{\vee}\rangle > 0$ (resp. $= 0$) if and only if $r_\beta(\alpha) < 0$ (resp. $= \alpha$), which is equivalent to $r_\beta(\alpha^{\vee}) < 0$ (resp. $= \alpha^{\vee}$). Furthermore, if $m\alpha + n\beta \in \Delta$ for some $m, n \in \mathbb{Z}_+$, then $\langle m\alpha + n\beta, \beta^{\vee}\rangle \geq 2n$, hence $m\alpha \in \Delta$ and $m \leq 1$; similarly, $n \leq 1$; in particular, $2(\alpha+\beta)$ is not a root.]

5.20. Let $\alpha = \sum_i k_i\alpha_i \in K$ and let $\beta = \sum_i m_i\alpha_i \in Q_+$ be such that $\alpha - \beta \in Q_+$. Using the identity

$$(\alpha - \beta|\beta) = \sum_j m_j(k_j - m_j)k_j^{-1}(\alpha|\alpha_j) + \frac{1}{2}\sum_{i,j} b_{ij}\left(\frac{m_i}{k_i} - \frac{m_j}{k_j}\right)^2 k_i k_j,$$

show that $(\alpha - \beta|\beta) < 0$ unless $(\alpha|\alpha) = 0$ and β is proportional to α.

The rest of the exercises deal with Kac–Moody algebras $\mathfrak{g}(A)$ of rank 2, i.e.,

$$A = \begin{pmatrix} 2 & -a \\ -b & 2 \end{pmatrix}, \text{ where } a,\ b \text{ are positive integers, and } a \geq b.$$

We associate to $\mathfrak{g}(A)$ the field $\mathsf{F} = \mathbb{Q}(\sqrt{ab(ab-4)})$; when $ab \neq 4$ we denote by $\lambda \mapsto \lambda'$ the unique nontrivial involution of F. Fix the following symmetrization of A:

$$B = \begin{pmatrix} 2 & -a \\ -a & 2a/b \end{pmatrix}$$

and the corresponding standard form $(.|.)$ on $\mathfrak{h}^*$, so that $(\alpha_1|\alpha_1) = 2$, $(\alpha_2|\alpha_2) = 2a/b$, $(\alpha_1|\alpha_2) = -a$. Introduce the following numbers:

$$\eta = \frac{-ab + \sqrt{ab(ab-4)}}{2b}; \qquad \epsilon = -b\eta - 1;$$
$$\epsilon_0 = \epsilon \text{ if } a \neq b \quad \text{and} \quad = \eta \text{ if } a = b.$$

5.21. Assume that $ab \neq 4$. Show that in the basis α_1, α_2 of $\mathfrak{h}^*_{\mathbb{R}}$, one has

$$C^\vee = \{(x, y) \mid 2x \geq ay,\ 2y \geq bx\}.$$

Show that if $ab > 4$, then

$$\overline{X}^\vee = \{(x, y) \mid -\eta' y \leq x \leq -\eta y\}.$$

5.22. Show that ϵ and ϵ' are eigenvalues of $r_1 r_2$, and that

$$(r_1 r_2)^k = \frac{\epsilon^k - \epsilon'^k}{\epsilon - \epsilon'} r_1 r_2 - \frac{\epsilon^{k-1} - \epsilon'^{k-1}}{\epsilon - \epsilon'} I, \quad k \in \mathbb{Z}.$$

[Use the fact that for a 2×2 matrix a with eigenvalues λ_1 and λ_2, one has

$$a^k = \frac{\lambda_1^k - \lambda_2^k}{\lambda_1 - \lambda_2} a - \frac{\lambda_1^k \lambda_2 - \lambda_2^k \lambda_1}{\lambda_1 - \lambda_2} I \quad .]$$

5.23. Assume that $ab \neq 4$. Let $xy' + x'y$ be the trace form of the field $\mathbb{F}$. Show that the map

$$\phi : k_1\alpha_1 + k_2\alpha_2 \mapsto k_1 + k_2\eta$$

is an isometry of the lattices Q and $\mathbb{Z}[\eta] \subset \mathsf{F}$, so that the fundamental reflections r_1 and r_2 of Q induce automorphisms of the lattice $\mathbb{Z}[\eta]$:

$$r_1(\lambda) = -\lambda', \qquad r_2(\lambda) = -\epsilon\lambda',$$

and the group from Corollary 5.10 a) maps isomorphically onto the group generated by multiplication by the unit ϵ_0, the involution $'$ and -1. Show that

$$\phi(Q_+) = \{x + y\sqrt{ab(ab-4)} \in \mathbf{Z}[\eta] \mid y \geq 0 \text{ and } x \geq 0 \text{ if } y = 0\},$$

$$\phi(\Delta_+^{re}) = \{\epsilon^n\eta, \epsilon^{-n}, -\epsilon^{-n}\eta', -\epsilon^{n+1}, \text{ where } n \geq 0\},$$

$$\phi(\Delta^{im}) = \{x \in \mathbf{Z}[\eta] \mid xx' \leq 0, x \neq 0\}.$$

5.24. Show that in the case when $a = b = m$, $m \neq 2$, we have $\mathbb{F} = \mathbb{Q}(\sqrt{m^2-4})$, $\eta = \epsilon_0$, $\epsilon = \epsilon_0^2$; if in addition $m^2 - 4$ is square free except possibly for a factor of 4 when m is even, then $\mathbf{Z} + \frac{1}{2}(1 + \sqrt{m^2-4})\mathbf{Z}$ is the ring of integers of the field $\mathbf{F}$. Deduce that under the above hypotheses on m, the number $\frac{1}{2}(m + \sqrt{m^2-4})$ is a fundamental unit of the ring of integers of the field $\mathbf{Q}(\sqrt{m^2-4})$, i.e., ϵ_0 together with -1 generate the group of all integers of norm 1.

5.25. Show that

$$\Delta_+^{re}(A) = \{c_j\alpha_1 + d_{j+1}\alpha_2 \text{ and } c_{j+1}\alpha_1 + d_j\alpha_2 \ (j \in \mathbf{Z}_+)\},$$

where the sequences c_j and d_j are defined by the following recurrent formulas for $j \geq 0$:

$$\begin{aligned} c_{j+2} + c_j &= ad_{j+1} \\ d_{j+2} + d_j &= bc_{j+1} \end{aligned}$$

and $c_0 = d_0 = 0$, $c_1 = d_1 = 1$.

5.26. Show that all integral solutions (x, y) with relatively prime x and y of the equation

$$bx^2 - abxy + ay^2 = a \text{ or } b$$

are $\pm(c_j, d_{j+1})$ with j odd or $\pm(c_{j+1}, d_j)$ with j even, $j \geq 0$. Show that more solutions exist only if the right-hand side is a and $a/b = r^2$, where r is an integer > 1, and they are (rx, ry), where $bx^2 - abxy + ay^2 = b$.

5.27. Show that

$$\Delta_+^{re}\begin{pmatrix} 2 & -2 \\ -2 & 2 \end{pmatrix} = \{j\alpha_1 + (j+1)\alpha_2 \text{ and } (j+1)\alpha_1 + j\alpha_2 \quad (j \in \mathbf{Z}_+)\}.$$

$$\Delta_+^{re}\begin{pmatrix} 2 & -4 \\ -1 & 2 \end{pmatrix} = \{2j\alpha_1 + (j+1)\alpha_2 \text{ and } (j+1)\alpha_1 + \tfrac{1}{2}j\alpha_2$$

for even $j \in \mathbf{Z}_+$; $j\alpha_1 + \frac{1}{2}(j+1)\alpha_2$ and $2(j+1)\alpha_1 + j\alpha_2$ for odd $j \in \mathbf{Z}_+\}$.

5.28. Show that

$$\Delta_+^{re}\begin{pmatrix} 2 & -3 \\ -3 & 2 \end{pmatrix} = \{\phi_{2j}\alpha_1 + \phi_{2j+2}\alpha_2 \text{ and } \phi_{2j+2}\alpha_1 + \phi_{2j}\alpha_2 \quad (j \in \mathbb{Z}_+)\},$$

where ϕ_j is the jth Fibonacci number:

$$\phi_0 = 0, \qquad \phi_1 = 1, \qquad \phi_{j+2} = \phi_{j+1} + \phi_j \text{ for } j \in \mathbb{Z}_+.$$

§5.14. Bibliographical notes and comments.

The notion of real and imaginary roots were introduced in Kac [1968 B], where Propositions 5.2 and 5.5 and Theorem 5.6 were proved. (Moody [1968] introduced, independently, real roots; he called them Weyl roots.) Lemma 5.3, Theorem 5.4, and Proposition 5.7 are proved in Kac [1980 A]; the exposition of §§5.1–5.7 and §5.10 is taken from this paper. The strengthening of Proposition 5.10 b) as compared to previous editions, is due to Mark Kruelle.

Proposition 5.10 c) was obtained by Moody [1979], where he initiated a detailed study of hyperbolic root systems. The material of §5.8 is taken from Kac–Peterson [1984 A]. Proposition 5.9 is proved in Kac [1978 A]. Most of the results of §5.11 are due to Vinberg [1972], [1975], who developed a general algorithm allowing to compute the subgroup W generated by reflections in the automorphism group of a Lorentzian lattice. In particular, he showed that W has finite index in $\operatorname{Aut}\Lambda_n$ if and only if $n \leq 20$ (Vinberg [1972], [1975], Vinberg–Kaplinskaya [1978]). Recall that $H_2 \oplus nE_8$ are the only even unimodular lattices (of dimension $8n+2$) (see Serre [1970]). For $n = 1$ and 2, their automorphism groups were computed by Vinberg [1975]; and for $n = 3$, by Conway [1983]. See Vinberg [1985] for a survey of hyperbolic reflection groups.

As shown in Kac [1980 A], given a symmetric generalized Cartan matrix A, the set of positive roots $\Delta_+(A)$ describes the set of dimensions of indecomposable representations of the graph $S(A)$, equipped with some orientation. Moreover, the number of absolutely indecomposable representations of dimension $\alpha \in \Delta_+(A)$ over a finite field $\mathbb{F}_q$ is given by a polynomial $q^N + a_1 q^{N-1} + \cdots + a_N$, where $a_i \in \mathbb{Z}$, $N = 1 - (\alpha|\alpha)$ (see Kac [1982 A]). In these papers several conjectures are posed; the most intriguing of them suggests that $a_N = \operatorname{mult}\alpha$. Many of these conjectures (but not the latter) have been solved by Schofield [1988A-C]. Ringel [1989] found a connection of this field to quantized Kac–Moody algebras via Hall algebras.

The nature of the root multiplicities in the indefinite case still remains mysterious: there is no single case when the answer is known explicitly. Asymptotic behavior of root multiplicities was studied in Kac–Peterson [1984 A]; in some cases upper-bounds were found by Frenkel [1985] and Borcherds [1986].

Exercises 5.10 and 5.12 are taken from Kac [1978 A]; and 5.11 and 5.20, from Kac [1980 A]. Exercise 5.19 is taken from Peterson–Kac [1983]; it is one of the key lemmas in the structure theory of groups attached to Kac–Moody algebras. In Lepowsky–Moody [1979] one can find a detailed study of the root systems in the hyperbolic rank 2 case by making use of the map ϕ; Exercise 5.23 is due to them. Exercise 5.28 is taken from Feingold [1980]. The remaining exercises are either new or standard.

Chapter 6. Affine Algebras: the Normalized Invariant Form, the Root System, and the Weyl Group

§6.0. The results of Chapter 4 show that a Kac–Moody algebra $\mathfrak{g}(A)$ is finite-dimensional if and only if all principal minors of A are positive. These Lie algebras are semisimple; moreover, by the classical structure theory, they exhaust all finite-dimensional semisimple Lie algebras. So, the classical Killing–Cartan theory of simple Lie algebras is, in our terminology, the theory of Kac–Moody algebras associated to a matrix of finite type.

In this chapter we consider the next case, when the matrix A is of affine type. Recall that this is a generalized Cartan matrix A, all of whose proper principal minors are positive, but $\det A = 0$ (A is then automatically indecomposable). A Kac–Moody algebra associated to a generalized Cartan matrix of affine type is called an *affine* (Kac–Moody) algebra.

We describe in detail the standard bilinear form, the root system, and the Weyl group of an affine algebra $\mathfrak{g}$ in terms of the "underlying" simple finite-dimensional Lie algebra $\mathring{\mathfrak{g}}$. In particular, we show that the Weyl group of $\mathfrak{g}$ is the so-called affine Weyl group of $\mathring{\mathfrak{g}}$; this explains the term "affine" algebra.

At the end of the chapter we give an explicit construction of the root lattice, the root system and the Weyl group of all simple finite-dimensional Lie algebras.

§6.1. Let A be a generalized Cartan matrix of affine type of order $\ell + 1$ (and rank ℓ), and let $S(A)$ be its Dynkin diagram from Table Aff. Let $a_0, a_1, \ldots, a_\ell$ be the numerical labels of $S(A)$ in Table Aff. Then $a_0 = 1$ unless A is of type $A_{2\ell}^{(2)}$, in which case $a_0 = 2$.

We denote by $a_i^\vee$ ($i = 0, \ldots, \ell$) the labels of the Dynkin diagram $S({}^tA)$ of the dual algebra which is obtained from $S(A)$ by reversing the directions of all arrows and keeping the same enumeration of vertices. Note that in all cases

$$a_0^\vee = 1. \tag{6.1.1}$$

The numbers

$$h = \sum_{i=0}^{\ell} a_i \quad \text{and} \quad h^\vee = \sum_{i=0}^{\ell} a_i^\vee$$

are called, respectively, the *Coxeter number* and the *dual Coxeter number* of the matrix A. We list these important numbers below:

A	h	$h^\vee$	A	h	$h^\vee$
$A_l^{(1)}$	$l+1$	$l+1$	$A_{2l}^{(2)}$	$2l+1$	$2l+1$
$B_l^{(1)}$	$2l$	$2l-1$	$A_{2l-1}^{(2)}$	$2l-1$	$2l$
$C_l^{(1)}$	$2l$	$l+1$	$D_{l+1}^{(2)}$	$l+1$	$2l$
$D_l^{(1)}$	$2l-2$	$2l-2$	$E_6^{(2)}$	9	12
$E_6^{(1)}$	12	12	$D_4^{(3)}$	4	6
$E_7^{(1)}$	18	18			
$E_8^{(1)}$	30	30			
$F_4^{(1)}$	12	9			
$G_2^{(1)}$	6	4			

Another important number is r, the number of the Table Aff r containing A.

Remark 6.1. The dual Coxeter number of the affine matrix $X_N^{(r)}$ is independent of r.

The matrix A is symmetrizable by Lemma 4.6. Moreover, we have

$$(6.1.2) \qquad A = \operatorname{diag}(a_0 a_0^{\vee -1}, a_1 a_1^{\vee -1}, \dots, a_\ell a_\ell^{\vee -1})B, \text{ where } B = {}^tB.$$

Indeed, let $\delta = {}^t(a_0, \dots, a_\ell)$ and $\delta^\vee = {}^t(a_0^\vee, \dots, a_\ell^\vee)$; if $A = DB$ where D is diagonal invertible and $B = {}^tB$, then $B\delta = 0$ and hence ${}^t\delta B = 0$. On the other hand, ${}^t\delta^\vee A = 0$ implies $({}^t\delta^\vee)DB = 0$, and we use the fact that $\dim \ker B = 1$.

§6.2. Let $\mathfrak{g} = \mathfrak{g}(A)$ be the affine algebra associated to a matrix A of affine type from Table Aff r, let $\mathfrak{h}$ be its Cartan subalgebra, $\Pi = \{\alpha_0, \dots, \alpha_\ell\} \subset \mathfrak{h}^*$ the set of simple roots, $\Pi^\vee = \{\alpha_0^\vee, \dots, \alpha_\ell^\vee\} \subset \mathfrak{h}$ the set of simple coroots, Δ the root system, Q and $Q^\vee$ the root and coroot lattices, etc. It follows from Proposition 1.6 that the center of $\mathfrak{g}(A)$ is 1-dimensional and is spanned by

$$K = \sum_{i=0}^{\ell} a_i^\vee \alpha_i^\vee.$$

The element K is called the *canonical central element.*

Recall the definition of the element δ (cf. Theorem 5.6):

$$\delta = \sum_{i=0}^{\ell} a_i \alpha_i \in Q.$$

Fix an element $d \in \mathfrak{h}$ which satisfies the following conditions:

$$\langle \alpha_i, d \rangle = 0 \text{ for } i = 1, \ldots, \ell; \qquad \langle \alpha_0, d \rangle = 1.$$

(Such an element is defined up to a summand proportional to K.) The element d is called the *scaling element.* It is clear that the elements $\alpha_0^\vee, \ldots, \alpha_\ell^\vee, d$ form a basis of $\mathfrak{h}$. Note that

$$\mathfrak{g} = [\mathfrak{g}, \mathfrak{g}] + \mathbf{C}d.$$

We define a nondegenerate symmetric bilinear $\mathbf{C}$-valued form $(.|.)$ on $\mathfrak{h}$ as follows (cf. (2.1.2, 3, and 4) and (6.1.1)):

$$\text{(6.2.1)} \qquad \begin{cases} (\alpha_i^\vee | \alpha_j^\vee) = a_j a_j^{\vee -1} a_{ij} & (i, j = 0, \ldots, \ell); \\ (\alpha_i^\vee | d) = 0 & (i = 1, \ldots, \ell); \\ (\alpha_0^\vee | d) = a_0; & (d|d) = 0. \end{cases}$$

By Theorem 2.2 this form can be uniquely extended to a bilinear form $(.|.)$ on the whole Lie algebra $\mathfrak{g}$ such that all the properties described by this theorem hold. From now on we fix this form on $\mathfrak{g}$. This is, clearly, a standard form. We call it the *normalized invariant form.*

To describe the induced bilinear form on $\mathfrak{h}^*$, we define an element $\Lambda_0 \in \mathfrak{h}^*$ by

$$\langle \Lambda_0, \alpha_i^\vee \rangle = \delta_{0i} \text{ for } i = 0, \ldots, \ell; \qquad \langle \Lambda_0, d \rangle = 0.$$

Then $\{\alpha_0, \ldots, \alpha_\ell, \Lambda_0\}$ is a basis of $\mathfrak{h}^*$ and we have

$$\text{(6.2.2)} \qquad \begin{cases} (\alpha_i | \alpha_j) = a_i^\vee a_i^{-1} a_{ij} & (i, j = 0, \ldots, \ell); \\ (\alpha_i | \Lambda_0) = 0 & (i = 1, \ldots, \ell); \\ (\alpha_0 | \Lambda_0) = a_0^{-1}; & (\Lambda_0 | \Lambda_0) = 0. \end{cases}$$

The map $\nu : \mathfrak{h} \mapsto \mathfrak{h}^*$ defined by $(.|.)$ looks as follows:

$$\text{(6.2.3)} \qquad a_i^\vee \nu(\alpha_i^\vee) = a_i \alpha_i; \qquad \nu(K) = \delta; \qquad \nu(d) = a_0 \Lambda_0.$$

We also record some other simple formulas:

(6.2.4) $(\delta|\alpha_i) = 0 \ (i = 0, \ldots, \ell)$; $(\delta|\delta) = 0$; $(\delta|\Lambda_0) = 1$;

(6.2.5) $(K|\alpha_i^\vee) = 0 \ (i = 0, \ldots, \ell)$; $(K|K) = 0$; $(K|d) = a_0$.

Denote by $\mathring{\mathfrak{h}}$ (resp. $\mathring{\mathfrak{h}}_{\mathbf{R}}$) the linear span over $\mathbf{C}$ (resp. $\mathbf{R}$) of $\alpha_1^\vee, \ldots, \alpha_\ell^\vee$. The dual notions $\mathring{\mathfrak{h}}^*$ and $\mathring{\mathfrak{h}}_{\mathbf{R}}^*$ are defined similarly. Then we have an orthogonal direct sum of subspaces:

$$\mathfrak{h} = \mathring{\mathfrak{h}} \oplus (\mathbf{C}K + \mathbf{C}d); \qquad \mathfrak{h}^* = \mathring{\mathfrak{h}}^* \oplus (\mathbf{C}\delta + \mathbf{C}\Lambda_0).$$

We set $\mathfrak{h}_{\mathbf{R}} = \mathring{\mathfrak{h}}_{\mathbf{R}} + \mathbf{R}K + \mathbf{R}d$, $\mathfrak{h}_{\mathbf{R}}^* = \mathring{\mathfrak{h}}_{\mathbf{R}}^* + \mathbf{R}\Lambda_0 + \mathbf{R}\delta$.

Note that the restriction of the bilinear form $(.|.)$ to $\mathring{\mathfrak{h}}_{\mathbf{R}}^*$ and $\mathring{\mathfrak{h}}_{\mathbf{R}}$ (resp. $\mathring{\mathfrak{h}}_{\mathbf{R}}^* + \mathbf{R}\delta$ and $\mathring{\mathfrak{h}}_{\mathbf{R}} + \mathbf{R}K$) is positive-definite (resp. positive-semidefinite with kernels $\mathbf{R}\delta$ and $\mathbf{R}K$) by Proposition 4.7 a) and b).

For a subset S of $\mathfrak{h}^*$ denote by $\overline{S}$ the orthogonal projection of S on $\mathring{\mathfrak{h}}^*$. (This should not be confused with the sign of closure in metric topology.) Then we have the following useful formula for $\lambda \in \mathfrak{h}^*$ such that $\lambda(K) \neq 0$:

$$\lambda - \overline{\lambda} = \langle \lambda, K \rangle \Lambda_0 + \langle 2\lambda, K \rangle^{-1} (|\lambda|^2 - |\overline{\lambda}|^2)\delta. \tag{6.2.6}$$

Indeed, $\lambda - \overline{\lambda} = b_1 \Lambda_0 + b_2 \delta$. Taking inner product with δ, we obtain, by (6.2.2, 3, and 4) that $b_1 = \langle \lambda, K \rangle$. As $|\lambda|^2 = |\overline{\lambda}|^2 + 2b_1 b_2$, we are done. We also have another useful formula:

$$\lambda = \overline{\lambda} + \langle \lambda, K \rangle \Lambda_0 + (\lambda|\Lambda_0)\delta. \tag{6.2.7}$$

Define $\rho \in \mathfrak{h}^*$ by $\langle \rho, \alpha_i^\vee \rangle = 1 \ (i = 0, \ldots, \ell)$ and $\langle \rho, d \rangle = 0$ (cf. §2.5). Then (6.2.7) gives

$$\rho = \overline{\rho} + h^\vee \Lambda_0. \tag{6.2.8}$$

§6.3. Denote by $\mathring{\mathfrak{g}}$ the subalgebra of $\mathfrak{g}$ generated by the e_i and f_i with $i = 1, \ldots, \ell$. By the results of Chapter 1 it is clear that $\mathring{\mathfrak{g}}$ is a Kac–Moody algebra associated to the matrix $\mathring{A}$ obtained from A by deleting the 0th row and column. The elements $e_i, f_i \ (i = 1, \ldots, \ell)$ are the Chevalley generators of $\mathring{\mathfrak{g}}$, and $\mathring{\mathfrak{h}} = \mathring{\mathfrak{g}} \cap \mathfrak{h}$ is its Cartan subalgebra; $\mathring{\Pi} = \{\alpha_1, \ldots, \alpha_\ell\}$ is the root basis, and $\mathring{\Pi}^\vee = \{\alpha_1^\vee, \ldots, \alpha_\ell^\vee\}$ is the coroot basis for $\mathring{\mathfrak{g}}$. Futhermore, by

Proposition 4.9, $\mathring{\mathfrak{g}} = \mathfrak{g}(\mathring{A})$ is a simple finite-dimensional Lie algebra whose Dynkin diagram $S(\mathring{A})$ is obtained from $S(A)$ by removing the 0th vertex.

The set $\mathring{\Delta} = \Delta \cap \mathring{\mathfrak{h}}^*$ is the root system of $\mathring{\mathfrak{g}}$; it is finite and consists of real roots (by Proposition 4.9), the set $\mathring{\Delta}_+ = \mathring{\Delta} \cap \Delta_+$ being the set of positive roots. Denote by $\mathring{\Delta}_s$ and $\mathring{\Delta}_\ell$ the sets of short and long roots, respectively, in $\mathring{\Delta}$. Put $\mathring{Q} = \mathbb{Z}\mathring{\Delta}$. Let $\mathring{W}$ be the Weyl group of $\mathring{\Delta}$.

Recall that the sets of imaginary and positive imaginary roots of $\mathfrak{g}$ are as follows (Theorem 5.6 b):

$$\Delta^{im} = \{\pm\delta, \pm 2\delta, \dots\}, \qquad \Delta_+^{im} = \{\delta, 2\delta, \dots\}.$$

The following proposition describes the set of real roots Δ^{re} and positive real roots Δ_+^{re} in terms of $\mathring{\Delta}$ and δ.

PROPOSITION 6.3. a) $\Delta^{re} = \{\alpha + n\delta \mid \alpha \in \mathring{\Delta}, n \in \mathbb{Z}\}$ *if* $r = 1$.

b) $\Delta^{re} = \{\alpha + n\delta \mid \alpha \in \mathring{\Delta}_s,\ n \in \mathbb{Z}\} \cup \{\alpha + nr\delta \mid \alpha \in \mathring{\Delta}_\ell,\ n \in \mathbb{Z}\}$ *if* $r = 2$ *or* 3, *but A is not of type* $A_{2\ell}^{(2)}$.

c) $\Delta^{re} = \{\frac{1}{2}(\alpha + (2n-1)\delta) \mid \alpha \in \mathring{\Delta}_\ell,\ n \in \mathbb{Z}\} \cup \{\alpha + n\delta \mid \alpha \in \mathring{\Delta}_s, n \in \mathbb{Z}\}$ $\cup \{\alpha + 2n\delta \mid \alpha \in \mathring{\Delta}_\ell, n \in \mathbb{Z}\}$ *if A is of type* $A_{2\ell}^{(2)}$.

d) $\Delta^{re} + r\delta = \Delta^{re}$.

e) $\Delta_+^{re} = \{\alpha \in \Delta^{re}$ with $n > 0\} \cup \mathring{\Delta}_+$.

Proof. It is clear that d) and e) follow from a), b), c). The proof of a), b), and c) is based on Proposition 5.10. Let a and b denote the square lengths of a short and a long root, respectively, and let Δ_s^{re}, Δ_ℓ^{re} be the sets of short and long real roots.

First, suppose that A is not of type $A_{2\ell}^{(2)}$. Then $\delta = \alpha_0 + a_1\alpha_1 + \cdots$. If now $\alpha = \sum_{i=0}^{\ell} k_i\alpha_i \in \Delta_s^{re}$, then $a = |\alpha|^2 = |\alpha - k_0\delta|^2$ and hence $\alpha - k_0\delta \in \mathring{\Delta}_s$ by Proposition 5.10 a), which gives the inclusion $\subset$ in the relation

$$\Delta_s^{re} = \{\alpha + n\delta \mid \alpha \in \mathring{\Delta}_s, n \in \mathbb{Z}\} \text{ if } A \neq A_{2\ell}^{(2)}. \tag{6.3.1}$$

The reverse inclusion also follows from Proposition 5.10 a).

If $r = 1$, then α_0 is a long root and the same argument as above gives, using Proposition 5.10 b):

$$\Delta_\ell^{re} = \{\alpha + n\delta \mid \alpha \in \mathring{\Delta}_\ell, n \in \mathbb{Z}\} \text{ if } r = 1. \tag{6.3.2}$$

If $r = 2$ or 3, then α_0 is a short root, hence, by Proposition 5.10b), $\alpha = k_0\alpha_0 + \cdots \in \Delta_\ell^{re}$ only if k_0 is divisible by r. Therefore we obtain, by Proposition 5.10b),

(6.3.3) $$\Delta_\ell^{re} = \{\alpha + nr\delta \mid \alpha \in \mathring{\Delta}_\ell, n \in \mathbb{Z}\} \text{ if } r = 2, 3; A \neq A_{2\ell}^{(2)}.$$

Formulas (6.3.1, 2, and 3) prove a) and b).

Finally, let A be of type $A_{2\ell}^{(2)}$. Then short (resp. long) real roots have square length 1 (resp. 4), and the roots from $\Delta_m^{re} := \Delta^{re} \setminus (\Delta_s^{re} \cup \Delta_\ell^{re})$ have square length 2 ($\Delta_m^{re} \neq \emptyset$ if $\ell > 1$). We have to show that

(6.3.4) $$\Delta_s^{re} = \{\tfrac{1}{2}(\alpha + (2n-1)\delta) \mid \alpha \in \mathring{\Delta}_\ell, n \in \mathbb{Z}\},$$

(6.3.5) $$\Delta_m^{re} = \{\alpha + n\delta \mid \alpha \in \mathring{\Delta}_s, n \in \mathbb{Z}\},$$

(6.3.6) $$\Delta_\ell^{re} = \{\alpha + 2n\delta \mid \alpha \in \mathring{\Delta}_\ell, n \in \mathbb{Z}\}.$$

By Proposition 5.10 b), $\alpha = k_0\alpha_0 + \cdots \in \Delta_\ell^{re}$ only if k_0 is divisible by 4. Now the same argument as above proves (6.3.6). A similar argument gives (6.3.4 and 5).

□

Note that Proposition 6.3 will also follow from the explicit construction of affine algebras given in the next chapters. Note also that (6.1.1) implies:

(6.3.7) $$Q^\vee = \mathring{Q}^\vee \oplus \mathbb{Z}K \text{ (orthogonal direct sum)},$$

and that (due to (6.2.1) and (6.2.2)) we have the isomorphism of lattices equipped with bilinear forms

(6.3.8) $$Q^\vee(A) \simeq Q({}^tA).$$

Warning. $\mathring{\Delta} = \overline{\Delta}\setminus\{0\}$ in all cases except $A_{2\ell}^{(2)}$, in which case $\overline{\Delta}\setminus\{0\}$ is a non-reduced root system, and $\mathring{\Delta}$ is the associated reduced root system.

§6.4. Introduce the following important element:

$$\theta = \delta - a_0\alpha_0 = \sum_{i=1}^{\ell} a_i\alpha_i \in \mathring{Q}.$$

It is easy to deduce from the formulas of §6.2 that

$$|\theta|^2 = 2a_0. \tag{6.4.1}$$

Hence $|\theta|^2$ is equal to the square length of a long root if A is from Table Aff 1 or is of type $A_{2\ell}^{(2)}$, and it is equal to the square length of a short root otherwise. One deduces now from Proposition 5.10 a), b) that in all cases $\theta \in \mathring{\Delta}_+$. Again, from the formulas of §6.2 we deduce:

$$\theta = a_0\nu(\theta^\vee); \qquad |\theta^\vee|^2 = 2a_0^{-1}; \qquad \alpha_0^\vee = \nu^{-1}(\delta - \theta) = K - a_0\theta^\vee.$$

Furthermore, one has

PROPOSITION 6.4. a) *If A is from* Table Aff 1 *or is of type $A_{2\ell}^{(2)}$, then $\theta \in (\mathring{\Delta}_+)_\ell$ and θ is the unique root in $\mathring{\Delta}$ of maximal height ($= h - a_0$).*
b) *If A is from* Table Aff 2 *or* 3 *and is not of type $A_{2\ell}^{(2)}$, then $\theta \in (\mathring{\Delta}_+)_s$ and is the unique root in $\mathring{\Delta}_s$ of maximal height ($= h - 1$).*

Proof. It is easy to check that all simple roots of the same square length in $\mathring{\Delta}$ are $\mathring{W}$-equivalent, hence both $\mathring{\Delta}_s$ and $\mathring{\Delta}_\ell$ are orbits of $\mathring{W}$. Also, $\langle\theta, \alpha_i^\vee\rangle = -a_0\langle\alpha_0, \alpha_i^\vee\rangle \geq 0$ for all $i = 1, \ldots, \ell$. Now a) and b) follow from Proposition 3.12 b).

□

Unless otherwise stated, in the case of a finite type matrix A, we shall normalize the standard invariant form $(.|.)$ on $\mathfrak{g}(A)$ by the condition

$$(\alpha|\alpha) = 2 \quad \text{if } \alpha \in \Delta_\ell, \tag{6.4.2}$$

and shall call it the ***normalized invariant form***. We deduce from (6.4.1) and Proposition 6.4 the following:

COROLLARY 6.4. *Let $\mathfrak{g}$ be an affine algebra of type $X_N^{(r)}$. Then the ratio of the normalized invariant form of $\mathfrak{g}$ restricted to $\mathring{\mathfrak{g}}$ to the normalized invariant form of $\mathring{\mathfrak{g}}$ is equal to r.*

□

Note that we have the following description of Π and $\Pi^\vee$:

$$\Pi = \{\alpha_0 = a_0^{-1}(\delta - \theta), \alpha_1, \ldots, \alpha_\ell\},$$
$$\Pi^\vee = \{\alpha_0^\vee = K - a_0\theta^\vee, \alpha_1^\vee, \ldots, \alpha_\ell^\vee\}.$$

§6.5. Now we turn to the description of the Weyl group W of the affine algebra $\mathfrak{g}$. Recall that W is generated by fundamental reflections r_0, r_1, ..., r_ℓ, which act on $\mathfrak{h}^*$ by

$$r_i(\lambda) = \lambda - \langle \lambda, \alpha_i^\vee \rangle \alpha_i, \quad \lambda \in \mathfrak{h}^*.$$

As $\langle \delta, \alpha_i^\vee \rangle = 0$ $(i = 0, \dots, \ell)$, we have

$$w(\delta) = \delta \text{ for all } w \in W.$$

Recall also that the invariant standard form is W-invariant.

Denote by $\mathring{W}$ the subgroup of W generated by $r_1, \dots, r_\ell$. As $r_i(\Lambda_0) = \Lambda_0$ for $i = 1, \dots, \ell$, we deduce that $\mathring{W}$ operates trivially on $\mathbf{C}\Lambda_0 + \mathbf{C}\delta$; it is also clear that $\mathring{\mathfrak{h}}^*$ is $\mathring{W}$-invariant. We conclude that $\mathring{W}$ operates faithfully on $\mathring{\mathfrak{h}}^*$, and we can identify $\mathring{W}$ with the Weyl group of the Lie algebra $\mathring{\mathfrak{g}}$, operating on $\mathring{\mathfrak{h}}^*$. Hence (by Proposition 3.12 e)) the group $\mathring{W}$ is finite.

Recall that for a real root α we have a reflection $r_\alpha \in W$ defined by

$$r_\alpha(\lambda) = \lambda - \langle \lambda, \alpha^\vee \rangle \alpha, \quad \lambda \in \mathfrak{h}^*.$$

LEMMA 6.5. *Let $\alpha \in \Delta_+^{re}$ be such that $\beta := \delta - a\alpha \in \Delta_+^{re}$ for some a. Then*

$$r_\alpha r_\beta(\lambda) = \lambda + \langle \lambda, K \rangle \nu(\beta^\vee) - (\langle \lambda, \beta^\vee \rangle + \tfrac{1}{2}|\beta^\vee|^2 \langle \lambda, K \rangle)\delta$$

for $\lambda \in \mathfrak{h}^*$.

Proof. First, we compute mod $\mathbf{C}\delta$:

$$\begin{aligned} r_\alpha r_\beta(\lambda) \bmod \mathbf{C}\delta &= r_\alpha(\lambda + (\lambda|\delta)2a\alpha|a\alpha|^{-2} - \langle \lambda, \alpha^\vee \rangle \alpha) \bmod \mathbf{C}\delta \\ &= \lambda - \langle \lambda, \alpha^\vee \rangle \alpha + \langle \lambda, K \rangle \nu(\beta^\vee) + \langle \lambda, \alpha^\vee \rangle \alpha \bmod \mathbf{C}\delta \\ &= \lambda + \langle \lambda, K \rangle \nu(\beta^\vee) \bmod \mathbf{C}\delta. \end{aligned}$$

To compute the coefficient of δ, we use the equality $|r_\alpha r_\beta(\lambda)|^2 = |\lambda|^2$. □

Applying Lemma 6.5 to $\theta = \delta - a_0\alpha_0 \in \mathring{\Delta}_+ \subset \Delta_+^{re}$, we get

$$r_{\alpha_0} r_\theta(\lambda) = \lambda + \langle \lambda, K \rangle \nu(\theta^\vee) - (\langle \lambda, \theta^\vee \rangle + \tfrac{1}{2}|\theta^\vee|^2 \langle \lambda, K \rangle)\delta. \tag{6.5.1}$$

Motivated by this formula, we introduce the following endomorphism t_α of the vector space $\mathfrak{h}^*$ for $\alpha \in \mathring{\mathfrak{h}}^*$:

$$t_\alpha(\lambda) = \lambda + \langle \lambda, K \rangle \alpha - ((\lambda|\alpha) + \tfrac{1}{2}|\alpha|^2 \langle \lambda, K \rangle)\delta. \tag{6.5.2}$$

In the case when $m := \langle \lambda, K \rangle \neq 0$ we can rewrite this as follows using (6.2.6):

$$t_\alpha(\lambda) = m\Lambda_0 + (\overline{\lambda} + m\alpha) + \frac{1}{2m}(|\lambda|^2 - |\overline{\lambda} + m\alpha|^2)\delta. \tag{6.5.3}$$

As $\langle \Lambda_0, K \rangle = 1$, we obtain, in particular:

$$t_\alpha(\Lambda_0) = \Lambda_0 + \alpha - \tfrac{1}{2}|\alpha|^2 \delta. \tag{6.5.4}$$

Note also that (6.5.2) implies

$$t_\alpha(\lambda) = \lambda - (\lambda|\alpha)\delta, \text{ if } \langle \lambda, K \rangle = 0. \tag{6.5.5}$$

Now we can easily deduce the additivity property of t_α:

$$t_\alpha t_\beta = t_{\alpha+\beta}. \tag{6.5.6}$$

Indeed, by (6.5.5) it is sufficient to check (6.5.6) for $\lambda = \Lambda_0$. But then, by (6.5.4 and 5) we have $t_\alpha t_\beta(\Lambda_0) = t_\alpha(\Lambda_0 + \beta - \frac{1}{2}|\beta|^2\delta) = t_\alpha(\Lambda_0) + t_\alpha(\beta - \frac{1}{2}|\beta|^2\delta) = \Lambda_0 + \alpha - \frac{1}{2}|\alpha|^2\delta + \beta - \frac{1}{2}|\beta|^2\delta - ((\beta - \frac{1}{2}|\beta|^2\delta)|\alpha)\delta = \Lambda_0 + \alpha + \beta - \frac{1}{2}|\alpha + \beta|^2\delta = t_{\alpha+\beta}(\Lambda_0)$.

We also have

$$t_{w(\alpha)} = w t_\alpha w^{-1} \text{ for } w \in \mathring{W}. \tag{6.5.7}$$

Indeed, $w t_\alpha(w^{-1}(\lambda)) = w(w^{-1}(\lambda) + \langle w^{-1}(\lambda), K \rangle \alpha - ((w^{-1}(\lambda)|\alpha) + \frac{1}{2}|\alpha|^2 \langle w^{-1}(\lambda), K \rangle)\delta)$. Now (6.5.7) follows since $w(K) = K$ and $(.|.)$ is W-invariant.

Now we introduce the following important lattice $M \subset \mathring{\mathfrak{h}}^*_{\mathbb{R}}$. Let $\mathbb{Z}(\mathring{W} \cdot \theta^\vee)$ denote the lattice in $\mathring{\mathfrak{h}}_{\mathbb{R}}$ spanned over $\mathbb{Z}$ by the (finite) set $\mathring{W} \cdot \theta^\vee$, and set $M = \nu(\mathbb{Z}(\mathring{W} \cdot \theta^\vee))$.

Here is a description of the lattice M:

$$\begin{aligned} &M = \overline{Q} = \mathring{Q} \text{ if } A \text{ is symmetric or } r > a_0; \\ &M = \nu(\overline{Q^\vee}) = \nu(\mathring{Q}^\vee) \text{ otherwise.} \end{aligned} \tag{6.5.8}$$

Indeed, if $r = 1$, then $\theta^\vee$ is a short root of $\mathring{\Delta}^\vee$, hence $\mathring{W} \cdot \theta^\vee = (\mathring{\Delta}^\vee)_s$. It is well known (cf. Exercise 6.9) that in the finite type case the root lattice is spanned over $\mathbb{Z}$ by the short roots, hence in this case $M = \nu(\mathring{Q}^\vee)$, giving (6.5.8) for $r = 1$. Equivalently,

$$M = \mathring{Q} \text{ (resp. } M = \mathbb{Z}\mathring{\Delta}_\ell) \\ \text{if } r = 1 \text{ and } A \text{ is symmetric (resp. nonsymmetric).}$$

Similarly, if $a_0 r = 2$ or 3, then $\theta^\vee$ is a long root of $\mathring{\Delta}^\vee$, hence $\mathring{W} \cdot \theta^\vee = (\mathring{\Delta}^\vee)_\ell$, and we get $M = \mathring{Q}$. Finally for $A_{2\ell}^{(2)}$ one has $\nu(\theta^\vee) = \frac{1}{2}\theta$, hence $M = \frac{1}{2}\mathbb{Z}\mathring{\Delta}_\ell$, which is equivalent to (6.5.8) in this case also (see (6.3.6)). Note also the following useful fact:

$$(6.5.9) \qquad \mathring{Q} \supset \nu(\mathring{Q}^\vee) \text{ if } r = 1 \quad \text{and} \quad \mathring{Q} \subset \nu(\mathring{Q}^\vee) \text{ if } r > 1.$$

The lattice M, considered as an abelian group, operates faithfully on $\mathfrak{h}^*$ by formula (6.5.2). We denote the corresponding subgroup of $GL(\mathfrak{h}^*)$ by T, and call it the ***group of translations*** (formula (6.6.3) below explains why). Now we can prove

PROPOSITION 6.5. $W = \mathring{W} \ltimes T$.

Proof. By (6.5.1), $t_\alpha \in W$ for $\alpha = \nu(\theta^\vee)$, hence $t_{w(\alpha)} \in W$ for $w \in \mathring{W}$ by (6.5.7). Now it follows from (6.5.6) that $t_\alpha \in W$ for all $\alpha \in M$. Since $\mathring{W}$ is finite and M is a free abelian group, $\mathring{W} \cap T = 1$. It follows from (6.5.7) that T is a normal subgroup in W. Finally $r_{\alpha_0} = t_{\nu(\theta^\vee)} r_\theta$, and therefore the subgroup of W generated by T and $\mathring{W}$ contains all r_i ($i = 0, \ldots, \ell$) and hence coincides with W.

□

Since $t_{\nu(\theta^\vee)} = r_{\alpha_0} r_\theta$ and T is generated by $w t_{\nu(\theta^\vee)} w^{-1}$ ($w \in \mathring{W}$), we have

$$(6.5.10) \qquad \det_{\mathfrak{h}^*} w = 1 \text{ for } w \in T.$$

Remark 6.5. A vertex i of the diagram $S(A)$ is called ***special*** if $\theta^{(i)} := \delta - a_i\alpha_i$ is a positive root. For example, $i = 0$ is a special vertex. Denote by $W^{(i)}$ the subgroup in W generated by all r_j ($j \neq i$), and denote by $M^{(i)}$ the lattice spanned over $\mathbb{Z}$ by the set $\nu(W^{(i)}(\theta^{(i)\vee}))$. Then the same argument as above shows that $T = \{t_\alpha \mid \alpha \in M^{(i)}\}$ and $W = W^{(i)} \ltimes T$. Most of the results of this chapter hold if we take a special vertex i in place of the vertex 0.

§6.6. For $s \in \mathbf{R}$ set

$$\mathfrak{h}_s^* = \{\lambda \in \mathfrak{h}_{\mathbf{R}}^* \mid \langle \lambda, K \rangle = s\}.$$

Note that $\mathfrak{h}_0^* = \sum_{i=0}^{\ell} \mathbf{R}\alpha_i$ and that the hyperplanes $\mathfrak{h}_s^*$ are W-invariant. Furthermore, the action of W on $\mathfrak{h}_0^*$ is faithful by (3.12.1).

Consider now the affine space $\mathfrak{h}_1^*$ mod $\mathbf{R}\delta$. Since the action of W on $\mathfrak{h}_0^*$ is faithful, its action on $\mathfrak{h}_{\mathbf{R}}^*/\mathbf{R}\delta \simeq \mathfrak{h}_0^*$ and thus on $\mathfrak{h}_1^*$ mod $\mathbf{R}\delta$ is also faithful.

The affine action of W on the affine space $\mathfrak{h}_1^* \bmod \mathbf{R}\delta$ has the following simple geometrical meaning. We identify $\mathfrak{h}_1^* \bmod \mathbf{R}\delta$ with $\mathring{\mathfrak{h}}_{\mathbf{R}}^*$ by projection, thus obtaining an isomorphism from W onto a group of affine transformations W_{af} of $\mathring{\mathfrak{h}}_{\mathbf{R}}{}^*$. We denote this isomorphism by af, so that

$$\text{af}(w)(\overline{\lambda}) = \overline{w(\lambda)} \text{ for } \lambda \in \mathfrak{h}_1^*.$$

Now we describe the action of W_{af} on $\mathring{\mathfrak{h}}_{\mathbf{R}}^*$. It is clear that

$$\text{(6.6.1)} \qquad \text{af}(w) = w \text{ for } w \in \mathring{W}.$$

Furthermore,

$$\text{(6.6.2)} \qquad \text{af}(r_{\alpha_0})(\lambda) = r_\theta(\lambda) + \nu(\theta^\vee) \quad (\lambda \in \mathring{\mathfrak{h}}_{\mathbf{R}}^*).$$

Indeed, if $\mu \in \mathfrak{h}_1^*$, then $\langle \mu, K \rangle = 1$, hence $\langle \mu, \alpha_0^\vee \rangle = \langle \mu, K - a_0\theta^\vee \rangle = 1 - a_0\langle \mu, \theta^\vee \rangle$, and $r_{\alpha_0}(\mu) = \mu - (1 - a_0\langle \mu, \theta^\vee \rangle)\alpha_0 = \mu + (1 - a_0\langle \mu, \theta^\vee \rangle)a_0^{-1}\theta$ $= \mu - \langle \mu, \theta^\vee \rangle(\theta^\vee) + \nu(\theta^\vee) \bmod \mathbf{R}\delta$. So, $\text{af}(r_{\alpha_0})$ is a reflection in the hyperplane

$$\theta = 1 \text{ (i.e.,}\{\lambda \in \mathring{\mathfrak{h}}_{\mathbf{R}}^* \mid (\lambda|\theta) = 1\}).$$

Also (6.6.2) implies $\text{af}(t_{\nu(\theta^\vee)})(\lambda) = \text{af}(r_{\alpha_0})(r_\theta(\lambda)) = \lambda + \nu(\theta^\vee)(\lambda \in \mathring{\mathfrak{h}}_{\mathbf{R}}^*)$. Hence, by (6.5.6 and 7), we obtain

$$\text{(6.6.3)} \qquad \text{af}(t_\alpha)(\lambda) = \lambda + \alpha \text{ for } \lambda \in \mathring{\mathfrak{h}}_{\mathbf{R}}^*,\ \alpha \in M.$$

So, the group W_{af} is none other than the so-called affine Weyl group of $\mathring{\mathfrak{g}}$.

Introduce the *fundamental alcove:*

$$C_{\text{af}} = \{\lambda \in \mathring{\mathfrak{h}}_{\mathbf{R}}^* \mid (\lambda|\alpha_i) \geq 0 \text{ for } 1 \leq i \leq \ell, \text{ and } (\lambda|\theta) \leq 1\}.$$

PROPOSITION 6.6. a) *Every point of* $\mathring{\mathfrak{h}}^*_{\mathbb{R}}$ *is* $\mathring{W}$*-equivalent* mod M *to a unique point of* $C_{\rm af}$.

b) *The stabilizer of every point of* $C_{\rm af}$ *under the action of* $\mathring{W}$ *on* $\mathring{\mathfrak{h}}^*_{\mathbb{R}}/M$ *is generated by its intersection with* $\{r_\theta, r_{\alpha_1}, \ldots, r_{\alpha_\ell}\}$.
c) *For every* $\lambda \in \mathfrak{h}^*_1$ *one has*

$$\mathrm{af}(W_\lambda) = (W_{\rm af})_{\overline{\lambda}},$$

and $W_\lambda \cap T = e$, *where* W_λ *denotes the stabilizer of* λ.

Proof. Consider the projection map $\pi : \mathfrak{h}^*_1 \to \mathring{\mathfrak{h}}^*_{\mathbb{R}}$. Then it is clear that π is surjective and that $\mathrm{af}(w) \circ \pi = \pi \circ w$ for $w \in W$. Furthermore, (cf. §3.12):

$$\pi^{-1}(C_{\rm af}) = C^\vee \cap \mathfrak{h}^*_1.$$

But by Proposition 3.12b), $C^\vee \cap \mathfrak{h}^*_1$ is a fundamental domain for the action of W on $\mathfrak{h}^*_1 \subset X^\vee$. This together with (6.6.1 and 3) proves a). We also deduce from the above that $\mathrm{af}(W_\lambda) = (W_{\rm af})_{\overline{\lambda}}$. Now b) follows from Proposition 3.12a). Finally $W_\lambda \cap T = 1$, since $\mathrm{af}(W_\lambda) = (W_{\rm af})_{\overline{\lambda}}$ contains no nontrivial translation.

□

§6.7. As we have seen, the root and coroot lattices, the root system and the Weyl group of an affine algebra can be expressed in terms of the corresponding objects for the "underlying" simple finite-dimensional Lie algebra.

For the convenience of the reader, we give below an explicit construction of all these objects for all simple finite-dimensional Lie algebras.

Let $\mathbf{R}^n$ be the n-dimensional real Euclidean space with the standard basis $v_1, \ldots, v_n$ and the bilinear form:

$$(v_i|v_j) = \delta_{ij}.$$

All the lattices below will be sublattices of $\mathbf{R}^n$ with the inherited bilinear form $(.|.)$. All indices are assumed to be distinct.

$$A_\ell : Q = Q^\vee = \{\sum_i k_i v_i \in \mathbf{R}^{\ell+1} \mid k_i \in \mathbf{Z}, \sum_i k_i = 0\},$$

$$\Delta = \{v_i - v_j\},$$

$$\Pi = \{\alpha_1 = v_1 - v_2, \alpha_2 = v_2 - v_3, \ldots, \alpha_\ell = v_\ell - v_{\ell+1}\},$$

$\theta = v_1 - v_{\ell+1}$,

$W = \{\text{all permutations of the } v_i\}$.

$D_\ell : Q = Q^\vee = \{\sum_i k_i v_i \in \mathbf{R}^\ell \mid k_i \in \mathbf{Z}, \sum_i k_i \in 2\mathbf{Z}\}$,

$\Delta = \{\pm v_i \pm v_j\}$,

$\Pi = \{\alpha_1 = v_1 - v_2, \ldots, \alpha_{\ell-1} = v_{\ell-1} - v_\ell, \alpha_\ell = v_{\ell-1} + v_\ell\}$,

$\theta = v_1 + v_2$,

$W = \{\text{all permutations and even number of sign changes of the } v_i\}$.

$B_\ell : Q = \{\sum_i k_i v_i \in \mathbf{R}^\ell \mid k_i \in \mathbf{Z}\},\ Q^\vee = Q(D_\ell)$,

$\Delta = \{\pm v_i \pm v_j, \pm v_i\}$,

$\Pi = \{\alpha_1 = v_1 - v_2, \ldots, \alpha_{\ell-1} = v_{\ell-1} - v_\ell, \alpha_\ell = v_\ell\}$,

$\theta = v_1 + v_2$,

$W = \{\text{all permutations and sign changes of the } v_i\} = \operatorname{Aut} Q$.

$C_\ell : \Delta = \{\frac{1}{\sqrt{2}}(\pm v_i \pm v_j), \quad \pm\sqrt{2}v_i\}$,

$\Delta = \{\frac{1}{\sqrt{2}}(\pm v_i, \pm v_j), \quad \pm\sqrt{2}v_i\}$,

$\Pi = \{\alpha_1 = \frac{1}{\sqrt{2}}(v_1 - v_2), \ldots, \alpha_{\ell-1} = \frac{1}{\sqrt{2}}(v_{\ell-1} - v_\ell), \alpha_\ell = \sqrt{2}v_\ell\}$,

$\theta = \sqrt{2}v_1$,

$W = W(B_\ell)$.

$G_2 : Q = \frac{1}{\sqrt{3}}Q(A_2), \quad Q^\vee = Q(A_2)$,

$\Delta = \{\frac{1}{\sqrt{3}}(v_i - v_j),\ \pm\frac{1}{\sqrt{3}}(v_i + v_j - 2v_k)\}$,

$\Pi = \{\alpha_1 = \frac{1}{\sqrt{3}}(-v_1 + 2v_2 - v_3),\ \alpha_2 = \frac{1}{\sqrt{3}}(v_1 - v_2)\}$,

$\theta = \frac{1}{\sqrt{3}}(v_1 + v_2 - 2v_3)$,

$W = \pm\{\text{all permutations of the } v_i\} = \operatorname{Aut} Q$.

$F_4 :\ Q = \{\sum_i k_i v_i \in \mathbb{R}^4 \mid \text{ all } k_i \in \mathbb{Z} \text{ or all } k_i \in \frac{1}{2} + \mathbb{Z}\},\ Q^\vee = Q(D_4)$,

$\Delta = \{\pm v_i, \pm v_i \pm v_j, \frac{1}{2}(\pm v_1 \pm v_2 \pm v_3 \pm v_4)\}$,

$\Pi = \{\alpha_1 = v_2 - v_3, \alpha_2 = v_3 - v_4, \alpha_3 = v_4, \alpha_4 = \frac{1}{2}(v_1 - v_2 - v_3 - v_4)\}$,

$\theta = v_1 + v_2,$

$W = \operatorname{Aut} Q.$

$E_8 : Q = Q^\vee = \{\sum_i k_i v_i \in \mathbf{R}^8 \mid \text{ all } k_i \in \mathbf{Z} \text{ or all } k_i \in \frac{1}{2} + \mathbf{Z}, \sum_i k_i \in 2\mathbf{Z}\},$

$\Delta = \{\pm v_i \pm v_j, \frac{1}{2}(\pm v_1 \pm v_2 \pm \cdots \pm v_8) \text{ (even number of minuses)}\},$

$\Pi = \{\alpha_1 = v_2 - v_3, \alpha_2 = v_3 - v_4, \alpha_3 = v_4 - v_5,$

$\alpha_4 = v_5 - v_6, \alpha_5 = v_6 - v_7, \alpha_6 = v_7 - v_8,$

$\alpha_7 = \frac{1}{2}(v_1 - v_2 - \ldots - v_7 + v_8), \alpha_8 = v_7 + v_8\},$

$\theta = v_1 + v_2,$

$W = \operatorname{Aut} Q.$

$E_7 : Q = Q^\vee = \{\sum_i k_i v_i \in \mathbf{R}^8 \mid \text{ all } k_i \in \mathbf{Z} \text{ or all } k_i \in \frac{1}{2} + \mathbf{Z}, \sum_i k_i = 0\},$

$\Delta = \{v_i - v_j, \ \frac{1}{2}(\pm v_1 \pm \cdots \pm v_8) \text{ (four minuses)}\},$

$\Pi = \{\alpha_1 = v_2 - v_3, \alpha_2 = v_3 - v_4, \alpha_3 = v_4 - v_5,$

$\alpha_4 = v_5 - v_6, \alpha_5 = v_6 - v_7, \alpha_6 = v_7 - v_8,$

$\alpha_7 = \frac{1}{2}(-v_1 - v_2 - v_3 - v_4 + v_5 + v_6 + v_7 + v_8)\},$

$\theta = v_2 - v_1,$

$W = \operatorname{Aut} Q.$

$E_6 : Q = Q^\vee = \{k_1 v_1 + \cdots + k_6 v_6 + \sqrt{2} k_7 v_7 \in \mathbf{R}^7 | \text{ all } k_i \in \mathbf{Z}$

$\text{or all } k_i \in \frac{1}{2} + \mathbf{Z}, k_1 + \cdots + k_6 = 0\},$

$\Delta = \{v_i - v_j (i, j \leq 6), \frac{1}{2}((\varepsilon_1 v_1 + \cdots + \varepsilon_6 v_6) \pm \sqrt{2} v_7)(\varepsilon_i = \pm 1,$

$\sum \varepsilon_i = 0), \pm\sqrt{2} v_7\},$

$\Pi = \{\alpha_1 = v_1 - v_2, \alpha_2 = v_2 - v_3, \alpha_3 = v_3 - v_4, \alpha_4 = v_4 - v_5,$

$\alpha_5 = v_5 - v_6, \alpha_6 = \frac{1}{2}(-v_1 - v_2 - v_3 + v_4 + v_5 + v_6 + \sqrt{2} v_7)\},$

$\theta = \sqrt{2} v_7,$

$W \times \{\pm 1\} = \operatorname{Aut} Q.$

In order to prove these facts we first check directly that Π is a basis of Q over $\mathbb{Z}$, and that the matrix $(2(\alpha_i|\alpha_j)/(\alpha_i|\alpha_i))$ is the Cartan matrix of the corresponding type. Furthermore, one easily checks that $\Delta = \{\alpha \mid (\alpha|\alpha) = 2\}$ for A_ℓ, D_ℓ and E_ℓ, and $\Delta =$ (resp. $\subset$) $\{\alpha \mid (\alpha|\alpha) = 1 \text{ or } r\}$ for B_ℓ, G_2, and F_4 (resp. C_ℓ), where $r = 2$ for B_ℓ, C_ℓ, F_4, and $r = 3$ for G_2.

Using Proposition 5.10 a), b), we see that Δ is the set of all roots. The computation of W follows easily from Corollary 5.10 and Remark 5.9 b).

Remark 6.7. It follows from the above discussion that if A is of finite type, then $\mathrm{Aut}Q = \mathrm{Aut}(A) \ltimes W$, except for A of type C_4. In the latter case, W is a group of index 3 in $\mathrm{Aut}Q$ (cf. Exercise 5.8).

We can list now the lattices M for all affine algebras:

$$\begin{aligned} X_\ell^{(1)}: &\quad M = Q^\vee(X_\ell); \\ A_{2\ell-1}^{(2)}: &\quad M = Q(D_\ell); \\ D_{\ell+1}^{(2)}: &\quad M = \sqrt{2}Q(B_\ell); \\ E_6^{(2)}: &\quad M = Q(D_4); \\ D_4^{(3)}: &\quad M = Q(A_2); \\ A_{2l}^{(2)}: &\quad M = Q(B_\ell). \end{aligned}$$

§6.8. Exercises.

6.1. Check that the square length $(\alpha|\alpha)$ of a root α from $\mathring{\Delta}_\ell$ (resp. $\mathring{\Delta}_s$) is equal to $2r$ (resp. $2r/s$, where $s = \max\limits_{a_{ij}\neq 0} a_{ji}/a_{ij}$).

6.2. Let $r = 1$. Show that h is the Coxeter number of the root system $\mathring{\Delta}$ and that $h^\vee = \phi(\theta,\theta)^{-1} = 1 + (\mathring{\rho}|\theta)$, where ϕ is the Killing form of $\mathring{\mathfrak{g}}$ and $\mathring{\rho}$ is the half sum of the roots from $\mathring{\Delta}_+$. Show that $\phi(x,y) = 2h^\vee(x|y)$ for $x, y \in \mathring{\mathfrak{g}}$.
[Note that $\phi(\theta + 2\mathring{\rho}, \theta)$ is the eigenvalue of the Casimir operator associated to the Killing form ϕ, hence it equals 1, and use (6.2.3).]

6.3. Let A be a Cartan matrix of type $X_N^{(r)}$ from Table Aff r, let $\ell = \mathrm{rank}\, A$, and let h be the Coxeter number. Let Δ_0 be the finite root system of type X_N. Check that

$$r\ell h = |\Delta_0|.$$

6.4. Let A be of type $A_1^{(1)}$, i.e., $A = \begin{pmatrix} 2 & -2 \\ -2 & 2 \end{pmatrix}$. Then

$$\Delta_+ = \{(k-1)\alpha_0 + k\alpha_1, k\alpha_0 + (k-1)\alpha_1, k\alpha_0 + k\alpha_1, \text{ where } k = 1, 2, \dots\}.$$

6.5. Let A be of type $A_\ell^{(1)}$, $\ell > 1$, i.e.,

$$A = \begin{pmatrix} 2 & -1 & 0 & \dots & 0 & -1 \\ -1 & 2 & -1 & \dots & 0 & 0 \\ \dots & \dots & \dots & \dots & \dots & \dots \\ -1 & 0 & 0 & \dots & -1 & 2 \end{pmatrix}.$$

Show that

$$\Delta_+ = \{k(\alpha_0 + \cdots + \alpha_{i-1}) + (k \pm 1)(\alpha_i + \cdots + \alpha_{j-1}) + k(\alpha_j + \cdots + \alpha_\ell)\},$$

where $k = 0, 1, 2, \ldots$; $0 \leq i \leq j \leq \ell + 1$, and only $+$ is allowed if $k = 0$.

6.6. Let A be of type $A_2^{(2)}$, i.e., $A = \begin{pmatrix} 2 & -4 \\ -1 & 2 \end{pmatrix}$. Then

$$\Delta_+ = \{4n\alpha_0 + (2n-1)\alpha_1,\ 4(n-1)\alpha_0 + (2n-1)\alpha_1, (2n-1)\alpha_0 + n\alpha_1,\\ (2n-1)\alpha_0 + (n-1)\alpha_1,\ 2n\alpha_0 + n\alpha_1;\ n = 1, 2, \ldots\}.$$

6.7. Let $s_0, s_1, \ldots, s_\ell$ be the reflections in the vector space $\mathring{\mathfrak{h}}$ with respect to the hyperplanes $\theta = 1$, $\alpha_1 = 0, \ldots, \alpha_\ell = 0$, respectively, and let W_a be the group generated by $s_0, \ldots, s_\ell$. Show that $W_a = \mathring{W} \ltimes \nu^{-1}(M)$ and that the map $s_i \mapsto r_{\alpha_i}$ defines an isomorphism $\phi : W_a \to W$. Show that the image under ϕ of a translation by $\alpha \in \nu^{-1}(M)$ is t_α.

6.8. Let $\mathring{\mathfrak{h}}^a$ denote the vector space of affine linear functions on $\mathring{\mathfrak{h}}$. Consider the isomorphism $\psi : \mathfrak{h}_0^* \to \mathring{\mathfrak{h}}^a$ defined by

$$\psi(\alpha_i) = \overline{\alpha_i} \text{ for } i = 1, \ldots, \ell; \qquad \psi(\delta) = 1.$$

The action of W_a on $\mathring{\mathfrak{h}}$ defined in Exercise 6.7 induces a linear action of $W_a = W$ on $\mathring{\mathfrak{h}}^a$. Show that the morphism $(\phi, \psi^{-1}) : (W_a, \mathring{\mathfrak{h}}^a) \to (W, \mathfrak{h}_0^*)$ is equivariant.

6.9. Show that for a finite or affine type matrix A, the set of short real roots $\Delta_s^{re}(A)$ span the lattice Q over $\mathbb{Z}$.

[The set $Q' = \mathbb{Z}\Delta_s^{re}$ is W-invariant. We can assume that A is not symmetric. If $A \neq A_{2\ell}^{(2)}$, then there exists a short simple root α and a long simple root β such that $(\alpha|\beta) \neq 0$; then $r_\beta(\alpha) = \alpha + \beta \in Q'$. Hence, $\beta \in Q'$ and $\Delta_\ell^{re} \subset Q'$. So $\Pi \subset \Delta^{re} \subset Q'$ and $Q = Q'$. The same argument works in the case $A_{2\ell}^{(2)}$ as well.]

6.10. Let $r = 1$ and $\gamma \in M$. Show that

$$\ell(t_\gamma) = \sum_{\alpha \in \mathring{\Delta}_+} |(\gamma|\alpha)|.$$

[Use Exercise 3.6.]

6.11. Denote by Γ_n the lattice in $\mathbf{R}^n$ defined by the same conditions as the root lattice of E_8 in $\mathbf{R}^8$. Show that if n is divisible by 8, then Γ_n is an even unimodular lattice.

6.12. Show that an integral lattice Q (i.e., $(x|y) \in \mathbb{Z}$ if $x, y \in Q$) is spanned over $\mathbb{Z}$ by its vectors of square length 2 if and only if Q is an orthogonal direct sum of root lattices of type A_ℓ, D_ℓ, E_6, E_7, E_8.

6.13. Show that the group of all automorphisms of $\mathfrak{h}$ preserving the bilinear form $(.|.)$ and fixing K is the semidirect product of the group of all orthogonal automorphisms of $\overset{\circ}{\mathfrak{h}}$ (fixing K and d) and the group of all t_α, $\alpha \in \overset{\circ}{\mathfrak{h}}{}^*$.

§6.9. Bibliographical notes and comments.

The study of affine algebras was started by Kac [1968 B] and Moody [1969]. The material of this chapter is fairly standard. The exposition is based on papers Macdonald [1972], Kac [1978 A], Kac–Peterson [1984 A]. The "quadratic" action (6.5.2) "explains" the appearance of theta functions in the theory of affine algebras (see Chapter 12). Exercise 6.10 is due to Haddad and Peterson.

Chapter 7. Affine Algebras as Central Extensions of Loop Algebras

§7.0. In this chapter we describe in detail a "concrete" construction of all "nontwisted" affine algebras. It turns out that such an algebra $\mathfrak{g}$ can be realized entirely in terms of an "underlying" simple finite-dimensional Lie algebra $\mathring{\mathfrak{g}}$. Namely, its derived algebra $[\mathfrak{g}, \mathfrak{g}]$ is the universal central extension (the center being 1-dimensional) of the Lie algebra of polynomial maps from $\mathbf{C}^\times$ into $\mathring{\mathfrak{g}}$. The corresponding algebra of "currents" plays an important role in quantum field theory.

At the end of the chapter we give an explicit construction of finite-dimensional simple Lie algebras.

§7.1. Let $\mathcal{L} = \mathbf{C}[t, t^{-1}]$ be the algebra of Laurent polynomials in t. Recall that the residue of a Laurent polynomial $P = \sum_{k \in \mathbf{Z}} c_k t^k$ (where all but a finite number of c_k are 0) is defined by $\operatorname{Res} P = c_{-1}$. This is a linear functional on $\mathcal{L}$ defined by the properties:

$$\operatorname{Res} t^{-1} = 1; \qquad \operatorname{Res} \frac{dP}{dt} = 0.$$

Define a bilinear $\mathbf{C}$-valued function φ on $\mathcal{L}$ by

$$\varphi(P, Q) = \operatorname{Res} \frac{dP}{dt} Q.$$

Then it is easy to check the following two properties:

$$\varphi(P, Q) = -\varphi(Q, P), \tag{7.1.1}$$

$$\varphi(PQ, R) + \varphi(QR, P) + \varphi(RP, Q) = 0 \quad (P, Q, R \in \mathcal{L}). \tag{7.1.2}$$

§7.2. The affine algebra associated to a generalized Cartan matrix of type $X_\ell^{(1)}$ (from Table Aff 1) is called a *nontwisted affine algebra*. Here we describe an explicit construction of these Lie algebras. Note that the generalized Cartan matrix A of type $X_\ell^{(1)}$ (where $X = A, B, \ldots, G$) is nothing else but the so-called extended Cartan matrix of the simple finite-dimensional Lie algebra $\mathring{\mathfrak{g}} := \mathfrak{g}(\mathring{A})$, whose Cartan matrix $\mathring{A}$ is a matrix of finite type X_ℓ (obtained from A by removing the 0th row and column).

Consider the *loop algebra*

$$\mathcal{L}(\mathring{\mathfrak{g}}) := \mathcal{L} \otimes_{\mathbf{C}} \mathring{\mathfrak{g}}.$$

This is an infinite-dimensional complex Lie algebra with the bracket $[\ ,\ ]_0$ defined by

$$[P \otimes x, Q \otimes y]_0 = PQ \otimes [x, y] \quad (P, Q \in \mathcal{L};\ x, y \in \mathring{\mathfrak{g}}).$$

It may be identified with the Lie algebra of regular rational maps $\mathbf{C}^\times \to \mathring{\mathfrak{g}}$, so that the element $\sum_i (t^i \otimes x_i)$ corresponds to the mapping $z \mapsto \sum_i z^i x_i$.

Fix a nondegenerate invariant symmetric bilinear $\mathbf{C}$-valued form $(.|.)$ on $\mathring{\mathfrak{g}}$; such a form exists (e.g., by Theorem 2.2) and is unique up to a constant multiple. We extend this form by linearity to an $\mathcal{L}$-valued bilinear form $(.|.)_t$ on $\mathcal{L}(\mathring{\mathfrak{g}})$ by

$$(P \otimes x | Q \otimes y)_t = PQ(x|y).$$

Also, we extend every derivation D of the algebra $\mathcal{L}$ to a derivation of the Lie algebra $\mathcal{L}(\mathring{\mathfrak{g}})$ by

$$D(P \otimes x) = D(P) \otimes x.$$

Now we can define a $\mathbf{C}$-valued 2-cocycle on the Lie algebra $\mathcal{L}(\mathring{\mathfrak{g}})$ by

$$\psi(a, b) = \operatorname{Res}(\frac{da}{dt} \mid b)_t.$$

Recall that a $\mathbf{C}$-valued 2-cocycle on a Lie algebra $\mathfrak{g}$ is a bilinear $\mathbf{C}$-valued function ψ satisfying two conditions:

(Co 1) $$\psi(a, b) = -\psi(b, a)$$

(Co 2) $$\psi([a, b], c) + \psi([b, c], a) + \psi([c, a], b) = 0 \quad (a, b, c \in \mathfrak{g}).$$

It is sufficient to check these conditions for $a = P \otimes x$, $b = Q \otimes y$, $c = R \otimes z$, where P, Q, $R \in \mathcal{L}$ and x, y, $z \in \mathring{\mathfrak{g}}$. We have

$$\psi(a, b) = (x|y)\varphi(P, Q).$$

Hence, (Co 1) follows from (7.1.1) and the symmetry of $(.|.)$. Property (Co 2) follows from (7.1.2) and the symmetry and invariance of $(.|.)$. Indeed, the left-hand side of (Co 2) is

$$\begin{aligned} &([x, y]|z)\varphi(PQ, R) + ([y, z]|x)\varphi(QR, P) + ([z, x]|y)\varphi(RP, Q) \\ &= ([x, y]|z)(\varphi(PQ, R) + \varphi(QR, P) + \varphi(RP, Q)) = 0. \end{aligned}$$

Denote by $\tilde{\mathcal{L}}(\mathring{\mathfrak{g}})$ the extension of the Lie algebra $\mathcal{L}(\mathring{\mathfrak{g}})$ by a 1-dimensional center, associated to the cocycle ψ. Explicitly, $\tilde{\mathcal{L}}(\mathring{\mathfrak{g}}) = \mathcal{L}(\mathring{\mathfrak{g}}) \oplus \mathbb{C}K$ (direct sum of vector spaces) and the bracket is given by

$$[a + \lambda K, b + \mu K] = [a, b]_0 + \psi(a, b)K \quad (a, b \in \mathcal{L}(\mathring{\mathfrak{g}});\ \lambda, \mu \in \mathbb{C}). \tag{7.2.1}$$

Finally, denote by $\hat{\mathcal{L}}(\mathring{\mathfrak{g}})$ the Lie algebra that is obtained by adjoining to $\tilde{\mathcal{L}}(\mathring{\mathfrak{g}})$ a derivation d which acts on $\mathcal{L}(\mathring{\mathfrak{g}})$ as $t\frac{d}{dt}$ and which kills K (see §7.3). In other words, $\hat{\mathcal{L}}(\mathring{\mathfrak{g}})$ is a complex vector space

$$\hat{\mathcal{L}}(\mathring{\mathfrak{g}}) = \mathcal{L}(\mathring{\mathfrak{g}}) \oplus \mathbb{C}K \oplus \mathbb{C}d$$

with the bracket defined as follows ($x, y \in \mathring{\mathfrak{g}}$; λ, μ, λ_1, $\mu_1 \in \mathbb{C}$):

$$\begin{aligned} &[t^m \otimes x \oplus \lambda K \oplus \mu d, t^n \otimes y \oplus \lambda_1 K \oplus \mu_1 d] \\ &\quad = (t^{m+n} \otimes [x, y] + \mu n t^n \otimes y - \mu_1 m t^m \otimes x) \oplus m\delta_{m,-n}(x|y)K. \end{aligned} \tag{7.2.2}$$

We shall prove that $\hat{\mathcal{L}}(\mathring{\mathfrak{g}})$ is an affine algebra associated to the affine matrix A of type $X_\ell^{(1)}$.

§7.3. Here we check that d is a derivation of the Lie algebra $\tilde{\mathcal{L}}(\mathring{\mathfrak{g}})$. More generally, denote by d_s the endomorphism of the space $\tilde{\mathcal{L}}(\mathring{\mathfrak{g}})$ defined by

$$d_s|_{\mathcal{L}(\mathring{\mathfrak{g}})} = -t^{s+1}\frac{d}{dt}; \quad d_s(K) = 0, \tag{7.3.1}$$

so that $d_0 = -d$.

PROPOSITION 7.3. *d_s is a derivation of $\tilde{\mathcal{L}}(\mathring{\mathfrak{g}})$.*

Proof. Since $D := d_s$ is a derivation of $\mathcal{L}(\mathring{\mathfrak{g}})$, we deduce that

$$D([a + \lambda K, b + \mu K]) = D([a, b]_0) = [D(a), b]_0 + [a, D(b)]_0.$$

But

$$[D(a), b] = [D(a), b]_0 + \psi(D(a), b)K.$$

Hence, one has to check that

$$\psi(D(a), b) + \psi(a, D(b)) = 0. \tag{7.3.2}$$

Set $a = P \otimes x$, $b = Q \otimes y$; then the left-hand side of (7.3.2) is:

$$\begin{aligned}(x|y)(\varphi(D(P), Q) + \varphi(P, D(Q)))\\ &= (x|y)(-\varphi(Q, D(P)) + \varphi(P, D(Q)))\\ &= (x|y)\,\mathrm{Res}(-\frac{dQ}{dt}t^{s+1}\frac{dP}{dt} + \frac{dP}{dt}t^{s+1}\frac{dQ}{dt}) = 0.\end{aligned}$$

□

Note that

$$\mathfrak{d} := \bigoplus_{j \in \mathbb{Z}} \mathbb{C} d_j$$

is a $\mathbb{Z}$-graded subalgebra in $\mathrm{der}\tilde{\mathcal{L}}(\mathring{\mathfrak{g}})$ with the following commutation relations:

$$[d_i, d_j] = (i - j)d_{i+j}.$$

This is the Lie algebra of regular vector fields on $\mathbb{C}^\times$ (= derivations of $\mathcal{L}$).

The Lie algebra $\mathfrak{d}$ has a unique (up to isomorphism) nontrivial central extension by a 1-dimensional center, say $\mathbb{C}c$, called the *Virasoro algebra* Vir, which is defined by the following commutation relations (see Exercise 7.13):

$$[d_i, d_j] = (i - j)d_{i+j} + \frac{1}{12}(i^3 - i)\delta_{i,-j}c \quad (i, j \in \mathbb{Z}). \tag{7.3.3}$$

The semidirect product of Lie algebras Vir $+\tilde{\mathcal{L}}(\mathring{\mathfrak{g}})$ defined by (7.3.3), (7.2.1), (7.3.1), and the equation $[c, \tilde{\mathcal{L}}(\mathring{\mathfrak{g}})] = 0$, plays an important role in the representation theory of affine and Virasoro algebras and in the quantum field theory.

§7.4. Let $\mathring{\Delta} \subset \mathring{\mathfrak{h}}^*$ be the root system of the Lie algebra $\mathring{\mathfrak{g}}$, let $\{\alpha_1, \ldots, \alpha_\ell\}$ be the root basis, $\{H_1, \ldots, H_\ell\}$ the coroot basis, E_i, F_i $(i = 1, \ldots, \ell)$ the Chevalley generators. Let θ be the highest root of the finite root system $\mathring{\Delta}$ (see Remark 4.9). Let $\mathring{\mathfrak{g}} = \bigoplus_{\alpha \in \mathring{\Delta} \cup 0} \mathring{\mathfrak{g}}_\alpha$ be the root space decomposition of $\mathring{\mathfrak{g}}$. Recall that $(\alpha|\alpha) \neq 0$ and $\dim \mathring{\mathfrak{g}}_\alpha = 1$ for $\alpha \in \mathring{\Delta}$ (there are no imaginary roots). Let $\mathring{\omega}$ be the Chevalley involution of $\mathring{\mathfrak{g}}$. We choose $F_0 \in \mathring{\mathfrak{g}}_\theta$ such that $(F_0|\mathring{\omega}(F_0)) = -2/(\theta|\theta)$, and set $E_0 = -\mathring{\omega}(F_0)$. Then due to Theorem 2.2 e) we have

$$[E_0, F_0] = -\theta^\vee. \tag{7.4.1}$$

The elements E_i ($i = 0, \ldots, \ell$) generate the Lie algebra $\mathring{\mathfrak{g}}$, since in the adjoint representation we have $\mathring{\mathfrak{g}} = U(\mathring{\mathfrak{g}})(E_0) = U(\mathring{\mathfrak{n}}_+)(E_0)$ (recall that $\mathring{\mathfrak{g}}$ is simple).

Now we turn to the Lie algebra $\hat{\mathcal{L}}(\mathring{\mathfrak{g}})$. It is clear that $\mathbf{C}K$ is the (1-dimensional) center of the Lie algebra $\hat{\mathcal{L}}(\mathring{\mathfrak{g}})$, and that the centralizer of $\mathfrak{d}$ in $\hat{\mathcal{L}}(\mathring{\mathfrak{g}})$ is a direct sum of Lie algebras: $\mathbf{C}K \oplus \mathbf{C}d \oplus (1 \otimes \mathring{\mathfrak{g}})$. In particular, $1 \otimes \mathring{\mathfrak{g}}$ is a subalgebra of $\hat{\mathcal{L}}(\mathring{\mathfrak{g}})$; we identify $\mathring{\mathfrak{g}}$ with this subalgebra by $x \mapsto 1 \otimes x$. Furthermore,

$$\mathfrak{h} := \mathring{\mathfrak{h}} + \mathbf{C}K + \mathbf{C}d$$

is an $(\ell + 2)$-dimensional commutative subalgebra in $\hat{\mathcal{L}}(\mathring{\mathfrak{g}})$. We extend $\lambda \in \mathring{\mathfrak{h}}^*$ to a linear function on $\mathfrak{h}$ by setting $\langle \lambda, K \rangle = \langle \lambda, d \rangle = 0$, so that $\mathring{\mathfrak{h}}^*$ is identified with a subspace in $\mathfrak{h}^*$. We denote by δ the linear function on $\mathfrak{h}$ defined by $\delta|_{\mathring{\mathfrak{h}}+\mathbf{C}K} = 0$, $\langle \delta, d \rangle = 1$. Set

$$e_0 = t \otimes E_0, \qquad f_0 = t^{-1} \otimes F_0,$$
$$e_i = 1 \otimes E_i, \quad f_i = 1 \otimes F_i \quad (i = 1, \ldots, \ell).$$

We deduce from (7.4.1) that

$$[e_0, f_0] = \frac{2}{(\theta|\theta)} K - \theta^\vee. \tag{7.4.2}$$

Now we describe the root system and the root space decomposition of $\hat{\mathcal{L}}(\mathring{\mathfrak{g}})$ with respect to $\mathfrak{h}$:

$$\Delta = \{j\delta + \gamma, \text{ where } j \in \mathbf{Z},\ \gamma \in \mathring{\Delta}\} \cup \{j\delta, \text{ where } j \in \mathbf{Z}\backslash 0\},$$
$$\hat{\mathcal{L}}(\mathring{\mathfrak{g}}) = \mathfrak{h} \oplus \Big(\bigoplus_{\alpha \in \Delta} \mathcal{L}(\mathring{\mathfrak{g}})_\alpha\Big), \text{ where}$$
$$\mathcal{L}(\mathring{\mathfrak{g}})_{j\delta+\gamma} = t^j \otimes \mathring{\mathfrak{g}}_\gamma, \qquad \mathcal{L}(\mathring{\mathfrak{g}})_{j\delta} = t^j \otimes \mathring{\mathfrak{h}}.$$

We set

$$\Pi = \{\alpha_0 := \delta - \theta, \alpha_1, \ldots, \alpha_\ell\},$$
$$\Pi^\vee = \{\alpha_0^\vee := \frac{2}{(\theta|\theta)} K - 1 \otimes \theta^\vee,\ \alpha_1^\vee := 1 \otimes H_1, \ldots, \alpha_\ell^\vee := 1 \otimes H_\ell\}.$$

Note that our θ is the same as the one introduced in §6.4 (for $r = 1$). Then Proposition 6.4 a) implies

$$A = (\langle \alpha_j, \alpha_i^\vee \rangle)_{i,j=0}^{\ell}. \tag{7.4.3}$$

In other words, $(\mathfrak{h}, \Pi, \Pi^\vee)$ is a realization of the affine matrix A we started with. (Indeed, Π and $\Pi^\vee$ are linearly independent, i.e., (1.1.1) holds and $2n - \operatorname{rank} A = 2(\ell+1) - \ell = \ell + 2 = \dim \mathfrak{h}$, i.e., (1.1.3) holds.)

Now we can prove our first realization theorem.

THEOREM 7.4. *Let $\mathring{\mathfrak{g}}$ be a complex finite-dimensional simple Lie algebra, and let A be its extended Cartan matrix. Then $\hat{\mathcal{L}}(\mathring{\mathfrak{g}})$ is the affine Kac–Moody algebra associated to the affine matrix A, $\mathfrak{h}$ is its Cartan subalgebra, Π and $\Pi^\vee$ the root basis and the coroot basis, and $e_0, \ldots, e_\ell, f_0, \ldots, f_\ell$ the Chevalley generators. In other words, $(\hat{\mathcal{L}}(\mathring{\mathfrak{g}}), \mathfrak{h}, \Pi, \Pi^\vee)$ is the quadruple associated to A.*

Proof. We employ Proposition 1.4 a). Some of the hypotheses of this proposition have already been checked. The relations (1.4.2) are clear. As for relations (1.4.1), they evidently hold when $i, j = 1, \ldots, \ell$ because E_i, F_i ($i = 1, \ldots, \ell$) are Chevalley generators of $\mathring{\mathfrak{g}}$. The relations $[e_0, f_i] = 0$ and $[e_i, f_0] = 0$ for $i = 1, \ldots, \ell$ hold since θ is the highest root of $\mathring{\mathfrak{g}}$. This together with (7.4.2) proves all the relations (1.4.1).

Furthermore, $\hat{\mathcal{L}}(\mathring{\mathfrak{g}})$ has no ideals intersecting $\mathfrak{h}$ trivially. Indeed, if $\mathfrak{i}$ is a nonzero ideal of $\hat{\mathcal{L}}(\mathring{\mathfrak{g}})$ such that $\mathfrak{i} \cap \mathfrak{h} = 0$, then by Proposition 1.5, $\mathfrak{i} \cap \hat{\mathcal{L}}(\mathring{\mathfrak{g}})_\alpha \neq 0$, for some $\alpha \in \Delta$. Hence $t^j \otimes x \in \mathfrak{i}$ for some $j \in \mathbb{Z}$ and $x \in \mathring{\mathfrak{g}}_\gamma$, $x \neq 0$, $\gamma \in \mathring{\Delta} \cup 0$. Taking $y \in \mathring{\mathfrak{g}}_{-\gamma}$ such that $(x|y) \neq 0$, we obtain $[t^j \otimes x, t^{-j} \otimes y] = j(x|y)K + [x, y] \in \mathfrak{h} \cap \mathfrak{i}$. As $[x, y] \in \mathring{\mathfrak{h}}$, we deduce that $j = 0$. But then $\gamma \neq 0$, and hence $[x, y]$ is contained in $\mathring{\mathfrak{h}} \cap \mathfrak{i}$ and is different from zero. This is a contradiction.

Finally, it remains to show that e_i, f_i ($i = 0, \ldots, \ell$), and $\mathfrak{h}$ generate the Lie algebra $\hat{\mathcal{L}}(\mathring{\mathfrak{g}})$. For that purpose we denote by $\hat{\mathcal{L}}_1(\mathring{\mathfrak{g}})$ the subalgebra of $\hat{\mathcal{L}}(\mathring{\mathfrak{g}})$ generated by them. Since E_i, F_i ($i = 1, \ldots, \ell$) generate the Lie algebra $\mathring{\mathfrak{g}}$, we obtain that $1 \otimes \mathring{\mathfrak{g}} \subset \hat{\mathcal{L}}_1(\mathring{\mathfrak{g}})$. Furthermore, $t \otimes E_0 \in \hat{\mathcal{L}}_1(\mathring{\mathfrak{g}})$; since $[t \otimes x, 1 \otimes y] = t \otimes [x, y]$ for x, $y \in \mathring{\mathfrak{g}}$ and since $\mathring{\mathfrak{g}}$ is simple, we deduce that $t \otimes \mathring{\mathfrak{g}} \subset \hat{\mathcal{L}}_1(\mathring{\mathfrak{g}})$. Since $[t \otimes x, t^k \otimes y] = t^{k+1} \otimes [x, y]$, it follows by induction on k that $t^k \otimes \mathring{\mathfrak{g}} \subset \hat{\mathcal{L}}_1(\mathring{\mathfrak{g}})$ for all $k \geq 0$. A similar argument shows that $t^k \otimes \mathring{\mathfrak{g}} \subset \hat{\mathcal{L}}_1(\mathring{\mathfrak{g}})$ for all $k < 0$, completing the proof.

□

The following important corollary of Theorem 7.4 is immediate.

COROLLARY 7.4. *Let $\mathfrak{g}(A)$ be a nontwisted affine Lie algebra of rank $\ell+1$. Then the multiplicity of every imaginary root of $\mathfrak{g}(A)$ is ℓ.*

Remark 7.4. Given a simple finite dimensional Lie algebra $\mathfrak{g}$, the Lie algebra $\hat{\mathcal{L}}(\mathfrak{g})$ is usually referred to in the literature as the *affine algebra associated to* $\mathfrak{g}$, or the *affinization* of $\mathfrak{g}$.

§7.5. One can also describe explicitly the rest of the notions introduced in the previous chapters.

The normalized invariant form $(.|.)$ of $\mathfrak{g}$ (introduced in §6.2) can be described as follows. Take the normalized invariant form $(.|.)$ on $\mathring{\mathfrak{g}}$ and extend $(.|.)$ to the whole $\hat{\mathcal{L}}(\mathring{\mathfrak{g}})$ by

$$\begin{aligned}(P\otimes x|Q\otimes y) &= (\operatorname{Res} t^{-1}PQ)(x|y), \quad (x,y\in\mathring{\mathfrak{g}},\ P,Q\in\mathcal{L});\\ (\mathbb{C}K+\mathbb{C}d|\mathcal{L}(\mathring{\mathfrak{g}})) &= 0; \quad (K|K)=(d|d)=0; \quad (K|d)=1.\end{aligned} \tag{7.5.1}$$

It is clear that this is a nondegenerate symmetric bilinear form. We check the only nontrivial case of the invariance property:

$$([d,P\otimes x]|Q\otimes y) = (d|[P\otimes x, Q\otimes y]).$$

The left-hand side is $(t\frac{dP}{dt}\otimes x|Q\otimes y) = (\operatorname{Res}\frac{dP}{dt}Q)(x|y)$; the right-hand side is $(d|PQ\otimes[x,y]+(\operatorname{Res}\frac{dP}{dt}Q)(x|y)K) = (\operatorname{Res}\frac{dP}{dt}Q)(x|y)$.

Finally, the restriction of $(.|.)$ to $\mathfrak{h}$ coincides with the form given by (6.2.1). Indeed, for both forms, $(d|d) = 0$, and hence it is sufficient to compare them on one element. We have $(\alpha_0|\alpha_0) = (\delta-\theta|\delta-\theta) = (\theta|\theta) = 2$.

Note also that the element K of $\hat{\mathcal{L}}(\mathring{\mathfrak{g}})$ is then the canonical central element and that the element d is the scaling element.

§7.6. Let $\mathring{\mathfrak{g}} = \mathring{\mathfrak{n}}_-\oplus\mathring{\mathfrak{h}}\oplus\mathring{\mathfrak{n}}_+$ be the triangular decomposition (1.3.2) of the Lie algebra $\mathring{\mathfrak{g}}$. Then the triangular decomposition of $\hat{\mathcal{L}}(\mathring{\mathfrak{g}})$ can be expressed as follows:

$$\begin{aligned}\hat{\mathcal{L}}(\mathring{\mathfrak{g}}) &= \mathfrak{n}_-\oplus\mathfrak{h}\oplus\mathfrak{n}_+, \text{ where}\\ \mathfrak{n}_- &= (t^{-1}\mathbb{C}[t^{-1}]\otimes(\mathring{\mathfrak{n}}_+ + \mathring{\mathfrak{h}})) + \mathbb{C}[t^{-1}]\otimes\mathring{\mathfrak{n}}_-,\\ \mathfrak{n}_+ &= (t\mathbb{C}[t]\otimes(\mathring{\mathfrak{n}}_- + \mathring{\mathfrak{h}})) + \mathbb{C}[t]\otimes\mathring{\mathfrak{n}}_+.\end{aligned}$$

The Chevalley involution ω of $\hat{\mathcal{L}}(\mathring{\mathfrak{g}})$ can be expressed in terms of the Chevalley involution $\mathring{\omega}$ of $\mathring{\mathfrak{g}}$ as follows:

$$\omega(P(t)\otimes x+\lambda K+\mu d) = P(t^{-1})\otimes\mathring{\omega}(x) - \lambda K - \mu d,$$

where $P(t) \in \mathcal{L}$, $x \in \mathring{\mathfrak{g}}$; λ, $\mu \in \mathbb{C}$. Indeed, we obviously have $\omega(e_i) = -f_i$, $\omega(f_i) = -e_i$ $(i = 1, \ldots, \ell)$ and $\omega|_{\mathfrak{h}} = -\mathrm{id}$. Furthermore, $\omega(e_0) = \omega(t \otimes E_0) = t^{-1} \otimes \mathring{\omega}(E_0) = -t^{-1} \otimes F_0 = -f_0$ and similarly, $\omega(f_0) = -e_0$.

Analogously the compact involution ω_0 of $\hat{\mathcal{L}}(\mathring{\mathfrak{g}})$ can be expressed in terms of the compact involution $\mathring{\omega}_0$ of $\mathring{\mathfrak{g}}$:

$$\omega_0(P(t) \otimes x + \lambda K + \mu d) = \overline{P}(t^{-1}) \otimes \mathring{\omega}_0(x) - \overline{\lambda} K - \overline{\mu} d,$$

where for $P(t) = \sum c_j t^j$ we set $\overline{P}(t) = \sum \overline{c}_j t^j$ and $\overline{a}$ denotes the complex conjugate of $a \in \mathbb{C}$. Hence the compact form of $\hat{\mathcal{L}}(\mathring{\mathfrak{g}})$ = (space of polynomial maps from the unit circle to the compact form of $\mathring{\mathfrak{g}}$) + $i\mathbb{R}K + i\mathbb{R}d$.

Finally, we have

$$\begin{aligned}(P(t) \otimes x | P(t) \otimes x)_0 &= -(P(t) \otimes x | \overline{P}(t^{-1}) \otimes \mathring{\omega}_0(x)) \\ &= \mathrm{Res}(t^{-1} P(t) \overline{P}(t^{-1}))(-x|\mathring{\omega}_0(x)) \\ &= \sum_j |c_j|^2 \, (x|x)_0.\end{aligned}$$

One deduces that the Hermitian form $(.|.)_0$ is positive-definite on the subspace $\mathcal{L}(\mathring{\mathfrak{g}})$ of $\hat{\mathcal{L}}(\mathring{\mathfrak{g}})$, using the fact that it is positive-definite on $\mathring{\mathfrak{g}}$ (see Remark 7.9 d) below). A more general approach will be developed in Chapter 11.

§7.7. We briefly explain here the physicists' approach to affine algebras. Let $\mathfrak{g}$ be a simple finite-dimensional Lie algebra with the normalized invariant bilinear form $(.|.)$. Let x_i be a basis of $\mathfrak{g}$, and let

$$[x_i, x_j] = \sum_k f_{ij}^k x_k, \text{ where } f_{ij}^k \in \mathbb{C}.$$

We denote the element $t^n \otimes x$ of $\mathcal{L}(\mathfrak{g})$ by $x^{(n)}$. Then the elements $x_i^{(n)}$ $(n \in \mathbb{Z})$ and K form a basis of $\tilde{\mathcal{L}}(\mathfrak{g})$ and obey the following commutation relations (cf. (7.2.1)):

$$[x_i^{(m)}, x_j^{(n)}] = \sum_k f_{ij}^k x_k^{(m+n)} + m\delta_{m,-n}(x_i|x_j)K. \tag{7.7.1}$$

Physicists associate to $x \in \mathfrak{g}$ the generating series:

$$x(z) := \sum_{n \in \mathbb{Z}} x^{(n)} z^{-n-1},$$

called the *current* of x. Here z is a formal parameter. In order to write down the commutation relations between currents, we need to introduce the *formal δ-function*:

$$\delta(z_1 - z_2) = z_2^{-1} \sum_{n \in \mathbb{Z}} \left(\frac{z_1}{z_2}\right)^n .$$

Its basic property is

$$\operatorname{Res}_{z_1=0} f(z_1)\delta(z_1 - z_2) = f(z_2). \tag{7.7.2}$$

Here, for $f(z) = \sum_{j \in \mathbb{Z}} f_j z^j$, one defines $\operatorname{Res}_{z=0} f(z) = f_{-1}$. It is straightforward to check that

$$[x(z_1), y(z_2)] = [x, y](z_1)\delta(z_1 - z_2) - K(x|y)\delta'_{z_1}(z_1 - z_2). \tag{7.7.3}$$

In this form the algebra $\tilde{\mathcal{L}}(\mathfrak{g})$ is called the *current algebra*.

The following lemma is very useful for calculations:

LEMMA 7.7. *One has the following equality of formal power series in $z_1^{\pm 1}$ and $z_2^{\pm 1}$ whenever both sides make sense:*

$$f(z_1, z_2)\delta(z_1 - z_2) = f(z_2, z_2)\delta(z_1 - z_2), \tag{7.7.4}$$

(7.7.5)

$$f(z_1, z_2)\delta'_{z_1}(z_1 - z_2) = f(z_2, z_2)\delta'_{z_1}(z_1 - z_2) - f'_{z_1}(z_1, z_2)\,|_{z_1=z_2}\,\delta(z_1 - z_2).$$

Proof. Multiply both sides by z_1^n and check the equality of the $Res_{z_1=0}$ using (7.7.2).

□

One also easily checks the following relation:

$$d_m(x(z)) = z^m\left(z\frac{d}{dz} + (m+1)\right)x(z). \tag{7.7.6}$$

Given two Vir-modules V and V_1, and a sequence of operators $F_j : V \to V_1$, $j \in \mathbb{Z}$, the generating series $F(z) = \sum_{j \in \mathbb{Z}} F_j z^{-j-\Delta}$ is called a *primary field of conformal weight* Δ if

$$[d_m, F(z)] = z^m\left(z\frac{d}{dz} + \Delta(m+1)\right)F(z), \quad m \in \mathbb{Z}.$$

Equation (7.7.6) shows that currents are primary fields of conformal weight 1.

§7.8. As we have seen, the (nontwisted) affine algebras can be described entirely in terms of the corresponding simple finite-dimensional Lie algebras.

In this section we give an explicit construction of the "simply-laced" simple finite-dimensional Lie algebras, i.e., those of type A_ℓ, D_ℓ, and E_ℓ ($\ell = 6, 7, 8$). The remaining cases will be treated in §7.9.

Let Q be the root lattice of type A_ℓ, D_ℓ, or E_ℓ and let $(.|.)$ be the bilinear symmetric form on Q such that (see §6.7):

$$\Delta = \{\alpha \in Q \mid (\alpha|\alpha) = 2\}.$$

Let $\varepsilon : Q \times Q \to \{\pm 1\}$ be a function satisfying the bimultiplicativity condition

$$\varepsilon(\alpha + \alpha', \beta) = \varepsilon(\alpha, \beta)\varepsilon(\alpha', \beta),$$

$$(7.8.1) \qquad \varepsilon(\alpha, \beta + \beta') = \varepsilon(\alpha, \beta)\varepsilon(\alpha, \beta') \quad \text{for } \alpha, \alpha', \beta, \beta' \in Q,$$

and the condition

$$(7.8.2) \qquad \varepsilon(\alpha, \alpha) = (-1)^{\frac{1}{2}(\alpha|\alpha)} \quad \text{for } \alpha \in Q.$$

(Recall that $(\alpha|\alpha) \in 2\mathbb{Z}$ for $\alpha \in Q$.) We call such a function an *asymmetry function.* Since $(\alpha|\beta) \in \mathbb{Z}$ for $\alpha, \beta \in Q$, replacing α by $\alpha + \beta$ in (7.8.2), we obtain

$$(7.8.3) \qquad \varepsilon(\alpha, \beta)\varepsilon(\beta, \alpha) = (-1)^{(\alpha|\beta)} \quad \text{for } \alpha, \beta \in Q.$$

An asymmetry function ε can be constructed as follows: choose an orientation of the Dynkin diagram, let

$$(7.8.4) \qquad \begin{aligned} &\varepsilon(\alpha_i, \alpha_j) = -1 \ \text{ if } i = j \text{ or if } \overset{i}{\circ}\!\rightarrow\!\overset{j}{\circ} \\ &\varepsilon(\alpha_i, \alpha_j) = 1 \ \text{ otherwise, i.e., } \overset{i}{\circ}\ \overset{j}{\circ} \text{ or } \overset{i}{\circ}\!\leftarrow\!\overset{j}{\circ}, \end{aligned}$$

and then extend by bimultiplicativity. It is clear that then (7.8.2) holds for all $\alpha \in Q$. (More generally, choose a $\mathbb{Z}$-basis $\beta_1, \dots, \beta_\ell$ of Q, let $\varepsilon(\beta_i, \beta_i) = (-1)^{\frac{1}{2}(\beta_i|\beta_i)}$, $\varepsilon(\beta_i, \beta_j) = (-1)^{(\beta_i|\beta_j)}$ if $i < j$ and $= 1$ if $i > j$, and extend by bimultiplicativity. Thus, ε "breaks the symmetry" of $(.|.)$).

Now let $\mathfrak{h}$ be the complex hull of Q and extend $(.|.)$ from Q to $\mathfrak{h}$ by bilinearity. We shall identify $\mathfrak{h}$ with $\mathfrak{h}^*$ using this form. Take the direct sum of $\mathfrak{h}$ with 1-dimensional vector spaces $\mathbb{C}E_\alpha$, $\alpha \in \Delta$:

$$\mathfrak{g} = \mathfrak{h} \oplus \Big(\bigoplus_{\alpha \in \Delta} \mathbb{C}E_\alpha\Big).$$

Define a bracket on $\mathfrak{g}$ as follows:

$$(7.8.5)\qquad \begin{cases} [h,h'] = 0 & \text{if } h,h' \in \mathfrak{h}, \\ [h,E_\alpha] = (h|\alpha)E_\alpha & \text{if } h \in \mathfrak{h}, \alpha \in \Delta, \\ [E_\alpha, E_{-\alpha}] = -\alpha & \text{if } \alpha \in \Delta, \\ [E_\alpha, E_\beta] = 0 & \text{if } \alpha,\beta \in \Delta, \alpha+\beta \notin \Delta \cup \{0\}, \\ [E_\alpha, E_\beta] = \varepsilon(\alpha,\beta)E_{\alpha+\beta} & \text{if } \alpha,\beta,\alpha+\beta \in \Delta. \end{cases}$$

Define the symetric bilinear form $(.|.)$ on $\mathfrak{g}$ extending it from $\mathfrak{h}$ as follows:

$$(7.8.6)\quad (h|E_\alpha) = 0 \text{ if } h \in \mathfrak{h}, \alpha \in \Delta, \qquad (E_\alpha|E_\beta) = -\delta_{\alpha,-\beta} \text{ if } \alpha,\beta \in \Delta.$$

PROPOSITION 7.8. *If Q is the root lattice of type A_ℓ, D_ℓ, or E_ℓ and ε is an asymmetry function on $Q \times Q$, then $\mathfrak{g}$ is the simple Lie algebra of type A_ℓ, D_ℓ, or E_ℓ, respectively, the form $(.|.)$ being the normalized invariant form.*

Proof. Let $x, y, z \in \mathfrak{g}$ be from either $\mathfrak{h}$ or one of the E_α. If one of these elements lies in $\mathfrak{h}$, the Jacobi identity trivially holds. Let now $x = E_\alpha$, $y = E_\beta$, $z = E_\gamma$. If all $\alpha+\beta, \alpha+\gamma, \beta+\gamma \notin \Delta \cup \{0\}$, the Jacobi identity trivially holds. Thus, we may assume that $\alpha+\beta \in \Delta \cup \{0\}$.

Note that given $\alpha, \beta \in \Delta$ we have:

$$(7.8.7)\qquad \alpha \pm \beta \in \Delta \text{ iff } (\alpha|\beta) = \mp 1,$$

$$(7.8.8)\qquad \varepsilon(\alpha,\alpha) = -1, \qquad \varepsilon(\alpha,\beta)\varepsilon(\beta,\alpha) = (-1)^{(\alpha|\beta)}.$$

If $\alpha+\beta = 0$, consider four cases: 1) both $\alpha \pm \gamma \notin \Delta \cup \{0\}$, 2) $\alpha+\gamma = 0$ or $\alpha - \gamma = 0$, 3) $\alpha+\gamma \in \Delta$, 4) $\alpha - \gamma \in \Delta$; the Jacobi identity trivially holds in cases 1) and 2), and reduces to conditions $\varepsilon(\gamma,\alpha)\varepsilon(\gamma+\alpha,-\alpha) = (\alpha|\alpha)$ in case 3) and $\varepsilon(-\alpha,\gamma)\varepsilon(-\alpha+\gamma,\alpha) = (\alpha|\alpha)$ in case 4), which holds due to bimultiplicativity of ε, (7.8.7) and (7.8.8). Thus, we may assume that all $\alpha+\beta, \alpha+\gamma, \beta+\gamma \in \Delta$, i.e., that $(\alpha|\beta) = (\alpha|\gamma) = (\beta|\gamma) = -1$. Then $|\alpha+\beta+\gamma|^2 = 0$, hence $\alpha+\beta+\gamma = 0$ (this is the place where we use that $(.|.)$ is positive-definite). The Jacobi identity reduces to the condition:

$$\varepsilon(\alpha,\beta)(\alpha+\beta) + \varepsilon(\beta,\gamma)(\beta+\gamma) + \varepsilon(\gamma,\alpha)(\alpha+\gamma) = 0,$$

or

$$\varepsilon(\alpha,\beta)(\alpha+\beta) - \varepsilon(\beta,-\alpha-\beta)\alpha - \varepsilon(-\alpha-\beta,\alpha)\beta = 0,$$

which holds due to bimultiplicativity of ε, (7.8.6) and (7.8.7). Thus, $\mathfrak{g}$ is a Lie algebra.

Let (see §6.7)

$$\Pi = \Pi^\vee = \{\alpha_1, \ldots, \alpha_\ell\}, e_i = E_{\alpha_i},\ f_i = -E_{-\alpha_i}.$$

It is easy to see that $(\mathfrak{g}, \mathfrak{h}, \Pi, \Pi^\vee)$ is the quadruple associated to the matrix A of type A_ℓ, D_ℓ, or E_ℓ, respectively, hence, by Proposition 1.4 a), $\mathfrak{g}$ is of type A_ℓ, D_ℓ, or E_ℓ, respectively.

It remains to check that $(.|.)$ is invariant; this is straightforward.

□

§7.9. In this section we give an explicit construction of the non-simply-laced simple finite-dimensional Lie algebras, i.e., those of type B_ℓ, C_ℓ, F_4, and G_2.

Consider the following (simply-laced) root lattices $Q(X_N)$ and their automorphisms $\overline{\mu}$ of order $r = 2$ or 3. (We use the construction of $Q(X_N)$ given in §6.7.)

Case 1. $Q(D_{\ell+1}) : \overline{\mu}(v_i) = v_i$ for $1 \le i \le \ell$, $\overline{\mu}(v_{\ell+1}) = -v_{\ell+1}$;

Case 2. $Q(A_{2\ell-1}) : \overline{\mu}(v_i) = -v_{2\ell+1-i}$;

Case 3. $Q(E_6) : \overline{\mu}(v_i) = -v_{7-i}$ for $1 \le i \le 6$, $\overline{\mu}(v_7) = v_7$;

Case 4. $Q(D_4) : \overline{\mu}(\alpha_1) = \alpha_3$, $\overline{\mu}(\alpha_3) = \alpha_4$, $\overline{\mu}(\alpha_4) = \alpha_1$, $\overline{\mu}(\alpha_2) = \alpha_2$.

We denote here by $(.|.)'$ the normalized invariant form on $Q(X_N)$, by Δ' the set of elements of $Q(X_N)$ of square length 2, by $\Pi' = \{\alpha'_i, \ldots, \alpha'_N\}$ the set of simple roots, etc. Note that in all cases, $\overline{\mu}(\Pi') = \Pi'$, hence $\overline{\mu} \in \mathrm{Aut}(X_N)$ is the diagram automorphism of $Q(X_N)$.

In all cases, fix an orientation of the Dynkin diagram $S(X_N)$ which is invariant under $\overline{\mu}$, and let $\varepsilon(\alpha, \beta)$ be the corresponding asymmetry function (defined by (7.8.4)). Let

$$\mathfrak{g}(X_N) = \mathfrak{h}' \oplus \big(\bigoplus_{\alpha \in \Delta'} \mathbb{C}E'_\alpha \big)$$

be the corresponding simple Lie algebra, as defined in §7.8. It is clear that the automorphism $\overline{\mu}$ of $Q(X_N)$ induces an automorphism μ of $\mathfrak{g}(X_N)$ defined by

$$\mu(\alpha) = \overline{\mu}(\alpha), \quad \mu(E'_\alpha) = E'_{\overline{\mu}(\alpha)}. \tag{7.9.1}$$

This automorphism has the following characteristic property:

$$\mu(e_i') = e'_{\overline{\mu}(i)}, \quad \mu(f_i') = f'_{\overline{\mu}(i)}, \quad i = 1, \ldots, N. \tag{7.9.2}$$

Such an automorphism is called a *diagram automorphism* of $\mathfrak{g}(X_N)$.

Introduce the following notations:

$$\begin{aligned}
&\Delta_\ell = \{\alpha = \alpha' \in \Delta' \mid \overline{\mu}(\alpha') = \alpha'\};\\
&\Delta_s = \{\alpha = r^{-1}(\overline{\mu}(\alpha') + \cdots + \overline{\mu}^r(\alpha') \mid \alpha' \in \Delta', \overline{\mu}(\alpha') \neq \alpha'\};\\
&\Delta = \Delta_s \cup \Delta_\ell; \qquad Q = \mathbf{Z}\Delta;\\
&E_\alpha^{(j)} = \eta^j E'_{\overline{\mu}(\alpha')} + \cdots + \eta^{rj} E'_{\overline{\mu}^r(\alpha')}, \text{ where } \eta = \exp 2\pi i/r;\\
&E_\alpha = E'_\alpha \text{ if } \alpha \in \Delta_\ell, \qquad E_\alpha = E_\alpha^{(0)} \text{ if } \alpha \in \Delta_s;\\
&V^{(j)} = \bigoplus_{\alpha \in \Delta_s} \mathbf{C}E_\alpha^{(j)}; \qquad \mathfrak{h}^{(j)} = \{h \in \mathfrak{h}' \mid \mu(h) = \eta^j h\};\\
&\mathfrak{g}^{(j)} = \mathfrak{h}^{(j)} + V^{(j)}, \quad j = 1, \ldots, r-1;\\
&\mathfrak{h} = \mathfrak{h}^{(0)}; \qquad \mathfrak{g} = \mathfrak{h} \oplus (\bigoplus_{\alpha \in \Delta} \mathbf{C}E_\alpha).
\end{aligned}$$

We can now state the result.

PROPOSITION 7.9. *Let* $(X_N, r) = (D_{\ell+1}, 2), (A_{2\ell-1}, 2), (E_6, 2)$, *or* $(D_4, 3)$. *Then*

a) $\mathfrak{g}(X_N) = \mathfrak{g} \oplus \mathfrak{g}^{(1)}$ *(resp.* $= \mathfrak{g} \oplus \mathfrak{g}^{(1)} \oplus \mathfrak{g}^{(2)}$*) if* $r = 2$ *(resp.* $r = 3$*), where* $\mathfrak{g}$ *is the fixed point set of* μ *and* $\mathfrak{g}^{(1)}$ *(resp.* $\mathfrak{g}^{(j)}$, $j = 1$ *or* 2*) is the eigenspace of* μ *with eigenvalue* -1 *(resp.* $\exp 2\pi i j/3$*).*

b) $\mathfrak{g}$ *is the simple Lie algebra of type* B_ℓ, C_ℓ, F_4, *or* G_2, *respectively, and its commutation relations are as follows:*

$$\left\{\begin{array}{ll}
[h, h'] = 0 & \text{if } h, h' \in \mathfrak{h},\\
{[h, E_\alpha]} = (h|\alpha)' E_\alpha & \text{if } h \in \mathfrak{h}, \alpha \in \Delta,\\
{[E_\alpha, E_{-\alpha}]} = -\alpha (\text{resp. } -r\alpha) & \text{if } \alpha \in \Delta_\ell \ (\text{resp. } \alpha \in \Delta_s),\\
{[E_\alpha, E_\beta]} = 0 & \text{if } \alpha, \beta \in \Delta,\ \alpha + \beta \notin \Delta \cup \{0\},\\
{[E_\alpha, E_\beta]} = (p+1)\varepsilon(\alpha', \beta') E_{\alpha+\beta} & \text{if } \alpha, \beta, \alpha + \beta \in \Delta,\\
\text{where } p \text{ is the maximal positive integer such that } \alpha - p\beta \in \Delta. &
\end{array}\right. \tag{7.9.3}$$

c) *The normalized bilinear form* $(.|.)$ *on* $\mathfrak{g}$ *is given by the following formulas*

$$(h|h') = (h|h')'|_{\mathfrak{h}}, \quad (h|E_\alpha) = 0 \quad \text{if } h, h' \in \mathfrak{h}, \alpha \in \Delta,$$

$$(E_\alpha|E_\beta) = -\delta_{\alpha,-\beta} \ (\text{resp. } -r\delta_{\alpha,-\beta}) \quad \text{if } \alpha, \beta \in \Delta_\ell \ (\text{resp. } \in \Delta_s).$$

d) *Δ is the set of roots of $\mathfrak{g}$ with respect to the Cartan subalgebra $\mathfrak{h}$, Δ_s (resp. Δ_ℓ) being the set of short (resp. long) roots, and Q is its root lattice;* $\Delta^\vee = \Delta_\ell \cup r\Delta_s$.

e) *The sets of simple roots Π (numbered as in Table* Fin) *are these:*

Case 1: $\alpha_1 = \alpha'_1, \ldots, \alpha_{\ell-1} = \alpha'_{\ell-1}, \alpha_\ell = \frac{1}{2}(\alpha'_\ell + \alpha'_{\ell+1})$;

Case 2: $\alpha_1 = \frac{1}{2}(\alpha'_1 + \alpha'_{2\ell-1}), \ldots, \alpha_{\ell-1} = \frac{1}{2}(\alpha'_{\ell-1} + \alpha'_{\ell+1}), \alpha_\ell = \alpha'_\ell$;

Case 3: $\alpha_1 = \alpha'_6, \alpha_2 = \alpha'_3, \alpha_3 = \frac{1}{2}(\alpha'_2 + \alpha'_4), \alpha_4 = \frac{1}{2}(\alpha'_1 + \alpha'_5)$;

Case 4: $\alpha_1 = \alpha'_2, \alpha_2 = \frac{1}{3}(\alpha'_1 + \alpha'_3 + \alpha'_4)$.

f) *Letting $\Pi_\ell = \Pi \cap \Delta_\ell$, $\Pi_s = \Pi \cap \Delta_s$, we have: $\Pi^\vee = \Pi_\ell \cup r\Pi_s$.*

g) $[\mathfrak{g}, \mathfrak{g}^{(j)}] = \mathfrak{g}^{(j)}$, *and the $\mathfrak{g}$-module $\mathfrak{g}^{(j)}$ is irreducible with highest weight θ_0 ($= -$lowest weight) listed below:*

Case 1: $\theta_0 = \alpha_1 + \alpha_2 + \cdots + \alpha_\ell$;

Case 2: $\theta_0 = \alpha_1 + 2\alpha_2 + \cdots + 2\alpha_{\ell-1} + \alpha_\ell$;

Case 3: $\theta_0 = \alpha_1 + 2\alpha_2 + 3\alpha_3 + 2\alpha_4$;

Case 4: $\theta_0 = \alpha_1 + 2\alpha_2$.

Proof. The proof of a), c), d), e), f), and g) is straightforward; the first part of b) follows from d), e), and Proposition 1.4 a). All commutation relations (7.9.3) except for the last one are clear.

If one of the $\alpha, \beta \in \Delta$ is long and $\alpha + \beta \in \Delta$, we clearly have:

$$[E_\alpha, E_\beta] = \varepsilon(\alpha', \beta') E_{\alpha+\beta}, \tag{7.9.4}$$

and $\alpha - \beta \notin \Delta$, hence the last formula in b) holds in this case. It remains to consider the case when both α and β are short and $\alpha + \beta \in \Delta$. First, let $r = 2$; then $p = 1$, $\alpha + \beta \in \Delta_\ell$, and we have to show that

$$[E_\alpha, E_\beta] = 2\varepsilon(\alpha', \beta') E_{\alpha+\beta}. \tag{7.9.5}$$

This follows from the following fact:

(7.9.6) if $r = 2$ and $\alpha', \beta' \in \Delta' \backslash \Delta_\ell$, $\alpha' + \beta' \in \Delta'$, then $\alpha' + \overline{\mu}(\beta') \notin \Delta'$.

Since the $\mathbb{Z}$-span of $\alpha', \beta', \overline{\mu}(\alpha')$, and $\overline{\mu}(\beta')$ can be only one of the root lattices $Q(A_n)$, $n = 2, 3$, or 4, it suffices to check (7.9.6) in these cases only, which is straightforward. Hence, we have: $[E_\alpha, E_\beta] = [E'_{\alpha'} + E'_{\overline{\mu}(\alpha')}, E'_{\beta'} + E'_{\overline{\mu}(\beta')}] = \varepsilon(\alpha', \beta') E'_{\alpha'+\beta'} + \varepsilon(\overline{\mu}(\alpha'), \overline{\mu}(\beta')) E'_{\overline{\mu}(\alpha'+\beta')} = 2\varepsilon(\alpha', \beta') E'_{\alpha'+\beta'}$.

Finally, let $r = 3$. Then either $\alpha' + \beta' \in \Delta_s, p = 1$, and exactly one of $\alpha' + \overline{\mu}(\beta'), \alpha' + \overline{\mu}^2(\beta')$ lies in Δ', or $\alpha' + \beta' \in \Delta_\ell, p = 2$, and none of

$\alpha' + \overline{\mu}(\beta'), \alpha' + \overline{\mu}^2(\beta')$ lies in Δ. Hence, in both cases the last equation of (7.9.3) holds.

□

Remark 7.9. a) (7.8.5) is a special case of (7.9.3) for $r = 1$, since in the simply-laced case $p = 0$.

b) Commutation relations (7.9.3) can be extended to the whole $\mathfrak{g}(X_N)$ as follows. Let $E_\alpha^{(0)} = E_\alpha = E'_\alpha$ if $\alpha \in \Delta_\ell$ and let $\alpha^{(j)} = \eta^j \mu(\alpha') + \cdots + \eta^{rj} \mu^r(\alpha')$ if $\alpha \in \Delta_s$. Then $(0 \le i, j \le r-1)$:

$$
\begin{aligned}
&[h^{(i)}, h^{(j)}] = 0 && \text{if } h^{(i)} \in \mathfrak{h}^{(i)}, h^{(j)} \in \mathfrak{h}^{(j)}, \\
&[h^{(i)}, E_\alpha^{(j)}] = (h^{(i)}|\alpha)' E_\alpha^{(i+j)} && \text{if } h^{(i)} \in \mathfrak{h}^{(i)}, \alpha \in \Delta, \\
&[E_\alpha^{(i)}, E_{-\alpha}^{(j)}] = -\alpha^{(i+j)} (\text{resp. } -\alpha) && \text{if } \alpha \in \Delta_s (\text{resp. } \alpha \in \Delta_\ell), \\
&[E_\alpha^{(i)}, E_\beta^{(j)}] = 0 && \text{if } \alpha, \beta \in \Delta, \alpha + \beta \notin \Delta \cup \{0\}, \\
&[E_\alpha^{(i)}, E_\beta^{(j)}] = (p+1)\varepsilon(\alpha', \beta') E_{\alpha+\beta}^{(i+j)} && \text{if } \alpha, \beta, \alpha + \beta \in \Delta,
\end{aligned}
$$

where p is as in (7.9.3).

c) The Chevalley involution ω and the compact involution ω_0 of $\mathfrak{g}(X_N)$ are given by $(\alpha \in \Delta)$:

$$\omega(E_\alpha) = E_{-\alpha}, \quad \omega(\alpha) = -\alpha; \qquad \omega_0(E_\alpha) = E_{-\alpha}, \quad \omega_0(\alpha) = -\alpha.$$

d) It follows from c) and Proposition 4.9 that the Hermitian form $(.|.)_0$ is positive-definite on $\mathfrak{g}(X_\ell)$, where X_ℓ is of finite type.

§7.10. For applications in Chapter 8, we shall now take care of the remaining diagram automorphism:

Case 5: $Q(A_{2\ell}) : \overline{\mu}(v_i) = -v_{2\ell+2-i}$.

In this case there is no $\overline{\mu}$-invariant orientation; we consider the orientation o→o→ ⋯ →o instead. Then the diagram automorphism μ of $\mathfrak{g}(A_{2\ell})$ is defined by

$$\mu(\alpha) = \overline{\mu}(\alpha), \ \mu(E'_\alpha) = (-1)^{1+\mathrm{ht}\,\alpha} E'_{\overline{\mu}(\alpha)}. \tag{7.10.1}$$

Let

$$
\begin{aligned}
&\Delta_\ell \ (\text{resp. } \Delta_s) \\
&\quad = \{\tfrac{1}{2}(\alpha' + \overline{\mu}(\alpha')) \mid \alpha' \ne \overline{\mu}(\alpha') \text{ and } (\alpha'|\overline{\mu}(\alpha'))' = 0 \ (\text{resp. } \ne 0)\},
\end{aligned}
$$

$$\Delta = \Delta_\ell \cup \Delta_s,$$
$$E_\alpha = E'_{\alpha'} - (-1)^{\mathrm{ht}\,\alpha} E'_{\overline{\mu}(\alpha')} \text{ if } \alpha \in \Delta_\ell,$$
$$E_\alpha = \sqrt{2}(E'_{\alpha'} - (-1)^{\mathrm{ht}\,\alpha} E'_{\overline{\mu}(\alpha')}) \text{ if } \alpha \in \Delta_s,$$
$$\mathfrak{h} = \{h \in \mathfrak{h}' \mid \mu(h) = h\}, \qquad \mathfrak{h}^{(1)} = \{h \in \mathfrak{h}' \mid \mu(h) = -h\},$$
$$\mathfrak{g} = \mathfrak{h} \oplus (\bigoplus_{\alpha\in\Delta} \mathbb{C}E_\alpha),$$
$$\mathfrak{g}^{(1)} = \mathfrak{h}^{(1)} \oplus (\sum_{\substack{\alpha\in\Delta' \\ \alpha\neq\mu(\alpha)}} \mathbb{C}(E'_\alpha + (-1)^{\mathrm{ht}\,\alpha} E'_{\overline{\mu}(\alpha)})) \oplus (\sum_{\substack{\alpha\in\Delta' \\ \alpha=\mu(\alpha)}} \mathbb{C}E_\alpha).$$

Then we obtain the following extension of Proposition 7.9.

PROPOSITION 7.10. a) $\mathfrak{g}(A_{2\ell}) = \mathfrak{g} \oplus \mathfrak{g}^{(1)}$.

b) *$\mathfrak{g}$ is the simple Lie algebra of type B_ℓ, and its commutation relations are given by (7.9.3).*

c) *The normalized bilinear form $(.|.)$ on $\mathfrak{g}$ is given by*

$$(h|h') = 4(h|h')'|_{\mathfrak{h}}, \quad (h|E_\alpha) = 0 \quad \text{if } h, h' \in \mathfrak{h}, \alpha \in \Delta,$$
$$(E_\alpha|E_\beta) = -4\delta_{\alpha,-\beta} (\text{resp. } -8\delta_{\alpha,-\beta}) \quad \text{if } \alpha \in \Delta_\ell \ (\text{resp. } \in \Delta_s).$$

d) *Same as Proposition 7.9 d) (with $r = 2$).*

e) *The set of simple roots Π is this:*

$$\alpha_1 = \frac{1}{2}(\alpha'_1 + \alpha'_{2\ell}), \alpha_2 = \frac{1}{2}(\alpha'_2 + \alpha'_{2\ell-1}), \ldots, \alpha_\ell = \frac{1}{2}(\alpha'_\ell + \alpha'_{\ell+1}).$$

f) *Letting $\Pi_\ell = \Pi \cap \Delta_\ell, \Pi_s = \Pi \cap \Delta_s$, we have $\Pi^\vee = 2\Pi_\ell \cup 4\Pi_s$.*

g) *$[\mathfrak{g}, \mathfrak{g}^{(1)}] = \mathfrak{g}^{(1)}$ and the $\mathfrak{g}$-module $\mathfrak{g}^{(1)}$ is irreducible with highest weight (= −lowest weight) $\theta_0 = 2\alpha_1 + 2\alpha_2 + \cdots + 2\alpha_\ell$.*

□

Remark 7.10. Letting $E_\alpha^{(0)} = E_\alpha$, $E_\alpha^{(1)} = E'_{\alpha'} + (-1)^{\mathrm{ht}\,\alpha} E'_{\overline{\mu}(\alpha')}$ if $\alpha \in \Delta_\ell$, $E_\alpha^{(1)} = \sqrt{2}(E'_{\alpha'} + (-1)^{\mathrm{ht}\,\alpha} E'_{\overline{\mu}(\alpha')})$ if $\alpha \in \Delta_s$, $\Delta_m = \{\alpha \in \Delta' \mid \overline{\mu}(\alpha) = \alpha\}$, $E_\alpha^{(1)} = E'_\alpha$ if $\alpha \in \Delta_m$, formulas (7.9.3) extend in the same way as described in Remark 7.9.

§7.11. A generalized Cartan matrix of infinite order is called an *infinite affine matrix* if every one of its principal minors of finite order is positive. By Theorem 4.8a), it is clear that a complete list of infinite affine matrices is the following (see Exercise 4.14):

$$A_\infty, A_{+\infty}, B_\infty, C_\infty \text{ and } D_\infty.$$

Let A be an infinite affine matrix, and let $\mathfrak{g}'(A)$ be the associated Kac–Moody algebra (defined in Remark 1.5 for an infinite n). These Lie algebras are called ***infinite rank affine algebras.*** For these Kac–Moody algebras we have $\Delta = \Delta^{re}$; given $\alpha \in \Delta_+$, we denote by $\alpha^\vee$ the element of $\mathfrak{h}'$ such that $\alpha^\vee \in [\mathfrak{g}_\alpha, \mathfrak{g}_{-\alpha}]$ and $\langle \alpha, \alpha^\vee \rangle = 2$, and we put $\Delta^\vee = \{\alpha^\vee \in \mathfrak{h}' | \alpha \in \Delta\}$ to be the set of dual roots. Here we give an explicit construction of these Lie algebras, which generalizes the usual construction of classical finite-dimensional Lie algebras. We also construct certain completions and central extensions of them, which play an important role in the theory of completely integrable systems (see Chapter 14).

Denote by gl_∞ the Lie algebra of all complex matrices $(a_{ij})_{i,j\in\mathbb{Z}}$, such that the number of nonzero a_{ij} is finite, with the usual bracket. This Lie algebra acts in a usual way on the space $\mathbb{C}^\infty$ of all column vectors $(a_i)_{i\in\mathbb{Z}}$, such that all but a finite number of the a_i are zero. Let $E_{ij} \in gl_\infty$ be the matrix which has a 1 in the i,j-entry and 0 everywhere else, and let $v_i \in \mathbb{C}^\infty$ be the column vector which has a 1 in the i-th entry and 0 everywhere else, so that

$$E_{ij}(v_j) = v_i.$$

Let $A = A_\infty$; then $\mathfrak{g}'(A) = sl_\infty := \{a \in gl_\infty | \operatorname{tr} a = 0\}$. The Chevalley generators of $\mathfrak{g}'(A)$ are as follows:

$$e_i = E_{i,i+1}, \quad f_i = E_{i+1,i} \ (i \in \mathbb{Z}),$$

so that

$$\Pi^\vee = \{\alpha_i^\vee = E_{i,i} - E_{i+1,i+1} \ (i \in \mathbb{Z})\}$$

is the set of simple coroots. Then $\mathfrak{h}'$ consists of diagonal matrices, and $\mathfrak{n}_+$ (resp. $\mathfrak{n}_-$) of upper- (resp. lower-) triangular matrices. Denote by ϵ_i the linear function on $\mathfrak{h}'$ such that $\epsilon_i(E_{jj}) = \delta_{ij}$ $(j \in \mathbb{Z})$. Then the root system and the root spaces of $\mathfrak{g}'(A)$, attached to nonzero roots, are

$$\Delta = \{\epsilon_i - \epsilon_j \ (i \neq j, \ i,j \in \mathbb{Z})\}; \quad \mathfrak{g}_{\epsilon_i - \epsilon_j} = \mathbb{C}E_{ij},$$

$\epsilon_i - \epsilon_j$ being a positive root if and only if $i < j$.

The set of simple roots is

$$\Pi = \{\alpha_i = \epsilon_i - \epsilon_{i+1}\ (i \in \mathbb{Z})\}.$$

The set of positive dual roots is

$$\Delta_+^\vee = \Pi^\vee \cup \{\alpha_i^\vee + \alpha_{i+1}^\vee + \cdots + \alpha_j^\vee,\ \text{where } i < j;\ i, j \in \mathbb{Z}\}. \tag{7.11.1}$$

The description for $A = A_{+\infty}$ is similar, replacing $\mathbb{Z}$ by $\mathbb{Z}_+$.

The remaining infinite rank affine algebras, of type X_∞, where $X = B$, C or D, are subalgebras of the Lie algebra gl_∞, and consist of matrices which preserve the bilinear form X, i.e.,

$$\mathfrak{g}'(X_\infty) = \{a \in \mathfrak{g}'(A_\infty) | X(a(u), v) + X(u, a(v)) = 0 \text{ for all } u, v \in \mathbb{C}^\infty\}.$$

These bilinear forms are as follows:

$$B(v_i, v_j) = (-1)^i \delta_{i,-j} \quad (i, j \in \mathbb{Z})$$
$$C(v_i, v_j) = (-1)^i \delta_{i,-j+1} \quad (i, j \in \mathbb{Z})$$
$$D(v_i, v_j) = \delta_{i,-j+1} \quad (i, j \in \mathbb{Z}).$$

We describe below the Chevalley generators e_i, f_i $(i \in \mathbb{Z}_+)$, the set of simple coroots $\Pi^\vee$, the root system Δ, the set of simple roots Π, and the set of positive dual roots $\Delta_+^\vee$ of $\mathfrak{g}'(A)$, where $A = B_\infty$, C_∞ or D_∞. In all cases, $\mathfrak{h}'$ consists of diagonal matrices and $\mathfrak{n}_+$ (resp. $\mathfrak{n}_-$) consists of upper- (resp. lower-) triangular matrices.

$$\begin{aligned} B_\infty : e_0 &= E_{0,1} + E_{-1,0},\ e_i = E_{i,i+1} + E_{-i-1,-i}, \\ f_0 &= 2(E_{1,0} + E_{0,-1}),\ f_i = E_{i+1,i} + E_{-i,-i-1} (i = 1, 2, \ldots); \\ \Pi^\vee &= \{\alpha_0^\vee = 2(E_{-1,-1} - E_{1,1}),\ \alpha_i^\vee = E_{i,i} + E_{-i-1,-i-1} \\ &\qquad - E_{i+1,i+1} - E_{-i,-i}\ (i = 1, 2, \ldots)\}; \\ \Delta &= \{\pm\epsilon_i \pm \epsilon_j\ (i \neq j,\ i, j \in \mathbb{Z}_+),\ \pm\epsilon_i\ (i \in \mathbb{Z}_+)\}; \\ \mathfrak{g}_{\epsilon_i - \epsilon_j} &= \mathbb{C}(E_{i,j} - (-1)^{i+j} E_{-j,-i}),\ i \neq j,\ i, j \in \mathbb{Z}; \\ \Pi &= \{\alpha_0 = -\epsilon_1, \alpha_i = \epsilon_i - \epsilon_{i+1}\ (i = 1, 2, \ldots)\}; \\ \Delta_+^\vee &= \Pi^\vee \cup \{\alpha_i^\vee + \alpha_{i+1}^\vee + \cdots + \alpha_j^\vee\ (i < j,\ i, j \in \mathbb{Z}_+), \\ &\qquad \alpha_0^\vee + 2\alpha_1^\vee + \cdots + 2\alpha_i^\vee \\ &\qquad\quad + \alpha_{i+1}^\vee + \cdots + \alpha_j^\vee (i \leq j,\ i, j \in \mathbb{Z}_+)\}. \end{aligned}$$

Here the ϵ_i are viewed restricted to $\mathfrak{h}'$, so that $\epsilon_i = -\epsilon_{-i}$.

$$\begin{aligned} C_\infty : e_0 &= E_{0,1},\ e_i = E_{i,i+1} + E_{-i,-i+1}, \\ f_0 &= E_{1,0},\ f_i = E_{i+1,i} + E_{-i+1,-i}\ (i = 1, 2, \ldots); \\ \Pi^\vee &= \{\alpha_0^\vee = E_{00} - E_{11},\ \alpha_i^\vee = E_{i,i} + E_{-i,-i} - E_{i+1,i+1} \\ &\qquad - E_{-i+1,-i+1}\ (i = 1, 2, \ldots)\}; \end{aligned}$$

$$\Delta = \{\pm\epsilon_i \pm \epsilon_j,\ \pm 2\epsilon_i\ (i \neq j,\ i,j = 1,2,\dots)\};$$
$$\mathfrak{g}_{2\epsilon_i} = \mathbf{C}E_{i,-i+1},\quad \mathfrak{g}_{\epsilon_i-\epsilon_j} = \mathbf{C}(E_{ij} - (-1)^{i+j}E_{-j+1,-i+1});$$
$$\Pi = \{\alpha_0 = -2\epsilon_1,\quad \alpha_i = \epsilon_i - \epsilon_{i+1}\ (i = 1,2,\dots)\};$$
$$\Delta_+^\vee = \Pi^\vee \cup \{\alpha_i^\vee + \alpha_{i+1}^\vee + \dots + \alpha_j^\vee\ (i<j,\ i,j \in \mathbf{Z}_+),$$
$$2\alpha_0^\vee + \dots + 2\alpha_i^\vee + \alpha_{i+1}^\vee + \dots + \alpha_j^\vee\ (i<j,\ i,j\in\mathbf{Z}_+)\}.$$

Here the ϵ_i are viewed restricted to $\mathfrak{h}'$, so that $\epsilon_i = -\epsilon_{-i+1}$.

$$D_\infty : e_0 = E_{0,2} - E_{-1,1},\ e_i = E_{i,i+1} - E_{-i,-i+1},$$
$$f_0 = E_{2,0} - E_{1,-1},\ f_i = E_{i+1,i} - E_{-i+1,-i}\ (i = 1,2,\dots);$$
$$\Pi^\vee = \{\alpha_0^\vee = E_{0,0} + E_{-1,-1} - E_{2,2} - E_{1,1},$$
$$\alpha_i^\vee = E_{i,i} + E_{-i,-i} - E_{i+1,i+1}$$
$$- E_{-i+1,-i+1}\ (i = 1,2,\dots)\};$$
$$\Delta = \{\pm\epsilon_i \pm \epsilon_j\ (i \neq j;\ i,j = 1,2,\dots)\};$$
$$\mathfrak{g}_{\epsilon_i-\epsilon_j} = \mathbf{C}(E_{ij} - E_{-j+1,-i+1});$$
$$\Pi = \{\alpha_0 = -\epsilon_1 - \epsilon_2,\ \alpha_i = \epsilon_i - \epsilon_{i+1}\ (i = 1,2,\dots)\};$$
$$\Delta_+^\vee = \Pi^\vee \cup \{\alpha_i^\vee + \alpha_{i+1}^\vee + \dots + \alpha_j^\vee\ (1 \le i < j),$$
$$\alpha_0^\vee + \alpha_2^\vee + \dots + \alpha_j^\vee\ (j = 2,3,\dots),$$
$$\alpha_0^\vee + \alpha_1^\vee + 2\alpha_2^\vee$$
$$+ \dots + 2\alpha_i^\vee + \alpha_{i+1}^\vee + \dots + \alpha_j^\vee (1 \le i < j)\}.$$

Here the ϵ_i are viewed restricted to $\mathfrak{h}'$, so that $\epsilon_i = -\epsilon_{-i+1}$.

§7.12. Now we turn to the description of a completion and its central extension for an infinite rank affine algebra. More generally, let $A = (a_{ij})_{i,j\in I}$ be an infinite generalized Cartan matrix, such that every row (and hence column) contains only a finite number of nonzero entries. Let $\mathfrak{g}'(A) = \bigoplus_\alpha \mathfrak{g}'_\alpha$ be the associated Kac–Moody algebra. We denote by $\overline{\mathfrak{g}}(A)$ the subspace of $\prod_\alpha \mathfrak{g}'_\alpha$ consisting of the expressions of the form $u = \sum_\alpha a_\alpha e_\alpha$, where $a_\alpha \in \mathbf{C}$, $e_\alpha \in \mathfrak{g}'_\alpha$, and such that the set $\{j | j = \operatorname{ht}\alpha \text{ with } a_\alpha \neq 0\}$ is finite. It is clear that we can extend the bracket from $\mathfrak{g}'(A)$ to $\overline{\mathfrak{g}}(A)$ by linearity. The Lie algebra $\overline{\mathfrak{g}}(A)$ contains $\overline{\mathfrak{h}} := \prod_{i\in I} \mathbf{C}\alpha_i^\vee$, the completed Cartan subalgebra.

Denote by $\overline{gl}_\infty$ the Lie algebra of all complex matrices $(a_{ij})_{i,j\in\mathbf{Z}}$, such that $a_{ij} = 0$ for $|i-j|$ sufficiently large, with the usual bracket. It acts in a usual way on the vector space $\mathbf{C}^\infty$ of columns $(a_i)_{i\in\mathbf{Z}}$, where all but a finite number of the a_i are zero.

If A is an infinite affine matrix of type $X_\infty = A_\infty$ (resp. B_∞, C_∞ or D_∞), we denote $\overline{\mathfrak{g}}(A)$ by $\overline{x}_\infty$. Then, clearly, $\overline{x}_\infty$ is isomorphic to $\overline{gl}_\infty$ (resp. the subalgebra of $\overline{gl}_\infty$, which consists of matrices preserving the bilinear form B, C or D). *A completed infinite rank affine algebra,* denoted by x_∞, is the central extension of $\overline{x}_\infty$ defined as follows.

The Lie algebra $\overline{a}_\infty$ has a 2-cocycle ψ defined by:

$$\psi(E_{ij}, E_{ji}) = 1 = -\psi(E_{ji}, E_{ij}) \text{ if } i \leq 0 \text{ and } j \geq 1,$$
$$\psi(E_{ij}, E_{mn}) = 0 \text{ otherwise.} \tag{7.12.1}$$

One easily checks that this is a cocycle (see Exercise 7.17). Then if $A = A_\infty$ (resp. B_∞, C_∞ or D_∞), we put $r = 1$ (resp. $\frac{1}{2}$, 1 or $\frac{1}{2}$), and let $x_\infty = \overline{x}_\infty \oplus \mathbb{C}K$ be the Lie algebra with the following bracket:

$$[a \oplus \lambda K, b \oplus \mu K] = (ab - ba) \oplus r\psi(a,b)K \quad (a, b \in \overline{x}_\infty;\ \lambda, \mu \in \mathbb{C}).$$

The elements $e_i, f_i \in \mathfrak{g}'(X_\infty) \subset x_\infty$ (defined above) are called *Chevalley generators* of x_∞, and $\tilde{\mathfrak{h}} = \overline{\mathfrak{h}} + \mathbb{C}K$ is called the *Cartan subalgebra.* The elements of the set $\tilde{\Pi}^\vee = \{\tilde{\alpha}_0^\vee = \alpha_0^\vee + K, \tilde{\alpha}_i^\vee = \alpha_i^\vee \text{ for } i \neq 0\}$ are called *simple coroots.* It is easy to see that we have the usual relations:

$$[\tilde{\alpha}_i^\vee, \tilde{\alpha}_j^\vee] = 0,\ [e_i, f_j] = \delta_{ij}\tilde{\alpha}_i^\vee,\ [\tilde{\alpha}_i^\vee, e_j] = a_{ij}e_j,$$
$$[\tilde{\alpha}_i, f_j] = -a_{ij}f_j,\ (\operatorname{ad} e_i)^{1-a_{ij}}e_j = 0,\ (\operatorname{ad} f_i)^{1-a_{ij}}f_j = 0.$$

Remark 7.12. The Chevalley generators $\{e_i, f_i\}_{i \in I}$ generate a subalgebra $(\bigoplus_{i\in I} \mathbb{C}\tilde{\alpha}_i^\vee) \oplus (\bigoplus_{\alpha \in \Delta} \mathfrak{g}_\alpha)$ of x_∞, which is isomorphic to $\mathfrak{g}'(A)$. This follows from Corollary 5.11b). Note that the principal gradation of $\mathfrak{g}'(A)$ extends in a natural way to a gradation of x_∞, called its *principal* gradation.

We also have the following expressions for the *canonical central element*:

$$a_\infty : \quad K = \sum_{i \in \mathbb{Z}} \tilde{\alpha}_i^\vee;$$
$$b_\infty : \quad K = \tilde{\alpha}_0^\vee + 2\sum_{i \geq 1} \tilde{\alpha}_i^\vee;$$
$$c_\infty : \quad K = \sum_{i \geq 0} \tilde{\alpha}_i^\vee;$$
$$d_\infty : \quad K = \tilde{\alpha}_0^\vee + \tilde{\alpha}_1^\vee + 2\sum_{i \geq 2} \tilde{\alpha}_i^\vee.$$

§7.13. Exercises.

7.1. Let $\mathfrak{p}$ be a Lie algebra, $\mathfrak{h} \subset \mathfrak{p}$ a finite-dimensional diagonalizable subalgebra, $\mathfrak{p} = \bigoplus_\alpha \mathfrak{p}_\alpha$ the root space decomposition. Show that every $\mathbf{C}$-valued 2-cocycle ψ on $\mathfrak{p}$ is equivalent to a cocycle ψ_0 (i.e., $\psi(x,y) - \psi_0(x,y) = f([x,y])$ for some $f \in \mathfrak{p}^*$ and all $x, y \in \mathfrak{p}$) such that $\psi_0(\mathfrak{p}_\alpha, \mathfrak{p}_\beta) = 0$ for $\alpha + \beta \neq 0$.

[$\mathfrak{p}$ operates on the space of all 2-cocycles in a natural way, so that $\mathfrak{h}$ is diagonalizable. Show that an eigenvector with a nonzero eigenvalue is equivalent to 0.]

7.2. Let $\mathfrak{p}$ be a Lie algebra with an invariant symmetric bilinear $\mathbf{C}$-valued form $(.|.)$ and let d be a derivation of $\mathfrak{p}$ such that $(d(x)|y) = -(x|d(y))$, x, $y \in \mathfrak{p}$. Show that $\psi(x,y) := (d(x)|y)$ is a 2-cocycle on $\mathfrak{p}$. Let $\hat{\mathfrak{p}}'$ be the corresponding central extension. Show that d can be lifted to a derivation of the Lie algebra $\hat{\mathfrak{p}}'$, so that we obtain the Lie algebra $\hat{\mathfrak{p}} = \hat{\mathfrak{p}}' + \mathbf{C}d$.

7.3. Let $\mathfrak{p}$ be a Lie algebra with a nondegenerate invariant bilinear $\mathbf{C}$-valued form $(.|.)$. Let R be a commutative associative $\mathbf{C}$-algebra with unity, and f a linear functional on R. Extend the form $(.|.)$ to the complex Lie algebra $\overline{\mathfrak{p}} = R \otimes_{\mathbf{C}} \mathfrak{p}$ by

$$(r_1 \otimes p_1 | r_2 \otimes p_2) = f(r_1 r_2)(p_1|p_2).$$

Let a be a derivation of R; extend it to a derivation $D = a \otimes 1$ of $\overline{\mathfrak{p}}$. Show that

$$(D(x)|y) = -(x|D(y))$$

if and only if $f|_{a(R)} = 0$. Apply the construction of Exercise 7.2 to $\overline{\mathfrak{p}}$ with $R = \mathcal{L}$, $\mathfrak{p} = \mathfrak{g}$ a simple finite-dimensional Lie algebra, $f(r) = \operatorname{Res} r$, $a = \frac{d}{dt}$, and show that one obtains a nontwisted affine Lie algebra $\hat{\mathcal{L}}(\mathfrak{g})$.

7.4. Let R be as in Exercise 7.3, and let $\mathfrak{g}$ be a Lie algebra. Show that

$$\operatorname{der}(R \otimes \mathfrak{g}) = (\operatorname{der} R) \otimes I_{\mathfrak{g}} + R \otimes \operatorname{der} \mathfrak{g}.$$

[Choose a basis r_i of R and write: $d(1 \otimes x) = \sum_i r_i \otimes d_i(x)$, $x \in \mathfrak{g}$. Show that $d_i \in \operatorname{der} \mathfrak{g}$. Replacing d by $d - \sum_i r_i \otimes d_i$, we can assume that $d(1 \otimes x) = 0$, $x \in \mathfrak{g}$. Replacing $\mathfrak{g}$ by its associative envelope, we have $d(P \otimes x) = d(P \otimes 1)(1 \otimes x) = (1 \otimes x)d(P \otimes 1)$. Deduce that $d \in (\operatorname{der} R) \otimes 1$.]

7.5. Deduce from Exercise 7.4 that for a finite-dimensional simple Lie algebra $\mathfrak{g}$ one has:

$$\operatorname{der} \mathcal{L}(\mathfrak{g}) = \mathfrak{d} + \operatorname{ad} \mathcal{L}(\mathfrak{g}).$$

7.6. Let $\mathfrak{p}$ be a Lie algebra with a nondegenerate invariant bilinear form $(.|.)$ and let $u_1, u_2, \dots$ be its orthonormal basis. Let ψ be a $\mathbb{C}$-valued 2-cocycle on $\mathfrak{p}$ such that for each i only a finite number of $\psi(u_i, u_j)$ are nonzero. Show that ψ is of the form described in Exercise 7.2.

7.7. Prove that every 2-cocycle on the Lie algebra $\mathcal{L}(\mathfrak{g})$ is equivalent to a cocycle $\lambda\psi$, where $\lambda \in \mathbb{C}$, and ψ is described in §7.2.

[Use the action of d_0 on the space of 2-cocycles and apply Exercises 7.6, 7.4, 7.3.]

7.8. Show that $\tilde{\mathcal{L}}(\mathfrak{g})$ is the universal central extension of the Lie algebra $\mathcal{L}(\mathfrak{g})$ (cf. Exercise 3.14).

[Use Exercise 7.7.]

7.9. Show that a $\mathbb{C}$-valued 2-cocycle on the complex Lie algebra $\mathbb{C}[t_1, t_1^{-1}, \dots, t_s, t_s^{-1}] \otimes_{\mathbb{C}} \mathfrak{g}$, where $\mathfrak{g}$ is a finite-dimensional simple Lie algebra, is equivalent to a cocycle α_D, where $D = \sum_i P_i \frac{\partial}{\partial t_i}$, with $\sum_i \frac{\partial P_i}{\partial t_i} = 0$, defined by

$$\alpha_D(P \otimes x, Q \otimes y) = (\text{coefficient at } (t_1 \dots t_s)^{-1} \text{ in } D(P)Q)(x|y).$$

7.10. Consider the bilinear form

$$(x|y) = \operatorname{tr}(xy) \text{ if } \mathfrak{g} = s\ell_n \text{ or } sp_n, \text{ and}$$
$$(x|y) = \frac{1}{2}\operatorname{tr}(xy) \text{ if } \mathfrak{g} = so_n,\ n > 4.$$

Check that $|\theta|^2 = 2$, so that the "normalized" cocycle ψ on $\mathcal{L}(\mathfrak{g})$ is given by

$$\psi(a,b) = \operatorname{Res}\operatorname{tr}\left(\frac{da}{dt}b\right) \text{ if } \mathfrak{g} = s\ell_n \text{ or } sp_n,$$
$$\psi(a,b) = \frac{1}{2}\operatorname{Res}\operatorname{tr}\left(\frac{da}{dt}b\right) \text{ if } \mathfrak{g} = so_n,\ n > 4.$$

Show that the normalized standard bilinear form on $\mathcal{L}(\mathfrak{g})$ is given by

$$(a|b) = \text{constant term of } \operatorname{tr}(ab) \text{ if } \mathfrak{g} = s\ell_n \text{ or } sp_n,$$
$$(a|b) = \frac{1}{2} \text{ constant term of } \operatorname{tr}(ab) \text{ if } \mathfrak{g} = so_n,\ n > 4.$$

7.11. Let $\mathring{\mathfrak{g}} = s\ell_{\ell+1}(\mathbb{C})$. We keep all the notation of Exercise 1.13, and add $^\circ$ on the top: $\mathring{\mathfrak{g}}$, $\mathring{\mathfrak{h}}$, $\mathring{\Pi}$, and $\mathring{\Pi}^\vee$. We have $\mathcal{L}(\mathring{\mathfrak{g}}) = s\ell_{\ell+1}(\mathcal{L})$. Show that the space $\hat{\mathcal{L}}(\mathring{\mathfrak{g}}) = s\ell_{\ell+1}(\mathcal{L}) \oplus \mathbf{C}K \oplus \mathbf{C}d$ with the bracket

$$[a(t) + \lambda K + \mu d, a_1(t) + \lambda_1 K + \mu_1 d]$$
$$= \left(a(t)a_1(t) - a_1(t)a(t) + \mu t\frac{da_1(t)}{dt} - \mu_1 t\frac{da(t)}{dt}\right) + \operatorname{Res}\operatorname{tr}\left(\frac{da(t)}{dt}a_1(t)\right)K,$$

is the affine algebra of type $A_\ell^{(1)}$. Set $\mathfrak{h} = \mathring{\mathfrak{h}} \oplus \mathbf{C}K \oplus \mathbf{C}d$, extend ϵ_i from $\mathring{\mathfrak{h}}$ to $\mathfrak{h}$ by $\epsilon_i(K) = \epsilon_i(d) = 0$, and define $\delta \in \mathfrak{h}^*$ as in §7.4. Set

$$e_0 = tE_{\ell+1,1}, \qquad f_0 = t^{-1}E_{1,\ell+1}, \qquad \alpha_0^\vee = K - \sum_{i=1}^{\ell} \alpha_i^\vee .$$

Set $\Pi = \{\alpha_0, \mathring{\Pi}\}$, $\Pi^\vee = \{\alpha_0^\vee, \mathring{\Pi}^\vee\}$. Show that $(\mathfrak{h}, \Pi, \Pi^\vee)$ is the realization of the matrix (of type $A_\ell^{(1)}$):

$$A = \begin{pmatrix} 2 & -2 \\ -2 & 2 \end{pmatrix} \text{ if } \ell = 1,$$

and

$$A = \begin{pmatrix} 2 & -1 & 0 & \dots & 0 & -1 \\ -1 & 2 & -1 & \dots & 0 & 0 \\ 0 & -1 & 2 & \dots & 0 & 0 \\ \dots & \dots & \dots & \dots & \dots & \dots \\ 0 & 0 & 0 & \dots & 2 & -1 \\ -1 & 0 & 0 & \dots & -1 & 2 \end{pmatrix} \text{ if } \ell > 1.$$

The root space decomposition of $\hat{\mathcal{L}}(\mathring{\mathfrak{g}})$ is

$$\hat{\mathcal{L}}(\mathring{\mathfrak{g}}) = \mathfrak{h} \oplus \Big(\bigoplus_{\substack{i,j=1,\dots,\ell+1 \\ i \neq j,\ s \in \mathbf{Z}}} t^s E_{ij} \Big) \oplus \Big(\bigoplus_{\substack{s \in \mathbf{Z} \\ s \neq 0}} t^s \mathring{\mathfrak{h}} \Big).$$

Show that $\hat{\mathcal{L}}(\mathring{\mathfrak{g}}) = \mathfrak{g}(A)$, with the Chevalley generators e_i, f_i $(i = 0, \dots, \ell)$. Show that the set $\{\epsilon_i - \epsilon_j + s\delta \ (i \neq j,\ s \in \mathbf{Z});\ s\delta \ (s \in \mathbf{Z}\backslash 0)\}$ is the root system of $\mathfrak{g}(A)$. Show that its subalgebra $\mathfrak{n}_-$ (resp. $\mathfrak{n}_+$) consists of all matrices from $s\ell_{\ell+1}(\mathbf{C}[t])$ (resp. $s\ell_{\ell+1}(\mathbf{C}[t^{-1}])$) such that the entries on and under (resp. over) the diagonal are divisible by t (resp. t^{-1}). Show that the Chevalley involution ω of $\hat{\mathcal{L}}(\mathring{\mathfrak{g}})$ is defined by

$$\omega(a(t)) = -(\text{transpose of } a(t^{-1})) \text{ if } a(t) \in \mathcal{L}(\mathring{\mathfrak{g}}),$$
$$\omega(K) = -K, \qquad \omega(d) = -d.$$

7.12. Introduce the following basis of the Lie algebra $s\ell_2(\mathcal{L})$:

$$L_{3s} = \frac{1}{2}\begin{pmatrix} t^s & 0 \\ 0 & -t^s \end{pmatrix}, \quad L_{3s+1} = \frac{1}{2}\begin{pmatrix} 0 & t^s \\ 0 & 0 \end{pmatrix},$$
$$L_{3s-1} = \begin{pmatrix} 0 & 0 \\ t^s & 0 \end{pmatrix} \quad (s \in \mathbb{Z}).$$

Then $[L_i, L_j] = c_{ij} L_{i+j}$ $(i, j \in \mathbb{Z})$, where

$$c_{ij} = -1,\ 0, \text{ or } 1 \text{ according as } j - i \equiv -1,\ 0, \text{ or } 1 \mod 3.$$

7.13. Show that the Lie algebra $\mathfrak{d}$ does not admit a nonzero invariant bilinear form. Deduce from Exercise 7.1 that every C-valued 2-cocycle on $\mathfrak{d}$ is equivalent to a cocycle ψ such that $\psi(d_i, d_j) = 0$ whenever $i + j \neq 0$. Denote by C_2° the vector space (over $\mathbb{C}$) of such cocycles. Show that $\dim C_2^\circ = 2$ and that the cocycles ψ_1 and ψ_2, defined by

$$\psi_1(d_i, d_j) = \delta_{i,-j}\, j; \qquad \psi_2(d_i, d_j) = \delta_{i,-j}\, j^3$$

form a basis of C_2°. Deduce that Vir is the universal central extension of $\mathfrak{d}$.

7.14. Let E_γ $(\gamma \in \Delta)$, $\alpha_i^\vee$ $(i = 1, \ldots, \ell)$ be a Chevalley basis of a finite-dimensional simple Lie algebra $\mathfrak{g}$, i.e.,

$$[E_\gamma, E_{-\gamma}] = \gamma^\vee; \quad [\gamma^\vee, \gamma'^\vee] = 0; \quad [\gamma^\vee, E_\beta] = \langle \beta, \gamma^\vee \rangle E_\beta;$$
$$[E_\beta, E_\gamma] = n_{\beta,\gamma} E_{\beta+\gamma}, \quad \text{where } n_{\beta,\gamma} \in \mathbb{Z},\ n_{\beta,\gamma} = -n_{-\gamma,-\beta}.$$

(cf. §7.8 and §7.9 for an explicit construction of such a basis.) Then the elements

$$E_{\gamma+k\delta} = t^k \otimes E_\gamma \quad (k \in \mathbb{Z},\ \gamma \in \Delta);$$
$$E_{k\delta}^{(i)} = t^k \otimes \alpha_i^\vee \quad (k \in \mathbb{Z}\setminus\{0\},\ i = 1, \ldots, \ell);$$
$$\alpha_i^\vee \quad (i = 1, \ldots, \ell); \quad k \text{ and } d,$$

form a basis of $\hat{\mathcal{L}}(\mathfrak{g})$. Show that the $\mathbb{Z}$-span of this basis is closed under the bracket, by writing down all the commutation relations. This is the *Chevalley basis* of $\hat{\mathcal{L}}(\mathfrak{g})$.

7.15. Let $\mathfrak{g}$ be a simple finite-dimensional Lie algebra and V a finite-dimensional $\mathfrak{g}$-module. Then we may in a natural way define an $\mathcal{L}(\mathfrak{g})$-module $\overline{V} = \mathcal{L} \otimes_{\mathbb{C}} V$. Fix $\lambda \in \mathbb{C}$ and define an action π of the affine algebra $\hat{\mathcal{L}}(\mathfrak{g})$ on $\overline{V}$ by:

$$\pi|_{\mathcal{L}(\mathfrak{g})} \text{ unchanged}; \quad \pi(K) = 0; \quad \pi(d) = t\frac{d}{dt} \otimes 1 + \lambda 1_V.$$

Show that this is an integrable $\hat{\mathcal{L}}(\mathfrak{g})$-module.

7.16. Let $\mathring{\mathfrak{g}}$ be a simple finite-dimensional Lie algebra, $(.|.)$ the normalized invariant form, $\mathring{\mathfrak{h}}$ a Cartan subalgebra, $\mathring{\Delta}_+$ a system of positive roots, $\mathring{\rho}$ their half-sum, θ the highest root. Choose a basis $u_1, \dots, u_\ell$ of $\mathring{\mathfrak{h}}$ and the dual basis $u^1, \dots, u^\ell$. For each $\alpha \in \mathring{\Delta}_+$ choose root vectors e_α and $e_{-\alpha}$ such that $(e_\alpha | e_{-\alpha}) = 1$. Let $\mathring{\Omega}$ be the Casimir operator of $\mathring{\mathfrak{g}}$. Finally, let $\hat{\mathcal{L}}(\mathring{\mathfrak{g}})$ be the nontwisted affine Lie algebra associated to $\mathring{\mathfrak{g}}$, and set $\rho = h^\vee \Lambda_0 + \mathring{\rho}$ (where $h^\vee$ is the dual Coxeter number of $\hat{\mathcal{L}}(\mathring{\mathfrak{g}})$). Show that

$$\langle \rho, \alpha_i^\vee \rangle = 1 \quad (i = 0, \dots, \ell)$$

and that the Casimir operator Ω for the affine algebra $\hat{\mathcal{L}}(\mathring{\mathfrak{g}})$ (see Chapter 2) can be written as follows:

$$\Omega = 2d(K + h^\vee) + \mathring{\Omega}$$
$$+ 2 \sum_{n \geq 1} \{ \sum_{\alpha \in \mathring{\Delta}} (t^{-n} \otimes e_{-\alpha})(t^n \otimes e_\alpha) + \sum_{i=1}^{\ell} (t^{-n} \otimes u^i)(t^n \otimes u_i) \}.$$

7.17. Let $L_n = \sum_{i=1}^{n} \mathbb{Z} v_i$ be the standard lattice in the Eucildean space $\mathbb{R}^n$. Define on $L_n \times L_n$ a bimultiplicative function ε by letting

$$\varepsilon(v_j, v_k) = 1 \text{ if } j \leq k, \text{ and } = -1 \text{ if } j > k.$$

Show that by restricting to root lattices $Q(A_\ell) \subset L_{\ell+1}$ and $Q(D_\ell) \subset L_\ell$ (described in §6.7), we obtain an asymmetry function.

7.18. Let $\frac{1}{2} L_n = \sum_{i=1}^{n} \mathbb{Z} u_i$, where $u_i = \frac{1}{2} v_i$, and define on $(\frac{1}{2} L_n) \times (\frac{1}{2} L_n)$ a bimultiplicative function ε (extending the one from Exercise 7.17) by:

$$\varepsilon(u_j, u_k) = 1 \text{ if } j \leq k, \text{ and } = e^{\frac{\pi i}{4}} \text{ if } j > k.$$

Show that by restricting to the root lattices $Q(E_8)$ and $Q(E_7)$ $(\subset \frac{1}{2} L_8)$, we obtain an asymmetry function. Similarly, taking the lattice $\frac{1}{2} L_6 + \mathbb{Z} u$ in $\mathbb{R}^7$, where $u = \frac{1}{\sqrt{2}} v_7$, and extending ε by $\varepsilon(u, u_j) = \varepsilon(u, u) = e^{\frac{\pi i}{4}}, \varepsilon(u_j, u_7) = 1$, we obtain an asymmetry function for $Q(E_6)$.

7.19. Let $\mathfrak{p}$ be a subalgebra of $\mathcal{L}(\mathfrak{g})$ of finite codimension, where $\mathfrak{g}$ is a simple finite-dimensional Lie algebra. Then there exists a nonzero $Q \in \mathcal{L}$ such that $\mathfrak{p} \supset Q\mathcal{L} \otimes_{\mathbb{C}} \mathfrak{g}$.
[Show that for a subspace V of finite codimension in $\mathcal{L}$, the space V^2 contains a nonzero ideal of $\mathcal{L}$.]

7.20. Show that the adjoint representation of the affine Lie algebra $\mathfrak{g} = sl_n(\mathcal{L}) + \mathbf{C}K + \mathbf{C}d$ induces the adjoint representation of the loop group $G = SL_n(\mathcal{L})$ on $\mathfrak{g}$ given by the following formula:

$$\begin{aligned}&(\operatorname{Ad} a)(x + \lambda K + \mu d)\\ &\quad = (axa^{-1} - \mu t a' a^{-1}) + (\lambda + \operatorname{Res} \operatorname{tr}(a' x a^{-1} - \frac{1}{2}\mu t (a' a^{-1})^2) K + \mu d),\end{aligned}$$

where $a \in G$, a' denotes the derivative of a by t, $x \in sl_n(\mathcal{L})$, and $\lambda, \mu \in \mathbf{C}$. [Use the following facts: a) $\operatorname{Ad} a$ preserves $(.|.)$; b) $[a, \frac{d}{dt}] = -a'$; c) $(\operatorname{Ad} a)x \equiv axa^{-1} \mod \mathbf{C}K$; d) $(\operatorname{Ad} a)K = K$.]

7.21. Let $\mathring{G}$ be a connected simply-connected complex algebraic group operating faithfully on a finite-dimensional complex vector space $\mathring{V}$. This action extends to the action of the loop group $G := \mathring{G}(\mathcal{L})$ on $V := \mathcal{L} \otimes_{\mathbf{C}} \mathring{V}$. Let $\mathring{\mathfrak{g}}$ be the Lie algebra $\mathring{G}$, $\mathring{\mathfrak{h}}$ a Cartan subalgebra, $\mathring{Q}^\vee \subset \mathfrak{h}$ the dual root lattice; let $V = \bigoplus_\lambda V_\lambda$ be the weight space decomposition with respect to $\mathring{\mathfrak{h}}$. Given $\gamma \in \mathring{Q}^\vee$, define $t^\gamma \in \operatorname{End}_{\mathcal{L}} V$ by:

$$t^\gamma(v_\lambda) = t^{\langle \lambda, \gamma \rangle} v_\lambda.$$

Show that $t^\gamma \in G$, thus giving an injective homomorphism $f : \mathring{Q}^\vee \to G$. Show that, via the adjoint action of G on $\hat{\mathcal{L}}(\mathring{\mathfrak{g}})$, we have:

$$f(t^\gamma) \cdot (t^k \otimes e_\alpha) = t^{k + \langle \alpha, \gamma \rangle} \otimes e_\alpha \ (e_\alpha \in \mathring{\mathfrak{g}}_\alpha); \qquad f(t^\gamma)|_{\mathfrak{h}} = t_{-\gamma}.$$

(Thus, we obtain a canonical embedding of the group of translations T into G.) Similarly, define an injective homomorphism $f : \mathring{P}^\vee \to \operatorname{Ad} G$ (where $\mathring{P}^\vee = \{h \in \mathring{\mathfrak{h}} \mid \langle \alpha, h \rangle \in \mathbf{Z} \text{ for all } \alpha \in \mathring{\Delta}\}$). Show that the group $\widetilde{T} := f(\mathring{P}^\vee)$ acts on $\tilde{\mathcal{L}}(\mathring{\mathfrak{g}})$ by the same formulas. Show that the group $\widetilde{T}$ normalizes W and that $\widetilde{W} := W \ltimes \widetilde{T}$ preserves $Q^\vee$ and Q. Show that the group $\widetilde{W}_+ = \{w \in \widetilde{W} \mid w(\Delta_+) \subset \Delta_+\}$ acts simply transitively on $(\operatorname{Aut} A) \cdot \alpha_0^\vee$.

7.22. Let $\mathfrak{g}$ be a simple finite-dimensional Lie algebra and let R be a commutative associative algebra with unity. Let Ω_R^1 denote the space of all formal differentials over R, i.e., expressions of the form fdg, where $f, g \in R$, with relation $d(fg) = fdg + gdf$. Then the universal central extension of the complex Lie algebra $R \otimes_{\mathbf{C}} \mathfrak{g}$ is

$$0 \to \Omega_R^1/dR \to \tilde{\mathfrak{g}}_R := \mathfrak{g}_R \oplus (\Omega_R^1/dR) \to \mathfrak{g}_R \to 0,$$

where the bracket on $\tilde{\mathfrak{g}}_R$ is defined by:

$$[r_1 \otimes g_1, r_2 \otimes g_2] = r_1 r_2 \otimes [g_1, g_2] + (g_1|g_2) r_1 dr_2 \mod dR$$

(This is a generalization of Exercise 7.9.)

7.23. Let $T(z) = \sum_{n \in \mathbb{Z}} d_n z^{-n-2}$ be the generating series for Vir. Show that $[d_m, T(z)] = z^m(z\frac{d}{dz} + 2(m+1))T(z) + \frac{m^3-m}{12} z^{m-2} c$. ($T(z)$ is called the energy-momentum tensor. It is a primary field if and only if the "conformal anomaly" c is 0.)

7.24. Consider the representation of the Lie algebra $\mathfrak{d}$ (= representation of Vir with $c = 0$) on the space $U_{\alpha,\beta}$ of "densities" of the form $P(t)t^\alpha (dt)^\beta$, where α, β are some numbers and $P(t) \in \mathcal{L}$. Show that in the basis $v_k = t^{k+\alpha}(dt)^\beta$ this representation looks as follows:

$$d_n(v_k) = -(k + \alpha + \beta + \beta n) v_{n+k}.$$

Show that $F(z) = \sum_{k \in \mathbb{Z}} F_k z^{-k-\Delta}$, where $F_k : V \to V_1$, is a primary field of conformal weight Δ if and only if the map $F : U_{\Delta-1,1-\Delta} \otimes V \to V_1$ defined by $F(v_k \otimes v) = F_k(v)$ is a Vir-module homomorphism.

7.25. Let $\mathfrak{g}$ be a simply-laced simple Lie algebra as constructed in §7.8. Given $\alpha \in \Delta$, let

$$R_\alpha = (\exp - \operatorname{ad} E_{-\alpha})(\exp - \operatorname{ad} E_\alpha)(\exp - \operatorname{ad} E_{-\alpha}).$$

Show that $R_\alpha|_{\mathfrak{h}} = r_\alpha$, $R_\alpha(E_\beta) = -\varepsilon(\alpha,\beta) E_{r_\alpha(\beta)}$ if $(\alpha|\beta) \neq 0$ and $R_\alpha(E_\beta) = E_\beta$ otherwise.

7.26. Let $\varepsilon(\alpha, \beta)$ be a 2-cocycle on Q with values in $\{\pm 1\}$, i.e.,

$$\varepsilon(\alpha,\beta)\varepsilon(\alpha+\beta,\gamma) = \varepsilon(\beta,\gamma)\varepsilon(\alpha,\beta+\gamma) \text{ for } \alpha,\beta \in Q,$$
$$\varepsilon(0,0) = 1 \text{ (and hence } \varepsilon(0,\alpha) = \varepsilon(\alpha,0) = 1 \text{ for } \alpha \in Q);$$

suppose that in addition

$$\varepsilon(\alpha,-\alpha) = (-1)^{\frac{1}{2}(\alpha|\alpha)} \text{ and } \varepsilon(\alpha,\beta)\varepsilon(\beta,\alpha) = (-1)^{(\alpha|\beta)}, \; \alpha,\beta \in Q.$$

Then Proposition 7.8 still holds.

7.27. Let Q be the root lattice of a simply laced simple Lie algebra. Let $B : Q \times Q \longrightarrow \mathbb{Z}$ be a bilinear form such that

$$B(\alpha, \beta) + B(\beta, \alpha) = (\alpha|\beta).$$

Then $\varepsilon(\alpha, \beta) = (-1)^{B(\alpha,\beta)}$ is an asymmetry function. Write $B(\alpha, \beta) = (R\alpha|\beta)$. Then the conditions on B are equivalent to $R(Q) \subset Q^*$ and $R + R^* = I$. Show that if $w \in \operatorname{Aut} Q$ is such that $Q \subset (1-w)Q^*$, then $R = 1-w$ satisfies these conditions. Show that the Coxeter element has the above property.

7.28. Let $V = V_- \bigoplus V_+$ be a vector space over $\mathbb{C}$ (in general infinite dimensional) represented as a direct sum of two subspaces. Let $\bar{gl}_*(V)$ denote the Lie algebra of endomorphisms of V which have the form $a = \begin{pmatrix} a_1 & a_2 \\ a_3 & a_4 \end{pmatrix}$ with respect to the above decomposition, where $a_3 : V_- \longrightarrow V_+$ has a finite rank. Show that

$$f(a, b) := tr_{V_-} a_2 b_3 - tr_{V_-} b_2 a_3$$

is a 2-cocycle on the Lie algebra $\bar{gl}_*(V)$. Show that the restriction of this cocycle to the subalgebra of finite rank endomorphisms is trivial. Show that for $V = \bigoplus_{i \in \mathbb{Z}} \mathbb{C}v_i, V_- = \sum_{i>0} \mathbb{C}v_i, V_+ = \sum_{i \le 0} \mathbb{C}v_i$, the cocycle f restricted to $\bar{gl}_\infty$ coincides with the cocycle ψ defined in §7.12.

7.29. Let $V = \mathbb{C}[t, t^{-1}]^n$ be the natural $s\ell_n(\mathbb{C}[t, t^{-1}])$-module. This extends to the module over the Lie algebra of differential operators $\mathfrak{a} = s\ell_n(\mathbb{C}[t, t^{-1}, \frac{d}{dt}])$. Set $V_+ = (t\mathbb{C}[t])^n, V_- = (\mathbb{C}[t^{-1}]^n \subset V$. Show that the restriction of the cocycle f of Exercise 7.28 to the subalgebra $\mathfrak{a} \subset \bar{gl}_*(V)$ is given by the following formula ($\ell \in \mathbb{Z}_+; m, m' \in \mathbb{Z}; a, a' \in g\ell_n(\mathbb{C})$):

$$f\left(t^{\ell+m}\left(\frac{d}{dt}\right)^{\ell} \otimes a, t^{\ell'+m'}\left(\frac{d}{dt}\right)^{\ell'} \otimes a'\right) = \delta_{m,-m'}(tr\, aa')(-1)^{\ell}\ell!\ell'!\binom{m+\ell}{\ell+\ell'+1}.$$

§7.14. Bibliographical notes and comments.

Except for the explicit formula for the central extension, the realization of nontwisted affine Lie algebras was given by Kac [1968 B] and Moody [1969]. The formula for the cocycle has been known to physicists for such a long time that it is difficult to trace the original source. Proposition 7.8 (in the form given by Exercise 7.26) is due to Frenkel–Kac [1980].

The Virasoro algebra (first studied by Virasoro [1970]) plays a prominent role in the dual strings theory (see, e.g., Mandelstam [1974], Schwartz [1973], [1982]). Mathematicians started to develop a representation theory of the Virasoro algebra quite recently (Kac [1978 B], [1979], [1982 B],

Frenkel–Kac [1980], Segal [1981], Feigin–Fuchs [1982], [1983 A, B], [1984 A, B], and others). A survey of some of these results may be found in the book Kac–Raina [1987].

This has recently become a topic of interest among physicists in connection with statistical mechanics and a revival of the dual strings theory. The conformally invariant field theory, which is the foundation of both of these remarkable developments, originated in papers by Belavin–Polyakov–Zamolodchikov [1984 A, B]. The notion of a primary field, briefly discussed in §7.7, is a key notion of this theory. A survey of recent developments in string theory may be found in the two volume book by Green–Schwartz–Witten [1987].

Exercise 7.2 is due to Kupershmidt [1984] and Zuckerman (unpublished). The first published proof of Exercises 7.7 and 7.8 that I know is in Garland [1980]. Exercise 7.14 is due to Garland [1978]. In this paper Garland studies in detail the $\mathbb{Z}$-form of the universal enveloping algebra of an affine Lie algebra. Exercises 7.28 and 7.29 are taken from Kac–Peterson [1981]. Exercise 7.19 is due to R. Coley [1983]. Exercise 7.20 is taken from Frenkel [1984], Segal [1981], and Kac–Peterson [1984 B]. Exercise 7.22 is due to Kassel [1984]. The rest of the material of Chapter 7 is fairly standard.

There has been recently a number of papers dealing with the groups associated to affine algebras. Such a group is a central extension by $\mathbb{C}^\times$ of the group of polynomial (or analytic, etc.) maps of $\mathbb{C}^\times$ to a complex simple finite-dimensional Lie group. The corresponding "compact form" is a central extension by a circle of the group of polynomial (or analytic, or C^∞, etc.) loops on a connected simply-connected compact Lie group. Thus, there is a whole range of groups associated to an affine Lie algebra (or rather a certain completion of it). The group of polynomial maps is, naturally, the minimal associated group; this is a special case of groups discussed in §3.15. Various aspects of the theory of loop groups may be found in Garland [1980], Frenkel [1984], Pressley [1980], Segal [1981], Atiyah–Pressley [1983], Goodman–Wallach [1984 A], Kac–Peterson [1984 B], [1985 B], [1987], Freed [1985], Pressley–Segal [1986], Mitchell [1987], [1988], Kazhdan–Lusztig [1988], Mickelsson [1987], and others.

Chapter 8. Twisted Affine Algebras and Finite Order Automorphisms

§8.0. Here we describe a realization of the remaining, "twisted" affine algebras. This turns out to be closely related to the Lie algebra of equivariant polynomial maps from $\mathbb{C}^\times$ to a simple finite-dimensional Lie algebra with the action of a finite cyclic group. As a side result of this construction we deduce a nice description of the finite order automorphisms of a simple finite-dimensional Lie algebra, and in particular the classification of symmetric spaces.

§8.1. Let $\mathfrak{g}$ be a simple finite-dimensional Lie algebra and let σ be an automorphism of $\mathfrak{g}$ satisfying $\sigma^m = 1$ for a positive integer m. Set $\epsilon = \exp \frac{2\pi i}{m}$. Then each eigenvalue of σ has the form ϵ^j, $j \in \mathbb{Z}/m\mathbb{Z}$, and since σ is diagonalizable, we have the decomposition

$$\mathfrak{g} = \bigoplus_{j \in \mathbb{Z}/m\mathbb{Z}} \mathfrak{g}_j, \tag{8.1.1}$$

where $\mathfrak{g}_j$ is the eigenspace of σ for the eigenvalue ϵ^j. Clearly, (8.1.1) is a $\mathbb{Z}/m\mathbb{Z}$-gradation of $\mathfrak{g}$. Conversely, if a $\mathbb{Z}/m\mathbb{Z}$-gradation (8.1.1) is given, the linear transformation of $\mathfrak{g}$ given by multiplying the vectors of $\mathfrak{g}_j$ by ϵ^j is an automorphism σ of $\mathfrak{g}$ which satisfies $\sigma^m = 1$.

Let $\overline{\mathfrak{h}}_0$ be a maximal ad-diagonalizable subalgebra of the Lie algebra $\mathfrak{g}_{\overline{0}}$. We first prove the following:

LEMMA 8.1. a) *Let $(.|.)$ be a nondegenerate invariant bilinear form on $\mathfrak{g}$. Then: $(\mathfrak{g}_i|\mathfrak{g}_j) = 0$ if $i + j \not\equiv 0 \mod m$, and $\mathfrak{g}_i$ and $\mathfrak{g}_j$ are nondegenerately paired if $i + j \equiv 0 \mod m$.*
b) *The centralizer $\mathfrak{z}$ of $\mathfrak{h}_{\overline{0}}$ in $\mathfrak{g}$ is a Cartan subalgebra of $\mathfrak{g}$.*
c) *$\mathfrak{g}_{\overline{0}}$ is a reductive subalgebra of $\mathfrak{g}$.*

Proof. Given $x \in \mathfrak{g}_i$, $y \in \mathfrak{g}_j$, we have $(x|y) = (\sigma(x)|\sigma(y)) = \epsilon^{i+j}(x|y)$ (the form $(.|.)$ is Aut $\mathfrak{g}$-invariant being a multiple of the Killing form), which proves the first part of a). The second part follows since $(.|.)$ is nondegenerate.

In order to prove b), note that

$$\mathfrak{z} = \mathfrak{h} + \sum_\alpha \mathfrak{g}_\alpha,$$

where $\mathfrak{h}$ is a Cartan subalgebra of $\mathfrak{g}$ containing $\mathfrak{h}_{\bar{0}}$, $\mathfrak{g}_\alpha$ are root spaces with respect to $\mathfrak{h}$ and the summation is taken over $\alpha \in \Delta$ such that $\alpha|_{\mathfrak{h}_{\bar{0}}} = 0$. It follows that $\mathfrak{z} = \mathfrak{h} + \mathfrak{s}$, where $\mathfrak{s}$ is a σ-invariant semisimple subalgebra (= the derived subalgebra of $\mathfrak{z}$), whose intersection with $\mathfrak{g}_{\bar{0}}$ is trivial. This, in particular, proves c). Thus (8.1.1) induces a $\mathbb{Z}/m\mathbb{Z}$ gradation $\mathfrak{s} = \oplus_j \mathfrak{s}_j$ such that $\mathfrak{s}_{\bar{0}} = \{0\}$. Numbering the elements of $\mathbb{Z}/m\mathbb{Z}$ by the corresponding integers in the set $N_m = \{0, 1, \ldots, m-1\}$ and defining $\mathfrak{s}_a = \mathfrak{s}_b$ if $b \in N_m$ and $a \equiv b \mod m$, we shall prove $\mathfrak{s}_n = \mathfrak{s}_{-n} = 0$ by induction on n.

We know $\mathfrak{s}_0 = 0$; let $n > 0$ and $x \in \mathfrak{s}_n$. Then $(\operatorname{ad} x)^r \mathfrak{s}_i \subset \mathfrak{s}_{nr+i}$. Select a positive integer r such that $n(r-1) < m-i \leq nr$. Then $nr+i = m+t$ with $0 \leq t < n$, so by the inductive assumption, $\mathfrak{s}_{nr+i} = \mathfrak{s}_t = 0$. Thus $\operatorname{ad} x|_{\mathfrak{s}}$ is nilpotent; similarly, $\operatorname{ad} y|_{\mathfrak{s}}$ is nilpotent if $y \in \mathfrak{s}_{-n}$. But $[\mathfrak{s}_n, \mathfrak{s}_{-n}] \subset \mathfrak{s}_0 = 0$, so $\operatorname{ad}x$ and $\operatorname{ad}y$ commute on $\mathfrak{s}$ and by the nilpotency, $\operatorname{tr}_{\mathfrak{s}}(\operatorname{ad} x \operatorname{ad} y) = 0$. Now Lemma 8.1a) (applied to $\mathfrak{s}$) implies $\mathfrak{s} = 0$, proving b).

□

It follows from Lemma 8.1b) that $\mathfrak{h}_{\bar{0}}$ contains a regular element, say x, of $\mathfrak{g}$. Hence, the centralizer $\mathfrak{h}$ of x in $\mathfrak{g}$ is a σ-invariant Cartan subalgebra, and the sum of eigenspaces of $\operatorname{ad}x$ with positive eigenvalues (we say that $a \in \mathbb{C}$ is positive if either $\operatorname{Re} a > 0$ or $\operatorname{Re} a = 0$ and $\operatorname{Im} a > 0$) is a maximal nilpotent σ-invariant subalgebra $\mathfrak{n}_+$. Let Δ_+ be the corresponding set of positive roots. Thus, σ induces an automorphism of Δ_+; let μ be the corresponding diagram automorphism of $\mathfrak{g}$. It is clear that $\sigma\mu^{-1} = \exp(\operatorname{ad} h)$, where $h \in \mathfrak{h}_{\bar{0}}$. Since Cartan subalgebras of $\mathfrak{g}$ are conjugate, we have proved the following

PROPOSITION 8.1. *Let $\mathfrak{g}$ be a simple finite-dimensional Lie algebra, let $\mathfrak{h}$ be its Cartan subalgebra and let $\Pi = \{\alpha'_1, \ldots, \alpha'_N\}$ be a set of simple roots. Let $\sigma \in \operatorname{Aut} \mathfrak{g}$ be such that $\sigma^m = 1$. Then σ is conjugate to an automorphism of $\mathfrak{g}$ of the form*

$$\mu \exp(\operatorname{ad} \frac{2\pi i}{m} h), \quad h \in \mathfrak{h}_{\bar{0}}, \tag{8.1.2}$$

where μ is a diagram automorphism preserving $\mathfrak{h}$ and Π', $\mathfrak{h}_{\bar{0}}$ is the fixed point set of μ in $\mathfrak{h}$, and $\langle \alpha'_i, h \rangle \in \mathbb{Z}, i = 1, \ldots, N$.

□

§8.2. We associate a subalgebra $\mathcal{L}(\mathfrak{g}, \sigma, m)$ of $\mathcal{L}(\mathfrak{g})$ to the automorphism σ of $\mathfrak{g}$ as follows:

(8.2.1) $\mathcal{L}(\mathfrak{g},\sigma,m)=\bigoplus_{j\in\mathbb{Z}}\mathcal{L}(\mathfrak{g},\sigma,m)_j$, where $\mathcal{L}(\mathfrak{g},\sigma,m)_j=t^j\otimes\mathfrak{g}_{j \bmod m}$.

The decomposition (8.2.1) is clearly a $\mathbb{Z}$-gradation of $\mathcal{L}(\mathfrak{g},\sigma,m)$.

Note that $\mathcal{L}(\mathfrak{g},\sigma,m)$ is the fixed point set of the automorphism $\tilde{\sigma}$ of $\mathcal{L}(\mathfrak{g})$ defined by

$$\tilde{\sigma}(t^j\otimes x)=(\epsilon^{-j}t^j)\otimes\sigma(x),\quad (j\in\mathbb{Z},\ x\in\mathfrak{g}).$$

Hence, $\mathcal{L}(\mathfrak{g},\sigma,m)$ may be identified with the Lie algebra of equivariant maps (with respect to the action of $\mathbb{Z}/m\mathbb{Z}$):

$$(\mathbb{C}^\times;\ \text{multiplication by } \epsilon^{-1})\to(\mathfrak{g};\ \text{action of } \sigma).$$

Recall the Lie algebra $\hat{\mathcal{L}}(\mathfrak{g})=\mathcal{L}(\mathfrak{g})\oplus\mathbb{C}K'\oplus\mathbb{C}d'$ defined in §7.2 (here we write K' and d' instead of K and d for reasons which will become clear later on). We set

$$\hat{\mathcal{L}}(\mathfrak{g},\sigma,m)=\mathcal{L}(\mathfrak{g},\sigma,m)\oplus\mathbb{C}K'\oplus\mathbb{C}d'.$$

This is a subalgebra of $\hat{\mathcal{L}}(\mathfrak{g})$, which is the fixed point set of an automorphism $\hat{\sigma}$ of $\hat{\mathcal{L}}(\mathfrak{g})$ defined by $\hat{\sigma}|_{\mathcal{L}(\mathfrak{g},\sigma,m)}=\tilde{\sigma}$, $\hat{\sigma}(K')=K'$, $\hat{\sigma}(d')=d'$. Its derived Lie algebra is $\tilde{\mathcal{L}}(\mathfrak{g},\sigma,m):=\mathcal{L}(\mathfrak{g},\sigma,m)\oplus\mathbb{C}K'$. Note also that $\mathcal{L}(\mathfrak{g},1,1)=\mathcal{L}(\mathfrak{g})$, $\tilde{\mathcal{L}}(\mathfrak{g},1,1)=\tilde{\mathcal{L}}(\mathfrak{g})$ and $\hat{\mathcal{L}}(\mathfrak{g},1,1)=\hat{\mathcal{L}}(\mathfrak{g})$.

Setting $\deg K'=\deg d'=0$ together with (8.2.1) defines a $\mathbb{Z}$-gradation of $\hat{\mathcal{L}}(\mathfrak{g},\sigma,m)$:

$$\hat{\mathcal{L}}(\mathfrak{g},\sigma,m)=\bigoplus_{j\in\mathbb{Z}}\hat{\mathcal{L}}(\mathfrak{g},\sigma,m)_j. \tag{8.2.2}$$

§8.3. The structure of the Lie algebra $\hat{\mathcal{L}}(\mathfrak{g},\sigma,m)$ for arbitrary σ will be studied in the following sections.

Here we consider some very special examples, which give us an explicit construction of all *twisted* affine algebras, i.e., those listed in Tables Aff 2 and Aff 3.

Let $\mathfrak{g}=\mathfrak{h}'\oplus(\oplus_{\alpha\in\Delta}\mathbb{C}E'_\alpha)$ be a simple finite-dimensional Lie algebra of type X_N, as constructed in §7.8 and 7.9 with the normalized invariant form $(.|.)$. Let $\overline{\mu}$ be an automorphism of the Dynkin diagram of $\mathfrak{g}$ of order r ($=1$, 2 or 3), and let μ be the corresponding diagram automorphism of $\mathfrak{g}$.

Case 0. For $X_N=A_\ell$, B_ℓ, $\dots$, E_8 and $r=1$ let $\Pi=\{\alpha_1,\dots,\alpha_\ell\}$ be the set of simple roots of $\mathfrak{g}$ enumerated as in Table Fin. Let $E_i=E'_{\alpha_i}$, $F_i=-E'_{-\alpha_i}$, $H_i=\alpha_i^\vee$ $(i=1,\dots,\ell)$, $E_0=E'_{-\theta}$, $F_0=-E'_\theta$, $H_0=\theta=\theta_0$.

Cases 1–5. Let now $X_N = D_{\ell+1}$, $A_{2\ell-1}$, E_6, D_4 or $A_{2\ell}$ with the ordering of the index set as in Table Fin, and let $r = 2$, 2, 2, 3 or 2 respectively (these are, clearly, all the cases when $\mu \neq 1$). We have the corresponding $\mathbb{Z}/r\mathbb{Z}$-gradations (described by Propositions 7.9a) and 7.10a) in a slightly different notation):

(8.3.1) $$\mathfrak{g} = \mathfrak{g}_{\bar{0}} \oplus \mathfrak{g}_{\bar{1}} \text{ if } r = 2, \text{ and } \mathfrak{g} = \mathfrak{g}_{\bar{0}} \oplus \mathfrak{g}_{\bar{1}} \oplus \mathfrak{g}_{\bar{2}} \text{ if } r = 3;$$

(8.3.2) $$\mathfrak{h}' = \mathfrak{h}_{\bar{0}} \oplus \mathfrak{h}_{\bar{1}} \text{ if } r = 2, \text{ and } \mathfrak{h}' = \mathfrak{h}_{\bar{0}} \oplus \mathfrak{h}_{\bar{1}} \oplus \mathfrak{h}_{\bar{2}} \text{ if } r = 3.$$

Here and further, $\bar{s} \in \mathbb{Z}/r\mathbb{Z}$ stands for the residue of $s \bmod r$. Let $\Pi' = \{\alpha'_1, \dots, \alpha'_N\} \in \mathfrak{h}$ be the set of simple roots of $\mathfrak{g}$ (enumerated as in Table Fin) and let $E'_i = E'_{\alpha_i}$, $F'_i = -E'_{-\alpha_i}$ $(i = 1, \dots, N)$ be its Chevalley generators and $H'_i = \alpha'^{\vee}_i$ simple coroots. Introduce the following elements $\theta^0 \in \mathfrak{h}$ and E_i, F_i, H_i $(i = 0, \dots, \ell) \in \mathfrak{g}$:

Case 1: $X_N = D_{\ell+1}$, $r = 2$:

$$\theta^0 = \alpha'_1 + \dots + \alpha'_\ell;$$
$$H_i = H'_i (1 \le i \le \ell - 1), H_\ell = H'_\ell + H'_{\ell+1}, H_0 = -\theta^0 - \bar{\mu}(\theta^0);$$
$$E_i = E'_i (1 \le i \le \ell - 1), E_\ell = E'_\ell + E'_{\ell+1}, E_0 = E'_{-\theta^0} - E'_{-\bar{\mu}(\theta^0)};$$
$$F_i = F'_i (1 \le i \le \ell - 1), F_\ell = F'_\ell + F'_{\ell+1}, F_0 = -E'_{\theta^0} + E'_{\bar{\mu}(\theta^0)}.$$

Case 2: $X_N = A_{2\ell-1}$, $r = 2$:

$$\theta^0 = \alpha'_1 + \dots + \alpha'_{2\ell-2};$$
$$H_i = H'_i + H'_{2\ell-i}\ (1 \le i \le \ell - 1),\ H_\ell = H'_\ell,\ H_0 = -\theta^0 - \bar{\mu}(\theta^0);$$
$$E_i = E'_i + E'_{2\ell-i}\ (1 \le i \le \ell - 1),\ E_\ell = E'_\ell,\ E_0 = E'_{-\theta^0} - E'_{-\bar{\mu}(\theta^0)};$$
$$F_i = F'_i + F'_{2\ell-i}\ (1 \le i \le \ell - 1),\ F_\ell = F'_\ell,\ F_0 = -E'_{\theta^0} + E'_{\bar{\mu}(\theta^0)}.$$

Case 3: $X_N = E_6$, $r = 2$:

$$\theta^0 = \alpha'_1 + 2\alpha'_2 + 2\alpha'_3 + \alpha'_4 + \alpha'_5 + \alpha'_6;$$
$$H_1 = H'_1 + H'_5;, H_2 = H'_2 + H'_4, H_3 = H'_3, H_4 = H'_6, H_0 = -\theta^0 - \bar{\mu}(\theta^0);$$
$$E_1 = E'_1 + E'_s, E_2 = E'_2 + E'_4, E_3 = E'_3, E_4 = E'_4, E_0 = E'_{-\theta^0} - E'_{-\bar{\mu}(\theta^0)};$$
$$F_1 = F'_1 + F'_5, F_2 = F'_2 + F'_4, F_3 = F'_3, F_4 = F'_6, F_0 = -E'_{\theta^0} + E_{\bar{\mu}(\theta^0)}.$$

Case 4: $X_N = D_4$, $r = 3, \eta = \exp 2\pi i/3$:

$$\theta^0 = \alpha'_1 + \alpha'_2 + \alpha'_3;$$

$$H_1 = H_1' + H_3' + H_4', H_2 = H_2', H_0 = -\theta^0 - \overline{\mu}(\theta^0) - \overline{\mu}^2(\theta^0);$$
$$E_1 = E_1' + E_3' + E_4', E_2 = E_2', E_0 = E'_{-\theta^0} + \eta^2 E'_{-\overline{\mu}(\theta^0)} + \eta E'_{-\overline{\mu}^2(\theta^0)};$$
$$F_1 = F_1' + F_3' + F_4', F_2 = F_2', F_0 = -E'_{\theta^0} - \eta E'_{\overline{\mu}(\theta^0)} - \eta^2 E'_{\overline{\mu}^2(\theta^0)}.$$

Case 5: $X_N = A_{2\ell}$, $r = 2$:

$$\theta^0 = \alpha_1' + \cdots + \alpha_{2\ell}';$$
$$H_i = H_i' + H_{2\ell-i+1}' (1 \leq i \leq \ell - 1), H_0 = 2(H_\ell' + H_{\ell+1}'), H_\ell = -\theta^0;$$
$$E_i = E_i' + E_{2\ell-i+1}' \ (1 \leq i \leq \ell - 1), E_0 = \sqrt{2}(E_\ell' + E_{\ell+1}')m, E_\ell = E'_{-\theta^0}.$$
$$F_i = F_i' + F_{2\ell-i+1}' \ (1 \leq i \leq \ell - 1), F_0 = \sqrt{2}(F_\ell' + F_{\ell+1}'), F_\ell = -E'_{\theta^0}.$$

Let $\theta_0 = \frac{1}{r}(\overline{\mu}(\theta^0) + \cdots + \overline{\mu}^r(\theta^0))$ in Cases 1–4, and $\theta_0 = \theta^0$ in Case 5. Let $\varepsilon = 0$ in cases 0–4, and $\varepsilon = \ell$ in case 5, and let $I = \{0, 1, \ldots, \ell\} \backslash \{\varepsilon\}$.

PROPOSITION 8.3. a) *The elements* E_i $(i = 0, \ldots, \ell)$ *generate the Lie algebra* $\mathfrak{g}$.
b) *Elements* E_i, F_i *with* $i \in I$ *are Chevalley generators of the Lie algebras* $\mathfrak{g}_{\overline{0}}$, *elements* $\alpha_i = 2H_i/(H_i \,|\, H_i)$, $i \in I$, *being simple roots.*
c) $[E_\varepsilon, F_\varepsilon] = H_\varepsilon$, $(E_\varepsilon \,|\, F_\varepsilon) = r/a_0$, *and* $(\theta_0 \,|\, \theta_0) = 2a_0/r$, *where* $a_0 = 1$ *in Cases 1–4 and* $a_0 = 2$ *in Case 5.*
d) *The representation of* $\mathfrak{g}_{\overline{0}}$ *on* $\mathfrak{g}_{\overline{1}}$ *is irreducible, and is equivalent to the representation on* $\mathfrak{g}_{-\overline{1}}$.
e) F_0 *(resp.* E_0*) is the highest (resp. lowest) weight vector of the* $\mathfrak{g}_{\overline{0}}$*-module* $\mathfrak{g}_{\overline{1}}$ *(resp.* $\mathfrak{g}_{-\overline{1}}$*) with weight* θ_0 *(resp.* $-\theta_0$*) (i.e.* $\theta_0 + \alpha_i$ *is not a weight). The types of* $\mathfrak{g}_{\overline{0}}$ *and the decompositions* $\theta_0 = \sum_{i \in I} a_i \alpha_i$ *are listed in the following table:*

r	$\mathfrak{g}$	$\mathfrak{g}_{\overline{0}}$	a_i
2	$A_{2\ell},\ \ell \geq 2$	B_ℓ	$\overset{2}{\circ} - \overset{2}{\circ} - \cdots - \overset{2}{\circ} \Rightarrow \overset{2}{\circ}$
2	$A_{2\ell-1},\ \ell \geq 3$	C_ℓ	$\overset{1}{\circ} - \overset{2}{\circ} - \cdots - \overset{2}{\circ} \Leftarrow \overset{1}{\circ}$
2	$D_{\ell+1},\ \ell \geq 2$	B_ℓ	$\overset{1}{\circ} - \overset{1}{\circ} - \cdots - \overset{1}{\circ} \Rightarrow \overset{1}{\circ}$
2	A_2	A_1	$\overset{2}{\circ}$
2	E_6	F_4	$\overset{1}{\circ} - \overset{2}{\circ} \Rightarrow \overset{3}{\circ} - \overset{2}{\circ}$
3	D_4	G_2	$\overset{1}{\circ} \Rrightarrow \overset{2}{\circ}$

Proof follows immediately from the results of §§7.9 and 7.10.

□

Remark 8.3. Letting $\alpha_\epsilon = -\theta_0$ and $a_\epsilon = 1$, we can write:

$$\sum_{i=0}^{\ell} a_i \alpha_i = 0,$$

where the a_i are given by Table Aff.

The restriction of $(.|.)$ to $\mathfrak{h}_{\bar{0}} = \mathfrak{h}' \cap \mathfrak{g}_{\bar{0}}$ is nondegenerate, and hence defines an isomorphism $\nu : \mathfrak{h}_{\bar{0}} \to \mathfrak{h}_{\bar{0}}^*$. Let $\Delta_{\bar{s}}$ $(s = 0, \dots, r-1)$ be the set of nonzero weights of $\mathfrak{h}_{\bar{0}}$ on $\mathfrak{g}_{\bar{s}}$, and let $\mathfrak{g}_{\bar{s}} = \bigoplus_{\alpha \in \Delta_{\bar{s}} \cup \{0\}} \mathfrak{g}_{\bar{s},\alpha}$ be the weight space decomposition. Proposition 8.2 implies that $(\alpha|\alpha) \neq 0$, $\dim \mathfrak{g}_{\bar{s},\alpha} = 1$ and $[\mathfrak{g}_{\bar{s},\alpha}, \mathfrak{g}_{-\bar{s},-\alpha}] = \mathbf{C}\nu^{-1}(\alpha)$ if $\alpha \in \Delta_{\bar{s}}$.

Now we turn to the Lie algebra $\hat{\mathcal{L}}(\mathfrak{g}, \mu, r)$, which we denote by $\hat{\mathcal{L}}(\mathfrak{g}, \mu)$ for short.. Set $\mathfrak{h} = \mathfrak{h}_{\bar{0}} + \mathbf{C}K' + \mathbf{C}d'$ and define $\delta \in \mathfrak{h}^*$ by $\delta|_{\mathfrak{h}_{\bar{0}}+\mathbf{C}K'} = 0$, $\langle \delta, d' \rangle = 1$. Set $e_\epsilon = t \otimes E_\epsilon$, $f_\epsilon = t^{-1} \otimes F_\epsilon$, $e_i = 1 \otimes E_i$, $f_i = 1 \otimes F_i$ $(i \in I)$. Then we have:

$$[e_i, f_i] = 1 \otimes H_i \ (i \in I); \quad [e_\epsilon, f_\epsilon] = r a_0^{-1} K' + 1 \otimes H_\epsilon.$$

We describe the root system and the root space decomposition of $\hat{\mathcal{L}}(\mathfrak{g}, \mu)$ with respect to $\mathfrak{h}$:

(8.3.3)
$$\Delta = \{j\delta + \gamma, \text{ where } j \in \mathbf{Z}, \gamma \in \Delta_{\bar{s}}, \ j \equiv s \mod r, \ s = 0, \dots, r-1\} \cup \{j\delta, \text{ where } j \in \mathbf{Z}, \ j \neq 0\};$$

(8.3.4)
$$\hat{\mathcal{L}}(\mathfrak{g}, \mu) = \mathfrak{h} \oplus (\bigoplus_{\alpha \in \Delta} \mathcal{L}(\mathfrak{g}, \mu)_\alpha),$$

where

(8.3.5)
$$\mathcal{L}(\mathfrak{g}, \mu)_{s\delta+\gamma} = t^s \otimes \mathfrak{g}_{\bar{s},\gamma}, \quad \mathcal{L}(\mathfrak{g}, \mu)_{s\delta} = t^s \otimes \mathfrak{g}_{\bar{s},0}.$$

We set

(8.3.6)
$$\Pi = \{\alpha_\epsilon := \delta - \theta_0, \ \alpha_i \ (i \in I)\},$$

(8.3.7)
$$\Pi^\vee = \{\alpha_\epsilon^\vee := r a_0^{-1} K' + 1 \otimes H_\epsilon, \ \alpha_i^\vee := 1 \otimes H_i \ (i \in I)\}.$$

Using Proposition 8.3 we obtain that if $\mathfrak{g}$ is of type X_N and r $(= 2$ or $3)$ is the order of μ, then the matrix

$$A = (\langle \alpha_j, \alpha_i^\vee \rangle)_{i,j=0}^{\ell}$$

is of type $X_N^{(r)}$ and the integers $a_0, \dots, a_\ell$ are the labels at the diagram of this matrix in Tables Aff2 and Aff3.

Now we can state the second realization theorem. Its proof is similar to that of Theorem 7.4.

THEOREM 8.3. *Let $\mathfrak{g}$ be a complex simple finite dimensional Lie algebra of type $X_N = D_{\ell+1}$, $A_{2\ell-1}$, E_6, D_4 or $A_{2\ell}$ and let $r = 2, 2, 2, 3$, or 2, respectively. Let μ be a diagram automorphism of $\mathfrak{g}$ of order r.*[1] *Then the Lie algebra $\hat{\mathcal{L}}(\mathfrak{g}, \mu)$ is a (twisted) affine Kac–Moody algebra $\mathfrak{g}(A)$ associated to the affine matrix A of type $X_N^{(r)}$ from Table Aff r ($r = 2$ or 3), $\mathfrak{h}$ is its Cartan subalgebra, Δ the root system, Π and $\Pi^\vee$ the root basis and the coroot basis, and $e_0, \dots, e_\ell, f_0, \dots, f_\ell$ the Chevalley generators. In other words, $(\hat{\mathcal{L}}(\mathfrak{g}, \mu), \mathfrak{h}, \Pi, \Pi^\vee)$ is the quadruple associated to A.*

□

We can summarize the results of Theorems 7.4 and 8.3 as follows. Let A be an affine matrix of type $X_N^{(r)}$, let $\mathfrak{g}$ be a simple finite-dimensional Lie algebra of type X_N and let μ be a diagram automorphism of $\mathfrak{g}$ of order r (= 1, 2 or 3). Then the Lie algebra $\hat{\mathcal{L}}(\mathfrak{g}, \mu)$ is isomorphic to the affine Lie algebra $\mathfrak{g}(A)$. Note that $\tilde{\mathcal{L}}(\mathfrak{g}, \mu)$ is isomorphic to $\mathfrak{g}'(A)$ and $\mathcal{L}(\mathfrak{g}, \mu)$ to $\mathfrak{g}'(A)/\mathbb{C}K$.

COROLLARY 8.3. *Let $\mathfrak{g}(A)$ be an affine algebra of rank $\ell + 1$ and let A be of type $X_N^{(r)}$. Then the multiplicity of the root $jr\delta$ is equal to ℓ, and the multiplicity of the root $s\delta$ for $s \not\equiv 0 \mod r$ is equal to $(N - \ell)/(r - 1)$.*

Proof. If $r = 1$, then $\operatorname{mult} j\delta = \ell$ (for $j \neq 0$) by Theorem 7.4. If $r = 2$ or 3, then by Theorem 8.3, mult $s\delta$ (for $s \neq 0$) is equal to the multiplicity of the eigenvalue $\exp 2\pi i s/r$ of μ operating on $\mathfrak{h}'$, which gives the result.

□

The Chevalley involution ω, the compact involution ω_0 and the triangular decomposition of the Lie algebra $\hat{\mathcal{L}}(\mathfrak{g}, \mu) \subset \hat{\mathcal{L}}(\mathfrak{g})$ are induced by those from $\hat{\mathcal{L}}(\mathfrak{g})$.

The normalized invariant form on $\hat{\mathcal{L}}(\mathfrak{g}, \mu)$ is given by

$$\begin{aligned} &(P \otimes x \,|\, Q \otimes y) = r^{-1} \operatorname{Res}(t^{-1}PQ)(x \,|\, y) \ (x, y \in \mathfrak{g},\ P, Q \in \mathcal{L}); \\ &(\mathbb{C}K' + \mathbb{C}d' \,|\, \mathcal{L}(\mathfrak{g}, \mu)) = 0;\ (K'|K') = (d'|d') = 0;\ (K'|d') = 1, \end{aligned} \tag{8.3.8}$$

where $(.|.)$ is the normalized invariant form on $\mathfrak{g}$. The proof is similar to that of (7.5.1).

It is also easy to see that $K = rK'$ is the canonical central element, and that $d = a_0 r^{-1} d'$ is the scaling element.

---

[1] For $r = 3$ there are two such automorphisms which are equivalent. We choose one of them.

Warning. The Lie algebra $\mathfrak{g}_{\bar{0}}$ is isomorphic to the Lie algebra $\overset{\circ}{\mathfrak{g}}$ introduced in §6.3 in all cases except $A_{2\ell}^{(2)}$; in the latter case $\overset{\circ}{\mathfrak{g}}$ is of type C_ℓ whereas $\mathfrak{g}_{\bar{0}}$ is of type B_ℓ.

§8.4. Here we present another application of realization theorems.

PROPOSITION 8.4. *Let $\mathfrak{g}(A)$ be an affine algebra.*

a) *Set* $\mathfrak{t} = \mathbb{C}K + \sum_{\substack{s\in\mathbb{Z}\\ s\neq 0}} \mathfrak{g}_{s\delta}$. *Then $\mathfrak{t}$ is isomorphic to the infinite-dimensional Heisenberg algebra (= Heisenberg Lie algebra of order ∞; see §2.9) with center $\mathbb{C}K$.*

b) *The Hermitian form $(x|y)_0 = -(\omega_0(x)|y)$ is positive semidefinite on $\mathfrak{g}'(A)$ with kernel $\mathbb{C}K$.*

Proof. By the realization theorem, $\mathfrak{g}'(A)/\mathbb{C}K \simeq \mathcal{L}(\mathfrak{g},\mu)$. The gradation of $\mathfrak{g}$ which corresponds to μ induces the gradation of the Cartan subalgebra $\mathfrak{h}'$ of $\mathfrak{g}$ (see §8.2): $\mathfrak{h}' = \bigoplus_{j\in\mathbb{Z}/r\mathbb{Z}} \mathfrak{h}'_j$. We obtain the following isomorphism:

$$\mathfrak{t}/\mathbb{C}K \simeq \bigoplus_{\substack{s\in\mathbb{Z}\\ s\neq 0}} t^s \otimes \mathfrak{h}'_{s \bmod r}.$$

It follows that $\mathfrak{t}/\mathbb{C}K$ is a commutative subalgebra. It is easy to see that the restriction of the cocycle ψ to this subalgebra is nondegenerate. This proves a); b) follows from the remarks at the end of §7.6.

□

The subalgebra $\mathfrak{t}$ is called the ***homogeneous Heisenberg subalgebra*** of the affine algebra $\mathfrak{g}(A)$. It plays an important role in representation theory of affine algebras.

§8.5. Let $\mathfrak{g}$ be a simple finite-dimensional Lie algebra, let m be a positive integer and let $\varepsilon = \exp\frac{2\pi i}{m}$. Let $\gamma(t) : \mathbb{C}^\times \to \operatorname{Aut}\mathfrak{g}$ be a regular map (we view $\operatorname{Aut}\mathfrak{g}$ as an algebraic group over $\mathbb{C}$); we may regard $\gamma(t)$ as an element of $\operatorname{Aut}\mathcal{L}(\mathfrak{g})$ (by pointwise action of $\gamma(t)$ on $a(t) \in \mathcal{L}(\mathfrak{g})$). We shall view $\sigma \in \operatorname{Aut}\mathfrak{g}$ as an element of $\operatorname{Aut}\mathcal{L}(\mathfrak{g})$ by letting $= \sigma(t^k \otimes a) = t^k \otimes \sigma(a)$. The following lemma follows from an equivalent definiton of $\mathcal{L}(\mathfrak{g},\sigma,m)$ ($\sigma^m = 1$):

$$\mathcal{L}(\mathfrak{g},\sigma,m) = \{a(t) \in \mathcal{L}(\mathfrak{g}) \mid \sigma(a(\varepsilon^{-1}t)) = a(t)\}.$$

LEMMA 8.5. *Let $\alpha,\beta \in \operatorname{Aut}\mathfrak{g}$ be such that $\alpha^m = \beta^m = 1$, and let $\gamma(t) \in \operatorname{Aut}\mathcal{L}(\mathfrak{g})$. Then*

$$\gamma(t)\mathcal{L}(\mathfrak{g},\alpha,m) = \mathcal{L}(\mathfrak{g},\beta,m)$$

if and only if

$$\beta\gamma(\varepsilon^{-1}t) = \gamma(t)\alpha \quad \text{for all } t \in \mathbf{C}^\times. \tag{8.5.1}$$

□

If $\alpha, \beta \in \operatorname{Aut} \mathfrak{g}$ and $\gamma : \mathbf{C}^\times \to \operatorname{Aut} \mathfrak{g}$ (a regular map) are such that (8.5.1) holds, we write $\alpha \xrightarrow{\gamma} \beta$. The following relations are immediate:

$$\alpha \xrightarrow{\gamma} \beta \text{ implies } \beta \xrightarrow{\gamma^{-1}} \alpha; \tag{8.5.2}$$

$$\alpha \xrightarrow{\gamma} \beta \xrightarrow{\gamma_1} \beta_1 \text{ imply } \alpha \xrightarrow{\gamma\gamma_1} \beta_1, \tag{8.5.3}$$

in particular

$$\alpha \xrightarrow{\gamma} \beta \text{ implies } \alpha \xrightarrow{g\gamma} g\beta g^{-1} \text{ for } g \in \operatorname{Aut} \mathfrak{g}. \tag{8.5.4}$$

PROPOSITION 8.5. *Let σ be an automorphism of a simple finite-dimensional Lie algebra $\mathfrak{g}$ of the form (8.1.2). Denote by t^h the regular map $\mathbf{C}^\times \to \operatorname{Aut} \mathfrak{g}$ such that t^h on $\mathfrak{g}_\alpha$ is an operator of multiplication by $t^{\langle\alpha,h\rangle}$. Then*

$$t^h(\mathcal{L}(\mathfrak{g}, \mu, m)) = \mathcal{L}(\mathfrak{g}, \sigma, m). \tag{8.5.5}$$

Proof follows immediately from Lemma 8.5, since $t^h\mu = \mu t^h$.

□

Remark 8.5. It follows from Proposition 8.5 that the isomorphism class of $\mathcal{L}(\mathfrak{g}, \sigma, m)$ depends only on the connected component of $\operatorname{Aut}\mathfrak{g}$ containing σ. It is not difficult to show that this statement still holds if $\mathfrak{g}$ is replaced by an arbitrary finite-dimensional algebra (not necessarily a Lie algebra).

We can now prove the following theorem.

THEOREM 8.5. *Let $\mathfrak{g}$ be a simple finite-dimensional Lie algebra of type X_N and let σ be an automorphism of $\mathfrak{g}$ such that $\sigma^m = 1$. Let r be the least positive integer such that σ^r is an inner automorphism; then $r = 1$, 2 or 3. Let $(.|.)$ be the normalized invariant form on $\mathfrak{g}$. Let μ be a diagram automorphism of $\mathfrak{g}$ of order r. Choose a Cartan subalgebra $\mathfrak{h}_{\bar{0}}$ of the fixed point set $\mathfrak{g}^\sigma$ of σ. Let A be the affine matrix of type $X_N^{(r)}$. Then there exists an isomorphism $\Phi : \hat{\mathcal{L}}(\mathfrak{g}, \sigma, m) \to \mathfrak{g}(A)$ such that:*

(i) *the bilinear form on* $\hat{\mathcal{L}}(\mathfrak{g},\sigma,m)$ *defined by (8.3.8) induces the normalized invariant form on* $\mathfrak{g}(A)$;

(ii) *the* $\mathbb{Z}$-*gradation (8.2.2) of* $\hat{\mathcal{L}}(\mathfrak{g},\sigma,m)$ *induces a* $\mathbb{Z}$-*gradation of* $\mathfrak{g}(A)$ *of type* $s = (s_0,\ldots,s_\ell)$, *where* s_j *are nonnegative integers which satisfy the relation*

$$r\sum_{j=0}^{\ell} a_j s_j = m; \tag{8.5.6}$$

(iii) $\Phi(1\otimes \mathfrak{h}_{\bar{0}} + \mathbb{C}K' + \mathbb{C}d')$ *is the Cartan subalgebra of* $\mathfrak{g}(A)$;

(iv) $m\Phi(K')$ *is the canonical central element* K *of* $\mathfrak{g}(A)$;

(v) $m^{-1}\Phi(d_0) = a_0^{-1}d + u - \frac{1}{2}(u|u)K$, *where* d *is the scaling element of* $\mathfrak{g}(A)$ *and* $u \in \mathring{\mathfrak{h}}$ *is defined by* $\langle \alpha_i, u\rangle = rs_i/m$ $(i = 1,\ldots,\ell)$.

Proof. Due to Proposition 8.1, Lemma 8.5 and (8.5.4), we may assume that σ is of the form (8.1.2). Note that r divides m (otherwise $am + br = 1$ for some $a,b \in \mathbb{Z}$ and hence $\sigma = (\sigma^r)^b$ would be an inner automorphism). By Proposition 8.5 we have an isomorphism

$$t^{\frac{m}{r}h} : \mathcal{L}(\mathfrak{g},\mu,r) \xrightarrow{\sim} \mathcal{L}(\mathfrak{g},\sigma,m).$$

In other words, according to Theorems 7.4 and 8.3, we have an isomorphism, denoted by φ for short,

$$\varphi : \mathfrak{g}'(A)/\mathbb{C}K \xrightarrow{\sim} \mathcal{L}(\mathfrak{g},\sigma,m).$$

The $\mathbb{Z}$-gradation (8.2.1) of $\mathcal{L}(\mathfrak{g},\sigma,m)$ induces, via φ^{-1}, the $\mathbb{Z}$-gradation of type $s' = (s'_0,\ldots,s'_\ell)$ of $\mathfrak{g}'(A)/\mathbb{C}K$, where $s'_i = \deg \bar{e}_i = -\deg \bar{f}_i$. Here and further in the proof, for $x \in \mathfrak{g}'(A)$ we denote by $\bar{x}$ the coset $x + \mathbb{C}K$. Note that $\deg \mathfrak{g}_\delta = \sum_{i=0}^{\ell} a_i s'_i$. On the other hand, multiplication by t^r increases the degree in $\mathcal{L}(\mathfrak{g},\sigma,m)$ by m. Since $\deg \mathfrak{h}_{\bar{0}} = 0$ and $\bar{\mathfrak{g}}_{r\delta} = t^r \mathfrak{h}_{\bar{0}}$, we deduce that $\deg \bar{\mathfrak{g}}_{r\delta} = m = r\sum_{i=0}^{\ell} a_i s'_i$, proving (8.5.6) for s'.

The isomorphism φ^{-1} can be lifted to a (unique) linear isomorphism $\Phi' : \hat{\mathcal{L}}(\mathfrak{g},\sigma,m) \to \mathfrak{g}(A)$, which satisfies (iv), (v) with s_i replaced by s'_i, and the following condition (and hence (iii)):

$$\Phi'^{-1}(\alpha_i^\vee) = \varphi(\overline{\alpha_i^\vee}) + r(a_i s'_i/a_i^\vee)K' \quad (i = 0,\ldots,\ell). \tag{8.5.7}$$

One easily checks that this is a Lie algebra isomorphism.

Furthermore, it is clear that the $\mathbb{Z}$-gradation (8.2.2) of $\hat{\mathcal{L}}(\mathfrak{g},\sigma,m)$ induces via the isomorphism Φ' the gradation of type s' of $\mathfrak{g}(A)$. The property (i) of Φ' is also straightforward. It remains to show that the s'_i can be made non-negative. For that pick $v \in \mathfrak{h}$ (the Cartan subalgebra of $\mathfrak{g}(A)$) such that $\langle \alpha_i, v\rangle = s'_i$; note that $\delta(v) = m/r > 0$. Hence, by Lemma 3.8, Proposition 3.12b) and Proposition 5.8b), there exists $w \in \widetilde{W}^{ad}$ (see Remark 3.8) such that $\langle \alpha_i, w(v)\rangle = s_i \in \mathbb{Z}_+$. Now $\Phi = w_0^{-1}\Phi'$ is the desired isomorphism.

□

§8.6. We deduce from Theorem 8.5 a classification of finite order automorphisms of a simple finite-dimensional Lie algebra $\mathfrak{g}$ of type X_N. Let μ be a diagram automorphism of $\mathfrak{g}$ of order r. Let E_i, F_i, H_i ($i = 0, \ldots, \ell$) be the elements of $\mathfrak{g}$ introduced in §8.3 and let $\alpha_0, \ldots, \alpha_\ell$ be the roots attached to the E_i. Recall that the elements E_i ($i = 0, \ldots, \ell$) generate $\mathfrak{g}$ (by Proposition 8.3a)) and that there exists a unique linear dependence $\sum_{i=0}^{\ell} a_i\alpha_i = 0$ such that the a_i are positive relatively prime integers (see Remark 8.3). Recall also that the vertices of the diagram $X_N^{(r)}$ are in one-to-one correspondence with the E_i and that the a_i are labels at this diagram.

In order to derive an application to finite order automorphisms, we need the following fact.

LEMMA 8.6. *Every ideal of the Lie algebra $\mathcal{L}(\mathfrak{g},\mu)$ is of the form $P(t^r)\mathcal{L}(\mathfrak{g},\mu)$, where $P(t) \in \mathcal{L}$. In particular, a maximal ideal is of the form $(1-(at)^r)\mathcal{L}(\mathfrak{g},\mu)$, where $a \in \mathbb{C}^\times$.*

Proof. Let $\mathfrak{i}$ be a nontrivial ideal of $\mathcal{L}(\mathfrak{g},\mu)$ and $x = \sum_{j,s} t^j P_{\bar{j},s}(t) \otimes a_{\bar{j},s} \in \mathfrak{i}$, where $0 \le j < r$ is such that $\bar{j} \equiv j \bmod r$, $P_{\bar{j},s}(t) \in \mathcal{L}$, $P_{\bar{j},s} \ne 0$ and $a_{j,s} \in \mathfrak{g}_j$ are linearly independent. We show that $Q(t^r)P_{\bar{j},s}(t)\mathcal{L}(\mathfrak{g},\mu) \subset \mathfrak{i}$ for all $Q(t) \in \mathcal{L}$.

Let $\mathfrak{h}_{\bar{0}}$ be a Cartan subalgebra of $\mathfrak{g}_{\bar{0}}$; we can assume that x is an eigenvector for ad $\mathfrak{h}_{\bar{0}}$ with weight $\alpha \in \mathfrak{h}_{\bar{0}}^*$. If $\alpha \ne 0$, taking $[x, t^j \otimes a_{-\bar{j}}]$ with $a_{-\bar{j}}$ of weight $-\alpha$, instead of x, we reduce the problem to the case

$\alpha = 0$ and $\bar{j} = 0$, i.e., $a_{\bar{j},s} \in \mathfrak{h}_{\bar{0}}$. Let $\gamma \in \mathfrak{h}_{\bar{0}}^*$ be a root of $\mathfrak{g}_{\bar{0}}$ such that $\langle \gamma, a_{\bar{j},s} \rangle \neq 0$. Then the element $y = [[x, Q(t^r) \otimes e_\gamma], e_{-\gamma}] \in \mathfrak{i}$, where $e_{\pm\gamma}$ is a root vector with root $\pm\gamma$, has the following form: $y = Q(t^r)(P \otimes h + tP_{\bar{1}} \otimes h_{\bar{1}} + \cdots + t^{r-1}P_{\overline{r-1}} \otimes h_{\overline{r-1}})$, where $P = P_{\bar{j},s}(t)$, $P_{\bar{i}} \in \mathcal{L}$, $h \in \mathfrak{h}'$, $h \neq 0$, and the $h_{\bar{i}} \in \mathfrak{g}_{\bar{i}}$ have zero weight with respect to $\mathfrak{h}$. Since $[y, e_j] \in \mathfrak{i}$ for all root vectors $e_j \in \mathfrak{g}_{\bar{0}}$, we conclude that $Q(t^r)P \otimes \mathfrak{h} \subset \mathfrak{i}$ and therefore $Q(t^r)P\mathcal{L}(\mathfrak{g}, \mu) \subset \mathfrak{i}$.

For $x \in \mathfrak{g}_{\bar{j}}$ let $\mathfrak{i}_0 = \{p(t) \in \mathcal{L} \,|\, t^j p(t) \otimes x \in \mathfrak{i}\}$. It follows from above that $\mathfrak{i}_0$ is an invariant ideal (with respect to transformation $t \to \eta^{-1}t$) of $\mathcal{L}$, independent of $\bar{j}$ and x, and that $\mathfrak{i} = \mathfrak{i}_0\mathcal{L}(\mathfrak{g}, \mu)$. Since all ideals of $\mathcal{L}$ are principal and $\mathfrak{i}_0$ is invariant, we deduce that $\mathfrak{i} = P\mathcal{L}(\mathfrak{g}, \mu)$ for some $P \in \mathbf{C}[t^r, t^{-r}]$.

□

THEOREM 8.6. *Let $s = (s_0, \ldots, s_\ell)$ be a sequence of nonnegative relatively prime integers; put $m = r \sum_{i=0}^{\ell} a_i s_i$. Then*

a) *The relations*

$$\sigma_{s;r}(E_j) = e^{2\pi i s_j/m} E_j \quad (j = 0, \ldots, \ell) \tag{8.6.1}$$

define (uniquely) an m-th order automorphism $\sigma_{s;r}$ of $\mathfrak{g}$.

b) *Up to conjugation by an automorphism of $\mathfrak{g}$, the automorphisms $\sigma_{s;r}$ exhaust all m-th order automorphisms of $\mathfrak{g}$.*

c) *The elements $\sigma_{s;r}$ and $\sigma_{s';r'}$ are conjugate by an automorphism of $\mathfrak{g}$ if and only if $r = r'$ and the sequence s can be transformed into the sequence s' by an automorphism of the diagram $X_N^{(r)}$.*

Proof. We use Theorem 8.5 and the *covering homomorphism* $\varphi_\sigma : \mathcal{L}(\mathfrak{g}, \sigma) \to \mathfrak{g}$ defined by $t \mapsto 1$, i.e., $\varphi_\sigma(\sum P_i \otimes g_i) = \sum P_i(1) g_i$. It is clear that

$$\operatorname{Ker} \varphi_\sigma = (1 - t^m)\mathcal{L}(\mathfrak{g}, \sigma).$$

To prove a) note that the root space decomposition (8.3.4) of $\hat{\mathcal{L}}(\mathfrak{g}, \mu)$ induces a gradation $\mathcal{L}(\mathfrak{g}, \mu) = \tilde{\mathcal{L}}(\mathfrak{g}, \mu)/\mathbf{C}K' = \bigoplus_\alpha \mathcal{L}_\alpha$. Define the automorphism $\tilde{\sigma}_s$ of $\mathcal{L}(\mathfrak{g}, \mu)$ by:

$$\tilde{\sigma}_s(e_\alpha) = \epsilon^{\Sigma k_i s_i} e_\alpha \text{ if } e_\alpha \in \mathcal{L}_\alpha, \quad \text{where } \alpha = \sum k_i \alpha_i.$$

If $\mathcal{L}(\mathfrak{g},\mu) = \bigoplus_{j\in\mathbb{Z}} \mathcal{L}(\mathfrak{g},\mu)_j$ is the gradation of type s, then $\mathcal{L}(\mathfrak{g},\mu)_j$ and $\mathcal{L}(\mathfrak{g},\mu)_{j+m}$ lie in the eigenspace of $\tilde{\sigma}_s$ with eigenvalue $\exp 2\pi i j/m$. Since $t^r\mathcal{L}(\mathfrak{g},\mu)_j \subset \mathcal{L}(\mathfrak{g},\mu)_{j+m}$, we deduce that the ideal $(1-t^r)\mathcal{L}(\mathfrak{g},\mu)$ is $\tilde{\sigma}_s$-invariant and hence $\tilde{\sigma}_s$ induces the automorphism of $\mathfrak{g}$ with the properties described in a).

Let now σ be an m-th order automorphism of $\mathfrak{g}$. Theorem 8.5 gives us an isomorphism

$$\overline{\Phi} : \mathcal{L}(\mathfrak{g},\sigma,m) \to \mathcal{L}(\mathfrak{g},\mu)$$

such that the $\mathbb{Z}$-gradation of $\mathcal{L}(\mathfrak{g},\sigma,m)$ induces a $\mathbb{Z}$-gradation of type s of $\mathcal{L}(\mathfrak{g},\mu)$ with $s_i \in \mathbb{Z}_+$ satisfying (8.5.6). Denote by τ_a the automorphism of $\mathcal{L}(\mathfrak{g},\mu)$, which corresponds to changing t to at, $a \in \mathbb{C}^\times$. Then, since by Lemma 8.6, any maximal ideal of $\mathcal{L}(\mathfrak{g},\mu)$ is of the form $\big(1-(at)^r\big)\mathcal{L}(\mathfrak{g},\mu)$, we have the following commutative diagram for a suitable automorphism ψ of $\mathfrak{g}$ and $a \in \mathbb{C}^\times$:

$$\begin{array}{ccccc} \mathcal{L}(\mathfrak{g},m,\sigma) & \xrightarrow{\overline{\Phi}} & \mathcal{L}(\mathfrak{g},\mu) & \xrightarrow{\tau_a} & \mathcal{L}(\mathfrak{g},\mu) \\ \downarrow{\scriptstyle\varphi_\sigma} & & & & \downarrow{\scriptstyle\varphi_\mu} \\ \mathfrak{g} & & \xrightarrow{\psi} & & \mathfrak{g} \end{array}$$

Hence $\psi\sigma\psi^{-1} = \sigma_{s,r}$, proving b).

The proof of c) also uses the covering map. Suppose that $\sigma = \sigma_{s;r}$ and $\sigma' = \sigma_{s';r'}$ are conjugate, i.e., $\tau\sigma\tau^{-1} = \sigma'$ for some $\tau \in \operatorname{Aut}\mathfrak{g}$. Note that by Propostion 8.6b) (see below), $\mathfrak{h}_{\overline{0}}$ (the Cartan subalgebra of $\mathfrak{g}^\mu$) is the Cartan subalgebra of $\mathfrak{g}^\sigma$ and $\mathfrak{g}^{\sigma'}$. Since $\tau(\mathfrak{g}^\sigma) = \mathfrak{g}^{\sigma'}$, $\tau(\mathfrak{h}_{\overline{0}})$ is another Cartan subalgebra of $\mathfrak{g}^{\sigma'}$. Let τ_1 be an inner automorphism of $\mathfrak{g}^{\sigma'}$ such that $\tau_1(\tau(\mathfrak{h}_{\overline{0}})) = \mathfrak{h}_{\overline{0}}$. Replacing τ by $\tau_1\tau$, we may assume that τ leaves $\mathfrak{h}_{\overline{0}}$ invariant and that the sets of positive roots of $\mathfrak{g}^\sigma$ and $\mathfrak{g}^{\sigma'}$ correspond to each other under τ (by Proposition 5.9 applied to $\mathfrak{g}^\sigma$). The extension $\tilde{\tau}$ of τ given by $\tilde{\tau}(t^j \otimes a) = t^j \otimes \tau(a)$, is an isomorphism of $\mathbb{Z}$-graded Lie algebras: $\mathcal{L}(\mathfrak{g},\sigma,m) \to \mathcal{L}(\mathfrak{g},\sigma',m)$, which maps positive roots onto positive roots. Hence the sequences s and s' correspond under an automorphism of the diagram $X_N^{(r)}$.

□

Given a nonzero sequence $s = (s_0,\ldots,s_\ell)$ of nonnegative integers and a number $r = 1$, 2 or 3, we call the automorphism $\sigma_{s;r}$ of $\mathfrak{g}$ defined by (8.6.1) the *automorphism of type* $(s;r)$. Let $\mathfrak{g} = \bigoplus_j \mathfrak{g}_j(s;r)$ be the $\mathbb{Z}/m\mathbb{Z}$-gradation associated to it. Here are some of its properties.

PROPOSITION 8.6. a) *r is the least positive integer for which $\sigma^r_{s;r}$ is an inner automorphism.*

b) *Let $i_1, \ldots, i_p$ be all the indices for which $s_{i_1} = \cdots = s_{i_p} = 0$. Then the Lie algebra $\mathfrak{g}_{\bar{0}}(s;r)$ is isomorphic to a direct sum of the $(\ell-p)$-dimensional center and a semisimple Lie algebra whose Dynkin diagram is the subdiagram of the affine diagram $X_N^{(r)}$ consisting of the vertices $i_1, \ldots, i_p$.*

c) *Let $j_1, \ldots, j_n$ be all the indices for which $s_{j_1} = \cdots = s_{j_n} = 1$. Then the $\mathfrak{g}_{\bar{0}}(s;r)$-module $\mathfrak{g}_{\bar{1}}(s;r)$ (resp. $\mathfrak{g}_{-\bar{1}}(s;r)$) is isomorphic to a direct sum of n irreducible modules with highest-weights $-\alpha_{j_1}, \ldots, -\alpha_{j_n}$ (resp. $\alpha_{j_1}, \ldots, \alpha_{j_n}$).*

Proof. a) follows from the (easy) fact that a finite order automorphism σ of $\mathfrak{g}$ is inner if and only if there exists a Cartan subalgebra which is pointwise fixed under σ. b) is immediate from the isomorphism $\mathfrak{g}^\sigma \simeq \mathcal{L}(\mathfrak{g},\sigma)_0$ and Corollary 5.12b). To prove c) note that the $\mathfrak{g}_{\bar{0}}(s;r)$-module $\mathfrak{g}_{\bar{1}}(s;r)$ is isomorphic to the $\mathcal{L}(\mathfrak{g},\sigma)_0$-module $\mathcal{L}(\mathfrak{g},\sigma)_1$. Furthermore, using the Jacobi identity, we see that $\mathcal{L}(\mathfrak{g},\sigma)_1$ is spanned by elements of the form $[\ldots[[e_{i_1}, e_{i_2}], e_{i_3}]\ldots e_{i_r}]$, such that $s_{i_1} = 1$ and $s_{i_t} = 0$ for $t > 1$. Using the Weyl complete reducibility theorem proves c).

□

§8.7. Later we will need the following reformulation of Theorem 8.5 (which is a generalization of Theorems 7.4 and 8.3).

THEOREM 8.7. *Let A be an affine matrix of type $X_N^{(r)}$. Let $\mathfrak{g}$ be a simple finite-dimensional Lie algebra of type X_N, and let $(.|.)$ be its normalized invariant form. Let μ be a diagram automorphism of order r of $\mathfrak{g}$ and E_i, F_i, H_i ($i = 0, \ldots, \ell$) the elements of $\mathfrak{g}$ introduced in §8.2. Let $\sigma_{s;r}$ be an automorphism of type $(s;r)$ of $\mathfrak{g}$, and let $\mathfrak{g} = \bigoplus_j \mathfrak{g}_j(s;r)$ be the associated $\mathbb{Z}/m\mathbb{Z}$-gradation, where $m = r\sum_{i=0}^{\ell} a_i s_i$. Define the Lie algebra structure on*

$(\sum_{j \in \mathbb{Z}} t^j \otimes \mathfrak{g}_{j \bmod m}(s;r)) \oplus \mathbb{C}K$ *by*

$$[(P_1(t) \otimes g_1) \oplus \lambda_1 K, (P_2(t) \otimes g_2) \oplus \lambda_2 K] \\ = P_1(t)P_2(t) \otimes [g_1, g_2] \oplus \frac{1}{m}(\operatorname{Res} \frac{dP_1(t)}{dt} P_2(t))(g_1|g_2)K.$$

Then this is isomorphic to the derived affine algebra $\mathfrak{g}'(A)$, *with Chevalley generators* $e_i = t^{s_i} \otimes E_i$, $f_i = t^{-s_i} \otimes F_i$ $(i = 0, \ldots, \ell)$, *the coroot basis* $\alpha_i^\vee = (1 \otimes H_i) \oplus (a_i s_i / a_i^\vee m)K$ $(i = 0, \ldots, \ell)$ *and canonical central element* K. *Extending this Lie algebra by* $\mathbb{C}d'$, *where* $d'(P(t) \otimes g) = \frac{a_0}{m} t \frac{dP(t)}{dt} \otimes g$ *and* $[d', K] = 0$, *we obtain a Lie algebra which is isomorphic to the affine algebra* $\mathfrak{g}(A)$ *with the scaling element* $d = a_0(d' - H - \frac{1}{2}(H|H)K)$, *where* $H \in \sum \mathbb{C}H_i$ *is defined by* $\langle \alpha_i, H \rangle = s_i/m$ $(i = 1, \ldots, \ell)$. *The normalized invariant form is defined by*

$$(P_1(t) \otimes g_1 | P_2(t) \otimes g_2) = r^{-1}(\operatorname{Res} t^{-1} P_1(t) P_2(t))(g_1|g_2),$$
$$(\mathbb{C}K + \mathbb{C}d' | P(t) \otimes g) = 0, \quad (K|K) = 0, \quad (d'|d') = 0 \text{ and } (K|d') = 1.$$

Finally, setting $\deg t = 1$, $\deg g = 0$ *for* $g \in \mathfrak{g}$ *and* $\deg K = \deg d' = 0$ *defines the* $\mathbb{Z}$*-gradation of type* s.

□

The realization of the affine algebra of type $X_N^{(r)}$ provided by Theorem 8.7 is called the *realization of type s*.

Using Corollary 6.4 and Exercise 6.2, we obtain

COROLLARY 8.7. *Let* $\hat{\mathcal{L}}(\mathfrak{g}, \mu, r)$ *be an affine algebra of type* $X_N^{(r)}$. *Then* $\phi(x, y) = 2rh^\vee(x|y)$ *for* $x, y \in \mathfrak{g}$, *where* ϕ *is the Killing form on* $\mathfrak{g}$. *In particular:*

$$\sum_{\alpha \in \Delta_{\mathfrak{g}}} (\lambda|\alpha)(\mu|\alpha) = 2rh^\vee(\lambda|\mu) \text{ for } \lambda, \mu \in \mathfrak{h}^*,$$

where $\mathfrak{h}$ *is a Cartan subalgebra of* $\mathfrak{g}$ *and* $\Delta_{\mathfrak{g}} \subset \mathfrak{h}^*$ *the root system of* $\mathfrak{g}$.

§8.8. Exercises.

In Exercises 8.1–8.4 we sketch another proof of Theorem 8.5.

8.1. Choose a Cartan subalgebra $\mathfrak{h}_{\bar{0}}$ of $\mathfrak{g}_{\bar{0}}$ and set $\mathfrak{h} = 1 \otimes \mathfrak{h}_{\bar{0}} + \mathbb{C}K' + \mathbb{C}d' \subset \hat{\mathcal{L}}(\mathfrak{g}, \sigma, m)$. With respect to $\mathfrak{h}$ we have the root space decomposition; denote by Δ the set of roots and by $\mathcal{L}_\alpha$ the root space attached to $\alpha \in \Delta$. Denote by $\bar{\lambda}$ the restriction of $\lambda \in \mathfrak{h}^*$ to $1 \otimes \mathfrak{h}_{\bar{0}}$ and set $(\lambda|\mu) = (\bar{\lambda}|\bar{\mu})$ for λ, $\mu \in \mathfrak{h}^*$. Set $\Delta^\circ = \{\alpha \in \Delta | \bar{\alpha} = 0\}$. Show that $\operatorname{ad} \mathcal{L}_\alpha$ is locally nilpotent provided that $\alpha \notin \Delta^\circ$.

***8.2.* Let $\alpha \in \Delta \backslash \Delta^\circ$. Then**

(i) dim $\mathcal{L}_\alpha = 1$ and $(\alpha|\alpha) \neq 0$;

(ii) for $\beta \in \Delta$, the set of $\beta + k\alpha \in \Delta \cup \{0\}$ is a string $\beta - p\alpha, \ldots, \beta + q\alpha$, where p and q are some nonnegative integers such that $p - q = 2(\alpha|\beta)/(\alpha|\alpha)$.

(iii) $[\mathcal{L}_\beta, \mathcal{L}_\gamma] \neq 0$ if $\beta, \gamma, \beta + \gamma \in \Delta$, $\beta \notin \Delta^\circ$.

8.3. Let $\Delta_0 \subset \Delta$ be the set of roots of $1 \otimes \mathfrak{g}_{\bar{0}} \subset \mathcal{L}(\mathfrak{g}, \sigma, m)$ and let Δ_{0+} be a subset of positive roots in Δ_0; let $\Delta_+ = \Delta_{0+} \cup \{\alpha \in \Delta | \langle \alpha, d \rangle > 0\}$. A root $\alpha \in \Delta_+$ is called simple if it is not a sum of two members of Δ_+. Let $\Pi = \{\alpha_1, \alpha_2, \ldots\}$ be the set of all simple roots. Then

(i) each $\alpha \in \Delta_+$ can be written in the form $\alpha = \sum_i k_i \alpha_i$, where $k_i \in \mathbb{Z}_+$;

(ii) $\Pi \subset \Delta \backslash \Delta^\circ$;

(iii) $\overline{\alpha_1}$, $\overline{\alpha_2}$, ... span $\mathfrak{h}_{\bar{0}}$ and the bilinear form $(. | .)$ is positive definite on $\sum_i \mathbb{R}\overline{\alpha_i}$;

(iv) there exists a nontrivial linear dependence of $\overline{\alpha_i}$ with nonnegative coefficients;

(v) for $i \neq j$ we have $a_{ij} := 2(\alpha_i|\alpha_j)/(\alpha_i|\alpha_i) \in -\mathbb{Z}_+$;

(vi) if $\alpha \in \Delta_+$ is not simple, then $\alpha - \alpha_i \in \Delta_+$ for some $\alpha_i \in \Pi$.

***8.4.* Deduce the existence of the isomorphism Φ.**

***8.5.* Let G be a connected semi-simple complex algebraic group and let $\widetilde{G}$ be the group of regular maps $\mathbb{C}^\times \to G$. Let $\alpha \in \operatorname{Aut} G$ be such that $\alpha^m = 1$;**

we extend α to an automorphism of $\tilde{G} \supset G$ by letting $\alpha(\gamma(t)) = \alpha(\gamma(\varepsilon^{-1}t))$. Show that, given $g \in G$, there exists $\gamma \in \tilde{G}$ such that

$$g = \gamma^{-1}\alpha(\gamma) \tag{8.8.1}$$

if and only if

$$g\alpha(g)\dots\alpha^{m-1}(g) = 1. \tag{8.8.2}$$

8.6. Taking for granted that any finite order automorphism of $\tilde{G}$ leaves invariant some connected semisimple subgroup G' of maximal rank (see Kac–Peterson [1987]), deduce from Exercise 8.5 that given $g \in \tilde{G}$, there exists $\gamma \in \tilde{G}$ satisfyng (8.8.1) if and only if g satisfies (8.8.2)

8.7. Show that every ideal of the Lie algebra $\mathcal{L}(\mathfrak{g}, \sigma)$ is of the form $P(t^m)\mathcal{L}(\mathfrak{g}, \sigma)$, where $P(t) \in \mathbb{C}[t]$. This is a maximal ideal if and only if $P(t) = a + bt$, where $a, b \in \mathbb{C}^\times$.

8.8. Taking for granted that all maximal ad-diagonalizable subalgebras of a Kac–Moody algebra $\mathfrak{g}$ are conjugate (see Peterson–Kac [1983]), show that every $\mathbb{Z}$-gradation $\mathfrak{g} = \oplus_{j \in \mathbb{Z}} \mathfrak{g}_j$ with finite-dimensional $\mathfrak{g}_0$ is conjugate to a $\mathbb{Z}$-gradation of type s with $s_i \in \mathbb{Z}_+$.

8.9. Let $\mathfrak{g}$ be a simple finite-dimensional Lie algebra and let e_i, f_i ($i = 1, \dots, \ell$) be its Chevalley generators. Let $\theta = \sum_{i=1}^{\ell} a_i\alpha_i$ be the decomposition of the highest root of $\mathfrak{g}$ via simple roots. Define an involution σ_i of $\mathfrak{g}$ by $\sigma_i(e_i) = -e_i$, $\sigma_i(f_i) = -f_i$, $\sigma_i(e_j) = e_j$, $\sigma_i(f_j) = f_j$ for $j \neq i$. Let μ be a diagram involution of $\mathfrak{g}$. Show that every involution σ of $\mathfrak{g}$ is conjugate to one from the following list: a) σ_i for $a_i = 1$; b) σ_i for $a_i = 2$; c) μ; d) $\mu \circ \sigma_i$ for $a_i = 1$ or 2 and $\mu(i) = i$. Show that the $\mathfrak{g}_{\bar{0}}$-module $\mathfrak{g}_{\bar{1}}$ is irreducible if and only if σ is of type b), c) or d), and that σ is inner if and only if it is of type a) or b). Show that in the case a), $\mathfrak{g}_{\bar{1}} = V_{-1} \oplus V_1$, where $V_{\pm 1}$ are irreducible $\mathfrak{g}_{\bar{0}}$-modules, and $\mathfrak{g} = V_{-1} \oplus \mathfrak{g}_{\bar{0}} \oplus V_1$ is a $\mathbb{Z}$-gradation of $\mathfrak{g}$; show that we thus obtain all up to conjugation $\mathbb{Z}$-gradations of the form $\mathfrak{g} = \mathfrak{g}_{-1} \oplus \mathfrak{g}_0 \oplus \mathfrak{g}_1$.

8.10. Show that an automorphism σ of order 2 or 3 of a simple Lie algebra is determined (up to conjugacy) by the isomorphism class of the fixed point subalgebra of σ. Give an example of two non-conjugate automorphisms of order 5 of A_2 with isomorphic fixed point subalgebras.

8.11. Show that the minimal order of a regular automorphism (i.e., an automorphism with an abelian fixed point set) of a simple finite-dimensional Lie algebra $\mathfrak{g}$ is the Coxeter number h (= (height of the highest root) +1). Show that such an automorphism is conjugate to the automorphism of type $(1,1,\ldots,1;1)$. Show that every regular automorphism of $\mathfrak{g}$ of order $h+1$ is conjugate to the authomorphism of type $(2,1,\ldots,1;1)$.

8.12. An automorphism σ of order m of a simple finite-dimensional Lie algebra $\mathfrak{g}$ is called *quasirational* if for the associated $\mathbb{Z}/m\mathbb{Z}$-gradation $\mathfrak{g} = \bigoplus_j \mathfrak{g}_j$ one has: $\dim \mathfrak{g}_i = \dim \mathfrak{g}_j$ if $(i,m) = (j,m)$, or, equivalently, the characteristic polynomial of σ on $\mathfrak{g}$ has rational coefficients. Show that the automorphisms of type $(1,1,\ldots,1;r)$ and $(2,1,\ldots,1;1)$ are quasirational. Classify all quasirational automorphisms of A_1 and A_2 up to conjugation.

8.13. An automorphism σ of order m of $\mathfrak{g}$ is called *rational* if σ^k is conjugate to σ for every k such that k and m are relatively prime. Show that a rational automorphism is quasirational. Find a counterexample to the converse statement.

8.14. Show that the automorphisms of Exercise 8.11 are rational.

8.15. Let μ be an automorphism of the Lie algebra $\mathfrak{g} = s\ell_n(\mathbb{C})$ defined by

$$\mu(E_{ij}) = -(-1)^{i+j} E_{n+1-j,n+1-i}.$$

Show that μ is the diagram automorphism of $\mathfrak{g}$. Show that the subalgebra $\hat{\mathcal{L}}(\mathfrak{g},\mu) \subset \hat{\mathcal{L}}(\mathfrak{g})$ is an affine Lie algebra of type $A_n^{(2)}$, K being the (resp. twice the) canonical central element if $n = 2\ell$ (resp. $n = 2\ell - 1$). Describe the root space decomposition.

8.16. Introduce the following basis of the Lie algebra $\mathcal{L}(s\ell_3(\mathbb{C}),\mu)$ $(s \in \mathbb{Z})$:

$$L_{8s} = \begin{pmatrix} t^{2s} & 0 & 0 \\ 0 & 0 & 0 \\ 0 & 0 & -t^{2s} \end{pmatrix}, \quad L_{8s+1} = \begin{pmatrix} 0 & t^{2s} & 0 \\ 0 & 0 & t^{2s} \\ 0 & 0 & 0 \end{pmatrix},$$

$$L_{8s+2} = \begin{pmatrix} 0 & 0 & 0 \\ 0 & 0 & 0 \\ t^{2s+1} & 0 & 0 \end{pmatrix}, \quad L_{8s+3} = \begin{pmatrix} 0 & 0 & 0 \\ t^{2s+1} & 0 & 0 \\ 0 & -t^{2s+1} & 0 \end{pmatrix},$$

$$L_{8s+4} = \begin{pmatrix} t^{2s+1} & 0 & 0 \\ 0 & -2t^{2s+1} & 0 \\ 0 & 0 & t^{2s+1} \end{pmatrix}, \quad L_{8s+5} = \begin{pmatrix} 0 & t^{2s+1} & 0 \\ 0 & 0 & -t^{2s+1} \\ 0 & 0 & 0 \end{pmatrix},$$

$$L_{8s+6} = \begin{pmatrix} 0 & 0 & t^{2s+1} \\ 0 & 0 & 0 \\ 0 & 0 & 0 \end{pmatrix}, \quad L_{8s+7} = \begin{pmatrix} 0 & 0 & 0 \\ t^{2s+2} & 0 & 0 \\ 0 & t^{2s+2} & 0 \end{pmatrix}.$$

Show that $[L_i, L_j] = d_{ij} L_{i+j}$ $(i, j \in \mathbb{Z})$, where $d_{ij} \in \mathbb{Z}$ depend on i, j mod 8; compute them.

8.17. Construct a "Chevalley basis" of $\hat{\mathcal{L}}(\mathfrak{g}, \mu)$ (cf. Exercise 7.14).

§8.9. Bibliographical notes and comments.

The realization of the "centerless" twisted affine Lie algebras was given in Kac [1968 A, B]; so was the application to the classification of symmetric spaces. The application to the classification of all finite order automorphisms was found in Kac [1969 A]; a detailed exposition of this is given in Helgason [1978]. The present proof of Theorem 8.5 is simpler than the original proof. It resulted from conversations with G. Segal and D. Peterson.

Exercises 8.1–8.4 are taken from Kac [1969 A] and Helgason [1978]. As B. Weisfeiler pointed out, Exercise 8.6 means that $H^1(\mathbb{Z}/m\mathbb{Z}, \widetilde{G}) = \{1\}$; note also that $H^0(\mathbb{Z}/m\mathbb{Z}, G) = G^\alpha$. Steinberg [1965] proved that $H^1(\mathbb{Z}/m\mathbb{Z}, G(\mathbb{C}((t))) = 1$. Exercises 8.12–8.14 are taken from Kac [1978 A].

Levstein [1988] has classified the involutions of all affine algebras. For a classification of finite order automorphisms and conjugate linear automorphisms see Bausch–Rousseau [1989], Kac–Peterson [1987], Rousseau [1989].

There is an intriguing connection of the material of Chapter 8 with invariant theory. Namely, given a $\mathbb{Z}/m\mathbb{Z}$-gradation of a simple Lie algebra $\mathfrak{g} = \bigoplus_j \mathfrak{g}_j$ corresponding to an m-th order automorphism σ, we get a $\mathfrak{g}_0$-module $\mathfrak{g}_1$; the corresponding connected reductive linear group G acting on the space $\mathfrak{g}_1$ is called a σ-group. These groups have many nice properties (see Kac [1975], Popov [1976], Vinberg [1976], Kac–Popov–Vinberg [1976]), but the most remarkable thing is that σ-groups almost exhaust all "nice" irreducible linear groups (see Kac [1980 D] for a review of these theories).

I believe that this is an indication of a deep connection between the theory of infinite-dimensional Lie algebras and groups, and invariant theory. We have already discussed one aspect of such a connection in §5.14.

Chapter 9. Highest-Weight Modules over Kac–Moody Algebras

§**9.0.** In this chapter we begin to develop the representation theory of Kac–Moody algebras. Here we introduce the so-called category $\mathcal{O}$, which is roughly speaking the category of restricted $\mathfrak{h}$-diagonalizable modules (the precise definition is given below). We study the "elementary" objects of this category, the so-called Verma modules, and their connection with irreducible modules. We discuss the problems of irreducibility and complete reducibility in the category $\mathcal{O}$. We find, as an application of the representation theory, the defining relations of Kac–Moody algebras with a symmetrizable Cartan matrix. At the end of the chapter we discuss the category σ for the infinite-dimensional Heisenberg algebra and the Virasoro algebra.

§**9.1.** As in Chapters 1 and 2, we start with an arbitrary complex $n \times n$ matrix A and consider the associated Lie algebra $\mathfrak{g}(A)$. Recall the triangular decomposition:

$$\mathfrak{g}(A) = \mathfrak{n}_- \oplus \mathfrak{h} \oplus \mathfrak{n}_+.$$

We have the corresponding decomposition of the universal enveloping algebra:

$$U(\mathfrak{g}(A)) = U(\mathfrak{n}_-) \otimes U(\mathfrak{h}) \otimes U(\mathfrak{n}_+). \tag{9.1.1}$$

Recall that a $\mathfrak{g}(A)$-module V is called $\mathfrak{h}$-diagonalizable if it admits a weight space decomposition $V = \bigoplus_{\lambda \in \mathfrak{h}^*} V_\lambda$ by weight spaces V_λ (see §3.6). A nonzero vector from V_λ is called a *weight vector* of *weight* λ. Let $P(V) = \{\lambda \in \mathfrak{h}^* | V_\lambda \neq 0\}$ denote the set of weights of V. Finally, for $\lambda \in \mathfrak{h}^*$ set $D(\lambda) = \{\mu \in \mathfrak{h}^* | \mu \leq \lambda\}$.

The category $\mathcal{O}$ is defined as follows. Its objects are $\mathfrak{g}(A)$-modules V which are $\mathfrak{h}$-diagonalizable with finite-dimensional weight spaces and such that there exists a finite number of elements $\lambda_1, \ldots, \lambda_s \in \mathfrak{h}^*$ such that

$$P(V) \subset \bigcup_{i=1}^{s} D(\lambda_i).$$

The morphisms in $\mathcal{O}$ are homomorphisms of $\mathfrak{g}(A)$-modules.

Note that (by Proposition 1.5) any submodule or quotient module of a module from the category $\mathcal{O}$ is also in $\mathcal{O}$. Also, it is clear that a sum or tensor product of a finite number of modules from $\mathcal{O}$ is again in $\mathcal{O}$. Finally, remark that every module from $\mathcal{O}$ is restricted (see §2.5).

§9.2. Important examples of modules from the category $\mathcal{O}$ are highest-weight modules. A $\mathfrak{g}(A)$-module V is called a ***highest-weight module*** with ***highest weight*** $\Lambda \in \mathfrak{h}^*$ if there exists a nonzero vector $v_\Lambda \in V$ such that

(9.2.1) $$\mathfrak{n}_+(v_\Lambda) = 0; \ h(v_\Lambda) = \Lambda(h)v_\Lambda \text{ for } h \in \mathfrak{h}; \text{and}$$

(9.2.2) $$U(\mathfrak{g}(A))(v_\Lambda) = V.$$

The vector v_Λ is called a ***highest-weight vector***. Note that by (9.1.1), condition (9.2.2) can be replaced by

(9.2.3) $$U(\mathfrak{n}_-)(v_\Lambda) = V.$$

It follows from (9.2.1 and 3) that

(9.2.4) $$V = \bigoplus_{\lambda \leq \Lambda} V_\lambda; \quad V_\Lambda = \mathbf{C}v_\Lambda; \quad \dim V_\lambda < \infty.$$

Hence a highest-weight module lies in $\mathcal{O}$, and every two highest-weight vectors are proportional.

A $\mathfrak{g}(A)$-module $M(\Lambda)$ with highest weight Λ is called a *Verma module* if every $\mathfrak{g}(A)$-module with highest weight Λ is a quotient of $M(\Lambda)$.

PROPOSITION 9.2. a) *For every $\Lambda \in \mathfrak{h}^*$ there exists a unique up to isomorphism Verma module $M(\Lambda)$.*

b) *Viewed as a $U(\mathfrak{n}_-)$-module, $M(\Lambda)$ is a free module of rank 1 generated by a highest-weight vector.*

c) *$M(\Lambda)$ contains a unique proper maximal submodule $M'(\Lambda)$.*

Proof. If $M_1(\Lambda)$ and $M_2(\Lambda)$ are two Verma modules, then by definition there exists a surjective homomorphism of $\mathfrak{g}(A)$-modules $\psi : M_1(\Lambda) \to M_2(\Lambda)$. In particular, $\psi(M_1(\Lambda)_\lambda) = M_2(\Lambda)_\lambda$ and hence $\dim M_1(\Lambda)_\lambda \geq \dim M_2(\Lambda)_\lambda$. Exchanging $M_1(\Lambda)$ and $M_2(\Lambda)$ proves that ψ is an isomorphism.

To prove the existence of a Verma module, consider the left ideal $J(\Lambda)$ in $U(\mathfrak{g}(A))$ generated by $\mathfrak{n}_+$ and the elements $h - \Lambda(h)$ $(h \in \mathfrak{h})$, and set

$$M(\Lambda) = U(\mathfrak{g}(A))/J(\Lambda).$$

The left multiplication on $U(\mathfrak{g}(A))$ induces a structure of $U(\mathfrak{g}(A))$-module on $M(\Lambda)$. It is clear that $M(\Lambda)$ is a $\mathfrak{g}(A)$-module with highest weight Λ, the highest-weight vector being the image of $1 \in U(\mathfrak{g}(A))$. If now V is a $\mathfrak{g}(A)$-module with highest weight Λ, then the annihilator of $V_\Lambda \subset V$ is a left ideal J_1 which contains $J(\Lambda)$. Hence, $V \simeq U(\mathfrak{g}(A))/J_1$ and we have an epimorphism of $\mathfrak{g}(A)$-modules $M(\Lambda) \to V$,which proves a).

b) follows from the explicit construction of $M(\Lambda)$ given above and the Poincaré–Birkhoff–Witt theorem. c) follows from the fact that a sum of proper submodules in $M(\Lambda)$ is again a proper submodule (since, by Proposition 1.5, every submodule of $M(\Lambda)$ is graded with respect to the weight space decomposition and does not contain $M(\Lambda)_\Lambda$).

□

Remark 9.2. One can also obtain $M(\lambda)$ via the construction of an induced module. Let V be a left module over a Lie algebra $\mathfrak{a}$, and suppose we are given a Lie algebra homomorphism $\psi : \mathfrak{a} \to \mathfrak{b}$. Recall that the induced $\mathfrak{b}$-module is defined by

$$U(\mathfrak{b}) \otimes_{U(\mathfrak{a})} V := (U(\mathfrak{b}) \otimes_{\mathbf{C}} V)/\sum_{a,b,v} \mathbf{C}(b\psi(a) \otimes v - b \otimes a(v)),$$

where the summation is over $b \in U(\mathfrak{b})$, $a \in \mathfrak{a}$, $v \in V$, and the action of $\mathfrak{b}$ is induced by left multiplication in $U(\mathfrak{b})$. Define the $(\mathfrak{n}_+ + \mathfrak{h})$-module $\mathbf{C}_\lambda$ with underlying space $\mathbf{C}$ by $\mathfrak{n}_+(1) = 0$, $h(1) = \langle\lambda, h\rangle 1$ for $h \in \mathfrak{h}$. Then

$$M(\lambda) = U(\mathfrak{g}(A)) \otimes_{U(\mathfrak{n}_+ + \mathfrak{h})} \mathbf{C}_\lambda.$$

§9.3. It follows from Proposition 9.2c) that among the modules with highest weight Λ there is a unique irreducible one, namely the module

$$L(\Lambda) = M(\Lambda)/M'(\Lambda).$$

Clearly, $L(\Lambda)$ is a quotient of any module with highest weight Λ.

To show that the $L(\Lambda)$ exhaust all irreducible modules from the category $\mathcal{O}$, as well as for some other purposes, we introduce the following notion. Let V be a $\mathfrak{g}(A)$-module. A vector $v \in V_\lambda$ is called ***primitive*** if there exists a submodule U in V such that

$$v \notin U; \quad \mathfrak{n}_+(v) \subset U.$$

Then λ is called a ***primitive weight***. Similarly, one defines primitive vectors and weights for a $\mathfrak{g}'(A)$-module. A weight vector v such that $\mathfrak{n}_+(v) = 0$ is obviously primitive.

PROPOSITION 9.3. *Let V be a nonzero module from the category $\mathcal{O}$. Then*
a) *V contains a nonzero weight vector v such that $\mathfrak{n}_+(v) = 0$.*
b) *The following conditions are equivalent:*
(i) *V is irreducible;*
(ii) *V is a highest-weight module and any primitive vector of V is a highest-weight vector;*
(iii) *$V \simeq L(\Lambda)$ for some $\Lambda \in \mathfrak{h}^*$.*
c) *V is generated by its primitive vectors as a $\mathfrak{g}(A)$-module.*

Proof. To prove a), take a maximal $\lambda \in P(V)$ (with respect to the ordering $\geq$). Then one can take v to be a weight vector of weight λ.

Let V be an irreducible module; then a weight vector v is primitive if and only if $\mathfrak{n}_+(v) = 0$. Take a primitive vector v of weight λ. Then $U(\mathfrak{g})(v)$ is a submodule of V, hence $V = U(\mathfrak{g})(v)$ and V is a module with highest weight λ. In particular, $P(V) \leq \lambda$ and $\dim V_\lambda = 1$. Hence every primitive vector is proportional to v, which proves the implication (i) $\Rightarrow$ (ii) of b).

If V is a highest-weight module and $U \subset V$ is a proper submodule, then U contains a primitive vector by a). This proves the implication (ii) $\Rightarrow$ (iii) and the assertion b).

Let V' be the submodule in V generated by all primitive vectors. If $V' \neq V$, then the $\mathfrak{g}(A)$-module V/V' contains a primitive vector v by a). But a weight vector in V which is a preimage of v is a primitive vector.

□

Thus we have a bijection between $\mathfrak{h}^*$ and irreducible modules from the category $\mathcal{O}$, given by $\Lambda \mapsto L(\Lambda)$. Note that $L(\Lambda)$ can also be defined as an irreducible $\mathfrak{g}(A)$-module, which admits a nonzero vector v_Λ such that

$$\mathfrak{n}_+(v_\Lambda) = 0 \text{ and } h(v_\Lambda) = \Lambda(h)v_\Lambda \text{ for } h \in \mathfrak{h}. \tag{9.3.1}$$

Remark 9.3. A module V from the category $\mathcal{O}$ is generated by its primitive vectors even as a $\mathfrak{n}_-$-module. Indeed, a weight vector $v \in V$ is not primitive if and only if $v \in U(\mathfrak{n}_-)U_0(\mathfrak{n}_+)v$ (= the submodule generated by $\mathfrak{n}_+(v)$).

Here and further $U_0(\mathfrak{g})$ denotes the augmentation ideal $\mathfrak{g}U(\mathfrak{g})$ of $U(\mathfrak{g})$. We have the following "Schur lemma."

LEMMA 9.3. $\operatorname{End}_{\mathfrak{g}(A)} L(\Lambda) = \mathbf{C} I_{L(\Lambda)}$.

Proof. If a is an endomorphism of the module $L(\Lambda)$ and v_Λ is a highest-weight vector, then by Proposition 9.3b), we have $a(v_\Lambda) = \lambda v_\Lambda$ for some $\lambda \in \mathbf{C}$. But then $a(u(v_\Lambda)) = \lambda u(v_\Lambda)$ for every $u \in U(\mathfrak{g})$, hence $a = \lambda I_{L(\Lambda)}$.

□

§9.4. Let $L(\Lambda)^*$ be the $\mathfrak{g}(A)$-module contragredient to $L(\Lambda)$. Then $L(\Lambda)^* = \prod_\lambda (L(\Lambda)_\lambda)^*$. The subspace

$$L^*(\Lambda) := \bigoplus_\lambda (L(\Lambda)_\lambda)^*$$

is a submodule of the $\mathfrak{g}(A)$-module $L(\Lambda)^*$. It is clear that the module $L^*(\Lambda)$ is irreducible and that for $v \in \big(L(\Lambda)_\Lambda\big)^*$ one has:

$$\mathfrak{n}_-(v) = 0; \; h(v) = -\langle \Lambda, h \rangle v \text{ for } h \in \mathfrak{h}.$$

Such a module is called an *irreducible module with lowest weight* $-\Lambda$. As in §9.3 we have a bijection between $\mathfrak{h}^*$ and irreducible lowest-weight modules: $\Lambda \mapsto L^*(-\Lambda)$.

Denote by π_Λ the action of $\mathfrak{g}(A)$ on $L(\Lambda)$, and introduce the new action π_Λ^* on the space $L(\Lambda)$ by

$$\pi_\Lambda^*(g)v = \pi_\Lambda(\omega(g))v, \tag{9.4.1}$$

where ω is the Chevalley involution of $\mathfrak{g}(A)$. It is clear that $(L(\Lambda), \pi_\Lambda^*)$ is an irreducible $\mathfrak{g}(A)$-module with lowest weight $-\Lambda$. By the uniqueness theorem, this module can be identified with $L^*(\Lambda)$, and the pairing between $L(\Lambda)$ and $L^*(\Lambda)$ gives us a nondegenerate bilinear form B on $L(\Lambda)$ such that

$$B(g(x), y) = -B(x, \omega(g)(y)) \text{ for all } g \in \mathfrak{g}(A), \; x, y \in L(\Lambda). \tag{9.4.2}$$

A bilinear form on $L(\Lambda)$ which satisfies (9.4.2) is called a *contravariant bilinear form.*

PROPOSITION 9.4. *Every $\mathfrak{g}(A)$-module $L(\Lambda)$ carries a unique up to constant factor nondegenerate contravariant bilinear form B. This form is symmetric and $L(\Lambda)$ decomposes into an orthogonal direct sum of weight spaces with respect to this form.*

Proof. The existence of B was proved above, the uniqueness follows from Lemma 9.3. The symmetry follows from the uniqueness. The fact that $B(L(\Lambda)_\lambda, \; L(\Lambda)_\mu) = 0$ if $\lambda \neq \mu$ follows from (9.4.2) for $g \in \mathfrak{h}$, $x \in L(\Lambda)_\lambda$, $y \in L(\Lambda)_\mu$.

□

A more explicit way to introduce the contravariant bilinear form is the following. Let V be a highest-weight $\mathfrak{g}(A)$-module with a fixed highest-weight vector v_Λ. Given $v \in V$, we define its *expectation value* $\langle v \rangle \in \mathbf{C}$ by

$$v = \langle v \rangle v_\Lambda + \sum_{\alpha \in Q_+ \setminus \{0\}} v_{\Lambda-\alpha}, \text{ where } v_{\Lambda-\alpha} \in V_{\Lambda-\alpha}.$$

Extend the negative Chevalley involution $-\omega$ to an anti-involution $\hat{\omega}$ of $U(\mathfrak{g}(A))$. Due to (9.1.1) we immediately see that

$$\langle \hat{\omega}(a)v_\Lambda \rangle = \langle av_\Lambda \rangle, \quad a \in U(\mathfrak{g}(A)). \tag{9.4.3}$$

It follows that $\langle \hat{\omega}(a)a'v_\Lambda \rangle$ is symetric in $a, a' \in U(\mathfrak{g}(A))$, hence the formula

$$B(av_\Lambda, a'v_\Lambda) = \langle \hat{\omega}(a)a'v_\Lambda \rangle$$

gives a well-defined symmetric bilinear form on V, which is contravariant, and normalized by $B(v_\Lambda, v_\Lambda) = 1$.

§9.5. The underlying statement of most of the complete reducibility theorems is contained in the following lemma.

LEMMA 9.5. *Let V be a $\mathfrak{g}(A)$-module from the category $\mathcal{O}$. If for any two primitive weights λ and μ of V the inequality $\lambda \geq \mu$ implies $\lambda = \mu$, then the module V is completely reducible (i.e., V decomposes into a direct sum of irreducible modules).*

Proof. Set $V^0 = \{v \in V | \mathfrak{n}_+(v) = 0\}$. This is $\mathfrak{h}$-invariant, hence we have the weight space decomposition $V^0 = \bigoplus_{\lambda \in L} V^0_\lambda$, where all elements from L are primitive weights. Let $\lambda \in L$ and $v \in V^0_\lambda$, $v \neq 0$. Then the $\mathfrak{g}(A)$-module $U(\mathfrak{g})(v)$ is irreducible (and hence isomorphic to $L(\lambda)$). Indeed, if this is not the case, then by Proposition 9.3b), we have $U(\mathfrak{n}_-)(v) \cap V^0_\mu \neq 0$ for some $\mu < \lambda$. This contradicts the assumption of the lemma. Therefore, the $\mathfrak{g}(A)$-submodule V' of V generated by V^0 is completely reducible.

It remains to show that $V' = V$. If this is not the case, we consider the $\mathfrak{g}(A)$-module V/V'. Then there exists a weight vector $v \in V$ of weight μ such that $v \notin V'$ but $e_i(v) \in V'$ and $\neq 0$ for some i. But then, since V is from the category $\mathcal{O}$, there exists $\lambda \in L$ such that $\lambda \geq \mu + \alpha_i$, and hence $\lambda > \mu$, which contradicts the assumption of the lemma.

□

§9.6. Unfortunately, a module $V \in \mathcal{O}$ does not always admit a composition series (i.e., a sequence of submodules $V \supset V_1 \supset V_2 \supset \ldots$ such that each V_i/V_{i+1} is irreducible) (see Exercises 10.3 and 10.4). However, one can manage with the following substitute for it.

LEMMA 9.6. *Let $V \in \mathcal{O}$ and $\lambda \in \mathfrak{h}^*$. Then there exists a filtration by a sequence of submodules $V = V_t \supset V_{t-1} \supset \cdots \supset V_1 \supset V_0 = 0$ and a subset $J \subset \{1, \ldots, t\}$ such that:*
(i) if $j \in J$, then $V_j/V_{j-1} \simeq L(\lambda_j)$ for some $\lambda_j \geq \lambda$;
(ii) if $j \notin J$, then $(V_j/V_{j-1})_\mu = 0$ for every $\mu \geq \lambda$.

Proof. Let $a(V, \lambda) = \sum_{\mu \geq \lambda} \dim V_\mu$. We prove the lemma by induction on $a(V, \lambda)$. If $a(V, \lambda) = 0$, then $0 = V_0 \subset V_1 = V$ is the required filtration with $J = \emptyset$. Let $a(V, \lambda) > 0$. Choose a maximal element $\mu \in P(V)$ such that $\mu \geq \lambda$, choose a weight vector $v \in V_\mu$, and let $U = U(\mathfrak{g})(v)$. Then, clearly, U is a highest-weight module. Proposition 9.2c) implies that U contains a maximal proper submodule $\overline{U}$. We have

$$0 \subset \overline{U} \subset U \subset V; \quad U/\overline{U} \simeq L(\mu), \ \mu \geq \lambda.$$

Since $a(\overline{U}, \lambda) < a(V, \lambda)$ and $a(V/U, \lambda) < a(V, \lambda)$, we use induction to get a suitable filtration for $\overline{U}$ and V/U. Combining them we get the required filtration of V.

□

Let $V \in \mathcal{O}$ and $\mu \in \mathfrak{h}^*$. Fix $\lambda \in \mathfrak{h}^*$ such that $\mu \geq \lambda$ and construct a filtration given by Lemma 9.6. Denote by $[V : L(\mu)]$ the number of times μ appears among $\{\lambda_j | j \in J\}$. It is clear that $[V : L(\mu)]$ is independent of the filtration furnished by Lemma 9.6 and of the choice of λ; this number is called the ***multiplicity*** of $L(\mu)$ in V. Note that $L(\mu)$ has a nonzero multiplicity in V if and only if μ is a primitive weight of V.

§9.7. Now we will introduce and study the formal characters of modules from $\mathcal{O}$. For that purpose, we define a certain algebra $\mathcal{E}$ over $\mathbf{C}$. The elements of $\mathcal{E}$ are series of the form

$$\sum_{\lambda \in \mathfrak{h}^*} c_\lambda e(\lambda),$$

where $c_\lambda \in \mathbb{C}$ and $c_\lambda = 0$ for λ outside the union of a finite number of sets of the form $D(\mu)$. The sum of two such series and the multiplication by a number are defined in the usual way. $\mathcal{E}$ becomes a commutative associative algebra if we decree that $e(\lambda)e(\mu) = e(\lambda + \mu)$ and extend by linearity; here the identity element is $e(0)$. The elements $e(\lambda)$ are called formal exponentials. They are linearly independent and are in one-to-one correspondence with the elements λ of $\mathfrak{h}^*$.

Let now V be a module from the category $\mathcal{O}$ and let $V = \bigoplus_{\lambda \in \mathfrak{h}^*} V_\lambda$ be its weight space decomposition. We define the *formal character* of V by

$$\operatorname{ch} V = \sum_{\lambda \in \mathfrak{h}^*} (\dim V_\lambda) e(\lambda).$$

Clearly, $\operatorname{ch} V \in \mathcal{E}$.

First, we prove the following

PROPOSITION 9.7. *Let V be a $\mathfrak{g}(A)$-module from the category $\mathcal{O}$. Then*

$$\operatorname{ch} V = \sum_{\lambda \in \mathfrak{h}^*} [V : L(\lambda)] \operatorname{ch} L(\lambda). \tag{9.7.1}$$

Proof. Denote by ϕ the map which associates to each $V \in \mathcal{O}$ the difference $\phi(V) \in \mathcal{E}$ between the left- and right-hand sides of (9.7.1). Then $\phi(L(\lambda)) = 0$, and given an exact sequence of modules $0 \to V_1 \to V_2 \to V_3 \to 0$ we have $\phi(V_2) = \phi(V_1) + \phi(V_3)$.

Using Lemma 9.6, we deduce that given $\lambda \in \mathfrak{h}^*$ there exist modules $M_1, \ldots, M_r \in \mathcal{O}$ such that $(M_i)_\mu = 0$ when $\mu \geq \lambda$, and $\phi(V) = \sum_{i=1}^{r} \phi(M_i)$. In particular, for every $\lambda \in \mathfrak{h}^*$ the coefficient at $e(\lambda)$ in $\phi(V)$ is zero.

□

Let us compute the formal character of a Verma module $M(\Lambda)$. Using Proposition 9.2b) and the Poincaré–Birkhoff–Witt theorem, one can construct a basis of the space $M(\Lambda)_\lambda$ as follows. Let $\beta_1, \beta_2, \ldots$ be all the positive roots of the Lie algebra $\mathfrak{g}(A)$, and let $e_{-\beta_s, i_s}$ be a basis of $\mathfrak{g}_{-\beta_s} (1 \leq i_s \leq \operatorname{mult} \beta_s = m_s)$. Let v_Λ be a highest-weight vector of $M(\Lambda)$. Then the vectors

$$e_{-\beta_1,1}^{n_{1,1}} \cdots e_{-\beta_1,m_1}^{n_{1,m_1}} \; e_{-\beta_2,1}^{n_{2,1}} \cdots e_{-\beta_2,m_2}^{n_{2,m_2}} \cdots (v_\Lambda),$$

such that $(n_{1,1} + \cdots + n_{1,m_1})\beta_1 + (n_{2,1} + \cdots + n_{2,m_2})\beta_2 + \cdots = \Lambda - \lambda$ and $n_{i,j} \in \mathbb{Z}_+$, form a basis of $M(\Lambda)_\lambda$. Therefore

$$\operatorname{ch} M(\Lambda) = e(\Lambda) \prod_{\alpha \in \Delta_+} (1 + e(-\alpha) + e(-2\alpha) + \ldots)^{\operatorname{mult} \alpha}.$$

Hence, we have

$$\operatorname{ch} M(\Lambda) = e(\Lambda) \prod_{\alpha \in \Delta_+} (1 - e(-\alpha))^{-\operatorname{mult} \alpha}. \tag{9.7.2}$$

§9.8. Assume now that A is a symmetrizable matrix and let $(.|.)$ be a bilinear form on $\mathfrak{g}(A)$ provided by Theorem 2.2. Then the generalized Casimir operator Ω acts on each module from the category $\mathcal{O}$ (see §2.6).

LEMMA 9.8. a) *If V is a $\mathfrak{g}(A)$-module with highest weight Λ, then*

$$\Omega = (|\Lambda + \rho|^2 - |\rho|^2) I_V .$$

b) *If V is a module from the category $\mathcal{O}$ and v is a primitive vector with weight λ, then there exists a submodule $U \subset V$ such that $v \notin U$ and*

$$\Omega(v) = (|\lambda + \rho|^2 - |\rho|^2) v \mod U.$$

Proof. follows immediately from Corollary 2.6.

□

PROPOSITION 9.8. *Let V be a $\mathfrak{g}(A)$-module with highest weight Λ. Then*

$$\text{ch}\, V = \sum_{\substack{\lambda \leq \Lambda \\ |\lambda+\rho|^2 = |\Lambda+\rho|^2}} c_\lambda \,\text{ch}\, M(\lambda), \text{ where } c_\lambda \in \mathbb{Z},\ c_\Lambda = 1. \tag{9.8.1}$$

Proof. Using (9.7.1), it suffices to prove (9.8.1) for $V = L(\Lambda)$. Set $B(\Lambda) = \{\lambda \leq \Lambda \,||\lambda + \rho|^2 = |\Lambda + \rho|^2\}$, and order the elements of this set, $\lambda_1, \lambda_2, \ldots$ so that the inequality $\lambda_i \geq \lambda_j$ implies $i \leq j$. Then the proof of Proposition 9.7 and Lemma 9.8 imply the following system of equations:

$$\text{ch}\, M(\lambda_i) = \sum_{\lambda_j \in B(\Lambda)} c_{ij} \,\text{ch}\, L(\lambda_j).$$

The matrix (c_{ij}) of this system is triangular with ones on the diagonal. Solving this system proves (9.8.1).

□

§9.9. Now, using the Casimir operator, we can investigate irreducibility and complete reducibility in $\mathcal{O}$.

PROPOSITION 9.9. *Let A be a symmetrizable matrix.*

a) *If $2(\Lambda+\rho|\beta) \neq (\beta|\beta)$ for every $\beta \in Q_+, \beta \neq 0$, then the $\mathfrak{g}(A)$-module $M(\Lambda)$ is irreducible.*

b) *If V is a $\mathfrak{g}(A)$-module from the category $\mathcal{O}$ such that for any two primitive weights λ and μ of V, such that $\lambda-\mu=\beta>0$, one has $2(\lambda+\rho|\beta) \neq (\beta|\beta)$, then V is completely reducible.*

Proof. Proposition 9.3b) implies that if $M(\Lambda)$ is not irreducible, then there exists a primitive weight $\lambda = \Lambda-\beta$, where $\beta>0$. But then Lemma 9.8a) gives: $2(\Lambda+\rho|\beta) = (\beta|\beta)$, proving a).

To prove b) we may assume that the $\mathfrak{g}(A)$-module V is indecomposable. Since, clearly, Ω is locally finite on V, i.e., every $v \in V$ lies in a finite-dimensional Ω-invariant subspace, we obtain that there exists $a \in \mathbb{C}$ such that $\Omega - aI$ is locally nilpotent on V. Hence Lemma 9.8b) implies $|\lambda+\rho|^2 = |\mu+\rho|^2$ for any two primitive weights λ and μ. Now b) follows from Lemma 9.5.

□

§9.10. Here we consider the subalgebra $\mathfrak{g}'(A) = [\mathfrak{g}(A), \mathfrak{g}(A)]$ of $\mathfrak{g}(A)$ instead of $\mathfrak{g}(A)$. Recall that $\mathfrak{g}(A) = \mathfrak{g}'(A) + \mathfrak{h}$, and that $\mathfrak{h}' = \sum_i \mathbb{C}\alpha_i^\vee = \mathfrak{g}'(A) \cap \mathfrak{h}$. Recall the free abelian group Q and the decomposition $U(\mathfrak{g}'(A)) = \bigoplus_{\alpha \in Q} U'_\alpha$ (see §2.6 for details).

A $\mathfrak{g}'(A)$-module V is called a ***highest-weight module*** with ***highest weight*** $\Lambda \in (\mathfrak{h}')^*$ if V admits a Q_+-gradation $V = \bigoplus_{\alpha\in Q_+} V_{\Lambda-\alpha}$ such that $U'_\beta(V_{\Lambda-\alpha}) \subset V_{\Lambda-\alpha+\beta}$, $\dim V_\Lambda = 1$, $h(v) = \Lambda(h)v$ for $h \in (\mathfrak{h}')^*$, $v \in V_\Lambda$, and $V = U(\mathfrak{g}'(A))(V_\Lambda)$. In other words, this is a restriction of a highest-weight module over $\mathfrak{g}(A)$ to $\mathfrak{g}'(A)$. In the same way as in §9.2, we define the Verma module $M(\Lambda)$ over $\mathfrak{g}'(A)$ and show that it contains a unique proper maximal graded submodule $M'(\Lambda)$. We put $L(\Lambda) = M(\Lambda)/M'(\Lambda)$. This is, of course, a restriction of an irreducible $\mathfrak{g}(A)$-module to $\mathfrak{g}'(A)$.

LEMMA 9.10. *The $\mathfrak{g}'(A)$-module $L(\Lambda)$ is irreducible.*

Proof. Let $V \subset L(\Lambda)$ be a nonzero $\mathfrak{g}'(A)$-submodule. We choose $v = \sum_{i=1}^m v_i \in V$ such that $v_i \in L(\Lambda)_{\lambda_i}$, $v_i \neq 0$ and $\sum_i ht(\Lambda-\lambda_i)$ is minimal. If $\lambda_i \neq \Lambda$ for some i, then $e_j(v_{\lambda_i}) \neq 0$ for some j (by Proposition 9.3b). Hence $v \in L(\Lambda)_\Lambda$ and $V = L(\Lambda)$.

□

We shall sometimes describe $\Lambda \in \mathfrak{h}^*$ by its *labels* $\langle \Lambda, \alpha_i^\vee \rangle$ $(i = 1, \ldots, n)$. If $\Lambda, M \in \mathfrak{h}^*$ have the same labels, they may differ only off $\mathfrak{h}'$. Then it is clear from Lemma 9.10 that the modules $L(\Lambda)$ and $L(M)$ when restricted to $\mathfrak{g}'(A)$ are isomorphic as (irreducible) $\mathfrak{g}'(A)$-modules, and the actions of elements of $\mathfrak{g}(A)$ on them differ only by scalar operators. Note that

$$\dim L(\Lambda) = 1 \text{ if and only if } \Lambda|_{\mathfrak{h}'} = 0. \tag{9.10.1}$$

Indeed, if $\Lambda|_{\mathfrak{h}'} = 0$, we can consider the 1-dimensional $\mathfrak{g}(A)$-module $\mathbf{C}$ which is trivial on $\mathfrak{g}'(A)$ and is defined by $h \to \langle \Lambda, h \rangle$ on $\mathfrak{h}$; by the uniqueness of $L(\Lambda)$ (see §9.3), $\mathfrak{g}(A)$-modules $L(\Lambda)$ and $\mathbf{C}$ are isomorphic. The "only if" part follows from the computation: $0 = e_i f_i(v) = \langle \Lambda, \alpha_i^\vee \rangle v$ if v is a highest-weight vector.

If A is a symmetrizable matrix, then $\mathfrak{g}'(A)$ carries a symmetric invariant bilinear form $(.|.)$ which is defined on $\mathfrak{h}'$ by (2.1.2), and whose kernel is $\mathfrak{c}$ (by Theorem 2.2). We put $(\beta|\gamma) = (\nu^{-1}(\beta)|\nu^{-1}(\gamma))$ for for β, $\gamma \in Q$ (see §2.6). We can also define $\rho \in (\mathfrak{h}')^*$ by $\langle \rho, \alpha_i^\vee \rangle = \frac{1}{2} a_{ii}$ $(i = 1, \ldots, n)$.

In the sequel we will use the following version of Proposition 9.9 for the Lie algebra $\mathfrak{g}'(A)$, where A may be infinite.

PROPOSITION 9.10. *Let A be a symmetrizable matrix, possibly infinite.*

a) *If $2\langle \Lambda+\rho, \nu^{-1}(\beta)\rangle \neq (\beta|\beta)$ for every $\beta \in Q_+ \backslash \{0\}$, then the $\mathfrak{g}'(A)$-module $M(\Lambda)$ is irreducible.*

b) *Let V be a $\mathfrak{g}'(A)$-module such that the following three conditions are satisfied:*

(i) *for every $v \in V, e_i(v) = 0$ for all but a finite number of the e_i;*

(ii) *for every $v \in V$ there exists $k > 0$ such that $e_{i_1} \cdots e_{i_s}(v) = 0$ whenever $s > k$;*

(iii) $V = \bigoplus_{\lambda \in (\mathfrak{h}')^*} V_\lambda$, *where $V_\lambda = \{v \in V | h(v) = \langle \lambda, h \rangle v$ for all $h \in \mathfrak{h}'\}$;*

(iv) *if λ and $\mu \in (\mathfrak{h}')^*$ are primitive weights such that $\lambda - \mu = \beta|_{\mathfrak{h}'}$ for some $\beta \in Q_+ \backslash \{0\}$, then $2\langle \lambda + \rho, \nu^{-1}(\beta)\rangle \neq (\beta|\beta)$.*

Then V is completely reducible, i.e., is isomorphic to a direct sum of $\mathfrak{g}'(A)$-modules of the form $L(\Lambda)$, $\Lambda \in (\mathfrak{h}')^*$.

Proof. To prove the proposition, we employ the operator Ω_0 instead of Ω (see §2.5). It is clear that Ω_0 is locally finite on V as long as conditions (i) and (ii) of b) hold. Furthermore, (2.6.1) implies the following fact. Let $v \in V_\lambda$ be such that $(\Omega_0 - aI_V)^k(v) = 0$ for some $k \in \mathbf{Z}_+$ and $a \in \mathbf{C}$, and let $v' \in U'_{-\beta}(v)$ $(\beta \in Q)$; then (see (3.4.1))

$$(\Omega_0 - (a + 2\langle \lambda + \rho, \nu^{-1}(\beta)\rangle - (\beta|\beta))I_V)^k v' = 0. \tag{9.10.2}$$

To prove a), suppose that J is a proper nonzero submodule of $M(\Lambda)$. Take a nonzero element $v' = \sum_i v_i \in J$, where $v_i \in U'_{-\beta_i}(v)$, v is a highest-weight vector of $M(\Lambda)$ and $\beta_i \in Q_+\backslash\{0\}$, such that $\sum_i \operatorname{ht}\beta_i$ is minimal. Then $\mathfrak{n}_+(v_i) = 0$ and therefore $\Omega_0(v_i) = 0$. Note also that $\Omega_0(v) = 0$. Applying (9.10.2) to $V = M(\Lambda)$ and $v' = v_i$, we arrive to a contradiction with the hypothesis of a).

To prove b), put $V^0 = \{v \in V | \mathfrak{n}_+(v) = 0\}$ and $V' = U(\mathfrak{g}'(A))V^0$. Note that V^0 is graded with respect to the decomposition (iii). Using (9.10.2) and (iv) we show that every $v \in V^0 \cap V_\lambda$ generates a $\mathfrak{g}'(A)$-module $L(\lambda)$. Indeed, first we show that the $\mathfrak{g}'(A)$-module $U(\mathfrak{g}'(A))(v)$ is Q_+-graded. In the contrary case, there exists a linear combination of nonzero vectors $\sum_i v_i = 0$, where $v_i \in U'_{-\beta_i}(v), \beta_i \in Q_+$ and $\beta_i \neq \beta_j$ for $i \neq j$, with $\sum_i \operatorname{ht}\beta_i$ minimal. By the same argument as in proof of a) we arrive at a contradiction with (iv). Second, we show that this module is irreducible. By Lemma 9.10 it suffices to show that it has no Q_+-graded ideals. In the contrary case, there exists a nonzero $v' \in U'_{-\beta}(v)$, where $\beta \in Q_+\backslash\{0\}$, such that $\mathfrak{n}_+(v') = 0$. Again, (9.10.2) gives us a contradiction with (iv). Hence the $\mathfrak{g}'(A)$-module V' is completely reducible.

Now we prove that $V' = V$. Suppose the contrary; then there exists a vector $v \in V_\lambda \backslash V'$ such that $\mathfrak{n}_+(v) \subset V'$ and $(\Omega_0 - aI_V)^k(v) = 0$, for some $k \in \mathbb{Z}_+$ and $a \in \mathbb{C}$. Since, clearly, $\Omega_0(v) \in V'$, we have $a = 0$ and hence $\Omega_0^k(v) = 0$. But then there exists $\beta \in Q_+\backslash\{0\}$ and $u \in U'_\beta$ such that $\mathfrak{n}_+(u(v)) = 0$ (by (ii)). Since $\Omega_0(u(v)) = 0$, using again (9.10.2), we arrive at a contradiction with (iv).

□

COROLLARY 9.10. *Let A be a symmetrizable matrix with nonpositive real entries and let V be a $\mathfrak{g}'(A)$-module satisfying conditions* (i), (ii) *and* (iii) *of Proposition 9.10b). Suppose that for every weight λ of V one has $\langle \lambda, \alpha_i^\vee \rangle > 0$ for all i. Then V is a direct sum of irreducible $\mathfrak{g}'(A)$-modules, which are free of rank 1 when viewed as $U(\mathfrak{n}_-)$-modules.*

Proof. We may assume that $A = (a_{ij})$ is a symmetric matrix (replacing the elements $\alpha_i^\vee$ by proportional ones). For $\beta = \sum_i k_i\alpha_i \in Q_+\setminus\{0\}$ and a weight λ of V we have

$$2\langle\lambda+\rho,\nu^{-1}(\beta)\rangle - (\beta|\beta) = 2\sum_i k_i\langle\lambda,\alpha_i^\vee\rangle - \sum_{i\neq j} a_{ij}k_ik_j - \sum_i a_{ii}(k_i^2-k_i) > 0.$$

Now we apply b) and then a) of Proposition 9.10.

□

§**9.11.** Here we give an unexpected application of the results of the representation theory developed in this chapter, to the defining relations of the Lie algebras $\mathfrak{g}(A)$ with a symmetrizable Cartan matrix A. As before, $(.|.)$ denotes a bilinear form on $\mathfrak{g}(A)$ provided by Theorem 2.2. Let $\tilde{\mathfrak{g}}(A)$ be the Lie algebra introduced in §1.2, so that $\mathfrak{g}(A) = \tilde{\mathfrak{g}}(A)/\mathfrak{r}$ and $\mathfrak{r} = \mathfrak{r}_- \oplus \mathfrak{r}_+$. Set $\mathfrak{r}_\alpha = \tilde{\mathfrak{g}}_\alpha \cap \mathfrak{r}$.

PROPOSITION 9.11. *The ideal $\mathfrak{r}_+$ (resp. $\mathfrak{r}_-$) is generated as an ideal in $\tilde{\mathfrak{n}}_+$ (resp. $\tilde{\mathfrak{n}}_-$) by those $\mathfrak{r}_\alpha$ (resp. $\mathfrak{r}_{-\alpha}$) for which $\alpha \in Q_+\setminus\Pi$ and $2(\rho|\alpha) = (\alpha|\alpha)$.*

Proof. We define a Verma module $\widetilde{M}(\lambda)$ over $\tilde{\mathfrak{g}}(A)$ by $U(\tilde{\mathfrak{g}}(A))/\tilde{J}$, where $\tilde{J}$ is the left ideal generated by $\tilde{\mathfrak{n}}_+$ and $h-\lambda(h)$ $(h\in\mathfrak{h})$ (cf. §9.2). Let $\widetilde{M}'(\lambda)$ be the unique (proper) maximal submodule of $\widetilde{M}(\lambda)$. Then we have an isomorphism of $\tilde{\mathfrak{g}}(A)$-modules

$$\widetilde{M}'(0) \simeq \bigoplus_{i=1}^n \widetilde{M}(-\alpha_i). \tag{9.11.1}$$

This is due to the fact that $\tilde{\mathfrak{n}}_-$ is a free Lie algebra (see Theorem 1.2b)) and hence $U(\tilde{\mathfrak{n}}_-)$ is freely generated by $f_1,\ldots,f_n$. The isomorphism (9.11.1) gives us the following isomorphism of $\tilde{\mathfrak{g}}(A)$-modules:

$$\begin{aligned} &U(\mathfrak{g})\otimes_{U(\tilde{\mathfrak{g}})}\widetilde{M}'(0)\\ &\simeq U(\mathfrak{g})\otimes_{U(\tilde{\mathfrak{g}})}\Big(\bigoplus_{i=1}^n\widetilde{M}(-\alpha_i)\Big)\simeq\bigoplus_{i=1}^n M(-\alpha_i).\end{aligned} \tag{9.11.2}$$

Here and further we write $U(\mathfrak{g}),\ldots$ in place of $U(\mathfrak{g}(A)),\ldots$ for short. Let $\pi:\tilde{\mathfrak{g}}(A)\to\mathfrak{g}(A)$ be the canonical homomorphism. We define a map

$\lambda_1 : \mathfrak{r}_- \to U(\mathfrak{g}) \otimes_{U(\tilde{\mathfrak{g}})} \widetilde{M}'(0)$ by $\lambda_1(a) = 1 \otimes a(\tilde{v})$, where $\tilde{v}$ is a highest-weight vector of $\widetilde{M}(0)$. This is a $\tilde{\mathfrak{g}}$-module homomorphism; indeed, for $x \in \tilde{\mathfrak{g}}$, $a \in \mathfrak{r}_-$ we have

$$\lambda_1([x,a]) = 1 \otimes (xa - ax)\tilde{v} = \pi(x) \otimes a(\tilde{v}) - \pi(a) \otimes x(\tilde{v})$$
$$= \pi(x) \otimes a(\tilde{v}) = x(\lambda_1(a)) \text{ since } \pi(a) = 0.$$

Similarly, we get $\lambda_1([\mathfrak{r}_-, \mathfrak{r}_-]) = 0$, so that we have a $\mathfrak{g}(A)$-module homomorphism:

$$\lambda : \quad \mathfrak{r}_-/[\mathfrak{r}_-, \mathfrak{r}_-] \to \bigoplus_{i=1}^{n} M(-\alpha_i)$$

by (9.11.2). More explicitly, λ is described as follows: write $a \in \mathfrak{r}_-$ in the form $a = \sum_i u_i f_i$, where $u_i \in U_0(\tilde{\mathfrak{n}}_-)$; then $\lambda(a + [\mathfrak{r}_-, \mathfrak{r}_-]) = \sum_i \pi(u_i) v_i$, where v_i is a highest-weight vector of $M(-\alpha_i)$. We deduce that λ is injective. Indeed, $\lambda(a + [\mathfrak{r}_-, \mathfrak{r}_-]) = 0$ implies $\pi(u_i) = 0$ for all i, hence $u_i \in \mathfrak{r}_- U_0(\tilde{\mathfrak{n}}_-)$, hence $a \in \mathfrak{r}_- U_0(\tilde{\mathfrak{n}}_-) \cap \mathfrak{r}_-$. Therefore, $a \in [\mathfrak{r}_-, \mathfrak{r}_-]$ by the following general fact: given a Lie algebra $\mathfrak{n}$ and a subalgebra $\mathfrak{r} \subset \mathfrak{n}$, one has

$$\text{(9.11.3)} \qquad \mathfrak{r} \cap \mathfrak{r} U_0(\mathfrak{n}) = [\mathfrak{r}, \mathfrak{r}].$$

Using the Poincaré–Birkhoff–Witt theorem, we reduce (9.11.3) to another fact about an arbitrary Lie algebra $\mathfrak{r}$:

$$\text{(9.11.4)} \qquad \mathfrak{r} \cap U_0(\mathfrak{r})^2 = [\mathfrak{r}, \mathfrak{r}].$$

This follows by passing to $U(\mathfrak{r}/[\mathfrak{r}, \mathfrak{r}])$, which is a polynomial ring.

As a result we have an imbedding $\lambda : \mathfrak{r}_-/[\mathfrak{r}_-, \mathfrak{r}_-] \to \oplus M(-\alpha_i)$ in the category $\mathcal{O}$ of $\mathfrak{g}(A)$-modules. Now let $-\alpha$ ($\alpha \in Q_+$) be a primitive weight of the $\mathfrak{g}(A)$-module $\mathfrak{r}_-/[\mathfrak{r}_-, \mathfrak{r}_-]$.

Note that $\alpha \notin \Pi$ since no f_i lies in $\mathfrak{r}$. Using the embedding λ we conclude by Lemma 9.8 that $|-\alpha_i + \rho|^2 = |-\alpha + \rho|^2$ for some i and therefore $2(\rho|\alpha) = (\alpha|\alpha)$ (since $2(\rho|\alpha_i) = (\alpha_i|\alpha_i)$). Applying Remark 9.3 we deduce that $\mathfrak{r}_-$ is generated as an ideal in $\tilde{\mathfrak{n}}_-$ by those $\mathfrak{r}_{-\alpha}$ for which $\alpha \in Q_+ \setminus \Pi$ and $2(\rho|\alpha) = (\alpha|\alpha)$. This completes the proof of the proposition for $\mathfrak{r}_-$. The result for $\mathfrak{r}_+$ follows by applying the involution $\tilde{\omega}$ of $\tilde{\mathfrak{g}}(A)$.

□

In the case of a Kac–Moody algebra we deduce the following theorem.

THEOREM 9.11. *Let $\mathfrak{g}(A) = \tilde{\mathfrak{g}}(A)/\mathfrak{r}$ be a Kac–Moody algebra with a symmetrizable Cartan matrix A. Then the elements*

$$(9.11.5) \qquad (\operatorname{ad} e_i)^{1-a_{ij}} e_j, \ i \neq j \ (i,j = 1, \dots, n),$$

$$(9.11.6) \qquad (\operatorname{ad} f_i)^{1-a_{ij}} f_j, \ i \neq j \ (i,j = 1, \dots, n)$$

generate the ideals $\mathfrak{r}_+$ and $\mathfrak{r}_-$, respectively.

Proof. Denote by $\mathfrak{g}_1(A)$ the quotient of $\tilde{\mathfrak{g}}(A)$ by the ideal generated by all elements (9.11.5 and 6) (these relations hold by (3.3.1)). We have the induced Q-gradation $\mathfrak{g}_1(A) = \bigoplus_{\alpha \in Q} \mathfrak{g}'_\alpha$. Let $\mathfrak{r}'_\pm$ denote the image of $\mathfrak{r}_\pm$ in $\mathfrak{g}_1(A)$. Suppose the contrary to the statement of the theorem: $\mathfrak{r}'_+ \neq 0$ (the case of $\mathfrak{r}'_-$ is obtained by applying ω). Choose a root α of minimal height among the roots $\alpha \in Q_+ \backslash \{0\}$ such that $\mathfrak{r}'_+ \cap \mathfrak{g}'_\alpha \neq 0$, and let $\alpha = \sum_i k_i \alpha_i$. It is clear that $\mathfrak{r}'_+ \cap \mathfrak{g}'_\alpha$ must occur in any system of homogeneous generators of $\mathfrak{r}'_+$.

It follows from the proofs of §3.8 (which used only the relations (3.3.1)) that there is $\tilde{r}_i \in \operatorname{Aut} \mathfrak{g}_1(A)$ such that $\tilde{r}_i(\mathfrak{g}'_\alpha) = \mathfrak{g}'_{r_i(\alpha)}$ and $\tilde{r}_i(\mathfrak{r}'_+) = \mathfrak{r}'_+$ for all i. Hence $\operatorname{ht} r_i(\alpha) \geq \operatorname{ht} \alpha$ for all $i = 1, \dots, n$, and hence $(\alpha|\alpha_i) \leq 0$ for all i, where $(.|.)$ is a standard form. Therefore $(\alpha|\alpha) \leq 0$. But $2(\rho|\alpha) = 2\sum_i k_i(\rho|\alpha_i) = \sum_i k_i(\alpha_i|\alpha_i) > 0$ and we arrive at a contradiction with Proposition 9.11.

□

Theorem 9.11 implies the following definition of a Kac–Moody algebra in terms of generators and relations. Let $A = (a_{ij})_{i,j=1}^n$ be a symmetrizable generalized Cartan matrix and let $(\mathfrak{h}, \Pi, \Pi^\vee)$ be a realization of A (see §1.1). Then $\mathfrak{g}(A)$ is a Lie algebra on generators e_i, f_i $(i = 1, \dots, n)$, $\mathfrak{h}$ and the following defining relations:

$$[e_i, f_j] = \delta_{ij} \alpha_i^\vee, \quad [h, e_i] = \langle \alpha_i, h \rangle e_i, \quad [h, f_i] = -\langle \alpha_i, h \rangle f_i,$$

$$[h, h'] = 0 \ (h, h' \in \mathfrak{h}), \quad (\operatorname{ad} e_i)^{1-a_{ij}} e_j = 0,$$

$$(\operatorname{ad} f_i)^{1-a_{ij}} f_j = 0 \ (i,j = 1, \dots, n; i \neq j).$$

Furthermore, $\mathfrak{g}'(A)$ has the presentation described in the Introduction (see §0.3).

§9.12. Given below are some other important applications of Proposition 9.11.

PROPOSITION 9.12. *Let $A = (a_{ij})$ be a symmetrizable $n \times n$-matrix, $\mathfrak{g}(A)$ the associated Lie algebra, Δ_+ the set of positive roots, $\mathfrak{g}(A) = \mathfrak{n}_- \oplus \mathfrak{h} \oplus \mathfrak{n}_+$ the triangular decomposition, etc.*

a) If a_{ij} are nonzero real numbers of the same sign for all $i, j = 1, \dots, n$, then $\mathfrak{n}_+$ (resp. $\mathfrak{n}_-$) is a free Lie algebra on generators $e_1, \dots, e_n$ (resp. $f_1, \dots, f_n$).

b) Let $L \subset \Delta_+$ be such that

(i) $(\alpha|\beta)$ are nonzero real numbers of the same sign for all $\alpha, \beta \in L$;

(ii) $\alpha, \beta \in L$ and $\alpha - \beta \in \Delta_+$ imply that $\alpha - \beta \in L$.

Let $\mathfrak{n}_\pm^L$ be the subalgebra of $\mathfrak{n}_\pm$ generated by $\bigoplus_{\alpha \in L} \mathfrak{g}_{\pm\alpha}$; set $\mathfrak{h}^L = \sum_{\alpha \in L} \mathbb{C}\nu^{-1}(\alpha)$, and

$$\mathfrak{g}^L = \mathfrak{n}_-^L \oplus \mathfrak{h}^L \oplus \mathfrak{n}_+^L. \tag{9.12.1}$$

For $\alpha \in L$ set $\mathfrak{g}_{\pm\alpha}^0 = \{x \in \mathfrak{g}_{\pm\alpha} | (x|y) = 0 \text{ for all } y \in [\mathfrak{n}_\mp^L, \mathfrak{n}_\mp^L]\}$. Let I be a (possibly infinite) set containing $\alpha \in L$ with multiplicity $\dim \mathfrak{g}_\alpha^0$, and consider the matrix $B = ((\alpha|\beta))_{\alpha,\beta \in I}$ and the associated Lie algebra $\mathfrak{g}'(B)$. Then $\mathfrak{g}^L$ is a Lie algebra isomorphic to a quotient of $\mathfrak{g}'(B)$ by a central ideal (which lies in the Cartan subalgebra of $\mathfrak{g}'(B)$). Furthermore, (9.12.1) is induced by the triangular decomposition of $\mathfrak{g}'(B)$, and $\mathfrak{n}_+^L$ (resp. $\mathfrak{n}_-^L$) is a free Lie algebra on a basis of the space $\bigoplus_{\alpha \in L} \mathfrak{g}_\alpha^0$ (resp. $\bigoplus_{\alpha \in L} \mathfrak{g}_{-\alpha}^0$).

Proof. First we prove a). Changing the elements of the dual root basis by proportional ones, we may assume that the matrix A is symmetric and has positive entries. But then for $\alpha = \sum_i k_i \alpha_i \in Q_+$ we have

$$2(\rho|\alpha) - (\alpha|\alpha) = \sum_i a_{ii}(k_i - k_i^2) - \sum_{i \neq j} a_{ij} k_i k_j < 0$$

if $\alpha \neq 0$ and $\alpha \notin \Pi$. Now we can apply Proposition 9.11.

To prove b) we may assume that L is finite. We use induction on $|L|$ and the statement a) to prove b) and also to prove that $(.|.)$ gives a non-degenerate pairing of $\mathfrak{n}_+^L$ and $\mathfrak{n}_-^L$.

□

COROLLARY 9.12. a) *Let A be a symmetrizable matrix and let $\alpha \in \Delta_+(A)$ be such that $(\alpha|\alpha) \neq 0$. Then $\bigoplus_{k\geq 1} \mathfrak{g}_{k\alpha}$ is a free Lie algebra on a basis of the space $\bigoplus_{k\geq 1} \mathfrak{g}^0_{k\alpha}$, where $\mathfrak{g}^0_{k\alpha} = \{x \in \mathfrak{g}_{k\alpha} | (x|y) = 0$ for all y from the subalgebra generated by $\mathfrak{g}_{-\alpha}, \ldots, \mathfrak{g}_{-(k-1)\alpha}\}$.*

b) *Let A be a generalized Cartan matrix and let $\alpha \in \Delta_+(A)$ be an isotropic root. Then $\mathbf{C}\nu^{-1}(\alpha) \oplus (\bigoplus_{j\neq 0} \mathfrak{g}_{j\alpha})$ is an infinite Heisenberg Lie algebra.*

Proof. a) follows from Proposition 9.12 b) by setting $L = \{k\alpha | k \in \mathbf{Z}_+\}$. If α is an imaginary root of an affine Lie algebra, then b) amounts to Proposition 8.4. Applying Proposition 5.7 and Lemma 3.8 proves b) in the general case.

□

§9.13. Following is an application of the results of §9.10 to representation theory of the infinite-dimensional Heisenberg algebra $\mathfrak{s}$ (= Heisenberg Lie algebra of order ∞; see §2.9). Recall that this is a Lie algebra with a basis $p_i, q_i (i = 1, 2, \ldots)$ and c, with the following commutation relations:

$$[p_i, q_i] = c (i = 1, 2, \ldots), \text{ all the other brackets are zero.} \tag{9.13.1}$$

This is a nilpotent Lie algebra with center $\mathbf{C}c$.

It is well known that for every $a \in \mathbf{C}^\times$, the Lie algebra $\mathfrak{s}$ has an irreducible representation σ_a, called the *canonical commutation relations representation*, on the space $R = \mathbf{C}[x_1, x_2, \ldots]$ of polynomials in infinitely many indeterminates x_i, defined by:

$$\sigma_a(p_i) = a\frac{\partial}{\partial x_i}, \quad \sigma_a(q_i) = x_i, \quad \sigma_a(c) = aI_R.$$

Here and further the operator of multiplication by a polynomial P is denoted by the same symbol. We denote this $\mathfrak{s}$-module by R_a.

Introduce the following commutative subalgebras of $\mathfrak{s}$:

$$\mathfrak{s}_+ = \sum_{j\geq 1} \mathbf{C}p_j; \quad \mathfrak{s}_- = \sum_{j\geq 1} \mathbf{C}q_j.$$

A vector v of an $\mathfrak{s}$-module is called a *vacuum vector* with *eigenvalue* $\lambda \in \mathbf{C}$ if $\mathfrak{s}_+(v) = 0$ and $c(v) = \lambda v$.

Note (cf. §2.9) that $\mathfrak{s}$ can be viewed as the Lie algebra $\mathfrak{g}'(0)/\mathfrak{r}_1$, where $\mathfrak{r}_1 \subset \mathfrak{r}$ is a central ideal, so that $\mathfrak{n}_+$ (resp. $\mathfrak{n}_-$) is identified with $\mathfrak{s}_+$ (resp. $\mathfrak{s}_-$). Since (R, σ_a) is a free $U(\mathfrak{s}_-)$-module of rank 1, it is nothing other than a Verma module over $\mathfrak{s}$, the vector 1 being the highest-weight vector = vacuum vector. Now the general Proposition 9.10 gives, in our special situation, the following result, which can be viewed as an algebraic version of the Stone–von Neumann theorem.

LEMMA 9.13. a) *Let V be an $\mathfrak{s}$-module such that $c = aI_V$, where $a \neq 0$, which has a vacuum vector $v_0 \neq 0$, such that $V = U(\mathfrak{s}_-)(v_0)$. Then the $\mathfrak{s}$-module V is isomorphic to R_a.*
b) *Let V be an $\mathfrak{s}$-module such that c is diagonalizable with nonzero eigenvalues and such that for every $v \in V$ there exists N such that $p_{i_1} \cdots p_{i_n}(v) = 0$ whenever $n > N$. Then V is isomorphic to a direct sum of $\mathfrak{s}$-modules of the form $R_a, a \neq 0$.*

Proof. We can assume in b) that $c = aI_V$ with $a \neq 0$. V may be viewed as a $\mathfrak{g}'(0)$-module for which $\alpha_i^\vee = aI_V$ for all i. But then for every weight λ and for $\beta \in Q$ we have $\langle \lambda, \nu^{-1}(\beta) \rangle = a\mathrm{ht}\beta$. Since $(\beta|\beta) = 0$ and $\rho = 0$, we have $2\langle \lambda + \rho, \nu^{-1}(\beta) \rangle = 2a\mathrm{ht}\beta \neq (\beta|\beta)$ for $\beta \in Q_+ \backslash \{0\}$. Now a) and b) follow from Proposition 9.10 a) and b).

□

COROLLARY 9.13. *Let V be an irreducible $\mathfrak{s}$-module which has a nonzero vacuum vector with a nonzero eigenvalue λ. Then the $\mathfrak{s}$-module V is isomorphic to R_λ.*

The Lie algebra $\mathfrak{s}$ is often extended by a derivation d_0 defined by

$$[d_0, q_j] = m_j q_j, \quad [d_0, p_j] = -m_j p_j,$$

where m_j are some positive integers. The Lie algebra $\mathfrak{a} = (\mathfrak{s} + \mathbb{C}d_0) \oplus \mathfrak{a}_0$, where $\mathfrak{a}_0$ is a finite-dimensional central ideal, is called an *oscillator algebra*. Given $b \in \mathbb{C}$ and $\lambda \in \mathfrak{a}_0^*$, we can extend the $\mathfrak{s}$-module R_a to the $\mathfrak{a}$-module $R_{a,b,\lambda}$ as follows:

$$d_0 \mapsto b + \sum_j m_j x_j \frac{\partial}{\partial x_j}, \quad a \mapsto \langle \lambda, a \rangle I \text{ for } a \in \mathfrak{a}_0.$$

Letting $\mathfrak{s}_0 = \mathbb{C}c + \mathbb{C}d_0 + \mathfrak{a}_0$, we have the trianglur decomposition

$$\mathfrak{r} = \mathfrak{s}_- \oplus \mathfrak{s}_0 \oplus \mathfrak{s}_+.$$

The following proposition is immediate by Lemma 9.13.

PROPOSITION 9.13. *Let V be an $\mathfrak{a}$-module such that $\mathfrak{s}_0$ is diagonalizable and c has only nonzero eigenvalues.*
a) *If there exists $v_0 \in V, v_0 \neq 0$, such that*

$$\mathfrak{s}_+(v_0) = 0, \quad U(\mathfrak{s}_-)v_0 = V,$$

than V is isomorphic to an $\mathfrak{a}$-module $R_{a,b,\lambda}$
b) If for every $v \in V$ there exists N such that $p_{i_1} \dots p_{i_n}(v) = 0$ whenever $n > N$, then V is isomorphic to a direct sum of $\mathfrak{a}$-modules $R_{a,b,\lambda}, a \neq 0$.

Note that the monomial $x_1^{j_1} \dots x_n^{j_n} \in R_{a,b,\lambda}$ is an eigenvector of d_0 with eigenvalue $\sum_k m_k j_k$. Hence, for the $\mathfrak{a}$-module $R = R_{a,b,\lambda}$ with $a \neq 0$, we have:

$$\operatorname{tr}_R q^{d_0} = q^b \prod_{j=1}^{\infty} (1 - q^{m_j})^{-1}. \tag{9.13.2}$$

Here, as usual, for a diagonalizable operator A on a vector space V with eigenvalues $\lambda_1, \lambda_2, \dots$ counting the multiplicities, one defines $\operatorname{tr}_V q^A = \sum_i q^{\lambda_i}$.

As in §9.4, it is easy to see that the $\mathfrak{s}$-module R_a carries a unique ("contravariant") bilinear form B such that $B(1,1) = 1$ and p_n is an operator adjoint to q_n, provided that $a \in \mathbf{R}$. It is also easy to see by induction that distinct monomials are orthogonal with respect to B and that

$$B(x_1^{k_1} \dots x_n^{k_n}, x_1^{k_1} \dots x_n^{k_n}) = a^{\Sigma k_i} \prod_j k_j!. \tag{9.13.3}$$

As in §9.4, B can be written also in the following form:

$$B(P, Q) = \left(P(a\frac{\partial}{\partial x_1}, a\frac{\partial}{\partial x_2}, \dots) Q(x_1, x_2, \dots) \right)(0). \tag{9.13.4}$$

Note finally that for the $\mathfrak{a}$-module $R_{a,b,\lambda}$ the operators d_0 and $s \in \mathfrak{s}_0$ are selfadjoint with respect to B.

§9.14. The following basic notions of the representaion theory of the Virasoro algebra Vir (see §7.3), very similar to that of Kac–Moody algebras, will be used later.

Define the *triangular decomposition* of Vir as follows:

$$\mathrm{Vir} = \mathrm{Vir}_- \oplus \mathrm{Vir}_0 \oplus \mathrm{Vir}_+, \text{ where } \mathrm{Vir}_\pm = \bigoplus_{j>0} \mathbf{C} d_{\pm j}, \mathrm{Vir}_0 = \mathbf{C} c \oplus \mathbf{C} d_0.$$

Given $c, h \in \mathbf{C}$, define a Vir-module V with highest weight (c, h) by the requirement that there exists a nonzero vector $v = v_{c,h}$ such that (cf. §9.2):

$$\mathrm{Vir}_+(v) = 0,\ U(\mathrm{Vir}_-)v = V,\ d_0(v) = hv,\ c(v) = cv.$$

(Here and further we follow the usual physicists practice to denote the eigenvalue of a central element by the same letter, hoping that this will not cause confusion.)

The definition of the Verma module $M(c, h)$ over Vir, as well as the proof of its existence and uniqueness are given in the same way as in §9.2.

It is clear that c acts on $M(c, h)$ as cI. The number c is called the *conformal central charge*. As in §9.7, we easily show that the elements

$$d_{-j_n} \dots d_{-j_2} d_{-j_1}(v_{c,h}), \text{ where } 0 < j_1 \le j_2 \le \dots, \tag{9.14.1}$$

form a basis of $M(c, h)$. Since $[d_0, d_{-n}] = nd_{-n}$, we see that d_0 is diagonalizable on $M(c, h)$ with spectrum $h + \mathbf{Z}_+$ and with the eigenspace decomposition

$$M(c,h) = \bigoplus_{j \in \mathbf{Z}_+} M(c,h)_{h+j}, \tag{9.14.2}$$

where $M(c, h)_{h+j}$ is spanned by elements of the form (9.14.1) with $j_1 + \dots + j_n = j$.

It follows that

$$\dim M(c,h)_{h+j} = p(j), \tag{9.14.3}$$

where $p(j)$ is the classical partition function. (Thus, the Kostant partition function for the Virasoro algebra is just the classical partition function.) Equation (9.14.3) can be rewritten as follows:

$$\mathrm{tr}_{M(c,h)}\, q^{d_0} := \sum_{\lambda} \dim M(c,h)_\lambda q^\lambda = q^h \prod_{j=1}^{\infty} (1 - q^j)^{-1}.$$

As in §9.7, the series $\mathrm{tr}_V\, q^{d_0}$ is called the *formal character* of the Vir-module V.

As in §9.2, one shows that there exists a unique irreducible Vir-module $L(c, h)$ with highest weight (c, h).

The Chevally involution ω of Vir is defined by

$$\omega(d_n) = -d_{-n}, \omega(c) = -c. \tag{9.14.4}$$

The contravariant bilinear form B on a Vir-module is defined as in §9.4. In other words this is a symmetric bilinear form with respect to which d_n and d_{-n} are adjoint operators. Its existence, uniqueness and construction are established in the same way as in §9.4.

§9.15. Exercises.

9.1. Show that every nonzero homomorphism $M(\lambda_1) \to M(\lambda_2)$ is an imbedding.

9.2. Prove that for $V \in \mathcal{O}$ there exists an increasing filtration (in general infinite) by submodules $0 = V_0 \subset V_1 \subset \dots$ such that V_{i+1}/V_i is a highest-weight module.

9.3. Let V be a module with highest weight Λ. Then $\operatorname{ch} V = \sum_{\lambda \leq \Lambda} c_\lambda \operatorname{ch} M(\lambda)$, where $c_\Lambda = 1, c_\lambda \in \mathbb{Z}$, and $c_\lambda = 0$ unless λ is a primitive weight of a Verma module $M(\mu)$, where μ is a primitive weight of V. Show also that if $c_\lambda \neq 0$, then $[M(\Lambda) : L(\lambda)] \neq 0$.

9.4. Let A be a generalized Cartan matrix of finite type, and $V \in \mathcal{O}$ a finitely generated module over $\mathfrak{g}(A)$. Show that V admits a Jordan–Hölder series, i.e., a filtration by submodules $V = V_0 \supset V_1 \supset \cdots \supset V_s = 0$ such that all the modules V_i/V_{i+1} are irreducible. Describe the Jordan–Hölder series for Verma modules over $sl_2(\mathbb{C})$.

9.5. Let $\mathfrak{g}'(0)$ be the derived algebra of the Lie algebra $\mathfrak{g}(0)$ associated to the $n \times n$ zero matrix. Given $c_1, \dots, c_n \in \mathbb{C}$, the following formulas define the structure of a $\mathfrak{g}'(0)$-module on the space $V = \mathbb{C}[x_1, \dots, x_n]$:

$$e_i \to c_i \frac{\partial}{\partial x_i}, \quad f_i \to \text{multiplication by } x_i, \quad \alpha_i^\vee \to c_i I_V.$$

Prove that $V \simeq M(\Lambda)$, where $\langle \Lambda, \alpha_i^\vee \rangle = c_i$ $(i = 1, \dots, n)$. Show that V is irreducible if and only if all c_i are nonzero.

9.6. Let $M(0)$ be the Verma module with highest weight 0 over $\mathfrak{g}(A)$. Then if v is a highest-weight vector, $f_i(v)$ is killed by $\mathfrak{n}_+$, and one has an imbedding $M(-\alpha_i) \subset M(0)$.

9.7. Assume that A is symmetrizable. Show that $M(-\alpha_i)$ is an irreducible module if $2(\rho|\gamma) \neq (\gamma|\gamma)$ for all nonzero $\gamma \in Q_+ \backslash \Pi$.

9.8. Using Exercise 9.7, show that if $2(\rho|\gamma) \neq (\gamma|\gamma)$ for all nonzero $\gamma \in Q_+ \backslash \Pi$, then the submodule $\sum_i M(-\alpha_i) \subset M(0)$ is, in fact, a direct sum $\bigoplus_i M(-\alpha_i)$. Deduce that in this case $\prod_{\alpha \in \Delta_+} (1 - e(-\alpha))^{\operatorname{mult} \alpha} = 1 - \sum_{i=1}^n e(-\alpha_i)$.

9.9. Use Exercise 9.8 to show that if A is symmetrizable and $2(\rho|\gamma) \neq (\gamma|\gamma)$ for all nonzero $\gamma \in Q_+ \backslash \Pi$, then the subalgebra $\mathfrak{n}_+$ (resp. $\mathfrak{n}_-$) of $\mathfrak{g}(A)$ is a free Lie algebra on generators e_i (resp. f_i) $i = 1, \dots, n$. (This is a special case of Proposition 9.11, but this alternative proof is simpler.)

9.10. Prove that the Lie algebra on generators e_i, f_i $(i = 1, \ldots, n)$ and h and defining relations $[e_i, f_j] = \delta_{ij} h$, $[h, e_i] = e_i$, $[h, f_i] = -f_i$ $(i, j = 1, \ldots, n)$ is simple, and therefore $e_1, \ldots, e_n$ generate a free Lie algebra.

9.11. Prove that if A is an indecomposable symmetrizable generalized Cartan matrix of indefinite type, then $\mathfrak{g}(A)$ contains a free Lie algebra of rank 2, and hence has exponential growth, i.e.,

$$\overline{\lim_{j \to \infty}} \log \dim \mathfrak{g}_j(1)/|j| > 0.$$

9.12. Let $\tilde{\mathfrak{p}}$ be a Lie algebra, $\mathfrak{a}$ an ideal in $\tilde{\mathfrak{p}}, \mathfrak{p} = \tilde{\mathfrak{p}}/\mathfrak{a}$. Show that the following sequence of $\mathfrak{p}$-modules is exact:

$$0 \to \mathfrak{a}/[\mathfrak{a}, \mathfrak{a}] \to U_0(\tilde{\mathfrak{p}})/\mathfrak{a}U_0(\tilde{\mathfrak{p}}) \to U_0(\mathfrak{p}) \to 0.$$

9.13. Using Exercise 9.12 construct the following exact sequence of $\mathfrak{g}(A)$-modules:

$$0 \to \mathfrak{r}_-/[\mathfrak{r}_-, \mathfrak{r}_-] \to \bigoplus_{i=1}^{n} M(-\alpha_i) \to M(0) \to \mathbf{C} \to 0,$$

where $\mathbf{C}$ is viewed as a trivial module.

9.14. Let $\mathcal{F}$ be the space of all functions on $\mathfrak{h}^*$ which vanish outside a finite union of sets of the form $D(\lambda)$. Then one can define product (convolution) of two functions $f, g \in \mathcal{F}$ by: $(f * g)(\lambda) = \sum_{\mu \in \mathfrak{h}^*} f(\mu) g(\lambda - \mu)$. Define the delta function $\delta_\lambda(\mu) = \delta_{\lambda,\mu}$. Show that the map $e(\lambda) \mapsto \delta_\lambda$ gives an algebra isomorphism $\mathcal{E} \xrightarrow{\sim} \mathcal{F}$.

9.15. Let A be an affine matrix of type $A_1^{(1)}$. Show that the $\mathfrak{g}'(A)$-module $M(-\rho)$ has an irreducible quotient which is not isomorphic to $L(-\rho)$.

[Let v be a highest-weight vector of $M(-\rho)$. Show that $v - (f_1 f_2 + f_2 f_1)v$ generates a proper submodule of $M(-\rho)$.]

9.16. Suppose that we have two filtrations $'$ and $''$ of $V \in \mathcal{O}$ satisfying conditions of Lemma 9.6. Then there exists a bijection $\sigma : J' \to J''$ such that $\lambda'_j = \lambda''_{\sigma(j)}$ for all $j \in J'$. Furthermore, if v_μ is a primitive vector of V, then $\mu = \lambda'_j$ where j is the least integer such that $v_\mu \in V'_j$.

9.17. Consider the associative algebra $\mathfrak{a}$ on generators $\alpha_n (n \in \mathbf{Z})$, and commutation relations

$$[\alpha_m, \alpha_n] = m\delta_{m,-n}.$$

Let V be an $\mathfrak{a}$-module such that $\alpha_0 \mapsto \mu I$ and for every $v \in V, \alpha_n(v) = 0$ for all but a finite number of $n > 0$. Let

$$L_0 = (\lambda^2 + \mu^2)/2 + \sum_{j>0} \alpha_{-j}\alpha_j,$$

$$L_n = \frac{1}{2}\sum_{j\in\mathbb{Z}} \alpha_{-j}\alpha_{j+n} + i\lambda n\alpha_n \text{ for } n \neq 0.$$

Show that

$$[L_m, L_n] = (m-n)L_{m+n} + \delta_{m,-n}\frac{m^3-m}{12}(1+12\lambda^2)I,$$

producing thereby a representation of Vir with conformal central charge $c = 1 + 12\lambda^2$.

9.18. Consider an associative algebra $\mathfrak{b}$ on generators a_m, a_m^* and $b_m (m \in \mathbb{Z})$ and the following defining relations:

$$[a_m^*, a_n] = \delta_{m,-n}, [b_m, b_n] = m\delta_{m,-n},$$

all other brackets are 0.

Let V be a $\mathfrak{b}$-module such that $b_0 \mapsto 0$ and for every $v \in V$, $a_n(v) = 0$, $a_n^*(v) = 0$ and $b_n(v) = 0$ for all but finitely many $n > 0$. The sign $::$ used below of normal product of generators, as usual, means that the generators with non-negative indices should be moved to the right of the rest of generators. Let

$$H_n = \sum_{j\in\mathbb{Z}} : a_{n-j}a_j^* :, \quad N = \sum_{j>0} ja_{-j}a_j^* + \frac{1}{2}\sum_{j>0} b_{-j}b_j.$$

Given numbers $\lambda, \mu \in \mathbb{C}$, let

$$\pi_{\lambda,\mu}(K) = \frac{1}{2}\mu^2 - 2, \quad \pi_{\lambda,\mu}(d) = -N, \quad \pi_{\lambda,\mu}(e^{(n)}) = a_n,$$

$$\pi_{\lambda,\mu}(f^{(n)}) = (n(\frac{1}{2}\mu^2 - 2) - \lambda)a_n^* - \sum_{j\in\mathbb{Z}} : H_j a_{n-j}^* : -\mu\sum_{j\in\mathbb{Z}} b_j a_{n-j}^*,$$

$$\pi_{\lambda,\mu}(h^{(n)}) = \lambda\delta_{n,0}I + 2H_n + \mu b_n.$$

Show that $\pi_{\lambda,\mu}$ is a restricted representation of the affine algebra $\hat{\mathcal{L}}(s\ell_2(\mathbb{C}))$ on V.

9.19. If in Exercise 9.17 (resp. 9.18) there exists a $\mathfrak{a}$- (resp. $\mathfrak{b}$-) cyclic vector $|0\rangle \in V$ such that $a_n|0\rangle = 0$ for $n > 0$ (resp. $a_n|0\rangle = 0 = b_n|0\rangle$ for $n \geq 0$ and $a_n^*|0\rangle = 0$ for $n > 0$) then the character of V is equal to the character of the Verma module $M(1 + 12\lambda^2, \frac{1}{2}(\lambda^2 + \mu^2))$(resp. $M((\frac{1}{2}\mu^2 - \lambda - 2)\Lambda_0 + \lambda\Lambda_1)$), but V is not necessarily isomorphic to a Verma module.

9.20. In Exercise 9.18 let $b_n = 0$ for all n and let $\mu = 0$. Show that if there exists a $\mathfrak{b}$-cyclic vector $|0\rangle \in V$ such that $a_n|0\rangle = 0$ for $n \geq 0$ and $a_n^*|0\rangle = 0$ for $n > 0$, then

$$\operatorname{ch} V = e((-\lambda - 2)\Lambda_0 - \lambda\Lambda_1) \prod_{\alpha \in \Delta_+^{re}} (1 - e(-\alpha))^{-1}.$$

9.21. (Open problem) Let V be an irreducible module over the Virasoro algebra such that d_0 is a diagonalizable operator whose eigenvalues have finite multiplicities. Then either these multiplicities are ≤ 1 (then V is an irreducible subquotient of a module $U_{\alpha,\beta}$ discussed in Exercise 7.24), or else V is a highest- or a lowest-weight module, i.e., there exists $v \neq 0$ such that $d_i(v) = 0$ for all $i > 0$ or for all $i < 0$.

§9.16. Bibliographical notes and comments.

The category $\mathcal{O}$ of modules over a finite-dimensional semisimple Lie algebra was introduced and studied in Bernstein–Gelfand–Gelfand [1971], [1975], [1976]. There is a vast literature on the subject, which is summarized in the books Dixmier [1974] and Jantzen [1979], [1983]. The first nontrivial results on Verma modules were obtained by Verma [1968]. One of the main techniques of the theory is the Jantzen filtration of a Verma module. The study of this filtration is based on a formula due to Shapovalov [1972] for the determinant of the contravariant form on $M(\lambda)$, lifted from $L(\lambda)$.

The study of the category $\mathcal{O}$ and the highest-weight modules over Kac–Moody algebras was started in Kac [1974]. There have been several developments of this in the papers Garland–Lepowsky [1976], Lepowsky [1979], Kac–Kazhdan [1979], Deodhar–Gabber–Kac [1982], Rocha–Wallach [1982], Ku [1988 A–C], [1989 A, B] and others. Again, the basic tool is the formula for the determinant of the contravariant form, proved in Kac–Kazhdan [1979] (generalizing Shapovalov's formula) and Jantzen's filtration.

This technique was applied by Kac [1978 B], [1979] to initiate a study of highest-weight modules over the Virasoro algebra by computing the determinant of the contravariant form on Verma modules $M(c, h)$. There have been several published proofs of the determinant formula: see Feigin–Fuchs [1982], Thorn [1984], Kac–Wakimoto [1986], and others. The computation of $chL(c, h)$ consists of two steps, based on this formula. First, one finds all possible inclusions of Verma modules, and all cases when $[M(c,h) : L(c,h')] \neq 0$, using the Jantzen filtration (see Kac [1978 B]). Second, one shows that $[M(c,h) : L(c,h')] \leq 1$. This more difficult fact, special for the Virasoro algebra, was conjectured by Kac [1982 B] and proved by Feigin–Fuchs [1983 A, B], [1984 A, B]. The explicit character formulas for all $L(c,h)$ are easily derived from these facts: see Feigin–Fuchs [1983 B], [1984]. (The cases $c \geq 1$ and $c = 0$ were examined previously by Kac [1979] and Rocha–Wallach [1983 A].) Using the determinant formula, Kac [1982 B] proved that $L(c,h)$ is unitarizable for $c \geq 1, h \geq 0$. By a detailed analysis of the determinant formula, the critical strip $0 \leq c < 1$ was examined by Friedan–Qiu–Shenker [1985] who found the (discrete) set of possible places of unitarity. The unitarizability for this set was proved by Goddard–Kent–Olive [1985], [1986], Kac–Wakimoto [1986] and Tsuchia–Kanie [1986 B]. The solution to the problem of integrability of unitary $L(c,h)$ was given by Segal [1981], Neretin [1983] and Goodman–Wallach [1985] (see Pressley–Segal [1986] for an account of this). As a result, the only problem from Kac [1982 B] that remains unsolved is Exercise 9.21. Kaplansky–Santaroubane [1985] solved this problem in the case when the spectrum of d_0 is simple and Chari–Pressley [1988 B] in the case when V is unitary.*

The remarkable link between the theory of highest-weight modules over the Virasoro algebra and conformal field theory and statistical mechanics was discovered by Belavin–Polyakov–Zamolodchikov [1984 A], [1984 B]. Conformal field theory has become by now a huge field with many remarkable ramifications to other fields of mathematics and mathematical physics: see the reviews Goddard–Olive [1986], Ginsparg [1988], Furlan–Sotkov–Todorov [1988], Goddard [1989], and many others. The basics of the representation theory of the Virasoro algebra may be found in the book Kac–Raina [1987].

One of the main problems in the theory of Verma modules is to compute the multiplicities $[M(\lambda) : L(\lambda')]$. In the finite-dimensional case, Kazhdan–Lusztig [1979] came up with a remarkable conjecture, which was soon after proved by Beilinson–Bernstein [1981] and Brylinski–Kashiwara [1981]. Kazhdan–Lusztig conjectures were generalized by Deodhar–Gabber–Kac [1982] to arbitrary Kac–Moody algebras. Quite recently these conjectures were proved by Casian [1989] and by Kashiwara [1989]. (Kazhdan–Lusztig's [1980] cohomological interpretation of the multiplicities was extended to arbitrary Kac–Moody algebras by Haddad [1984].)

The exposition of §§9.1–9.8 is a simplification of that in Deodhar–Gabber–Kac [1982]. A more complete statement than Proposition 9.9 a) can

---

* This problem has been solved recently by O. Mathieu "Classification of Harish-Chandra modules over the Virasoro algebra" and by C. Martin and A. Piard "Indecomposable modules over the Virasoro algebra and a conjecture of V. Kac."

be found in Kac–Kazhdan [1979]. The exposition of §§9.9, 9.10 and 9.12 is based on Kac–Peterson [1984 A]. The results of §9.11 on defining relations are due to Gabber–Kac [1981]. A simple cohomological proof of Theorem 9.11 was found by O. Mathieu (unpublished).

Exercise 9.2 is due to Garland–Lepowsky [1976]. Exercise 9.4 is taken from Bernstein–Gelfand–Gelfand [1971]. Exercises 9.6–9.10 follow Kac [1980 C]. Exercises 9.12 and 9.13 are taken from Gabber–Kac [1981].

Exercise 9.17 is taken from Chodos–Thorn [1974]; its special case goes back to Virasoro [1970]. Exercises 9.18 and 9.20 are due to Wakimoto [1986]. These "free field" constructions of representations of the Virasoro algebra and the affine algebra $\hat{\mathcal{L}}(s\ell_2(\mathbb{C}))$ (and, more generally, $\hat{\mathcal{L}}(s\ell_n(\mathbb{C}))$) play an important role both in the study of representations (Feigin–Frenkel [1988], [1989 A–C]) and in the conformal field theory (Feigin–Fuchs [1984 A], Dotsenko–Fateev [1984], Felder [1989], Bernard–Felder [1989], Bershadsky–Ooguri [1989], Distler–Qiu [1989]), and others.

One last comment concerns Exercises 9.10 and 9.11. The main result of the paper Kac [1968 B] is the following. Let $\mathfrak{g} = \bigoplus_{j\in\mathbb{Z}} \mathfrak{g}_j$ be an infinite-dimensional $\mathbb{Z}$-graded Lie algebra which satisfies the following conditions:
(i) $\overline{\lim}_{j\to\infty} \frac{\log \dim \mathfrak{g}_j}{\log |j|} < \infty$, i.e., $\dim \mathfrak{g}_j$ grows polynomially as $|j| \to \infty$;
(ii) there are no nontrivial graded ideals;
(iii) $\mathfrak{g}_{-1} + \mathfrak{g}_0 + \mathfrak{g}_1$ generate $\mathfrak{g}$ and the $\mathfrak{g}_0$-module $\mathfrak{g}_{-1}$ is irreducible.

Then $\mathfrak{g}$ is isomorphic either to $\mathfrak{g}'(A)/\mathbb{C}K$, where A is an affine matrix, with the gradation of type $(0, \dots, 1, \dots, 0)$, or to one of the simple $\mathbb{Z}$-graded Lie algebras of the polynomial vector fields on $\mathbb{C}^n$: W_n, S_n, H_n, and K_n.

It is actually proved that a Lie algebra satisfying (i)–(iii), which is outside this list, contains a subalgebra of Exercise 9.10 with $n = 2$, and hence has exponential growth, as shown by Kac [1980 C]. (By a different method, exponential growth was proved in the rank 2 hyperbolic case by Meurman [1982]). My conjecture is that if one drops condition (iii), the only algebra which should be added to the list is the "centerless" Virasoro algebra $\mathfrak{d}$ (see §7.3) and its subalgebra $\mathfrak{d}_+ = \sum_{j\geq -1} \mathbb{C}d_j$. A special case of this problem when $\dim \mathfrak{g}_j \leq 1$ has been solved by Mathieu [1986 A] (who showed that in this case the list consists of the $\mathbb{Z}$-graded Lie algebras defined in Exercises 7.12, 7.13 and 8.16). Quite recently Mathieu [1986 B] has solved this problem in the case $\dim \mathfrak{g}_j <$const.** He has shown that any $\mathbb{Z}$-graded Lie algebra $\mathfrak{g} = \oplus \mathfrak{g}_j$ with only trivial graded ideals, is either $\mathfrak{g}'(A)/\mathbb{C}K$ with gradation of type s, or $\mathfrak{d}$ or $\mathfrak{d}_+$.

** Recently Mathieu completely solved the problem in the paper "Classification of simple graded Lie algebras of finite growth."

Chapter 10. Integrable Highest-Weight Modules: the Character Formula

§10.0. The central result of this chapter is the character formula for an integrable highest-weight module $L(\Lambda)$ over a Kac–Moody algebra, which plays a key role in further considerations. We also study the region of convergence of characters, prove a complete reducibility theorem and find a product decomposition for the "q-dimension" of $L(\Lambda)$.

§10.1. Let $\mathfrak{g}(A)$ be a Kac–Moody algebra of rank n and let $\mathfrak{h}$ be its Cartan subalgebra. Set

$$P = \{\lambda \in \mathfrak{h}^* | \langle \lambda, \alpha_i^\vee \rangle \in \mathbf{Z}\ (i = 1, \dots, n)\},$$
$$P_+ = \{\lambda \in P | \langle \lambda, \alpha_i^\vee \rangle \geq 0\ (i = 1, \dots, n)\},$$
$$P_{++} = \{\lambda \in P | \langle \lambda, \alpha_i^\vee \rangle > 0\ (i = 1, \dots, n)\}.$$

The set P is called the *weight lattice*, elements from P (resp. P_+ or P_{++}) are called *integral weights* (resp. *dominant* or *regular dominant integral weights*). Note that P contains the root lattice Q.

Let V be a highest-weight module over $\mathfrak{g}(A)$, and v a highest-weight vector. It follows from Lemmas 3.4 b) and 3.5 that the module V is integrable if and only if $f_i^{N_i}(v) = 0$ for some $N_i > 0$ $(i = 1, \dots, n)$ (see §3.6 for the definition of an integrable module).

LEMMA 10.1. *The $\mathfrak{g}(A)$-module $L(\Lambda)$ is integrable if and only if $\Lambda \in P_+$.*

Proof. Formula (3.2.4) implies the "only if" part and the following formula:

$$e_i f_i^{\langle \Lambda, \alpha_i^\vee \rangle + 1}(v) = 0 \text{ if } \langle \Lambda, \alpha_i^\vee \rangle \in \mathbf{Z}_+,$$

where v is a highest-weight vector. It follows (since $[e_j, f_i] = 0$ for $j \neq i$) that if the vector $f_i^{\langle \Lambda, \alpha_i^\vee \rangle + 1}(v)$ is nonzero, it is a primitive vector of $L(\Lambda)$, which is impossible (by Proposition 9.3b). Hence, for $\Lambda \in P_+$, we have

$$f_i^{\langle \Lambda, \alpha_i^\vee \rangle + 1}(v) = 0 \quad (i = 1, \dots, n), \tag{10.1.1}$$

which proves the "if" part (by the remark preceding this lemma). □

Denote by $P(\Lambda)$ the set of weights of $L(\Lambda)$. It is clear that $P(\Lambda) \subset P$ if $\Lambda \in P$. The following proposition follows from Lemma 10.1 and Proposition 3.7a).

PROPOSITION 10.1. *If $\Lambda \in P_+$, then*

$$\text{mult}_{L(\Lambda)}\, \lambda = \text{mult}_{L(\Lambda)}\, w(\lambda) \text{ for } w \in W.$$

In particular, $P(\Lambda)$ is W-invariant.

□

COROLLARY 10.1. *If $\Lambda \in P_+$ then any $\lambda \in P(\Lambda)$ is W-equivalent to a unique $\mu \in P_+ \cap P(\Lambda)$.*

Proof. Take $\mu \in W \cdot \lambda$ such that $\text{ht}(\Lambda - \mu)$ is minimal; Lemma 3.12 b) implies the uniqueness.

□

§10.2. We let the Weyl group W act on the complex vector space $\tilde{\mathcal{E}}$ of all (possibly infinite) linear combinations of formal exponentials by

$$w(\sum_\lambda c_\lambda e(\lambda)) = \sum_\lambda c_\lambda e(w(\lambda)) \quad (w \in W).$$

The space $\tilde{\mathcal{E}}$ contains $\mathcal{E}$ as a subspace. However, product of two elements $P_1, P_2 \in \tilde{\mathcal{E}}$ doesn't always make sense, but if it does, then $w(P_1P_2) = w(P_1)w(P_2)$. Proposition 10.1 says that

$$w(\text{ch}\, L(\Lambda)) = \text{ch}\, L(\Lambda) \text{ for } w \in W \text{ and } \Lambda \in P_+ \tag{10.2.1}$$

Consider now the element (cf. §9.7)

$$R = \prod_{\alpha \in \Delta_+} (1 - e(-\alpha))^{\text{mult}\,\alpha} \in \mathcal{E}.$$

Fix an element $\rho \in \mathfrak{h}^*$ such that (cf. §2.5)

$$\langle \rho, \alpha_i^\vee \rangle = 1 \quad (i = 1, \ldots, n).$$

For $w \in W$ set $\epsilon(w) = (-1)^{\ell(w)}$. By (3.11.1) we have

$$\epsilon(w) = \det_{\mathfrak{h}^*} w.$$

Furthermore, one has

$$w(e(\rho)R) = \epsilon(w)e(\rho)R \text{ for } w \in W. \tag{10.2.2}$$

Indeed, it is sufficient to check (10.2.2) for each fundamental reflection r_i. Recall that by Lemma 3.7, the set $\Delta_+\backslash\{\alpha_i\}$ is r_i-invariant and, by Proposition 3.7, we have $\operatorname{mult} r_i(\alpha) = \operatorname{mult}\alpha$ for $\alpha \in \Delta_+$. Hence,

$$\begin{aligned} r_i(e(\rho)R) &= e(\rho - \alpha_i) r_i(1 - e(-\alpha_i)) r_i \prod_{\alpha\in\Delta_+\backslash\{\alpha_i\}} (1 - e(-\alpha))^{\operatorname{mult}\alpha} \\ &= e(\rho)e(-\alpha_i)(1 - e(\alpha_i)) \prod_{\alpha\in\Delta_+\backslash\{\alpha_i\}} (1 - e(-\alpha))^{\operatorname{mult}\alpha} \\ &= -e(\rho)R = \epsilon(r_i)e(\rho)R. \end{aligned}$$

§10.3. From now on we assume that the generalized Cartan matrix A is symmetrizable. Let $(.|.)$ be a standard invariant bilinear form on $\mathfrak{g}(A)$; recall that $(\alpha_i|\alpha_i) > 0$ $(i = 1, \dots, n)$. The following is the key fact in the proof of the character formula, complete reducibility theorem and other results.

LEMMA 10.3. *Let $\Lambda, \lambda \in P$ be such that $\lambda \le \Lambda$ and $\Lambda + \lambda \in P_+$. Then either $\langle \Lambda + \lambda, \alpha_i^\vee\rangle = 0$ for $i \in \operatorname{supp}(\Lambda - \lambda)$ or $(\Lambda|\Lambda) - (\lambda|\lambda) > 0$. In particular, if $\Lambda \in P_{++}, \lambda \in P_+$ and $\lambda < \Lambda$, then $(\Lambda|\Lambda) - (\lambda|\lambda) > 0$.*

Proof. We have $\lambda = \Lambda - \beta$, where $\beta = \sum_i k_i\alpha_i, k_i \ge 0$. Hence $(\Lambda|\Lambda) - (\lambda|\lambda) = (\Lambda + \lambda|\beta) = \sum_i \frac{1}{2}k_i(\alpha_i|\alpha_i)\langle\Lambda + \lambda, \alpha_i^\vee\rangle$. Since $(\alpha_i|\alpha_i) > 0$ for all i, the lemma follows.

□

§10.4. Now we can prove the following fundamental result of our representation theory.

THEOREM 10.4. *Let $\mathfrak{g}(A)$ be a symmetrizable Kac–Moody algebra, and let $L(\Lambda)$ be an irreducible $\mathfrak{g}(A)$-module with highest weight $\Lambda \in P_+$. Then*

$$\operatorname{ch} L(\Lambda) = \frac{\sum_{w\in W} \epsilon(w)e(w(\Lambda + \rho) - \rho)}{\prod_{\alpha\in\Delta_+} (1 - e(-\alpha))^{\operatorname{mult}\alpha}}. \tag{10.4.1}$$

Proof. Multiplying both sides of (9.8.1) by $e(\rho)R$ and using (9.7.2) we obtain

$$e(\rho)R \operatorname{ch} L(\Lambda) = \sum_{\substack{\lambda\le\Lambda \\ |\lambda+\rho|^2 = |\Lambda+\rho|^2}} c_\lambda e(\lambda + \rho), \tag{10.4.2}$$

where $c_\Lambda = 1$, $c_\lambda \in \mathbb{Z}$. By (10.2.1 and 2) the left-hand side of (10.4.2) is W-skew-invariant (an element $L \in \widetilde{\mathcal{E}}$ is called W-skew-invariant if $w(L) = \epsilon(w)L$). Hence the coefficients in the right-hand side of (10.4.2) have the following property:

(10.4.3) $$c_\lambda = \epsilon(w)c_\mu \text{ if } w(\lambda+\rho) = \mu+\rho \text{ for some } w \in W.$$

Let λ be such that $c_\lambda \neq 0$; then by (10.4.3) we have $c_{w(\lambda+\rho)-\rho} \neq 0$ for every $w \in W$; hence, it follows from (10.4.2) that $w(\lambda+\rho) \leq \Lambda+\rho$. Let $\mu \in \{w(\lambda+\rho)-\rho(w \in W)\}$ be such that $\mathrm{ht}(\Lambda-\mu)$ is minimal. Then, clearly, $\mu+\rho \in P_+$ and $|\mu+\rho|^2 = |\Lambda+\rho|^2$. Applying Lemma 10.3 to the elements $\Lambda+\rho \in P_{++}$ and $\mu+\rho$, we deduce that $\mu = \Lambda$. Thus, $c_\lambda \neq 0$ implies that $\lambda+\rho = w(\Lambda+\rho)$ for some $w \in W$ and in this case, $c_\lambda = \epsilon(w)$ (see (10.4.3)).

But $\Lambda+\rho \in P_{++}$, hence, by Proposition 3.12b), the equality $w(\Lambda+\rho) = \Lambda+\rho$ implies that $w = 1$. Hence, finally, we obtain

$$e(\rho)R\,\mathrm{ch}\,L(\Lambda) = \sum_{w \in W} \epsilon(w)e(w(\Lambda+\rho)),$$

which is (10.4.1).

□

Now set $\Lambda = 0$ in (10.4.1). Since $L(0)$ is the trivial 1-dimensional module over $\mathfrak{g}(A)$ we deduce the following "denominator identity":

(10.4.4) $$\prod_{\alpha \in \Delta_+} (1-e(-\alpha))^{\mathrm{mult}\,\alpha} = \sum_{w \in W} \epsilon(w)e(w(\rho)-\rho).$$

Substituting (10.4.4) in (10.4.1) we obtain another form of the character formula:

(10.4.5) $$\mathrm{ch}\,L(\Lambda) = \sum_{w \in W} \epsilon(w)e(w(\Lambda+\rho)) / \sum_{w \in W} \epsilon(w)e(w(\rho)).$$

Of course, in the case when $\mathfrak{g}(A)$ is a finite-dimensional (semisimple) Lie algebra, (10.4.5) is the classical Weyl character formula, formula (10.4.4) being the Weyl denominator identity.

Finally, remark that in the proof of Theorem 10.4 we used only the fact that $L(\Lambda)$ is an integrable highest-weight module (and never used its irreducibility). This fact is guaranteed by relations (10.1.1). Therefore, if V is a $\mathfrak{g}(A)$-module with highest weight $\Lambda \in P_+$ such that (10.1.1) hold, then $\mathrm{ch}\,V$ is given by formula (10.4.1) and hence $V = L(\Lambda)$. Thus we have

(10.4.6) $$L(\Lambda) = M(\Lambda) / \sum_i (U(\mathfrak{n}_-) f_i^{\langle\Lambda,\alpha_i^\vee\rangle+1} v_\Lambda) \text{ if } \Lambda \in P_+.$$

This can be stated also as follows:

COROLLARY 10.4. *Let A be a symmetrizable generalized Cartan matrix and let $\Lambda \in P_+$. Then the annihilator in $U(\mathfrak{g}(A))$ of a highest-weight vector of the $\mathfrak{g}(A)$-module $L(\Lambda)$ is a left ideal generated by the elements $e_i, f_i^{\langle\Lambda,\alpha_i^\vee\rangle+1}$ and $h-\langle\Lambda, h\rangle$, where $i = 1,\ldots,n$ and $h \in \mathfrak{h}$. In particular, an integrable highest-weight module over $\mathfrak{g}(A)$ is automatically irreducible.*

□

In the finite-dimensional case, (10.4.6) is a result of Harish–Chandra. Another corollary is the following

PROPOSITION 10.4. *Let A be a symmetrizable generalized Cartan matrix and let $\Lambda \in \mathfrak{h}'^*$ be such that $\langle\Lambda, \alpha_i^\vee\rangle \in \mathbb{Z}_+$ for $i = 1,\ldots,n$. Then the $\mathfrak{g}'(A)$-module $L(\Lambda)$ is characterized by the properties that it is irreducible and that there exists a nonzero vector $v \in L(\Lambda)$ such that*

$$\alpha_i^\vee(v) = \langle\Lambda, \alpha_i^\vee\rangle v \text{ and } e_i(v) = 0, \quad i = 1,\ldots,n. \tag{10.4.7}$$

In particular, the definitions of a $\mathfrak{g}'(A)$-module $L(\Lambda), \Lambda \in P_+$, given in the Introduction and in Chapter 9, are equivalent.

Proof. Let V be an irreducible $\mathfrak{g}'(A)$-module which has a nonzero element v satisfying (10.4.7). Considering the gradation of V by eigenspaces of $\alpha_i^\vee$, we see that the element $f_i^{\langle\Lambda,\alpha_i^\vee\rangle+1}(v)$ generates a proper submodule of V and hence is zero. Due to (10.4.6), we have a surjective $\mathfrak{g}'(A)$-module homomorphism $L(\Lambda) \to V$. Lemma 9.10 completes the proof.

□

§10.5. Consider the expansion

$$\prod_{\alpha\in\Delta_+} (1-e(\alpha))^{-\operatorname{mult}\alpha} = \sum_{\beta\in\mathfrak{h}^*} K(\beta)e(\beta), \tag{10.5.1}$$

defining a function K on $\mathfrak{h}^*$ called the (generalized) *partition function* (K in honor of Kostant). Note that $K(\beta) = 0$ unless $\beta \in Q_+$; furthermore, $K(0) = 1$, and $K(\beta)$ for $\beta \in Q_+$ is the number of partitions of β into a sum of positive roots, where each root is counted with its multiplicity. The last remark follows from another form of formula (10.5.1):

$$\sum_{\beta\in Q_+} K(\beta)e(\beta) = \prod_{\alpha\in\Delta_+} (1+e(\alpha)+e(2\alpha)+\ldots)^{\operatorname{mult}\alpha}.$$

Note that (9.7.2) can be rewritten as follows:

$$\text{mult}_{M(\Lambda)}\,\lambda = K(\Lambda - \lambda). \tag{10.5.2}$$

We proceed to rewrite formula (10.4.1) in terms of the partition function. Inserting (10.5.1) into (10.4.1), we have

$$\begin{aligned}\sum_{\lambda\le\Lambda}(\text{mult}_{L(\Lambda)}\,\lambda)e(\lambda) &= \sum_{w\in W}\epsilon(w)e(w(\Lambda+\rho)-\rho)\sum_{\beta\in\mathfrak{h}^*}K(\beta)e(-\beta)\\ &= \sum_{w\in W}\sum_{\beta\in\mathfrak{h}^*}\epsilon(w)K(\beta)e(-\beta+w(\Lambda+\rho)-\rho)\\ &= \sum_{w\in W}\sum_{\lambda\in\mathfrak{h}^*}\epsilon(w)K(w(\Lambda+\rho)-(\lambda+\rho))e(\lambda).\end{aligned}$$

Comparing the coefficients at $e(\lambda)$, we obtain the multiplicity formula:

$$\text{mult}_{L(\Lambda)}\,\lambda = \sum_{w\in W}\epsilon(w)K(w(\Lambda+\rho)-(\lambda+\rho)). \tag{10.5.3}$$

In the finite-dimensional case, this is Kostant's formula.

Warning. The series $e(-\rho)\sum_\beta K(\beta)e(\beta)$ is not W-skew-invariant, though this is the inverse of the W-skew-invariant element $e(\rho)R$. This is possible since the inverse in $\widetilde{\mathcal{E}}$ is not unique.

§10.6. Here we adopt a less formal point of view toward the characters, replacing the formal exponential $e(\lambda)$ by the function e^λ on $\mathfrak{h}$ defined by $e^\lambda(h) = e^{\langle\lambda,h\rangle}$ for $h \in \mathfrak{h}$. We define the *character* ch_V of a $\mathfrak{g}(A)$-module V from the category $\mathcal{O}$ to be the function

$$h \mapsto \text{ch}_V(h) = \sum_{\lambda\in P(V)} \text{mult}_V(\lambda)e^{\langle\lambda,h\rangle},$$

defined on the set $Y(V)$ of the elements $h \in \mathfrak{h}$ such that the series converges absolutely. Note that

$$\text{ch}_V(h) = \text{tr}_V \exp h \text{ for } h \in Y(V).$$

Let us introduce some additional notation. Define the *complexified Tits cone* $X_{\mathbb{C}}$ by

$$X_{\mathbb{C}} = \{x + iy \,|\, x \in X,\ y \in \mathfrak{h}_{\mathbb{R}}\}.$$

Set

$$Y = \{h \in \mathfrak{h} \mid \sum_{\alpha \in \Delta_+} (\text{mult}\,\alpha)|e^{-\langle\alpha,h\rangle}| < \infty\},$$

$$Y_N = \{h \in \mathfrak{h} \mid \text{Re}\langle\alpha_i, h\rangle > N \ (i = 1, \dots, n)\} \text{ for } N \in \mathbf{R}_+.$$

Note that Y lies in $X_{\mathbf{C}}$ (by Proposition 3.12 c)). We also have

$$X_{\mathbf{C}} = \bigcup_{w \in W} w(\overline{Y}_0). \tag{10.6.1}$$

LEMMA 10.6. *Let V be a highest-weight module over $\mathfrak{g}(A)$. Then*

a) *$Y(V)$ is a convex set.*

b) *$Y(V) \supset Y \cap Y_0$.*

c) *$Y(V) \supset Y_{\log n}$.*

Proof. a) is clear from the convexity of the function $|e^\lambda|$. Furthermore, since V is a quotient of a module $M(\Lambda)$, we deduce the following estimate from (10.5.2):

$$\text{mult}_V \lambda \leq K(\Lambda - \lambda),$$

which gives

$$\sum_{\lambda \in \mathfrak{h}^*} (\text{mult}_V \lambda)|e^{\langle\lambda,h\rangle}| \leq |e^{\langle\Lambda,h\rangle}| \sum_{\beta \in Q_+} K(\beta)|e^{-\langle\beta,h\rangle}|.$$

But (10.5.1) implies that for $h \in Y_0$ we have

$$\sum_{\beta \in Q_+} K(\beta)\left|e^{-\langle\beta,h\rangle}\right| = \prod_{\alpha \in \Delta_+} (1 - |e^{-\langle\alpha,h\rangle}|)^{-\text{mult}\,\alpha}.$$

The product part of this formula converges if $h \in Y$, proving b). Part c) follows from b) by the estimate (1.3.2).

□

For the proof of the proposition below we need the following

Remark 10.6. Let $T \subset X_{\mathbf{C}}$ be an open convex W-invariant set. Then

$$T \subset \text{convex hull}\,\Big(\bigcup_{w \in W} w(T \cap Y_0)\Big).$$

Indeed, $T_0 := \bigcup_{w \in W} w(\overline{Y}_0 \backslash Y_0)$ is nowhere dense in $X_{\mathbf{C}}$. Hence every $h \in T$ lies in the convex hull of $T \backslash T_0 = \bigcup_{w \in W} w(T \cap Y_0)$.

□

For a convex set R in a (real) vector space denote the interior of R in metric topology by Int R.

PROPOSITION 10.6. *Let $\mathfrak{g}(A)$ be a Kac–Moody algebra and let $L(\Lambda)$ be an irreducible $\mathfrak{g}(A)$-module with highest weight $\Lambda \in P_+$. Then*

a) *$Y(L(\Lambda))$ is a solid convex W-invariant set, which for every $x \in \mathrm{Int}\, X_{\mathbf{C}}$ contains tx for all sufficiently large $t \in \mathbf{R}_+$.*

b) *$\mathrm{ch}_{L(\Lambda)}$ is a holomorphic function on $\mathrm{Int}\, Y(L(\Lambda))$.*

c) *$Y(L(\Lambda)) \supset \mathrm{Int}\, Y$.*

d) *The series $\sum_{w \in W} \epsilon(w) e^{w(\Lambda+\rho)}$ converges absolutely on $\mathrm{Int}\, X_{\mathbf{C}}$ to a holomorphic function, and diverges absolutely on $\mathfrak{h} \setminus \mathrm{Int}\, X_{\mathbf{C}}$.*

e) *Provided that the Cartan matrix A is symmetrizable, $\mathrm{ch}_{L(\Lambda)}$ can be extended from $Y(L(\Lambda)) \cap X_{\mathbf{C}}$ to a meromorphic function on $\mathrm{Int}\, X_{\mathbf{C}}$.*

Proof. Set $T = \mathrm{Int}\, Y$; it is clear that T is open, convex (see the proof of Lemma 10.6a)), and W-invariant; by Lemma 10.6b), we have $Y(L(\Lambda)) \supset Y \cap Y_0$. Furthermore, Lemma 10.6a) and Proposition 10.1 imply that $Y(L(\Lambda))$ is a convex W-invariant set. Now c) follows from Remark 10.6.

In order to complete the proof of a) we have to show that $X' := \{x \in \mathrm{Int}\, X_{\mathbf{C}} | tx \in Y(L(\Lambda))$ for sufficiently large $t \in \mathbf{R}_+\}$ coincides with $\mathrm{Int}\, X_{\mathbf{C}}$. But again, X' is W-invariant and convex and contains Y_0 by Lemma 10.6c). Again, we can apply Remark 10.6.

The convexity of $|e^\lambda|$ implies that the absolute convergence is uniform on compact sets, which proves b).

In order to prove d) remark that $w(\Lambda+\rho)-(\Lambda+\rho)$ are distinct for distinct $w \in W$ by Proposition 3.12b). It is clear from the proof of Proposition 3.12d) that $(\Lambda+\rho) - w(\Lambda+\rho) \in Q_+$. Hence we have for $h \in Y_0$:

$$|\sum_{w \in W} \epsilon(w) e^{\langle w(\Lambda+\rho)-(\Lambda+\rho), h\rangle}| \leq \sum_{\alpha \in Q_+} |e^{-\langle \alpha, h\rangle}| < \infty.$$

Thus the region of absolute convergence of the series in question contains Y_0 and is clearly convex and W-invariant. Now we apply Remark 10.6. It follows that the series in question converges absolutely on $\mathrm{Int}\, X_{\mathbf{C}}$. On the other hand, let $h \in \mathfrak{h} \setminus \mathrm{Int}\, X_{\mathbf{C}}$. Then the set $\Delta_0 := \{\alpha \in \Delta_+^{re} | \mathrm{Re}\langle \alpha, h\rangle \leq 0\}$ is infinite by Proposition 3.12c) and f), and for every $\alpha \in \Delta_0$, we have $|e^{\langle r_\alpha(\Lambda+\rho), h\rangle}| \geq |e^{\langle \Lambda+\rho, h\rangle}|$, proving the divergence at h.

Finally, e) follows from d) and (10.4.5) (or rather (10.6.1))

□

We shall give a more precise description of $Y(L(\Lambda))$ in Chapter 11.

Denote by $\mathcal{O}_c$ the full subcategory of the category $\mathcal{O}$ of $\mathfrak{g}(A)$-modules V such that ch_V converges absolutely on Y_N for some $N > 0$. Also denote by $\mathcal{E}_c$ the subalgebra of the series from $\mathcal{E}$ which converge absolutely on Y_N for some N. By Lemma 10.6, every highest-weight module lies in $\mathcal{O}_c$. We have a homomorphism ψ of $\mathcal{E}_c$ into the algebra of functions, which are holomorphic on Y_N for some N, defined by $\psi : e(\lambda) \mapsto e^{\lambda}$. Applying ψ to both sides of formulas (10.4.5) and (10.4.4), we obtain on Y:

$$\mathrm{ch}_{L(\Lambda)} = \sum_{w\in W} \epsilon(w)e^{w(\Lambda+\rho)} / \sum_{w\in W} \epsilon(w)e^{w(\rho)}. \tag{10.6.2}$$

$$\prod_{\alpha\in\Delta_+} (1-e^{-\alpha})^{\mathrm{mult}\,\alpha} = \sum_{w\in W} \epsilon(w)e^{w(\rho)-\rho}. \tag{10.6.3}$$

§10.7. Here we prove the complete reducibility theorem.

THEOREM 10.7. *Let A be a symmetrizable generalized Cartan matrix.*
a) *Suppose that a $\mathfrak{g}'(A)$-module V satisfies the following two conditions:*

(10.7.1) for every $v \in V$ there exists N such that
$$e_{i_1}\cdots e_{i_k}(v) = 0 \text{ whenever } k > N.$$

(10.7.2) for every $v \in V$ and every i there exists N such that $f_i^N(v) = 0$.

Then V is isomorphic to a direct sum of $\mathfrak{g}'(A)$-modules $L(\Lambda)$ such that $\langle\Lambda, \alpha_i^\vee\rangle \in \mathbb{Z}_+$, for all i.
b) *Every integrable $\mathfrak{g}(A)$-module V from the category $\mathcal{O}$ is isomorphic to a direct sum of modules $L(\Lambda)$, $\Lambda \in P_+$.*

Proof. Thanks to Proposition 9.10 b) and Remark 3.6 in the case of statement a) (resp. Proposition 9.9b) in the case of statement b)), it suffices to check that if λ and μ are primitive weights and $\beta \in Q_+ \setminus \{0\}$ is such that $\lambda - \mu = \beta$, then
$$2\langle\lambda+\rho, \nu^{-1}(\beta)\rangle \neq (\beta|\beta).$$
Since the module V is integrable, we have by (3.2.4):

$$\langle\lambda, \alpha_i^\vee\rangle \geq 0 \ (i = 1, \ldots, n) \text{ for every primitive weight } \lambda. \tag{10.7.3}$$

But then we can write
$$\begin{aligned} 2\langle\lambda+\rho, \nu^{-1}(\beta)\rangle - (\beta|\beta) &= \langle\lambda + (\lambda-\beta) + 2\rho, \nu^{-1}(\beta)\rangle \\ &= \langle\lambda+\mu+2\rho, \nu^{-1}(\beta)\rangle > 0, \end{aligned}$$
using (10.7.3) and the fact that $\langle\rho, \nu^{-1}(\beta)\rangle > 0$ for all $\beta \in Q_+ \setminus \{0\}$.

□

COROLLARY 10.7. a) *A $\mathfrak{g}(A)$-module V from the category $\mathcal{O}$ is integrable if and only if V is a direct sum of modules $L(\Lambda)$ with $\Lambda \in P_+$.*

b) *Tensor product of a finite number of integrable highest-weight modules is a direct sum of modules $L(\Lambda)$ with $\Lambda \in P_+$.*

Theorem 10.7 contains as a special case the classical Weyl complete reducibility theorem for finite-dimensional representations of semi-simple Lie algebras.

§10.8. Let $s = (s_1, \ldots, s_n)$ be a sequence of integers. In §1.5 we introduced the $\mathbb{Z}$-gradation of type s:

$$\mathfrak{g}(A) = \bigoplus_{j \in \mathbb{Z}} \mathfrak{g}_j(s).$$

A particular case of this is the gradation of type $\mathbb{1} = (1, \ldots, 1)$ called the *principal gradation*. Note that if $s_i > 0$ $(i = 1, \ldots, n)$, we have

$$\dim \mathfrak{g}_j(s) < \infty.$$

Similarly we have the $\mathbb{Z}$-gradation of type s of the dual Kac–Moody algebra

$$\mathfrak{g}({}^tA) = \bigoplus_{j \in \mathbb{Z}} {}^t\mathfrak{g}_j(s).$$

Fix elements $\lambda_s \in \mathfrak{h}^*$ and $h^s \in \mathfrak{h}$ which satisfy

$$\langle \lambda_s, \alpha_i^\vee \rangle = s_i, \ \langle h^s, \alpha_i \rangle = s_i \ (i = 1, \ldots, n).$$

Note that $\lambda_{\mathbb{1}} = \rho$ and $h^{\mathbb{1}} = \rho^\vee$ are the elements ρ for the Kac–Moody algebras $\mathfrak{g}(A)$ and $\mathfrak{g}({}^tA)$, respectively.

Warning: $\nu(\rho^\vee) \neq 2\rho/(\rho|\rho)$.

Provided that all $s_i > 0$, the sequence s defines a homomorphism F_s : $\mathbb{C}[[e(-\alpha_1), \ldots, e(-\alpha_n)]] \to \mathbb{C}[[q]]$ by

$$F_s(e(-\alpha_i)) = q^{s_i} \quad (i = 1, \ldots, n). \tag{10.8.1}$$

This is called the *specialization of type s*. Note that

$$F_s(e(-\alpha)) = q^{\langle h^s, \alpha \rangle}. \tag{10.8.2}$$

PROPOSITION 10.8. *Let $\mathfrak{g}(A)$ be a symmetrizable Kac–Moody algebra. Then*

$$\dim \mathfrak{g}_j(\mathbb{1}) = \dim {}^t\mathfrak{g}_j(\mathbb{1}).$$

Proof. Note that both sides of identity (10.4.4) are elements from the algebra $\mathbf{C}[[e(-\alpha_1),\ldots,e(-\alpha_n)]]$. Applying the homomorphism $F_{\mathbb{1}}$ to both sides of this identity, we deduce

$$\prod_{j\geq 1}(1-q^j)^{\dim \mathfrak{g}_j(\mathbb{1})} = \sum_{w\in W(A)} \epsilon(w) q^{\langle \rho,\rho^\vee\rangle - \langle w(\rho),\rho^\vee\rangle}. \tag{10.8.3}$$

Similarly for $\mathfrak{g}({}^tA)$ we have

$$\prod_{j\geq 1}(1-q^j)^{\dim {}^t\mathfrak{g}_j(\mathbb{1})} = \sum_{w\in W({}^tA)} \epsilon(w) q^{\langle \rho^\vee,\rho\rangle - \langle w(\rho^\vee),\rho\rangle}. \tag{10.8.4}$$

Since $W(A)$ and $W({}^tA)$ are contragredient linear groups, the right-hand sides of (10.8.3 and 4) are equal.

□

Comparing (10.8.3 and 4), we also deduce

$$\prod_{\alpha\in\Delta_+}(1-q^{\langle \rho^\vee,\alpha\rangle})^{\operatorname{mult}\alpha} = \prod_{\alpha\in\Delta_+^\vee}(1-q^{\langle \rho,\alpha\rangle})^{\operatorname{mult}\alpha} \tag{10.8.5}$$

Remark 10.8. In the sequel we use the specialization of type s when some of the s_i are 0. Then F_s is not defined everywhere and we have to check that F_s is defined on a given power series.

§10.9. The specialization of type $\mathbb{1}$ is called the *principal specialization*. The following proposition gives a product decomposition of the principally specialized character.

Proposition 10.9. *Let $\mathfrak{g}(A)$ be a symmetrizable Kac–Moody algebra. Let $\Lambda\in P_+$ and set $s=(\langle\Lambda,\alpha_1^\vee\rangle,\ldots,\langle\Lambda,\alpha_n^\vee\rangle)$. Then*

$$F_{\mathbb{1}}(e(-\Lambda)\operatorname{ch} L(\Lambda)) = \prod_{j\geq 1}(1-q^j)^{\dim {}^t\mathfrak{g}_j(s+\mathbb{1}) - \dim {}^t\mathfrak{g}_j(\mathbb{1})}. \tag{10.9.1}$$

Proof. By (10.4.5) we have

$$e(-\Lambda)\operatorname{ch} L(\Lambda) = \frac{\sum_{w\in W}\epsilon(w)e(w(\Lambda+\rho)-(\Lambda+\rho))}{\sum_{w\in W}\epsilon(w)e(w(\rho)-\rho)}. \tag{10.9.2}$$

For $\lambda\in P_{++}$ set

$$N_\lambda = \sum_{w\in W}\epsilon(w)e(w(\lambda)-\lambda).$$

Note that $N_\lambda \in \mathbb{C}[[e(-\alpha_1), \ldots, e(-\alpha_n)]]$ (by Proposition 3.12 b), d)). We have

$$F_{\mathbb{1}}(N_\lambda) = \sum_{w \in W} \epsilon(w) q^{\langle \lambda, \rho^\vee \rangle - \langle w(\lambda), \rho^\vee \rangle} = \sum_{w \in W} \epsilon(w) q^{\langle \lambda, \rho^\vee - w(\rho^\vee) \rangle}$$
$$= F_r\Big(\sum_{w \in W} \epsilon(w) e(w(\rho^\vee) - \rho^\vee)\Big),$$

where $r = (\langle \lambda, \alpha_1^\vee \rangle, \ldots, \langle \lambda, \alpha_n^\vee \rangle)$. Applying identity (10.4.4) for $\mathfrak{g}({}^tA)$ we obtain

$$F_{\mathbb{1}}(N_\lambda) = F_r\Big(\prod_{\alpha \in \Delta_+^\vee} (1 - e(-\alpha))^{\mathrm{mult}\, \alpha}\Big).$$

Hence

$$F_{\mathbb{1}}(N_\lambda) = \prod_{\alpha \in \Delta_+^\vee} (1 - q^{\langle \lambda, \alpha \rangle})^{\mathrm{mult}\, \alpha}. \tag{10.9.3}$$

So, by (10.9.2 and 3) we have

$$F_{\mathbb{1}}(e(-\Lambda) \operatorname{ch} L(\Lambda)) = \prod_{\alpha \in \Delta_+^\vee} \Big(\frac{1 - q^{\langle \Lambda + \rho, \alpha \rangle}}{1 - q^{\langle \rho, \alpha \rangle}}\Big)^{\mathrm{mult}\, \alpha}. \tag{10.9.4}$$

It is clear that (10.9.4) is an equivalent form of (10.9.1).

□

§10.10. Let $V = \bigoplus_{\lambda \leq \Lambda} V_\lambda$ be a $\mathfrak{g}(A)$-module with highest weight Λ. Again, fix a sequence of nonnegative integers $s = (s_1, \ldots, s_n)$. Let

$$\deg(\Lambda - \sum_i k_i \alpha_i) = \sum_i k_i s_i.$$

Then setting

$$V_j(s) = \bigoplus_{\lambda : \deg \lambda = j} V_\lambda$$

defines the *gradation of* V *of type* s:

$$V = \bigoplus_{j \in \mathbb{Z}_+} V_j(s).$$

Note that $\dim V_j(s) < \infty$ if all $s_i > 0$. The gradation of type $\mathbb{1}$ of V is called the *principal gradation*. Provided that $\dim V_j(s) < \infty$, we have

$$F_s(e(-\Lambda) \operatorname{ch} V) = \sum_{j \in \mathbb{Z}_+} (\dim V_j(s)) q^j.$$

We call the formal power series $\sum_{j \geq 0} \dim V_j(\mathbb{1}) q^j$ the *q-dimension* of V and denote it by $\dim_q V$. Then Proposition 10.9 and formula (10.9.4) can be restated as follows (here we use also Proposition 10.8):

PROPOSITION 10.10. *Consider the principal gradation* $L(\Lambda) = \bigoplus_{j \in \mathbb{Z}_+} L_j(\mathbb{1})$. *Then under the hypotheses and notation of Proposition 10.9, one has:*

$$\begin{aligned} \dim_q L(\Lambda) &= \prod_{j \geq 1} (1 - q^j)^{\dim {}^t\mathfrak{g}_j(s + \mathbb{1}) - \dim \mathfrak{g}_j(\mathbb{1})} \\ &= \prod_{\alpha \in \Delta_+^\vee} \left(\frac{1 - q^{\langle \Lambda + \rho, \alpha \rangle}}{1 - q^{\langle \rho, \alpha \rangle}} \right)^{\operatorname{mult} \alpha} \end{aligned} \tag{10.10.1}$$

□

The following corollary is a classical result of Weyl.

COROLLARY 10.10. *Let A be a finite type matrix, so that $\mathfrak{g}(A)$ is a simple finite-dimensional Lie algebra. Let $\Lambda \in P_+$. Then $L(\Lambda)$ is finite-dimensional and*

$$\dim L(\Lambda) = \prod_{\alpha \in \Delta_+^\vee} \langle \Lambda + \rho, \alpha \rangle / \langle \rho, \alpha \rangle .$$

Proof. Let q tend to 1 in (10.10.1) and apply l'Hôpital's rule.

□

§10.11. Exercises.

10.1. Show that

$$\operatorname{ch} L(m\rho) = e(m\rho) \prod_{\alpha \in \Delta_+} (1 + e(-\alpha) + \cdots + e(-m\alpha))^{\operatorname{mult} \alpha}$$

[Replacing $e(-\alpha_i)$ by $e(-(m+1)\alpha_i)$ we deduce from (10.4.4):

$$\sum_{w \in W} \epsilon(w) e((m+1)w(\rho)) = e((m+1)\rho) \prod_{\alpha \in \Delta_+} (1 - e(-(m+1)\alpha))^{\operatorname{mult} \alpha}.]$$

10.2. Show that $\rho - w(\rho)$ $(w \in W)$ is equal to the sum of all positive real roots α such that $w^{-1}(\alpha) < 0$; the number of such roots equals $\ell(w)$. [See Exercises 3.6 and 3.12.]

10.3. Show that $w(\rho) - \rho$ $(w \in W)$ are primitive weights of the module $M(0)$ over a Kac–Moody algebra $\mathfrak{g}(A)$; deduce that this module has no irreducible submodules if $\dim \mathfrak{g}(A) = \infty$.

10.4. Let ω be the Chevalley involution of $\mathfrak{g}(A)$ and let V be a $\mathfrak{g}(A)$-module from the category $\mathcal{O}$. Set $V^\omega = \{f \in V^* | f(V_\lambda) = 0$ for all but finitely many $\lambda\}$. The algebra $\mathfrak{g}(A)$ acts on V^ω by $(x \cdot f)(v) = -f(\omega(x) \cdot v)(x \in \mathfrak{g}(A),\ v \in V,\ f \in V^\omega)$. Show that $V^\omega \in \mathcal{O}$, $\operatorname{ch} V = \operatorname{ch} V^\omega$, $L(\lambda) \simeq L(\lambda)^\omega$. Show that $M(0)^\omega$ has no proper maximal submodules if $\mathfrak{g}(A)$ is an infinite-dimensional Kac–Moody algebra.

10.5. Show that the adjoint representation of an affine Lie algebra is not completely reducible.

10.6. For $\lambda \in \mathfrak{h}^*$ such that $\langle \lambda, \alpha_i^\vee \rangle = s_i$ let F_λ stand for F_s. Show that $F_\mu(N_\lambda) = F_\lambda(N_\mu)$, $\lambda, \mu \in \mathfrak{h}^*$.

10.7. Let $A = (a_{ij})$ be a symmetrizable generalized Cartan matrix and let $\mathfrak{n}$ be a Lie algebra on generators e_i $(i = 1, \dots, n)$ and defining relations $(\operatorname{ad} e_i)^{1-a_{ij}} e_j = 0$ $(i \neq j)$. Setting $\deg e_i = \alpha_i$ defines a gradation $\mathfrak{n} = \bigoplus_{\alpha \in Q_+} \mathfrak{n}_\alpha$, where Q_+ is the semigroup generated by the α_i in the free abelian group Q on $\alpha_1, \dots, \alpha_n$. Define automorphisms r_i of Q by $r_i(\alpha_j) = \alpha_j - a_{ij}\alpha_i$, and let W be the group generated by $r_1, \dots, r_n$. Extend the action of W to the lattice $Q \oplus \mathbb{Z}\rho$ by $r_i(\rho) = \rho - \alpha_i$ and put $s(w) = \rho - w(\rho)$ for $w \in W$. Show that $s(w)$ is the sum of all $\alpha \in Q_+$ such that $\mathfrak{n}_\alpha \neq 0$ and $-w^{-1}(\alpha) \in Q_+$. Show that

$$\prod_{\alpha \in Q_+} (1 - e(\alpha))^{\dim \mathfrak{n}_\alpha} = \sum_{w \in W} \epsilon(w) e(s(w)).$$

10.8. Let $\mathfrak{g}(A)$ be a Kac–Moody algebra with the Cartan matrix $A = \begin{pmatrix} 2 & -a \\ -b & 2 \end{pmatrix}$, where $ab \geq 4$. For a given pair of nonnegative integers m and n define the sequences $a_j(m,n)$ and $b_j(m,n)$ $(j \in \mathbb{Z})$ by the following recurrent formulas $(j \in \mathbb{Z})$:

$$a_{j-1}(m,n) + a_{j+1}(m,n) = ab_j(m,n) + m$$
$$b_{j-1}(m,n) + b_{j+1}(m,n) = ba_j(m,n) + n$$

and $a_0(m,n) = a_1(m,n) = b_0(m,n) = b_1(m,n) = 0$. Show that for $\lambda \in \mathfrak{h}^*$ such that $\langle \lambda, \alpha_1^\vee \rangle = m$, $\langle \lambda, \alpha_2^\vee \rangle = n$, one has

$$\sum_{w \in W(A)} \epsilon(w) e(w(\lambda) - \lambda) = \sum_{j \in \mathbb{Z}} (-1)^j e(-a_j(m,n)\alpha_1 - b_{j+1}(m,n)\alpha_2).$$

10.9. Show that if we set $u = e(-\alpha_1)$, $v = e(-\alpha_2)$, the identity (10.4.4) for $A = \begin{pmatrix} 2 & -2 \\ -2 & 2 \end{pmatrix}$ turns into the famous Jacobi triple product identity:

$$\prod_{n=1}^{\infty}(1-u^n v^n)(1-u^{n-1}v^n)(1-u^n v^{n-1}) = \sum_{m\in\mathbb{Z}}(-1)^m u^{m(m-1)/2} v^{m(m+1)/2},$$

and if $A = \begin{pmatrix} 2 & -4 \\ -1 & 2 \end{pmatrix}$, the identity (10.4.4) becomes the important quintuple product identity:

$$\prod_{n=1}^{\infty}(1-u^{2n}v^n)(1-u^{2n-1}v^{n-1})(1-u^{2n-1}v^n)(1-u^{4n-4}v^{2n-1})(1-u^{4n}v^{2n-1})$$
$$= \sum_{m\in\mathbb{Z}}(u^{3m^2-2m}v^{(3m^2+m)/2} - u^{3m^2-4m+1}v^{(3m^2-m)/2}).$$

10.10. Let $L(\Lambda)$ be an integrable $A_1^{(1)}$-module, so that $k_i := \langle\Lambda, \alpha_i^\vee\rangle \in \mathbb{Z}_+$ $(i = 1, 2)$; put $s = k_1 + k_2 + 2$. Check that if $k_1 \neq k_2$, then

$$\prod_{n\geq 1}(1-q^{2n-1})\dim_q L(\Lambda) = \prod_{\substack{n\geq 1 \\ n\not\equiv 0,\pm(k_1+1) \bmod s}} (1-q^n)^{-1}.$$

(For $(k_1, k_2) = (2, 1)$ or $(3, 0)$ the right-hand side appears in the celebrated Rogers–Ramanujan identities.)

In Exercises 10.11–10.20, A is a finite type matrix, so that $\mathfrak{g}(A)$ is a simple finite-dimensional Lie algebra. Denote by θ its highest root and let $h = \operatorname{ht}\theta + 1$ be the Coxeter number. Let G be the associated complex Lie group.

10.11. Show that $\{L(\Lambda)\}_{\Lambda\in P_+}$ exhaust, up to isomorphism, all irreducible finite-dimensional modules over $\mathfrak{g}(A)$.

10.12. Set $E = \sum_i e_i$. Define constants $c_1, \ldots, c_n$ by $\rho^\vee = \sum_i c_i\alpha_i^\vee$, and put $H = 2\rho^\vee$ and $F = 2\sum_i c_i f_i$. Show that $\{E, H, F\}$ form a standard basis of $sl_2(\mathbb{C})$ (this is the so-called principal 3-dimensional subalgebra). Using the representation theory of $sl_2(\mathbb{C})$ show that for $\Lambda \in P_+$ the expression

$$\prod_{\alpha\in\Delta_+^\vee}\frac{1-q^{\langle\Lambda+\rho,\alpha\rangle}}{1-q^{\langle\rho,\alpha\rangle}}$$

is a polynominal in q with positive coefficients $d_0 = 1, d_1, \ldots, d_m$, and that this sequence is unimodal, i.e., it increases up to $d_{[m/2]}$, and $d_i = d_{m-i}$.

10.13. Show that the sequences of the coefficients of the following polynomials are unimodal:

a) $\frac{[k+n]!}{[n]![k]!}$ where $[j]! := (1-q^j)(1-q^{j-1})\dots(1-q)/(1-q)^j$;

b) $(1+q)(1+q^2)\dots(1+q^n)$.

[Apply Exercise 10.12 to the k-th symmetric power of the natural representation of A_n and the spin representation of B_n].

10.14. Let r be a real number such that $|r| \neq 0, 1, \dots, h-1$, and let $L(\Lambda)$ be an irreducible $\mathfrak{g}(A)$-module with highest weight $\Lambda \in P_+$. Show that

$$\operatorname{tr}_{L(\Lambda)} \exp(2\pi i \rho^\vee / r) = \prod_{\alpha \in \Delta_+^\vee} \sin \frac{\pi \langle \Lambda + \rho, \alpha \rangle}{r} \Big/ \sin \frac{\pi \langle \rho, \alpha \rangle}{r}.$$

Prove a similar formula for $\operatorname{tr}_{L(\Lambda)} \exp(2\pi i \nu^{-1}(\rho)/r)$.

10.15. Let $m = h$ (resp. $m = h+1$) and let $L = Q$ (resp. $L = P$). For $w \in W \ltimes mL$ put $\epsilon(w) = \epsilon(w')$, where w' is the canonical image of w in W. Show that for $\lambda \in P$, either $\langle \lambda, \alpha \rangle \equiv 0 \mod m$ for some $\alpha \in \Delta^\vee$ or else there exists a unique element $w_\lambda \in W \ltimes mL$ such that $w_\lambda(\lambda) = \rho$. Let $\Lambda \in P_+$, and let $s \in \mathbb{Z}$ be relatively prime to m. In the case $m = h+1$, assume that s is divisible by det A (which divides h). Show that

$$\operatorname{tr}_{L(\Lambda)} \exp(2\pi s i \rho^\vee / m) = \begin{cases} 0 \text{ if } \langle \Lambda + \rho, \alpha \rangle \equiv 0 \mod m \\ \qquad\qquad \text{for some } \alpha \in \Delta^\vee, \\ \epsilon(w_{\Lambda+\rho}) (= \pm 1) \text{ otherwise.} \end{cases}$$

[We have

$$\begin{aligned} &\operatorname{tr}_{L(\Lambda)} \exp(2\pi s i \rho^\vee / m) \\ &\quad = \sum_{w \in W} \epsilon(w) e^{2\pi s i (\frac{\Lambda+\rho}{m}, w(\rho^\vee))} \Big/ \sum_{w \in W} \epsilon(w) e^{2\pi s i (\frac{\rho}{m}, w(\rho^\vee))}, \end{aligned}$$

hence, this is unchanged (resp. changes only the sign) when $\Lambda+\rho$ is replaced by $\Lambda+\rho+m\alpha$ ($\alpha \in L$) (resp. by $r_j(\Lambda+\rho)$, $j = 1, \dots, \ell$).]

10.16. Let $\sigma_{h+1} = \exp \dfrac{2\pi i h}{h+1} \rho^\vee \in G$ and let C be the cyclic group generated by σ_{h+1}. Show that the order of C is $h+1$. Let $h+1$ be a prime integer. Deduce from Exercise 10.15 that the G-module $L(\Lambda)$ ($\Lambda \in P_+$), restricted to C, is a direct sum of several copies of the regular representation of C plus at most one 1-dimensional or h-dimensional representation of C.

[Show that C is defined over $\mathbb{Q}$.]

10.17. Let $h+1$ be a prime number and let $\Lambda \in P_+$. Show that

$$\dim L(\Lambda) \equiv 0 \text{ or } \pm 1 \mod (h+1),$$

(in particular, $\dim L(\Lambda) \equiv 0$ or ± 1 mod 7, 13, 13, 19, 31 if A is of type G_2, F_4, E_6, E_7, E_8, respectively), where 0, 1 or -1 appear according to Exercise 10.15 (in the case $m = h+1$).

10.18. For A of type A_{p-1}, p a prime number, and the element $\exp(2\pi i \rho^\vee/p) \in G$ one has statements similar to those of Exercises 10.16 and 10.17. Furthermore, show that:

$$\mathrm{tr}_{L((a-1)\rho)} \exp(2\pi i \rho^\vee/p) = \left(\frac{a}{p}\right) \text{ (the Legendre symbol).}$$

Deduce the quadratic reciprocity law. Compute $\mathrm{tr}\,\sigma_{h+1}$ in the spin representation of B_ℓ. Deduce that $2^{\frac{\ell-1}{2}} \equiv (-1)^{\frac{\ell^2-1}{8}} \mod \ell$ for an odd ℓ.

10.19. Let $M \subset \mathfrak{h}^*$ be the lattice spanned by long roots, $\Phi(.|.)$ the Killing form on $\mathfrak{g}(A)$, $\nu : \mathfrak{h} \to \mathfrak{h}^*$ the corresponding isomorphism, $g = \Phi(\theta|\theta)^{-1}$, and set $k = 2$ for type B_ℓ, C_ℓ, F_4, $k = 3$ for type G_2 and $k = 1$ for the rest of the types. Show that $\exp 4\pi i \nu^{-1}(\rho)$ is an element of order kg, which coincides with $\exp(2\pi i \rho^\vee/h)$ if $k = 1$. Show that for $\lambda \in P$, either $2(\lambda|\alpha) \in \mathbb{Z}$ for some $\alpha \in \Delta$ or else there exists a unique element $w'_\lambda \in W \ltimes gM$ such that $w'_\lambda(\lambda) = \nu^{-1}(\rho)$. Deduce that for $\lambda \in P_+$ one has

$$\mathrm{tr}_{L(\Lambda)} \exp 4\pi i \nu^{-1}(\rho) = \begin{cases} 0 \text{ if } \Phi(\Lambda+\rho|\alpha) \equiv 0 \mod kg \\ \qquad\qquad \text{for some } \alpha \in \Delta \\ \epsilon(w'_{\Lambda+\rho}) \text{ otherwise.} \end{cases}$$

10.20. Show that the conjugacy class of the element $\sigma_h = \exp 2\pi i \rho^\vee/h$ in G consists of all elements σ such that $\mathrm{Ad}\,\sigma$ is a conjugate of the automorphism of $\mathfrak{g}(A)$ of type $(\mathbb{1}\,;1)$. Show that the conjugacy class of the element σ_{h+1} in G consists of all regular elements of order $h+1$, and that $\mathrm{Ad}\,\sigma_{h+1}$ is an automorphism of $\mathfrak{g}(A)$ of type $(2,1,\ldots,1;1)$.

10.21. Let R be a countable-dimensional commutative associative algebra over $\mathbb{C}$ with a unique maximual ideal $\mathfrak{m}$. Show that any $x \in \mathfrak{m}$ is algebraic.

[If x is not algebraic then elements $(x-\lambda)^{-1}$ for $\lambda \in \mathbb{C}^\times$ are linearly independent elements of R.]

10.22. Consider the canonical filtration $U(\mathfrak{g}(A)) = \bigcup_{j\geq 0} U_j$ of the universal enveloping algebra of a Kac–Moody algebra $\mathfrak{g}(A)$, and let $\mathrm{Gr}\,U(\mathfrak{g}(A))$ be the associated graded algebra. Let Ann_Λ be the annihilator of a highest-weight vector of $L(\Lambda)$ in $U(\mathfrak{g}(A))$ and let $R_\Lambda = \mathrm{Gr}\,U(\mathfrak{g}(A))/\mathrm{Gr}\,\mathrm{Ann}_\Lambda$. Using Exercise 10.21, show that R_Λ has a unique maximal ideal if and only if $\Lambda \in P_+$.

10.23. (Open problem). Classify irreducible integrable modules over a Kac–Moody algebra.

§10.12. Bibliographical notes and comments.

Theorem 10.4 as well as other results of §§10.1–10.5 are due to Kac [1974]; the exposition closely follows this paper. Formula (10.4.1) is usually referred to as the Weyl–Kac character formula. The results of §10.6 are due to Kac–Peterson [1984 A] (see also Looijenga [1980]).

The first version of the complete reducibility theorem, which is Theorem 10.7 b), was obtained in Kac [1978 A]. In its present form, Theorem 10.7 is taken from Kac–Peterson [1984 A]. Note that this refinement is important for the proof of a Peter–Weyl type theorem in Kac–Peterson [1983].

The exposition of §§10.8–10.10 follows Kac [1978 A]. The trick employed in the proof of Proposition 10.9 goes back to Weyl and may be found in many textbooks (e.g. Bourbaki [1975], Jacobson [1962]). The only new thing is the use of the dual root system. The results of §§10.8 and 10.9 were generalized by Wakimoto [1983].

Exercises 10.3 and 10.4 are taken from Deodhar–Gabber–Kac [1982]. Exercise 10.10 is taken from Lepowsky–Milne [1978]. This observation eventually led Kac, Lepowsky and Wilson to a Lie algebraic interpretation and proof of the Rogers–Ramanujan identities, as announced at the 83rd AMS summer meeting in 1979, abstract 768-17-1, and at the 775th AMS meeting in 1980, abstract 775-A14 (see Lepowsky–Wilson [1982] for an exposition of these results). For further development see Lepowsky–Wilson [1984], Misra [1984 A,B], [1987], [1989 A,B], Meurman–Primc [1987], Lepowsky–Primc [1985], Primc [1989], and others.

Exercise 10.12 goes back to Dynkin, Exercise 10.13 is due to Hughes and Stanley (see Stanley [1980]). The fact that $\mathrm{tr}_{L(\Lambda)}(\exp 2\pi i\rho^\vee/h)$ and $\mathrm{tr}_{L(\Lambda)}(\exp 4\pi i\nu^{-1}(\rho))$ is $0, 1$ or -1 is due to Kostant [1976] (his proof is more complicated). The rest of the material of Exercises 10.14–10.19 is taken from Kac [1981]. Exercise 10.22 is essentially due to Feigin–Fuchs [1989].

Theorem 10.7 implies that $H^1(\mathfrak{g}'(A), L(\Lambda)) = 0$ for $\Lambda \in P_+$. Duflo (unpublished) and Kumar [1986 A] showed that $H^k(\mathfrak{g}'(A), L(\Lambda)) = 0$, $k \geq 1$, for every $\Lambda \in P_+$, $\Lambda \neq 0$. Kumar [1984] showed that $H^*(\mathfrak{g}'(A), \mathbb{C})$ is isomorphic to the cohomology of the associated topological group $G(A)$; Kac–Peterson [1984 B] computed previously the Poincaré series of $G(A)$.

The complexified Tits cone $X_{\mathbf{C}}$ is related to the theory of singularities of algebraic surfaces. Looijenga [1980] constructed a partial compactification of the space of orbits of $W \ltimes 2\pi i Q^\vee$ acting on the interior of $X_{\mathbf{C}}$, which plays an important role in the deformation theory of singularities. Further connections of the theory of Kac–Moody algebras and groups to the theory of singularities may be found in Slodowy [1981], [1985 A, B].

Chari [1986] and Chari–Pressley [1987], [1988 A], [1989] have made a significant progress in classification of irreducible integrable modules over affine algebras. In particular, they showed that such a module with finite-dimensional weight spaces is either $L(\Lambda)$ or $L^*(\Lambda)$ $(\Lambda \in P_+)$ or is of the type considered in Exercise 7.15.

In a remakable recent development, Kumar [1987 A] and Mathien [1987] proved Demazure's character formula for an arbitrary Kac–Moody algebra, using the geometry of Schubert varieties. (In the finite-dimensional case this is due to Demazure [1974] and Andersen [1985].) A corollary of this is the Weyl–Kac character formula for an arbitrary (not necessarily symmetrizable) Kac–Moody algebra.

The integrable module $L(\Lambda)$ over a symmetrizable Kac–Moody algebra $\mathfrak{g}(A)$ can be deformed to a module over the quantized Kac–Moody algebra $U_q(\mathfrak{g}(A))$, and it remains irreducible for generic values of q (Lusztig [1988]).

A remark on terminology. The integrable highest-weight modules are sometimes called standard modules in the literature. I object to this term since first, it carries no information, and second, it is already used in the representation theory of Lie groups for completely different representations. Also, the theory of modular invariant representations by Kac–Wakimoto [1988 B], [1989 A] shows that not only integrable $L(\Lambda)$ are of interest.

Chapter 11. Integrable Highest-Weight Modules: the Weight System and the Unitarizability

§11.0. In this chapter we describe in detail the set of weights $P(\Lambda)$ of an integrable highest-weight module $L(\Lambda)$ over a Kac–Moody algebra $\mathfrak{g}(A)$. We establish the existence of a positive-definite Hermitian form on $L(\Lambda)$, invariant with respect to the compact form $\mathfrak{k}(A)$ of $\mathfrak{g}(A)$. Finally, we study the decomposition of $L(\Lambda)$ with respect to various subalgebras of $\mathfrak{g}(A)$ and derive an explicit description of the region of convergence of $\mathrm{ch}_{L(\Lambda)}$.

§11.1. Fix $\Lambda \in P_+$. From the results of Chapter 3 one easily deduces the following statement.

PROPOSITION 11.1. *Let $\lambda \in P(\Lambda)$, $\alpha \in \Delta^{re}$ and $m_t = \mathrm{mult}_{L(\Lambda)}(\lambda + t\alpha)$. Then*

a) *The set of $t \in \mathbb{Z}$ such that $\lambda + t\alpha \in P(\Lambda)$ is the interval $\{t \in \mathbb{Z} | -p \leq t \leq q\}$, where p and q are nonnegative integers and*

$$p - q = \langle \lambda, \alpha^\vee \rangle.$$

b) *For $e_\alpha \in \mathfrak{g}_\alpha \setminus \{0\}$ the map $e_\alpha : L(\Lambda)_{\lambda+t\alpha} \to L(\Lambda)_{\lambda+(t+1)\alpha}$ is an injection if $-p \leq t < \frac{1}{2}(q-p)$; in particular, the function $t \mapsto m_t$ increases on this interval.*

c) *The function $t \mapsto m_t$ is symmetric with respect to $t = \frac{1}{2}(q-p)$.*

d) *If both λ and $\lambda + \alpha$ are weights, then $\mathfrak{g}_\alpha(L(\Lambda)_\lambda) \neq 0$.*

Proof. By Lemma 10.1 and the results of §3.6, the proposition holds for a simple root α. Applying Proposition 3.7a), we deduce that it holds for an arbitrary real root α.

□

§11.2. Fix $\Lambda \in P_+$. Recall that $P(\Lambda)$ is W-invariant (by Proposition 10.1).

An element $\lambda \in P$ is called *nondegenerate with respect to* Λ if either $\lambda = \Lambda$ or else $\lambda < \Lambda$ and for every connected component S of $\mathrm{supp}(\Lambda - \lambda)$ one has

$$S \cap \{i | \langle \Lambda, \alpha_i^\vee \rangle \neq 0\} \neq \emptyset. \qquad (11.2.1)$$

LEMMA 11.2. *Every weight λ of the $\mathfrak{g}(A)$-module $L(\Lambda)$ is nondegenerate with respect to Λ.*

Proof. Suppose that $\lambda \in P(\Lambda)\setminus\{\Lambda\}$. Let S be a connected component of $\mathrm{supp}(\Lambda - \lambda)$. Denote by $\mathfrak{n}_-(S)$ the subalgebra of $\mathfrak{n}_-$ generated by the f_i such that $i \in S$. Then

$$L(\Lambda)_\lambda \subset U(\mathfrak{n}_-)\mathfrak{n}_-(S)L(\Lambda)_\Lambda. \tag{11.2.2}$$

If (11.2.1) were false, then $\mathfrak{n}_-(S)L(\Lambda)_\Lambda = 0$ for some S and hence, by (11.2.2), $L(\Lambda)_\lambda = 0$, a contradiction.

□

Now we can describe explicitly the set of weights of the $\mathfrak{g}(A)$-module $L(\Lambda)$.

PROPOSITION 11.2. *Let $\Lambda \in P_+$. Then*

a) $P(\Lambda) = W \cdot \{\lambda \in P_+ | \lambda$ *is nondegenerate with respect to* $\Lambda\}$.

b) *If* $\{i|\langle\Lambda, \alpha_i^\vee\rangle = 0\} \subset S(A)$ *is a disjoint union of diagrams of finite type, then*

$$P(\Lambda) = W \cdot \{\lambda \in P_+ | \lambda \leq \Lambda\}.$$

Proof. The inclusion $\subset$ in a) and b) follows from Corollary 10.1 and Lemma 11.2. The other inclusion in b) follows from that of a). Indeed, if $\lambda \in P_+$ and $\lambda = \Lambda - \beta$, where $\beta \in Q_+$, then $\langle\beta, \alpha_i^\vee\rangle \leq 0$ for i such that $\langle\Lambda, \alpha_i^\vee\rangle = 0$, which implies that λ is nondegenerate with respect to Λ under the hypothesis of b).

It remains to show that if $\mu = \Lambda - \alpha \in P_+$, where $\alpha = \sum_i k_i\alpha_i$, $k_i \geq 0$, $\sum_i k_i > 0$, and $S \cap \{i|\langle\Lambda, \alpha_i^\vee\rangle \neq 0\} \neq \emptyset$ for every connected component S of $\mathrm{supp}\,\alpha$, then $\mu \in P(\Lambda)$. Let

$$\Omega_\alpha = \{\gamma \in Q_+ | \gamma \leq \alpha \text{ and } \Lambda - \gamma \in P(\Lambda)\}.$$

The set Ω_α is finite and the union of supports of its elements has a nonempty intersection with each connected component of $\mathrm{supp}\,\alpha$. Let $\beta = \sum_i m_i\alpha_i$ be an element of maximal height in Ω_α. It follows from Proposition 11.1a) that

$$\mathrm{supp}\,\beta = \mathrm{supp}\,\alpha. \tag{11.2.3}$$

Suppose that $\beta \neq \alpha$. Then

$$\Lambda - \beta - \alpha_i \notin P(\Lambda) \text{ if } k_i > m_i. \tag{11.2.4}$$

Set $S = \{j \in S(A) | k_j = m_j\}$. Let R be a connected component of $(\operatorname{supp} \alpha)\backslash S$. Since $\mu \in P_+$, we deduce from (11.2.4) and Proposition 11.1a)

$$\langle \beta, \alpha_i^\vee \rangle \geq \langle \Lambda, \alpha_i^\vee \rangle \text{ and } \langle \alpha, \alpha_i^\vee \rangle \leq \langle \Lambda, \alpha_i^\vee \rangle \text{ if } i \in R. \tag{11.2.5}$$

Set

$$\beta' = \sum_{i \in R} m_i \alpha_i, \quad \alpha' = \sum_{i \in R} (k_i - m_i) \alpha_i.$$

Then (11.2.3 and 5) imply

$$\langle \beta', \alpha_i^\vee \rangle \geq 0 \text{ if } i \in R, \tag{11.2.6}$$

$$\langle \alpha', \alpha_i^\vee \rangle \leq 0 \text{ if } i \in R. \tag{11.2.7}$$

It follows from (11.2.7) that R and hence $S(A)$ are not of finite type. In particular, for every $\lambda \in P(\Lambda)$ there exists α_i such that $\lambda - \alpha_i \in P(\Lambda)$. (Otherwise, $\dim L(\Lambda) < \infty$ and $\dim \mathfrak{g}(A) < \infty$). Hence $S \neq \emptyset$ and, by the properties of μ, we can choose R so that it is not a connected component of $\operatorname{supp} \alpha$. But then, in addition to (11.2.6), we have: $\langle \beta', \alpha_j^\vee \rangle > 0$ for some $j \in R$. Hence R is a diagram of finite type. This is a contradiction.

□

§11.3. We proceed to study the geometric properties of the set of weights $P(\Lambda)$ for $\Lambda \in P_+$.

PROPOSITION 11.3. a) *$P(\Lambda)$ coincides with the intersection of $\Lambda + Q$ with the convex hull of the orbit $W \cdot \Lambda$.*

b) *If $\lambda, \mu \in \mathfrak{h}^*$ are such that $\lambda - \mu \in Q$ and μ lies in the convex hull of $W \cdot \lambda$, then $\operatorname{mult}_{L(\Lambda)}(\mu) \geq \operatorname{mult}_{L(\Lambda)}(\lambda)$.*

Proof. First we prove by induction on $\operatorname{ht}(\Lambda - \lambda)$ that a weight λ of $L(\Lambda)$ lies in the convex hull of $W \cdot \Lambda$. If $\lambda = \Lambda$, there is nothing to prove. If $\lambda < \Lambda$, there exists i such that $\lambda + \alpha_i \in P(\Lambda)$. Take the maximal s such that $\mu := \lambda + s\alpha_i \in P(\Lambda)$. Then μ lies in the convex hull of $W \cdot \Lambda$ by the inductive assumption; since λ lies in the interval $[\mu, r_i(\mu)]$ (by Proposition 11.1), it also lies in the convex hull of $W \cdot \Lambda$.

Let now $\lambda = \sum_w c_w w(\Lambda) \in \Lambda + Q$, where $c_w \geq 0$ and $\sum_w c_w = 1$. Then

$$\Lambda - \lambda = \sum_w c_w(\Lambda - w(\Lambda)) \in Q_+. \tag{11.3.1}$$

Replacing λ by $w(\lambda)$ with minimal $\mathrm{ht}(\Lambda - \lambda)$, we may assume that $\lambda \in P_+$. Finally it is clear from (11.3.1) that λ is nondegenerate with respect to Λ. Hence, by Proposition 11.2a), $\lambda \in P(\Lambda)$, proving a).

To prove b) we may assume that $\lambda \in P_+$. Then we apply a) to $L(\lambda)$ to obtain $\mu \in P(\lambda)$. We prove b) by induction on $\mathrm{ht}(\lambda - \mu)$. If $\lambda = \mu$, there is nothing to prove. Otherwise, $\mu + \alpha_i \in P(\lambda)$ for some i. Let $s > 0$ be such that $\mu + s\alpha_i \in P(\lambda)$ but $\mu + (s+1)\alpha_i \notin P(\lambda)$. By a), $\mu + s\alpha_i$ lies in the convex hull of $W \cdot \lambda$ and in $P(\Lambda)$. Hence, by the inductive assumption, $\mathrm{mult}_{L(\Lambda)}(\lambda) \leq \mathrm{mult}_{L(\Lambda)}(\mu + s\alpha_i)$. On the other hand, μ lies in the interval $[\mu + s\alpha_i, r_i(\mu + s\alpha_i)]$ and hence $\mathrm{mult}_{L(\Lambda)}(\mu + s\alpha_i) \leq \mathrm{mult}_{L(\Lambda)}(\mu)$ by Proposition 11.1b), c). Combining these inequalities proves b).

□

§11.4. In the rest of the chapter we will assume that A is a symmetrizable generalized Cartan matrix; we fix a standard invariant bilinear form $(.|.)$ on $\mathfrak{g}(A)$.

PROPOSITION 11.4. *Let $\Lambda \in P_+$ and $\lambda, \mu \in P(\Lambda)$. Then*

a) $(\Lambda|\Lambda) - (\lambda|\mu) \geq 0$ *and equality holds if and only if* $\lambda = \mu \in W \cdot \Lambda$.

b) $|\Lambda + \rho|^2 - |\lambda + \rho|^2 \geq 0$ *and equality holds if and only if* $\lambda = \Lambda$.

Proof. Since both $(.|.)$ and $P(\Lambda)$ are W-invariant, we can assume in the proof of a) that $\lambda \in P_+$. Since $\beta := \Lambda - \lambda$ and $\beta_1 := \Lambda - \mu$ lie in Q_+, we have $(\Lambda|\Lambda) - (\lambda|\mu) = (\Lambda|\beta) + (\lambda|\beta_1) \geq 0$ (cf. the proof of Lemma 10.3). In the case of equality we have $(\Lambda|\beta) = (\lambda|\beta_1) = 0$. Since λ is nondegenerate with respect to Λ (by Lemma 11.2), we deduce that $\beta = 0$, i.e., $\lambda = \Lambda$. But then $(\Lambda|\beta_1) = 0$ and, by the same argument, $\mu = \Lambda$, proving a).

To prove b) we write:

$$(\Lambda + \rho|\Lambda + \rho) - (\lambda + \rho|\lambda + \rho) = ((\Lambda|\Lambda) - (\lambda|\lambda)) + 2(\Lambda - \lambda|\rho) \geq 0,$$

since $(\Lambda|\Lambda) - (\lambda|\lambda) \geq 0$ by a) and $\Lambda - \lambda \in Q_+$. Clearly, equality occurs if and only if $\Lambda = \lambda$.

□

§11.5. Let V be a $\mathfrak{g}(A)$-module. A Hermitian form H on V is called *contravariant* if

$$H(g(x), y) = -H(x, \omega_0(g)(y)) \text{ for all } g \in \mathfrak{g}(A),\ x, y \in V.$$

For example the Hermitian form $(.|.)_0$ on $\mathfrak{g}(A)$ is contravariant (see §2.7).

LEMMA 11.5. *Let $\Lambda \in \mathfrak{h}^*_{\mathbb{R}}$. Then the $\mathfrak{g}(A)$-module $L(\Lambda)$ carries a unique, up to a constant factor, nondegenerate contravariant Hermitian form. With respect to this form $L(\Lambda)$ decomposes into an orthogonal direct sum of weight spaces.*

Proof. Denote by $\mathfrak{g}(A)_{\mathbb{R}}$ the real subalgebra of $\mathfrak{g}(A)$ generated by e_i, f_i and $\mathfrak{h}_{\mathbb{R}}$, and let $L(\Lambda)_{\mathbb{R}} = U(\mathfrak{g}(A)_{\mathbb{R}})v_\Lambda$. Then H is the Hermitian extension of B (see Proposition 9.4) from $L(\Lambda)_{\mathbb{R}}$ to $L(\Lambda)$.

□

As in §9.4, it is easy to see that the fomula

$$H(gv_\Lambda, g'v_\Lambda) = \langle \hat{\omega}_0(g)g'v_\Lambda \rangle \tag{11.5.1}$$

where $g, g' \in U(\mathfrak{g}(A))$ and $\hat{\omega}_0$ is the extension of $-\omega_0$ to an antilinear anti-involution of $U(\mathfrak{g}(A))$, defines a contravariant Hermitian form on $L(\Lambda)$ (provided that $\Lambda \in \mathfrak{h}^*_{\mathbb{R}}$), normalized by $H(v_\Lambda, v_\Lambda) = 1$.

Note that, by definition, the operators g and $-\omega_0(g)$ are adjoint operators on $L(\Lambda)$ with respect to H. Thus, the compact form $\mathfrak{k}(A)$ is represented on $L(\Lambda)$ by skew-adjoint operators.

The $\mathfrak{g}(A)$-module $L(\Lambda)$ is called *unitarizable* if the Hermitian form H defined by (11.5.1) is positive definite.

§11.6. We proceed to prove the positivity of $(.|.)_0$ on $\mathfrak{n}_- + \mathfrak{n}_+$ and of H on $L(\Lambda)$. First we prove the inequality

$$2(\rho|\alpha) > (\alpha|\alpha) \text{ if } \alpha \in \Delta_+ \backslash \Pi. \tag{11.6.1}$$

If $\alpha \in \Delta_+^{im}$, this is clear, since then $(\alpha|\alpha) \le 0$, but $(\rho|\alpha) > 0$ for all $\alpha > 0$. If $\alpha \in \Delta_+^{re} \backslash \Pi$, then $\alpha^\vee \in (\Delta_+^\vee)^{re} \backslash \Pi^\vee$, and hence $2(\rho|\alpha)/(\alpha|\alpha) = \langle \rho, \alpha^\vee \rangle > 1$. This and $(\alpha|\alpha) > 0$ imply (11.6.1).

By analogy with the "partial" Casimir operator Ω_0, we define an operator Ω_1 on $\mathfrak{n}_-$ as follows:

$$\Omega_1(x) = \sum_{\alpha \in \Delta_+} \sum_i [e_{-\alpha}^{(i)}, [e_\alpha^{(i)}, x]_-] \quad (x \in \mathfrak{n}_-).$$

Here, as before, $\{e_\alpha^{(i)}\}$ and $\{e_{-\alpha}^{(i)}\}$ are dual bases of $\mathfrak{g}_\alpha$ and $\mathfrak{g}_{-\alpha}$ with respect to the bilinear form $(.|.)$, and the "minus" subscript denotes the projection on $\mathfrak{n}_-$ with respect to the triangular decomposition. Now we are in a position to prove the crucial lemma.

LEMMA 11.6. *If $\alpha \in \Delta_+$ and $x \in \mathfrak{g}_{-\alpha}$, then*

$$\Omega_1(x) = \big(2(\rho|\alpha) - (\alpha|\alpha)\big)x.$$

Proof. We calculate in $M(0)$ the expression $\Omega_0 x(v)$, where v is a highest-weight vector, in two different ways. By (2.6.1), we have

$$\Omega_0 x(v) = (2(\rho|\alpha) - (\alpha|\alpha))x(v). \tag{11.6.2}$$

On the other hand, by the definition of Ω_0, we have

$$\begin{aligned}\Omega_0 x(v) &= 2\sum_{\beta\in\Delta_+}\sum_i e_{-\beta}^{(i)} e_\beta^{(i)} x(v)\\ &= 2\sum_{\beta\in\Delta_+}\sum_i e_{-\beta}^{(i)} [e_\beta^{(i)}, x](v).\end{aligned}$$

Putting $S = \{\beta \in \Delta_+ | \beta < \alpha\}$, we may write

$$\begin{aligned}\Omega_0 x(v) &= 2\sum_{\beta\in S}\sum_i e_{-\beta}^{(i)}[e_\beta^{(i)}, x](v)\\ &= \sum_{\beta\in S}\sum_i ([e_{-\beta}^{(i)}, [e_\beta^{(i)}, x]] + [e_\beta^{(i)}, x]e_{-\beta}^{(i)} + e_{-\beta}^{(i)}[e_\beta^{(i)}, x])(v).\end{aligned}$$

Using (2.4.4), we obtain

$$\Omega_0 x(v) = \sum_{\beta\in S}\sum_i [e_{-\beta}^{(i)}, [e_\beta^{(i)}, x]](v).$$

Comparing this with (11.6.2) gives

$$\big((2(\rho|\alpha) - (\alpha|\alpha))x\big)(v) = \sum_{\beta\in\Delta_+}\sum_i [e_{-\beta}^{(i)}, [e_\beta^{(i)}, x]_-](v).$$

As $M(0)$ is a free $U(\mathfrak{n}_-)$-module (see Proposition 9.2 b)), the lemma follows.

□

§11.7. Now we are in a position to prove the following fundamental result.

THEOREM 11.7. *Let $\mathfrak{g}(A)$ be a symmetrizable Kac–Moody algebra. Then*
a) *The restriction of the Hermitian form $(.|.)_0$ to every root space $\mathfrak{g}_\alpha$ ($\alpha \in \Delta$) is positive-definite, i.e., $(.|.)_0$ is positive-definite on $\mathfrak{n}_- \oplus \mathfrak{n}_+$.*
b) *Every integrable highest-weight module $L(\Lambda)$ over $\mathfrak{g}(A)$ is unitarizable. Conversely, if $L(\Lambda)$ is unitarizable, then $\Lambda \in P_+$.*

Proof. We first prove a). Using ω_0, it suffices to show that $(.|.)_0$ is positive-definite on $\mathfrak{g}_{-\alpha}$ with $\alpha \in \Delta_+$. We do it by induction on $\operatorname{ht}\alpha$. The case $\operatorname{ht}\alpha = 1$ is clear by (2.2.1). Otherwise, put $S = \{\beta \in \Delta_+ | \beta < \alpha\}$ and use the inductive assumption to choose, for every $\beta \in S$, an orthonormal basis $\left\{e^{(i)}_{-\beta}\right\}$ of $\mathfrak{g}_{-\beta}$ with respect to $(.|.)_0$. Then, setting $e^{(i)}_\beta = -\omega_0(e^{(i)}_{-\beta})$, we have $(e^{(i)}_\beta | e^{(j)}_{-\beta}) = \delta_{ij}$. Now we apply Lemma 11.6 with this choice of $e^{(i)}_\beta$ and $e^{(i)}_{-\beta}$ (the choice for the $\beta \in \Delta_+ \backslash S$ is arbitrary). For $x \in \mathfrak{g}_{-\alpha}$ we have

$$\begin{aligned}(2(\rho|\alpha) - (\alpha|\alpha))(x|x)_0 &= (\Omega_1(x)|x)_0 \\ &= \sum_{\beta \in S}\sum_i ([e^{(i)}_{-\beta}, [e^{(i)}_\beta, x]]|x)_0 \\ &= \sum_{\beta \in S}\sum_i ([e^{(i)}_\beta, x]|[e^{(i)}_\beta, x])_0.\end{aligned}$$

By the inductive assumption, the last sum is nonnegative; using (11.6.1), we get $(x|x)_0 \geq 0$. Since $(.|.)_0$ is nondegenerate on $\mathfrak{g}_{-\alpha}$, we deduce that it is positive-definite, proving a).

Using Lemma 11.5, one has to show for b) that the restriction of H to $L(\Lambda)_\lambda$ is positive-definite. We prove this by induction on $\operatorname{ht}(\Lambda - \lambda)$. Let $\lambda \in P(\Lambda) \backslash \{\Lambda\}$ and $v \in L(\Lambda)_\lambda$. Thanks to a), we can choose a basis $\left\{e^{(i)}_\alpha\right\}$ of $\mathfrak{g}_\alpha$ such that $\left\{-\omega_0(e^{(i)}_\alpha)\right\}$ is the dual (with respect to $(.|.)$) basis of $\mathfrak{g}_{-\alpha}$. Then we have

$$\Omega = 2\nu^{-1}(\rho) + \sum_i u_i u^i - 2\sum_{\alpha \in \Delta_+}\sum_i \omega_0(e^{(i)}_\alpha)e^{(i)}_\alpha,$$

and hence:

$$\Omega(v) = (\lambda + 2\rho|\lambda)v - 2\sum_{\alpha \in \Delta_+}\sum_i \omega_0(e^{(i)}_\alpha)e^{(i)}_\alpha(v). \tag{11.7.1}$$

Computing $H(\Omega(v), v)$ in two different ways by making use of Corollary 2.6 and (11.7.1), and equating the results we obtain

$$(|\Lambda + \rho|^2 - |\lambda + \rho|^2)H(v, v) = 2 \sum_{\alpha \in \Delta_+} \sum_i H(e_\alpha^{(i)}(v), e_\alpha^{(i)}(v)).$$

By the inductive assumption, the right-hand side is nonnegative. Using Proposition 11.4 b) we deduce that $H(v, v) \geq 0$. Since H is nondegenerate on $L(\Lambda)_\lambda$, we conclude that it is positive-definite.

To prove the converse, note that by (3.2.4), we have:

$$\begin{aligned} 0 \leq H(f_i^k v_\Lambda, f_i^k v_\Lambda) &= H(f_i^{k-1} v_\Lambda, e_i f_i^k v_\Lambda) \\ &= k(\langle \Lambda, \alpha_i^\vee \rangle + 1 - k) H(f_i^{k-1} v_\Lambda, f_i^{k-1} v_\Lambda) \\ &= \cdots = \prod_{j=1}^{k} j(\langle \Lambda, \alpha_i^\vee \rangle + 1 - j). \end{aligned}$$

Hence $\langle \Lambda, a_i^\vee \rangle \in \mathbb{Z}_+$.

□

Warning. The restriction of $(\,.\,|\,.\,)_0$ to $\mathfrak{h}$ and even $\mathfrak{h}'$ is in general an indefinite Hermitian form: the matrix $((\alpha_i^\vee|\alpha_j^\vee)_0)$ is a symmetrization of the matrix A. In fact it is positive-definite (resp. positive-semidefinite) on $\mathfrak{h}'$ if and only if A is of finite (resp. affine) type.

As was just mentioned, if A is a matrix of finite type, then the restriction of $(\,.\,|\,.\,)_0$ to $\mathfrak{h}$ is positive-definite and hence, using Theorem 11.7a), the Hermitian form $(\,.\,|\,.\,)_0$ is positive-definite on $\mathfrak{g}(A)$, so that $\mathfrak{g}(A)$ carries a positive-definite $\mathfrak{k}(A)$-invariant Hermitian form. Thus, Theorem 11.7 is a generalization of a classical result of the finite-dimensional theory.

§11.8. We deduce from Theorem 11.7 b) another complete reducibility result. For that we first prove

LEMMA 11.8. *Let $h \in \operatorname{Int} X_{\mathbb{C}}$. Then for every $r \in \mathbb{R}$ the number of eigenvalues (counting multiplicities) λ of h in $L(\Lambda)$ ($\Lambda \in P_+$), such that $\operatorname{Re} \lambda > r$, is finite.*

Proof. This follows from Proposition 10.6d).

□

PROPOSITION 11.8. *Let $\mathfrak{a} \subset \mathfrak{g}(A)$ be an ω_0-invariant subalgebra which is normalized by an element $h \in \operatorname{Int} X_{\mathbf{C}}$ (i.e., $[h, \mathfrak{a}] \subset \mathfrak{a}$). Then with respect to $\mathfrak{a}$, the module $L(\Lambda)$ ($\Lambda \in P_+$) decomposes into an orthogonal (with respect to H) direct sum of irreducible h-invariant submodules.*

Proof. Put $\mathfrak{a}_1 = \mathfrak{a} + \mathbf{C}h$. By Theorem 11.7b) and Lemma 11.8, $L(\Lambda)$ decomposes into an orthogonal direct sum of finite-dimensional eigenspaces of h. It follows, using Theorem 11.7 b) and the ω_0-invariance of $\mathfrak{a}_1$, that for every $\mathfrak{a}_1$-submodule $V \subset L(\Lambda)$, the subspace $V^\perp$ is also an $\mathfrak{a}_1$-submodule and $L(\Lambda) = V \oplus V^\perp$. Hence $L(\Lambda)$ decomposes into an orthogonal direct sum of irreducible $\mathfrak{a}_1$-modules.

Let $U \subset L(\Lambda)$ be an irreducible $\mathfrak{a}_1$-submodule. It remains to show that U remains irreducible when restricted to $\mathfrak{a}$. Let U_λ denote the λ-eigenspace of h in U and let λ_0 be the eigenvalue of h with maximal real part. Let $\mathfrak{a}^\lambda$ denote the λ-eigenspace of $\operatorname{ad} h$ in $\mathfrak{a}$; we denote by $\mathfrak{a}_0$ (resp. $\mathfrak{a}_+$ or $\mathfrak{a}_-$) the sum of all $\mathfrak{a}^\lambda$ with $\operatorname{Re}\lambda = 0$ (resp. > 0 or < 0). Then $\mathfrak{a} = \mathfrak{a}_- \oplus \mathfrak{a}_0 \oplus \mathfrak{a}_+$, and it is clear that U_{λ_0} is an irreducible $\mathfrak{a}_0$-module and that $\{x \in U_\lambda | \mathfrak{a}_+(x) = 0\} = 0$ if $\operatorname{Re}\lambda < \operatorname{Re}\lambda_0$. Hence U is an irreducible $\mathfrak{a}$-module.

□

§11.9. Fix $\alpha \in \Delta_+$, and set

$$\mathfrak{n}_\pm^{(\alpha)} = \bigoplus_{j\geq 1} \mathfrak{g}_{\pm j\alpha}; \quad \mathfrak{g}^{(\alpha)} = \mathfrak{n}_-^{(\alpha)} \oplus \mathbf{C}\nu^{-1}(\alpha) \oplus \mathfrak{n}_+^{(\alpha)}.$$

Then $\mathfrak{g}^{(\alpha)}$ is a subalgebra of $\mathfrak{g}(A)$ (by Theorem 2.2e).

It follows from Proposition 5.1 and Chapter 3 that if α is a real root, then $\mathfrak{g}^{(\alpha)} \simeq sl_2(\mathbf{C})$ and module $L(\Lambda)$ restricted to $\mathfrak{g}^{(\alpha)}$ decomposes into a direct sum of irreducible finite-dimensional modules. If α is an imaginary root, then $\mathfrak{g}^{(\alpha)}$ is an infinite-dimensional Lie algebra as described in §9.12. Now we can describe the restriction of $L(\Lambda)$ to $\mathfrak{g}^{(\alpha)}$ for $\alpha \in \Delta_+^{im}$.

PROPOSITION 11.9. *Let $\alpha \in \Delta_+^{im}$ and $\Lambda \in P_+$. Introduce the following two subspaces of $L(\Lambda)$:*

$$L(\Lambda)_0^{(\alpha)} = \bigoplus_{\lambda:(\lambda|\alpha)=0} L(\Lambda)_\lambda; \quad L(\Lambda)_+^{(\alpha)} = \bigoplus_{\lambda:(\lambda|\alpha)>0} L(\Lambda)_\lambda.$$

a) *Considered as a $\mathfrak{g}^{(\alpha)}$-module, $L(\Lambda)$ decomposes into a direct sum of submodules $L(\Lambda) = L(\Lambda)_0^{(\alpha)} \oplus L(\Lambda)_+^{(\alpha)}$.*

b) $L(\Lambda)_0^{(\alpha)} = \{x \in L(\Lambda) | \mathfrak{g}^{(\alpha)}(x) = 0\}$.

c) $L(\Lambda)_+^{(\alpha)}$ *is a free* $U(\mathfrak{n}_-^{(\alpha)})$*-module on a basis of the subspace*

$$\{x \in L(\Lambda)_+^{(\alpha)} | \mathfrak{n}_+^{(\alpha)}(x) = 0\}.$$

d) *The* $\mathfrak{g}^{(\alpha)}$*-module* $L(\Lambda)$ *is completely reducible.*

Proof. Recall that, by Proposition 9.12 b), we can view $\mathfrak{g}^{(\alpha)}$ as a quotient of a Lie algebra $\mathfrak{g}'(B)$. Using Proposition 11.8, $L(\Lambda)$ decomposes into a direct sum of $\mathfrak{h}$-invariant irreducible $\mathfrak{g}^{(\alpha)}$-submodules. Each of these submodules is clearly generated by a nonzero vector $v_\lambda \in L(\Lambda)_\lambda$ such that $\mathfrak{n}_+^{(\alpha)}(v_\lambda) = 0$. If $(\lambda|\alpha) = 0$, then $\mathbf{C}v_\lambda$ is $\mathfrak{g}^{(\alpha)}$-stable by (9.10.1). If $(\lambda|\alpha) > 0$, then $\mathbf{C}v_\lambda$ generates a Verma module over $\mathfrak{g}^{(\alpha)}$ by Proposition 9.10 a).

To complete the proof note that

$$(\lambda|\alpha) \geq 0 \text{ if } \lambda \in P(\Lambda) \text{ and } \alpha \in \Delta_+^{im}. \tag{11.9.1}$$

Indeed, by Corollary 10.1 and because Δ_+^{im} is W-invariant, we may assume that $\lambda \in P_+$. But then (11.9.1) is obvious.

□

Statements a), b) and c) of Proposition 11.9 imply

COROLLARY 11.9. *Let* $\alpha \in \Delta_+^{im}$ *and* $\Lambda \in P_+$. *Let* λ *be a weight of the* $\mathfrak{g}(A)$*-module* $L(\Lambda)$. *Then either*

a) $(\lambda|\alpha) = 0$; *then* $\lambda - k\alpha$ *is not a weight of* $L(\Lambda)$ *for* $k \neq 0$; *or else*

b) $(\lambda|\alpha) \neq 0$; *then* $(\lambda|\alpha) > 0$ *and one has the following properties:*

(i) *the set of* $t \in \mathbf{Z}$ *such that* $\lambda - t\alpha \in P(\Lambda)$ *is an interval* $[-p, +\infty)$, *where* $p \geq 0$, *and* $t \mapsto \text{mult}_{L(\Lambda)}(\lambda - t\alpha)$ *is a nondecreasing function on this interval;*

(ii) *if* $x \in \mathfrak{g}_{-\alpha}$, $x \neq 0$, *then* $x : L(\Lambda)_{\lambda - t\alpha} \to L(\Lambda)_{\lambda-(t+1)\alpha}$ *is an injection.*

c)*If* $\lambda \in P_{++}$ *and* v *is a weight vector of weight* λ, *then the map* $\mathfrak{n}_- \to L(\Lambda)$ *defined by* $n \mapsto n(v)$ *is injective.*

§11.10. Here we use the results of the preceding section to describe explicitly the region of convergence of $\text{ch}_{L(\Lambda)}$.

PROPOSITION 11.10. *Let* A *be an indecomposable symmetrizable generalized Cartan matrix, and let* $L(\Lambda)$ *be an irreducible* $\mathfrak{g}(A)$*-module with highest weight* $\Lambda \in P_+$, *such that* $\langle \Lambda, \alpha_i^\vee \rangle \neq 0$ *for some* i. *Then the* *region* $Y(L(\Lambda)) \subset \mathfrak{h}$ *of absolute convergence of* $\mathrm{ch}_{L(\Lambda)}$ *is open and coincides with the set*

$$Y = \{h \in \mathfrak{h} | \sum_{\alpha \in \Delta_+} (\mathrm{mult}\, \alpha) \left| e^{-\langle \alpha, h \rangle} \right| < \infty\}.$$

Proof. By Proposition 10.6 c) it suffices to show that $Y(L(\Lambda)) \subset Y$ and that Y is open.

The inclusion in question is obvious if A is of finite type. If A is of affine type, we have

$$Y = \{h \in \mathfrak{h} | \operatorname{Re}\langle \delta, h \rangle > 0\}. \tag{11.10.1}$$

This follows from the description of the root system $\Delta(A)$ given by Proposition 6.3 and the fact that the multiplicities of roots are bounded (by ℓ) by Corollaries 7.4 and 8.3. Now the inclusion in question follows since $\mathrm{mult}_{L(\Lambda)}(\Lambda - s\delta) \neq 0$ for all $s \in \mathbb{Z}_+$ by Corollary 11.9b(i). Finally, if A is of indefinite type, then, by Theorem 5.6 c), there exists $\alpha \in \Delta_+^{im}$ such that $\mathrm{supp}\, \alpha = S(A)$ and $\langle \alpha, \alpha_i^\vee \rangle < 0$ for all i. But then $\Lambda - \alpha \in P(\Lambda)$ by Proposition 11.2a). Moreover, by Proposition 11.1b) and Corollary 11.9c), for every nonzero $v \in L(\Lambda)_{\Lambda - \alpha}$ the map $\psi : \mathfrak{n}_- \to L(\Lambda)$ defined by $\psi(y) = y(v)$ is injective. This completes the proof of the inclusion $Y(L(\Lambda)) \subset Y$.

To see that Y is open, note that

$$\alpha \in \Delta^{im} \text{ implies } \{h \in \operatorname{Int} X | \alpha(h) = 0\} = \emptyset. \tag{11.10.2}$$

This is clear since we can assume that $h \in C$ and then apply Proposition 3.12a) and f). Using Proposition 5.1a), we see that Y is the region of convergence of the infinite product $B := \prod_{\alpha \in \Delta_+^{im}} (1 - e^{-\alpha})^{\mathrm{mult}\, \alpha}$.

For $h \in \operatorname{Int} X$ consider the meromorphic function $f(z) = B(zh)^{-1}$. By a standard property of Dirichlet series with positive coefficients (see e.g. Serre [1970]), the set $\{t \in \mathbb{R} | th \in Y\}$ of convergence of the series obtained by multiplying out the product $B(zh)^{-1}$, which represents $f(z)$, is an open segment $(c, +\infty)$. It follows that if $th \in Y$, then $(t - \epsilon)h \in Y$ for some $\epsilon > 0$. But the argument in the proof of Proposition 10.6a) shows that Y is convex and that with every h' it contains $t'h'$ for sufficiently large $t' > 0$. Hence $Y \cap \mathfrak{h}_{\mathbb{R}}$ is open in $\mathfrak{h}_{\mathbb{R}}$, so that $Y = (Y \cap \mathfrak{h}_{\mathbb{R}}) + i\mathfrak{h}_{\mathbb{R}}$ is open in $\mathfrak{h}$.

□

§11.11. In this section we deduce a specialization formula of the denominator identity (10.4.4). Let $\mathfrak{h}_0$ be a subspace of $\mathfrak{h}$ such that

$$\mathfrak{h}_0 \cap \operatorname{Int} X_{\mathbf{C}} \neq \emptyset. \tag{11.11.1}$$

Let $\lambda \mapsto \overline{\lambda}$ denote the restriction map $\mathfrak{h}^* \to \mathfrak{h}_0^*$; denote by p the homomorphism of $\mathcal{E}$ to the completed group algebra of $\mathfrak{h}_0^*$ defined by $p(e(\lambda)) = e(\overline{\lambda})$. Put

$$\Delta_0 = \{\alpha \in \Delta | \overline{\alpha} = 0\}; \quad \Delta_{0+} = \Delta_0 \cap \Delta_+; \quad R_0 = \prod_{\alpha \in \Delta_{0+}} (1 - e(-\alpha)).$$

By Proposition 3.12, the set Δ_0 is finite; hence $\Delta_0 \subset \Delta^{re}$. It is clear that Δ_0 satisfies the usual axioms for a finite root system (see Bourbaki [1968]). Denote by W_0 the (finite) subgroup of W generated by reflections r_α ($\alpha \in \Delta_0$), and let ρ_0 (resp. $\rho_0^\vee$) be the half-sum of roots from Δ_{0+} (resp. $\Delta_{0+}^\vee$). Define a polynomial $D(\lambda)$ on $\mathfrak{h}^*$ by

$$D(\lambda) = \prod_{\alpha \in \Delta_{0+}} \langle \lambda, \alpha^\vee \rangle / \langle \rho_0, \alpha^\vee \rangle.$$

LEMMA 11.11. *For $\lambda \in \mathfrak{h}^*$ we have*

$$p\Big(R_0^{-1} \sum_{w \in W_0} \epsilon(w) e(w(\lambda))\Big) = D(\lambda) e(\overline{\lambda}).$$

Proof. Let $\Pi_0 \subset \Delta_{0+}$ be the set of simple roots of the root system Δ_0. Define the homomorphism $F : \mathbf{C}[e(-\alpha);\ \alpha \in \Pi_0] \to \mathbf{C}[q]$ by $F(e(-\alpha)) = q$ for all $\alpha \in \Pi_0$. Then

$$p(f) = \lim_{q \to 1} F(f); \quad F(e(-\alpha)) = q^{\langle \rho_0^\vee, \alpha \rangle}. \tag{11.11.2}$$

But by (10.9.3) we have

$$F\Big(\sum_{w \in W_0} \epsilon(w) e(w(\lambda) - \lambda)\Big) = \prod_{\alpha \in \Delta_{0+}} (1 - q^{\langle \lambda, \alpha^\vee \rangle}). \tag{11.11.3}$$

Formulas (11.11.2 and 3) together with (10.8.5) prove the lemma. □

Now we can deduce the specialization formula:

$$\prod_{\alpha \in \Delta_+ \setminus \Delta_0} (1 - e(-\overline{\alpha}))^{\operatorname{mult} \alpha} = \sum_{w \in W_0 \setminus W} \epsilon(w) D(w(\rho)) e(\overline{w(\rho)} - \overline{\rho}). \tag{11.11.4}$$

Here and further, $W_0 \backslash W$ denotes a set of representatives of left cosets of W_0 in W. Indeed, dividing both sides of (10.4.4) by R_0, we deduce that the left-hand side of (11.11.4), with $\overline{\alpha}$ replaced by α, is equal to

$$R_0^{-1} \sum_{w \in W} \epsilon(w) e(w(\rho) - \rho) = \sum_{v \in W_0 \backslash W} \epsilon(v) \Big(R_0^{-1} \sum_{u \in W_0} \epsilon(u) e(u(v(\rho)) - \rho) \Big).$$

Applying p and using Lemma 11.11, we get (11.11.4).

Now we consider a very special case of identity (11.11.4). Fix a sequence of nonnegative integers $s = (s_1, \ldots, s_n)$ such that the subdiagram $\{i \in S(A) | s_i = 0\} \subset S(A)$ is a union of diagrams of finite type. Fix an element $h^s \in \mathfrak{h}$ such that:

$$\langle \alpha_i, h^s \rangle = s_i \quad (i = 1, \ldots, n).$$

The subspace $\mathbb{C}h^s$ of $\mathfrak{h}$ satisfies condition (11.11.1) by Proposition 3.12.

Define $\lambda \in (\mathbb{C}h^s)^*$ by $\langle \lambda, h^s \rangle = 1$, and set $q = e(-\lambda)$. Then $p(e(-\alpha_i)) = q^{s_i}$ $(i = 1, \ldots, n)$. In other words, p is nothing else but the specialization of type s. Now (11.11.4) can be written as follows:

$$\prod_{j \geq 1} (1 - q^j)^{\dim \mathfrak{g}_j(s)} = \sum_{w \in W^s} \epsilon(w) D_s(w(\rho)) q^{\langle \rho - w(\rho), h^s \rangle}. \tag{11.11.5}$$

Here $D_s(\lambda) = \prod_{\alpha \in \Delta_{s+}} \langle \lambda, \alpha^\vee \rangle / \langle \rho_s, \alpha^\vee \rangle$, where $\Delta_{s+} = \{\alpha \in \Delta_+ | \langle \alpha, h^s \rangle = 0\}$ and ρ_s is the half-sum of roots from Δ_{s+}; W^s is a system of representatives of left cosets of the subgroup W_s generated by r_α, $\alpha \in \Delta_{s+}$, in W, so that $W = W_s W^s$; $\mathfrak{g}(A) = \bigoplus_j \mathfrak{g}_j(s)$ is the $\mathbb{Z}$-gradation of $\mathfrak{g}(A)$ of type s.

§11.12. In this section we shall discuss the unitarizability problem for the oscillator and the Virasoro algebras (cf. §§9.13 and 9.14).

A contravariant Hermitian form H on a $\mathfrak{a}$ (resp. Vir)-module V is a Hermitian form with respect to which the operators p_n and q_n are adjoint and the operators c and $a \in \mathfrak{a}_{0\mathbb{R}}$ are self adjoint (resp. d_n and d_{-n} are adjoint and c self adjoint). As in §11.5, it is clear that the $\mathfrak{a}$-module $R_{a,b,\lambda}$ (resp. the Vir-module $L(c,h)$) admits a contravariant Hermitian form if and only if $a, b \in \mathbb{R}$ and $\lambda \in \mathfrak{a}_{0\mathbb{R}}^*$ (resp. $c, h \in \mathbb{R}$), and that this form is uniquely determined by the condition

$$H(1,1) = 1 \quad (\text{resp. } H(v_{c,h}, v_{c,h}) = 1).$$

Recall that a module is called unitarizable if H is positive-definite.

It is immediate by (9.13.3) that the $\mathfrak{a}$-module $R_{a,b,\lambda}$ is unitarizable if and only if $a > 0$.

The unitarizability problem of the Vir-modules $L(c,h)$ is quite non-trivial. We shall make here only a few simple remarks (which will be used in Chapter 12).

PROPOSITION 11.12. a) *If the Vir-module $L(c,h)$ is unitarizable, then $h \geq 0$ and $c \geq 0$.*

b) *If $V = L(0,h)$ is unitarizable, then $h = 0$, and hence V is the trivial 1-dimensional Vir-module.*

c) *Let V be a unitarizable Vir-module such that d_0 is diagonalizable with finite-dimensional eigenspaces and with spectrum bounded below. Then V decomposes into an orthogonal direct sum of unitarizable Vir-modules $L(c,h)$, and the spectrum of d_0 is non-negative.*

Proof. Let v be the highest-weight vector of a unitarizable Vir-module $L(c,h)$. Then $H(d_{-1}(v), d_{-1}(v)) = H(v, d_1 d_{-1}(v)) = H(v, 2d_0(v)) + H(v, d_{-1}d_1(v)) = 2h$. Hence unitarizability implies $h \geq 0$. Looking at $H(d_{-j}(v), d_{-j}(v))$ as $j \to \infty$ similarly implies $c \geq 0$, proving a).

To prove b), consider the subspace U_n of $L(0,h)$ spanned by $d_{-2n}(v)$ and $d_{-n}^2(v)$. Since H is positive-definite on U_n we have: $\det_{U_n} H \geq 0$, which, after a simple calculation gives $4n^3h^2(8h - 5n) \geq 0$ for all integral $n > 0$. This implies $h = 0$.

To prove c) note that if U is a Vir-submodule of V, it is graded with respect to the eigenspace decomposition of d_0 (by Proposition 1.5), hence $V = U \oplus U^\perp$. Since $U^\perp$, the orthocomplement to U, is clearly also a Vir-submodule, we deduce that V is an orthogonal direct sum of unitarizable irreducible Vir-modules U_i such that d_0 is diagonalizable on U_i with specter bounded below. Thus the U_i are highest-weight modules $L(c,h)$ and the positivity of the spectrum of d_0 on V follows from a).

□

§11.13. In conclusion of this chapter, we shall sketch a theory of *generalized Kac–Moody algebras*. This is a Lie algebra $\mathfrak{g}(A)$ associated to a real $n \times n$ matrix $A = (a_{ij})$ satisfying the following properties (cf. §1.1):

(C1′) either $a_{ii} = 2$ or $a_{ii} \leq 0$;

(C2′) $a_{ij} \leq 0$ if $i \neq j$, and $a_{ij} \in \mathbb{Z}$ if $a_{ii} = 2$;

(C3′) $a_{ij} = 0$ implies $a_{ji} = 0$.

Let Π^{re} (resp. Π^{im}) $= \{\alpha_i \in \Pi \,|\, a_{ii} = 2 \text{ (resp. } \leq 0)\}$, let W be the subgroup of $GL(\mathfrak{h}^*)$ generated by the fundamental reflections r_i such that $\alpha_i \in \Pi^{re}$ (cf §3.7), and let $C = \{h \in \mathfrak{h}_{\mathbb{R}} \,|\, \langle \alpha_i, h\rangle \geq 0 \text{ if } a_{ii} = 2\}$, $C^\vee = \{\lambda \in \mathfrak{h}^*_{\mathbb{R}} \,|\, \langle \lambda, \alpha_i^\vee\rangle \geq 0 \text{ if } a_{ii} = 2\}$.

As in §3.3 we have:

$$(11.13.1)\qquad (\operatorname{ad} e_i)^{1-a_{ij}} e_j = 0;\ (\operatorname{ad} f_i)^{1-a_{ij}} f_i = 0 \quad \text{if } a_{ii} = 2 \text{ and } i \neq j;$$

$$(11.13.2)\qquad [e_i, e_j] = 0, [f_i, f_j] = 0 \quad \text{if } a_{ij} = 0.$$

A $\mathfrak{h}$-diagonalizable $\mathfrak{g}(A)$-module V is called *integrable* if e_i and f_i are locally nilpotent when $a_{ii} = 2$.

All the results of Chapter 3 hold with obvious modifications for generalized Kac–Moody algebras—one just restricts attention to the i with $a_{ii} = 2$.

The case of i with $a_{ii} \leq 0$ is treated using the following

LEMMA 11.13.1. *Let V be an integrable $\mathfrak{g}(A)$-module, let $a_{ii} \leq 0$, let $\lambda \in \mathfrak{h}^*$ be such that $\langle \lambda, \alpha_i^\vee\rangle < 0$ and let $v \in V_\lambda$ be a nonzero vector such that $f_i(v) = 0$. Then $e_i^j(v) \neq 0$ for all $j = 1, 2, \ldots$.*

Proof. This follows from the formula (cf. (3.6.1)):

$$f_i e_i^j(v) = -j(\langle \lambda, \alpha_i^\vee\rangle + \frac{1}{2}(j-1)a_{ii}) e_i^{j-1}(v).$$

□

COROLLARY 11.13.1. *If $a_{ii} < 0$, $\alpha \in \Delta_+ \backslash \{\alpha_i\}$ and $\operatorname{supp}(\alpha + \alpha_i)$ is connected, then $\alpha + j\alpha_i \in \Delta_+$ for all $j \in \mathbb{Z}_+$.*

Proof. We may assume that $\alpha - \alpha_i \notin \Delta_+$. Note that $\langle \alpha_i, \alpha_j^\vee\rangle \leq 0$ for all j. Since $\operatorname{supp}(\alpha + \alpha_i)$ is connected we conclude that $\langle \alpha, \alpha_i\rangle < 0$ and apply Lemma 11.13.1.

□

All the results of Chapter 4 hold for generalized Kac–Moody algebras, except that there is one additional affine matrix—the zero 1×1 matrix (cf. §2.9).

In order to describe the set of roots Δ, let $\Delta^{re} = W(\Pi^{re})$, $\Delta^{im} = \Delta \backslash \Delta^{re}$, $K = \{\alpha \in Q_+ \,|\, \alpha \in -C^\vee \text{ and } \operatorname{supp} \alpha \text{ is connected}\} \backslash \bigcup_{j \geq 2} j\Pi^{im}$.

Then we have

$$\Delta_+^{im} = \bigcup_{w \in W} w(K). \tag{11.13.3}$$

The proof is the same as that of Theorem 5.4 using Corollary 11.13.1. All other results of Chapter 5 hold for generalized Kac–Moody algebras with obvious modifications.

From now on in this section we will assume that $\mathfrak{g}(A)$ is a generalized Kac–Moody algebra with a symmetrizable Cartan matrix.

THEOREM 11.13.1. *$\mathfrak{g}(A)$ is a Lie algebra on generators e_i, f_i ($i = 1, \dots, n$) with defining relations (1.2.1), (11.13.1) and (11.13.2).*

Proof. Let $\mathfrak{g}_1(A)$ be the quotient of $\tilde{\mathfrak{g}}(A)$ by the ideal generated by all elements (11.13.1 and 2), and let Δ_1 be the set of roots of $\mathfrak{g}_1(A)$. By the argument proving Lemma 1.6, the support of any $\alpha \in \Delta_1$ is connected, and since Δ_1 is W-invariant, we obtain, using (11.13.3), that $\Delta_1 = \Delta$. Now the theorem follows from Proposition 9.11 and Lemma 11.13.2b) below.

□

LEMMA 11.13.2. a) *Let $\alpha = \sum_{i \in I} k_i \alpha_i$ be such that the k_i are positive integers and $\alpha \in -C^\vee$. Then $2(\rho \,|\, \alpha) \geq (\alpha \,|\, \alpha)$ with equality if and only if $(\alpha_i \,|\, \alpha_i) \leq 0$ and $(\alpha_i \,|\, \alpha_j) = 0$ when $i \neq j$ and $i, j \in I$, and $(\alpha_i \,|\, \alpha_i) = 0$ when $k_i > 1$.*

b) *Inequality (11.6.1) holds.*

Proof. $2(\rho|\alpha) - (\alpha|\alpha) = \sum_i k_i(\alpha_i|\alpha_i - \alpha)$. If $\alpha_i \in \Pi^{re}$, then $(\alpha_i|\alpha_i) > 0$ and $(\alpha_i| - \alpha) \geq 0$, so $k_i(\alpha_i|\alpha_i - \alpha) > 0$. If $\alpha_i \in \Pi^{im}$, then $k_i(\alpha_i|\alpha_i - \alpha) > 0$ since $\alpha - \alpha_i \in Q_+$. Hence $2(\rho|\alpha) - (\alpha|\alpha) \geq 0$, and if the equality holds then all the α_i with $i \in I$ are in Π^{im} and satisfy $(\alpha_i|\alpha_i - \alpha) = 0$. This completes the proof of a).

To prove b) note that we can keep on strictly reducing $(\rho|\alpha)$ by fundamental reflections while keeping α positive until either $\alpha \in \Pi$ or $\alpha \in -C^\vee$. In the second case $\alpha \in \Pi^{im}$ by a) since $\operatorname{supp} \alpha$ is connected.

□

It is clear by Lemma 11.13.2 b) that Theorem 11.7 a) holds for generalized symmetrizable Kac–Moody algebras. It turns out that this property characterizes them. More precisely, let A be a symmetrizable matrix, possibly infinite but countable, satisfying (C1′)–(C3′), and let $\mathfrak{g}'(A)$ be the associated algebra, i.e. the quotient of the Lie algebra defined in Remark 1.5

by relations (11.13.1 and 2). Let $\mathfrak{c}$ ($\subset \mathfrak{h}'$) be the center of $\mathfrak{g}'(A)$ and let $\mathfrak{g}$ be a central extension of $\mathfrak{g}'(A)/\mathfrak{c}$ (described explicitly by Borcherds [1989 A]). Define a $\mathbb{Z}$-gradation of $\mathfrak{g}$ by deg(center) $= 0$, $\deg e_i = -\deg f_i = s_i$, where s_i is a collection of positive integers with finite repetitions. Then $\mathfrak{g}$ has the following properties:

($\mathfrak{g}$1) $\mathfrak{g} = \bigoplus_j \mathfrak{g}_j$ is a $\mathbb{Z}$-graded Lie algebra, $[\mathfrak{g}_0, \mathfrak{g}_0] = 0$ and $\dim \mathfrak{g}_j < \infty$;

($\mathfrak{g}$2) $\mathfrak{g}$ has an antilinear involution ω which is -1 on $\mathfrak{g}_{0\mathbb{R}}$ and $\omega(\mathfrak{g}_m) = \mathfrak{g}_{-m}$;

($\mathfrak{g}$3) $\mathfrak{g}$ carries an invariant bilinear form $(\,.\,|\,.\,)$ such that $(\mathfrak{g}_i \,|\, \mathfrak{g}_j) = 0$ unless $i + j = 0$;

($\mathfrak{g}$4) the Hermitian form $(x\,|\,y)_0 = -(\omega(x)\,|\,y)$ is positive definite on $\mathfrak{g}_j$ if $j \neq 0$.

THEOREM 11.13.2. *Any Lie algebra $\mathfrak{g}$ satisfying ($\mathfrak{g}$1–4) can be obtained from a generalized Kac–Moody algebra $\mathfrak{g}'(A)$ as described above.*

Proof. The argument proving Proposition 9.12 b) shows that $\mathfrak{g}$ can be obtained from some $\mathfrak{g}'(B)$ as described above, where B is a symmetric matrix. It follows from the proof of Theorem 11.7a) that we have the inequality (11.6.1). This implies that DB satisfies (C1′)–(C3′) for an invertible diagonal matrix D (see Borcherds [1988] for details).

□

We turn now to the study of $\mathfrak{g}(A)$-modules $L(\Lambda)$ with Λ satisfying

$$(11.13.4) \qquad \langle \Lambda, \alpha_i^\vee \rangle \in \mathbb{Z}_+ \text{ if } a_{ii} = 2;\ \langle \Lambda, \alpha_i^\vee \rangle \geq 0 \text{ for all } i.$$

We denote by P_+ the set of such Λ's.

LEMMA 11.13.3. *Proposition 11.4 holds for a generalized Kac–Moody algebra.*

Proof. The proof of part a) of Proposition 11.4 for a generalized Kac–Moody algebra needs no modifications. The proof of its b) part is different. Let $\lambda \in P(\Lambda)$. It is clear that $(\lambda\,|\,\alpha_i) \geq 0$ for all $\alpha_i \in \Pi^{im}$. We can act on λ by fundamental reflections each time strictly increasing $|\lambda + \rho|^2$ until $\lambda \in C^\vee$, so we may assume that $(\lambda|\alpha_i) \geq 0$ for all $\alpha_i \in \Pi^{re}$. Thus:

$$(\lambda|\alpha_i) \geq 0 \text{ for all } \alpha_i \in \Pi.$$

If $\lambda \neq \Lambda$, let $i \in \operatorname{supp}(\Lambda - \lambda)$. If $\alpha_i \in \Pi^{re}$, then, of course $(\alpha_i|\Lambda+\lambda+2\rho) > 0$. If $\alpha_i \in \Pi^{im}$ we have: $(\alpha_i|\Lambda + \lambda + 2\rho) = (\alpha_i|\Lambda) + (\alpha_i|\lambda + \alpha_i) \geq 0$ since $(\Lambda|\alpha_i) \geq 0$ and $\lambda+\alpha_i = \Lambda-\beta$, where $\beta \in Q_+$. Hence $(\Lambda-\lambda\,|\,\Lambda+\lambda+2\rho) \geq 0$, which is equivalent to $|\Lambda+\rho|^2-|\lambda+\rho|^2 \geq 0$. Equality implies that $(\Lambda|\alpha_i)=0$ for any $i \in \text{supp}(\Lambda-\lambda)$, i.e. that $\Lambda=\lambda$.

□

It is clear by Lemma 11.13.3 that Theorem 11.7b) holds for generalized symmetrizable Kac–Moody algebras.

Finally, we find a character formula.

THEOREM 11.13.3. *Let $\Lambda \in P_+$, and let $S_\Lambda = e(\Lambda+\rho)\sum_s \varepsilon(s)e(-s)$, where s runs over all sums of elements from Π^{im} and $\varepsilon(s) = (-1)^m$ if s is the sum of m distinct pairwise perpendicular elements perpendicular to Λ, and $\varepsilon(s)=0$ otherwise. Then*

$$\text{ch}\, L(\Lambda) = \sum_{w\in W} \varepsilon(w)w(S_\Lambda)\Big/ e(\rho)\prod_{\alpha\in\Delta_+}(1-e(-\alpha))^{\text{mult}\,\alpha}.$$

Proof. Following the argument of §10.4 we find that

$$e(\rho)\prod_{\alpha\in\Delta_+}(1-e(-\alpha))^{\text{mult}\,\alpha}\,\text{ch}\, L(\Lambda) = \sum_\lambda c_\lambda e(\lambda+\rho), \tag{11.13.5}$$

where both sides are antisymmetric under W, the c_λ are integers and $\lambda \in \Lambda - Q_+$ satisfy $|\lambda+\rho|^2 = |\Lambda+\rho|^2$.

Let S_Λ be the sum of the terms on the right of (11.13.5) for which $\lambda+\rho \in C^\vee$. If $\lambda+\rho \notin \text{Int}\, C^\vee$, then $\sum_{w\in W}\varepsilon(w)e(w(\lambda+\rho)) = 0$, hence the right-hand side of (11.13.5) is equal to $\sum_{w\in W}\varepsilon(w)w(S_\Lambda)$, and it remains to evaluate S_Λ.

If $e(\lambda+\rho)$ occurs in S_Λ, we write $\lambda = \Lambda - \sum_{i\in I} k_i\alpha_i$ where k_i are positive integers. Then $|\lambda+\rho|^2 = |\Lambda+\rho|^2$ implies:

$$\sum_{i\in I} k_i(\Lambda|\alpha_i) + \sum_{i\in I} k_i(\lambda+2\rho|\alpha_i) = 0.$$

But $(\Lambda|\alpha_i) \geq 0$ and $(\lambda+2\rho|\alpha_i) \geq 0$ (the second inequality is proved in the same way as in Lemma 11.13.3). It follows that

$$(\Lambda|\alpha_i) = 0 \quad \text{and} \quad (\lambda+2\rho|\alpha_i) = 0. \tag{11.13.6}$$

Since $(\lambda+\rho|\alpha_i) \geq 0$ and $(\rho|\alpha_i) > 0$ for $\alpha_i \in \Pi^{re}$, it follows that $\alpha_i \in \Pi^{im}$ for $i \in I$. The second equality of (11.13.6) can be written as follows:

$$\sum_{i\in I}(k_i - \delta_{ij})(\alpha_i|\alpha_j) = 0, \quad j \in I.$$

Hence $(\alpha_i|\alpha_j) = 0$ for $i, j \in I$ unless $i = j$ and $k_i = 1$.

If $e(\Lambda - \sum_i k_i\alpha_i)$ occurs in $\mathrm{ch}\, L(\Lambda)$, then $(\Lambda|\alpha_i) \neq 0$ for some i, hence the only terms on the right-hand side of (11.13.5) which give a contribution to S_Λ are those coming from $e(\Lambda + \rho) \prod_{\alpha \in \Delta_+} (1 - e(-\alpha))^{\mathrm{mult}\, \alpha}$. If $(\alpha_i|\alpha_j) = 0$ for $i \neq j$, then the coefficient of $e(\Lambda + \rho - \sum k_i\alpha_i)$ in this expression is 0 if some of the k_i are greater than 1 and is $(-1)^{\Sigma k_i}$ otherwise.

□

Since $\mathrm{ch} L(0) = 1$, we deduce the following

COROLLARY 11.13.2. *Let $S = e(\rho) \sum_s \varepsilon(s) e(s)$, where s runs over all sums of elements from Π^{im} and $\varepsilon(s) = (-1)^m$ if s is the sum of m distinct pairwise perpendicular elements, and $\varepsilon(s) = 0$ otherwise. Then*

$$\prod_{\alpha \in \Delta_+} (1 - e(-\alpha))^{\mathrm{mult}\, \alpha} = e^{-\rho} \sum_{w \in W} \varepsilon(w) w(S).$$

□

§11.14. Exercises.

11.1. Show that if $\Lambda \in P_+$ and $\lambda \in P(\Lambda) \setminus \{\Lambda\}$, then

$$\sum_{w \in W} \epsilon(w)\, \mathrm{mult}_{L(\Lambda)}(\lambda + \rho - w(\rho)) = 0.$$

(This formula allows one to compute $\mathrm{mult}_{L(\Lambda)}$ by induction on $\mathrm{ht}(\Lambda - \lambda)$).

[Rewrite (10.4.5), multiplying by the denominator, as:

$$\sum_\lambda e(\lambda) \sum_{w \in W} \epsilon(w)\, \mathrm{mult}_{L(\Lambda)}(\lambda + \rho - w(\rho)) = \sum_{w \in W} \epsilon(w) e(w(\Lambda + \rho) - \rho),$$

and note that, by Proposition 11.4, the equality $\lambda + \rho = w(\Lambda + \rho)$ implies $\lambda = \Lambda$.]

11.2. If A is of finite, affine or strictly hyperbolic type, and $\Lambda \in P_+$, then every dominant $\lambda \leq \Lambda$ is nondegenerate with respect to Λ.

11.3. Show that if $\Lambda \in P_+$, then $P(\Lambda) \cap P_+ = \{\lambda \in P | \lambda \geq w(\lambda)$ for every $w \in W$ and λ is nondegenerate with respect to $\Lambda\}$.

11.4. Show that if $\Lambda \in P_+$, then the set of asymptotic rays for the set of rays through $\{\Lambda - \mu | \mu \in P(\Lambda)\}$ lies in $\overline{Z}$.

[Use Exercise 3.12 and Lemma 5.8.]

11.5. Let $\Lambda, \Lambda', \dots \in P_+$. Show that the $U(\mathfrak{g}(A))$-submodule generated by $v_\Lambda \otimes v_{\Lambda'} \otimes \dots$ of the $\mathfrak{g}(A)$-module $V = L(\Lambda) \otimes L(\Lambda') \otimes \dots$ coincides with the eigenspace of Ω corresponding to the eigenvalue $c_{\Lambda+\Lambda'+\dots}$, where $c_M := |M+\rho|^2 - |\rho|^2$. Moreover, $c_{\Lambda+\Lambda'+\dots} > c_M$ for every M such that $L(M) \subset V$.

11.6. Let $\mathfrak{g}(A)$ be an infinite-dimensional Kac–Moody algebra and let Λ, $M \in P_+$, $\Lambda|_{\mathfrak{h}'} \neq 0$. Show that $L(\Lambda) \otimes L(M) \not\supset L(0)$.

11.7. Let $\mathfrak{g}(A) = \bigoplus_{j \in \mathbb{Z}} \mathfrak{g}_j$ be a $\mathbb{Z}$-gradation of type $(\delta_{1,i}, \dots, \delta_{n,i})$ of a symmetrizable Kac–Moody algebra $\mathfrak{g}(A)$. Prove that the $\mathfrak{g}_0$-module $\mathfrak{g}_{-1}$ is isomorphic to the $\mathfrak{g}_0$-module $L(-\alpha_i)$.

[See the proof of Proposition 8.6c).]

11.8. Let A be a symmetrizable generalized Cartan matrix and $\alpha \in \Delta^{im}$. Show that $\psi(\alpha) := \lim\limits_{n\to\infty} \frac{\log \operatorname{mult} n\alpha}{n}$ exists. Show that $\psi(\alpha) = 0$ if and only if $(\alpha|\alpha) = 0$ and that $\psi(\alpha) > 0.48$ if $(\alpha|\alpha) \neq 0$. Show that $Y \subset \operatorname{Int} X_{\mathbb{C}}$ coincides with $\operatorname{Int} X_{\mathbb{C}}$ if and only if A is a direct sum of matrices of finite and affine type.

11.9. Let $\mathfrak{g}$ and $\mathfrak{g}^0$ be Kac–Moody algebras with symmetrizable Cartan matrices. Let $\mathfrak{h}$ and $\mathfrak{h}^0$ be their Cartan subalgebras, W and W^0 their Weyl groups, etc. Let $\pi : \mathfrak{g}^0 \to \mathfrak{g}$ be a homomorphism such that

$$\pi(\mathfrak{h}^0) \subset \mathfrak{h}; \quad \pi(\operatorname{Int} X^0_{\mathbb{C}}) \cap \operatorname{Int} X_{\mathbb{C}} \neq \emptyset; \quad (\pm\Pi) \cap \pi^*(Z) = \emptyset.$$

Let $\Lambda \in P_+$. Show that the $\mathfrak{g}$-module $L(\Lambda)$ is isomorphic as a $\mathfrak{g}^0$-module to a direct sum of integrable highest-weight modules $L^0(\mu)$ with finite multiplicities, which we denote by $(\Lambda : \mu)$. For $\mu = w(\mu' + \rho^0) - \rho^0$, where $w \in W^0$ and $\mu' \in P^0_+$, set $(\Lambda : \mu) = \epsilon(w)(\Lambda : \mu')$; for all other $\mu \in \mathfrak{h}^{0*}$ set $(\Lambda : \mu) = 0$. We define Δ_0, W_0, ρ_0, $D(\lambda)$, etc., for $\mathfrak{h}^0$ as in §11.11. Show that

$$\pi^*\Big(\prod_{\alpha \in \Delta_+ \setminus \Delta_{0+}} (1 - e(-\alpha))^{\operatorname{mult} \alpha}\Big) \sum_{\mu \in \mathfrak{h}^{0*}} (\Lambda : \mu) e(\mu)$$
$$= \sum_{w \in W_0 \setminus W} \epsilon(w) D(w(\Lambda+\rho)) e\big(\pi^*(w(\Lambda+\rho) - \rho)\big) \prod_{\alpha \in \Delta^0_+} (1 - e(-\alpha))^{\operatorname{mult}^0 \alpha}.$$

11.10. Let $\mathfrak{m}$ be a subalgebra of $\mathfrak{g}$ such that $\mathfrak{m} + \mathfrak{n}_-$ (resp. $\mathfrak{m} + \mathfrak{n}_+$) has finite codimension in $\mathfrak{g}(A)$ and let $\Lambda \in P_+$. Show that $\dim \{v \in L(\Lambda) | \mathfrak{m}(v) = 0\} < \infty$ (resp. $\dim L(\Lambda)/\mathfrak{m}L(\Lambda) < \infty$).

[Let $U_+ = \langle \exp t \operatorname{ad} e_i |\ t \in \mathbb{C},\ i = 1, \dots, n\rangle$. Show that $\bigcup_i U_+(e_i)$ span $\mathfrak{n}_+$.]

11.11. Let $\mathfrak{g}(A)$ be a symmetrizable Kac–Moody algebra. Set

$$R = \prod_{\alpha\in\Delta_+}(1-e(-\alpha))^{\operatorname{mult}\alpha};\quad F = -\log R.$$

Fix an orthonormal basis $\{u_i\}$ of $\mathfrak{h}$, and define derivations ∂_i of $\mathcal{E}$ by $\partial_i(e(\lambda)) = \langle\lambda, u_i\rangle e(\lambda)$. Show that

$$(\partial_i F)^2 - \partial_i^2 F = R^{-1}\partial_i^2 R;\quad \sum_i \partial_i^2 R = (\rho|\rho)R.$$

Deduce that

$$\sum_i((\partial_i F)^2 - \partial_i^2 F) = (\rho|\rho).$$

11.12. For $\beta \in Q_+$ set $c_\beta = \sum_{n\geq 1} n^{-1}\operatorname{mult}(\beta/n)$. Using Exercise 11.11 show that under the hypotheses of this exercise one has

$$(\beta|\beta-2\rho)c_\beta = \sum_{\substack{\beta'\beta''\in Q_+\\ \beta'+\beta''=\beta}} (\beta'|\beta'')c_{\beta'}c_{\beta''}.$$

(This formula allows one to compute the multiplicities of roots by induction on the height, thanks to (11.6.1). Note that if $\beta \in K$, then, due to Exercise 5.20, all summands on the right-hand side are non-positive.)

11.13. Consider the following generalized Cartan matrix of order m where $m \geq 3$:

$$A = \begin{pmatrix} 2 & -1 & -1 & \dots & -1 \\ -1 & 2 & 0 & \dots & 0 \\ -1 & 0 & 2 & \dots & 0 \\ \dots & \dots & \dots & \dots & \dots \\ -1 & 0 & 0 & \dots & 2 \end{pmatrix}$$

Set $\beta = 2\alpha_1 + \alpha_2 + \alpha_3 + \cdots + \alpha_m \in \Delta(A)$. Show that

$$\operatorname{mult}\beta = 2^{m-2} - (m-1).$$

11.14. Let $A : V_1 \to V_2$ and $B : V_2 \to V_1$ be linear maps of finite-dimensional spaces. Show that $\operatorname{tr}AB = \operatorname{tr}BA$. Let V be a module from the category $\mathcal{O}$ over a Kac–Moody algebra with a symmetrizable Cartan matrix. Deduce that for $e_\alpha \in \mathfrak{g}_\alpha$ and $e_{-\alpha} \in \mathfrak{g}_{-\alpha}$ such that $(e_\alpha|e_{-\alpha}) = 1$, $\alpha \in \Delta_{+j}$, one has

$$\operatorname{tr}_{V_\lambda} e_{-\alpha}e_\alpha = \sum_{j\geq 1}(\lambda + j\alpha|\alpha)\dim V_{\lambda+j\alpha}.$$

Let V be a module with highest weight Λ. Deduce the following generalization of Freudenthal's formula:

$$(|\Lambda+\rho|^2 - |\lambda+\rho|^2)\dim V_\lambda = 2\sum_{\alpha\in\Delta_+}\sum_{j\geq 1}(\operatorname{mult}\alpha)(\lambda+j\alpha|\alpha)\dim V_{\lambda+j\alpha}.$$

11.15. Let $V = \bigoplus_{i=1}^{n} L(\Lambda_i)$, where $\Lambda_i \in P_+$ satisfy $\langle\Lambda_i, \alpha_j^\vee\rangle = \delta_{ij}$ $(j = 1,\ldots,n)$. Show that the group associated to the $\mathfrak{g}(A)$-module V is a central extension of every group G^π constructed in Remark 3.8.

11.16. Show that if a module $L(\Lambda)$ over a generalized Kac–Moody algebra is unitarizable then $\Lambda \in P_+$.

11.17. Let $\Lambda, \Lambda' \in P_+$ and $\sigma \in W$ be such that $M := \sigma(\Lambda) + \Lambda' \in P_+$. Show that

$$\operatorname{mult}_{L(\Lambda)}(M+\rho-w(\Lambda'+\rho)) = \delta_{1,w}.$$

Deduce that the multiplicity of $L(M)$ in $L(\Lambda)\otimes L(\Lambda')$ is 1.
[Use Proposition 11.4 and the outer multiplicity formula: multiplicity of $L(M)$ in $L(\Lambda)\otimes L(\Lambda')$ is $\sum_{w\in W}\varepsilon(w)\operatorname{mult}_{L(\Lambda)}(M+\rho-w(\Lambda'+\rho))$.]

11.18. Using Exercise 9.17, show that the Vir-module $L(c,h)$ with $c \geq 1$ and $h \geq (c-1)/24$ is unitarizable.

11.19. Show that the fixed point set of a diagram automorphism of a generalized Kac–Moody algebra is a central extension of a generalized Kac–Moody algebra.

11.20. Prove an analogue of Corollary 10.4 for generalized Kac–Moody algebras.

11.21. Let $\mathfrak{g}(A)$ be a generalized Kac–Moody algebra. Show that $\Delta_+ = \Delta_+(A)$ is a subset of $Q_+\backslash\{0\}$ defined by the following properties:

(i) $\alpha_i \in \Delta_+$, $2\alpha_i \notin \Delta_+$, $i = 1,\ldots,n$;

(ii) if $\alpha \in \Delta_+$, $\alpha_i \in \Pi^{im}$ and $\operatorname{supp}(\alpha+\alpha_i)$ is connected, then $\alpha+\alpha_i \in \Delta_+$;

(iii) if $\alpha \in \Delta_+$, $\alpha_i \in \Pi^{re}$ and $\alpha \neq \alpha_i$, then $[\alpha, r_i(\alpha)] \cap Q_+ \subset \Delta_+$;

(iv) if $\alpha \in \Delta_+$ then $\operatorname{supp}\alpha$ is connected.

11.22. Let $\mathfrak{g}'(A)$ be a generalized Kac–Moody algebra. Show that the $U(\mathfrak{g}'(A))$-module $\bigoplus_{\Lambda\in P_+} L(\Lambda)$ is faithful. [Use Exercise 11.20 to show that $U(\mathfrak{n}_-)$ acts faithfully. Then $0 < H(u_-v_\Lambda, u_-v_\Lambda) = \langle\hat{\omega}_0(u_-)u_-v_\Lambda\rangle$ for some $\Lambda \in P_+$, depending on $u_- \in U(\mathfrak{n}_-)$. Hence $U(\mathfrak{n}_+)$ also acts faithfully.]

§11.15. Bibliographical notes and comments.

Proposition 11.2 was stated without a proof in Kac–Peterson [1984 A]. Propositions 11.3 and 11.4 are due to Kac–Peterson [1984 A]. Of course, Proposition 11.4 is a standard fact in the finite-dimensional case. Proposition 11.3 a) in the finite-dimensional case, is due to Steinberg (see Bourbaki [1975]). It seems that Proposition 11.3 b) was not previously known even in the finite-dimensional case.

Lemma 11.6 and the positivity Theorem 11.7 are due to Kac–Peterson [1984 B]. The proof of Theorem 11.7 b) is a direct generalization of that of Garland [1978] in the affine case (as soon as Theorem 11.7 a) is established). Theorem 11.7 is important since it allows one to apply powerful Hilbert space methods, see e.g. Goodman–Wallach [1984 A], Kac–Peterson [1984 B]. (Note that the Hilbert completion of $L(\Lambda)$ is the space of a unitary representation of the "compact form" of the group associated to a Kac–Moody algebra.) Jakobsen–Kac [1985], [1989] classified the unitarizable highest-weight modules over affine algebras for all natural choices of antilinear involutions and "Borel subalgebras"; it turned out that there exists only one essentially new family of unitarizable highest-weight representations, that of the Lie algebra of maps of the circle with a positive-finite measure into $su(n, 1)$. There is an interesting connection of these representations when $n = 1$ to the representations constructed in Exercise 9.20 (cf. Bernard–Felder [1989], Feigin–Frenkel [1989 B]).

The material of §§11.9–11.11 is taken from Kac–Peterson [1984 A]. The proof of Proposition 11.12 b) is taken from Goddard–Olive [1986]. Most of the material of §11.13 (on generalized Kac–Moody algebras) is taken from Borcherds [1988 A]. The new material is the description of the root system and Theorem 11.13.1 (that fills a slight gap in his paper).

Exercise 11.1 in the finite-dimensional case is attributed to Racah. Exercises 11.4, 11.8 and 11.9 are taken from Kac–Peterson [1984 A], and Exercises 11.5 and 11.6 are from Kac–Peterson [1983]. Exercise 11.10 is due to Peterson–Kac [1983]. This property has important applications to the conformal field theory. A similar "admissibility" property plays a prominent role in representation theory of p-adic groups.

Exercises 11.11, 11.12 and 11.14 are due to Peterson [1982], [1983]. The proofs indicated here were communicated to me by Peterson. These recurrent multiplicity formulas are very convenient for computations of root and weight multiplicities. Another formula for root multiplicities, which is also a formal consequence of identity (10.4.4), was found earlier by Berman–Moody [1979].

Exercise 11.13 is taken from Kac [1983 A]. It gives some nontrivial evidence in support of the conjecture mentioned in §5.13. Exercise 11.15 is taken from Peterson–Kac [1983]. Exercise 11.17 is taken from Kac–Wakimoto [1986]. This is a special case of the Parthasaraty–Rao–Varadarajan conjecture proved recently by Kumar [1988] and Mathieu [1988]. Exercise 11.19 is taken from Borcherds [1988 A]. The set of roots discussed in Exercise 11.22 is related to quivers with tadpoles (Kac [1983]).

The multiplicities of all roots of an indefinite-type Kac–Moody algebra are not known explicitly in any single case (in some cases explicit formulas are known for low "level" roots, see e.g. Feingold–Frenkel [1983], Kac–Moody–Wakimoto [1988]). Nevertheless Borcherds [1989 B], [1989 C] managed to find the root multiplicities of some generalized Kac–Moody algebras intimately related to the Monster simple group, and used this to establish the modularity properties of the Monster.

Given below are a few computations of the root multiplicities done on a computer by R. Gross, using Peterson's recurrent formula. Here $(k_1, k_2, \ldots)$ denotes the root $\alpha = k_1\alpha_1 + k_2\alpha_2 + \ldots$. We may assume that $\alpha \in -C^\vee$ and that $(\alpha\,|\,\alpha) := \frac{1}{2}\sum a_{ij}k_ik_j \neq 0$. If A is symmetric of rank 2, we can assume that $k_1 \geq k_2$, since then $\mathrm{mult}(k_1, k_2) = \mathrm{mult}(k_2, k_1)$. In Table H_2 (resp. H_3) all such roots are listed with $k_1 \leq 21$ (resp $k_1 \leq 14$). A part of Table H_3 was computed by Feingold–Frenkel [1983].

$$A = \begin{bmatrix} 2 & -3 \\ -3 & 2 \end{bmatrix}$$

TABLE H_2

α	$-(\alpha\|\alpha)$	mult α	α	$-(\alpha\|\alpha)$	mult α
(1, 1)	1	1	(15, 11)	149	23750
(2, 2)	4	1	(16, 11)	151	25923
(3, 2)	5	2	(13, 12)	155	30865
(3, 3)	9	3	(14, 12)	164	45271
(4, 3)	11	4	(13, 13)	169	55853
(4, 4)	16	6	(15, 12)	171	60654
(5, 4)	19	9	(16, 12)	176	74434
(6, 4)	20	9	(17, 12)	179	84121
(5, 5)	25	16	(18, 12)	180	87547
(6, 5)	29	23	(14, 13)	181	91257
(7, 5)	31	27	(15, 13)	191	135861
(6, 6)	36	39	(14, 14)	196	165173
(7, 6)	41	60	(16, 13)	199	185526
(8, 6)	44	73	(17, 13)	205	233487
(9, 6)	45	80	(15, 14)	209	271860
(7, 7)	49	107	(18, 13)	209	271702
(8, 7)	55	162	(19, 13)	211	292947
(9, 7)	59	211	(16, 14)	220	409725
(10, 7)	61	240	(15, 15)	225	492420
(8, 8)	64	288	(17, 14)	229	569358
(9, 8)	71	449	(18, 14)	236	732180
(10, 8)	76	600	(16, 15)	239	815214
(11, 8)	79	720	(19, 14)	241	874650
(12, 8)	80	758	(20, 14)	244	972117
(9, 9)	81	808	(21, 14)	245	1006994
(10, 9)	89	1267	(17, 15)	251	1242438
(11, 9)	95	1754	(16, 16)	256	1476973
(12, 9)	99	2167	(18, 15)	261	1752719
(10, 10)	100	2278	(19, 15)	269	2298090
(13, 9)	101	2407	(17, 16)	271	2458684
(11, 10)	109	3630	(20, 15)	275	2808958
(12, 10)	116	5130	(21, 15)	279	3207547
(11, 11)	121	6559	(22, 15)	281	3426450
(13, 10)	121	6555	(18, 16)	284	3783712
(14, 10)	124	7554	(17, 17)	289	4456255
(15, 10)	125	7936	(19, 16)	295	5411212
(12, 11)	131	10531	(20, 16)	304	7217527
(13, 11)	139	15204	(18, 17)	305	7453376
(12, 12)	144	19022	(21, 16)	311	9005900
(14, 11)	145	19902			

$$A = \begin{bmatrix} 2 & -2 & 0 \\ -2 & 2 & -1 \\ 0 & -1 & 2 \end{bmatrix}$$

TABLE H_3

α	$-(\alpha\|\alpha)$	mult α	α	$-(\alpha\|\alpha)$	mult α	α	$-(\alpha\|\alpha)$	mult α
(2, 2, 1)	1	2	(8, 8, 4)	16	297	(13, 13, 3)	30	6818
(3, 3, 1)	2	3	(10, 10, 2)	16	297	(10, 12, 5)	31	8326
(3, 4, 2)	3	5	(8, 9, 3)	17	385	(11, 12, 4)	31	8322
(4, 4, 1)	3	5	(10, 11, 2)	17	385	(10, 12, 6)	32	10111
(4, 4, 2)	4	7	(9, 9, 3)	18	490	(11, 13, 4)	32	10108
(5, 5, 1)	4	7	(11, 11, 2)	18	490	(12, 12, 4)	32	10107
(4, 5, 2)	5	11	(8, 9, 4)	19	627	(13, 14, 3)	32	10096
(6, 6, 1)	5	11	(11, 12, 2)	19	626	(10, 13, 6)	33	12266
(5, 5, 2)	6	15	(8, 10, 4)	20	792	(14, 14, 3)	33	12246
(7, 7, 1)	6	15	(9, 9, 4)	20	792	(11, 12, 5)	34	14821
(5, 6, 2)	7	22	(9, 10, 3)	20	792	(11, 12, 6)	35	17892
(8, 8, 1)	7	22	(12, 12, 2)	20	791	(12, 15, 5)	35	17893
(5, 6, 3)	8	30	(8, 10, 5)	21	1002	(12, 13, 4)	35	17886
(6, 6, 2)	8	30	(10, 10, 3)	21	1002	(14, 15, 3)	35	17861
(9, 9, 1)	8	30	(12, 13, 2)	21	1001	(11, 13, 5)	36	21525
(6, 6, 3)	9	42	(13, 13, 2)	22	1253	(12, 12, 6)	36	21526
(6, 7, 2)	9	42	(9, 10, 4)	23	1574	(12, 14, 4)	36	21514
(10, 10, 1)	9	42	(10, 11, 3)	23	1574	(13, 13, 4)	36	21515
(7, 7, 2)	10	56	(13, 14, 2)	23	1571	(11, 13, 6)	38	30993
(11, 11, 1)	10	56	(9, 10, 5)	24	1957	(11, 14, 6)	39	37083
(6, 7, 3)	11	77	(9, 11, 4)	24	1957	(12, 13, 5)	39	37080
(7, 8, 2)	11	77	(10, 10, 4)	24	1957	(13, 14, 4)	39	37053
(12, 12, 1)	11	77	(11, 11, 3)	24	1956	(11, 14, 7)	40	44258
(6, 8, 4)	12	101	(14, 14, 2)	24	1953	(13, 13, 5)	40	44247
(7, 7, 3)	12	101	(10, 10, 5)	25	2434	(14, 14, 4)	40	44217
(8, 8, 2)	12	101	(14, 15, 2)	25	2429	(13, 15, 4)	40	44219
(13, 13, 1)	12	101	(9, 11, 5)	26	3007	(12, 13, 6)	41	52753
(8, 9, 2)	13	135	(11, 12, 3)	26	3005	(12, 14, 5)	41	52741
(14, 14, 1)	13	135	(9, 12, 6)	27	3712	(13, 13, 6)	42	62719
(7, 8, 3)	14	176	(10, 11, 4)	27	3713	(12, 14, 6)	44	88255
(9, 9, 2)	14	176	(12, 12, 3)	27	3710	(13, 14, 5)	44	88230
(7, 8, 4)	15	231	(10, 12, 4)	28	4557	(12, 14, 7)	45	104456
(8, 8, 3)	15	231	(11, 11, 4)	28	4557	(14, 14, 5)	45	104415
(9, 10, 2)	15	231	(10, 11, 5)	29	5593	(12, 15, 6)	45	104450
(7, 9, 4)	16	297	(12, 13, 3)	29	5587	(13, 14, 6)	47	145513
			(11, 11, 5)	30	6826	(13, 14, 7)	48	171355
						(14, 14, 6)	48	171337
						(14, 14, 7)	49	201527

Chapter 12. Integrable Highest-Weight Modules over Affine Algebras. Application to η-Function Identities Sugawara Operators and Branching Functions

§12.0. In the last three chapters we developed a representation theory of arbitrary Kac–Moody algebras. From now on we turn to the special case of affine algebras.

We show that the denominator identity (10.4.4) for affine algebras is nothing else but the celebrated Macdonald identities. Historically this was the first application of the representation theory of Kac–Moody algebras. The basic idea is very simple: one gets an interesting identity by computing the character of an integrable representation in two different ways and equating the results. In particular, Macdonald identities are deduced via the trivial representation. Furthermore, we show that specializations (11.11.5) of the denominator identity turn into identities for q-series of modular forms, the simplest ones being Macdonald identities for the powers of the Dedekind η-function.

We study the structure of the weight system of an integrable highest-weight module over an affine algebra in more detail. This allows us to write its character in a different form to obtain the important theta function identity. This identity involves classical theta functions and certain modular forms called string functions, which are, essentially, generating functions for multiplicities of weights in "strings." Furthermore, we consider branching functions, which are a generalization of string functions when instead of the Cartan subalgebra a general reductive subalgebra is considered.

Finally, we introduce one of the most powerful tools of conformal field theory, the Sugawara construction and the coset construction, in relation to the study of general branching rules.

In the next chapter all this will be linked to the theory of modular forms and theta functions.

§12.1. Let $\mathfrak{g}(A)$ be an affine algebra of type $X_N^{(r)}$ (here $X = A, B, C, D, E, F$ or G; $r = 1, 2$ or 3; N as in Table Aff r). We keep the notation of Chapter 6. Since the multiplicities of the roots of $\mathfrak{g}(A)$ are known (by

Corollary 8.3), we can write down the denominator identity (10.4.4) explicitly.

Recall that the left-hand side of (10.4.4) is

$$R_{\text{left}} = \prod_{\alpha \in \Delta_+} (1 - e(-\alpha))^{\operatorname{mult} \alpha}.$$

Introduce the following polynomials in x (cf. Proposition 6.3):

$$L(x) = (1-x)^{\ell} \prod_{\alpha \in \mathring{\Delta}} (1 - xe(\alpha)) \text{ for } A \text{ of type } X_{\ell}^{(1)};$$

$$L(x) = (1-x)^{s}(1-x^{r})^{\ell - s} \prod_{\alpha \in \mathring{\Delta}_s} (1 - xe(\alpha)) \prod_{\alpha \in \mathring{\Delta}_\ell} (1 - x^{r} e(\alpha))$$

for A of type $X_N^{(r)}$ with $\alpha_0 r = 2$ or 3, where $s = \dfrac{N - \ell}{r - 1}$;

$$L(x) = (1-x)^{\ell} \prod_{\alpha \in \mathring{\Delta}_s} (1 - xe(\alpha)) \prod_{\alpha \in \mathring{\Delta}_\ell} (1 - xe(\tfrac{1}{2}(\delta - \alpha)))(1 - x^2 e(\alpha))$$

for A of type $A_{2\ell}^{(2)}$.

Set $\mathring{R} = \prod_{\alpha \in \mathring{\Delta}_+} (1 - e(-\alpha))$. Then, by Proposition 6.3 and Corollary 8.3 we have

$$R_{\text{left}} = \mathring{R} \prod_{n \geq 1} L(e(-n\delta)). \tag{12.1.1}$$

Furthermore, the right-hand side of (10.4.4) is

$$R_{\text{right}} := \sum_{w \in W} \epsilon(w) e(w(\rho) - \rho),$$

where ρ is as defined in §6.2. By Proposition 6.5, we can write $w = ut_{\alpha}$, where $u \in \mathring{W}$, $\alpha \in M$. Using the equalities $\langle \rho, K \rangle = h^{\vee}$ and $|\rho|^2 = |\overline{\rho}|^2$, we deduce from (6.5.3):

$$ut_{\alpha}(\rho) - \rho = u(\overline{\rho} + h^{\vee}\alpha) - \overline{\rho} + \frac{1}{2h^{\vee}}(|\overline{\rho}|^2 - |\overline{\rho} + h^{\vee}\alpha|^2)\delta. \tag{12.1.2}$$

Hence we obtain

$$(12.1.3)\quad R_{\text{right}} = e(\frac{|\overline{\rho}|^2}{2h^\vee}\delta) \sum_{\alpha\in M} \Big(\sum_{w\in \mathring{W}} \epsilon(w) e(w(\overline{\rho}+h^\vee\alpha)-\overline{\rho}) \Big) \times e(-\frac{1}{2h^\vee}|\overline{\rho}+h^\vee\alpha|^2\delta).$$

Comparing (12.1.1 and 3) we deduce

$$(12.1.4)\quad e(-\frac{|\overline{\rho}|^2}{2h^\vee}\delta)\mathring{R}\prod_{n\geq 1} L(e(-n\delta)) = \sum_{\alpha\in M}\Big(\sum_{w\in\mathring{W}} \epsilon(w)e(w(\overline{\rho}+h^\vee\alpha)-\overline{\rho})\Big)e(-\frac{1}{2h^\vee}|\overline{\rho}+h^\vee\alpha|^2\delta).$$

These are the Macdonald identities.

Example 12.1. Let A be of type $A_1^{(1)}$. Then $\Pi = \{\alpha_0,\alpha_1\}$; $\Delta_+ = \{(n-1)\alpha_0+n\alpha_1;\ n\alpha_0+(n-1)\alpha_1;\ n\alpha_0+n\alpha_1 (n=1,2,\ldots)\}$; $\operatorname{mult}\alpha = 1$ for $\alpha\in\Delta_+$; $M = \mathbb{Z}\alpha_1$; $\overline{\rho} = \frac{1}{2}\alpha_1$; $(\alpha_1|\alpha_1) = 2$; $h^\vee = 2$; $\mathring{W} = \{\pm 1\}$. Put $u = e(-\alpha_0)$, $v = e(-\alpha_1)$. Then using the expression for R_{left} in (10.4.4) and that given by (12.1.3) for R_{right}, we obtain:

$$(12.1.5)\quad \prod_{n\geq 1}(1-u^nv^n)(1-u^{n-1}v^n)(1-u^nv^{n-1}) = \sum_{j\in\mathbb{Z}}(-1)^j u^{\frac{1}{2}j(j+1)}v^{\frac{1}{2}j(j-1)}.$$

This is one of the forms of the Jacobi triple product identity.

For $\lambda\in\mathfrak{h}^*$ set

$$(12.1.6)\quad \mathring{\chi}(\lambda) = \frac{\sum_{w\in\mathring{W}} \epsilon(w)e(w(\lambda+\overline{\rho})-\overline{\rho})}{\prod_{\alpha\in\mathring{\Delta}_+}(1-e(-\alpha))}.$$

Note that if $\langle\lambda,\alpha_i^\vee\rangle\in\mathbb{Z}_+$ for $i=1,\ldots,\ell$, then $\mathring{\chi}(\lambda)$ is nothing other than the formal character of the $\mathring{\mathfrak{g}}$-module $\mathring{L}(\overline{\lambda})$ (where $\mathring{\mathfrak{g}}$ is a simple finite-dimensional Lie algebra with the root system $\mathring{\Delta}$). Dividing both sides of (12.1.4) by $\mathring{R}$ we obtain another form of Macdonald identities:

$$(12.1.7)\quad e(-\frac{|\overline{\rho}|^2}{2h^\vee}\delta)\prod_{n\geq 1}L(e(-n\delta)) = \sum_{\alpha\in M}\mathring{\chi}(h^\vee\alpha)e(-\frac{1}{2h^\vee}|\overline{\rho}+h^\vee\alpha|^2\delta).$$

Consider now the particular case of a nontwisted affine algebra $\mathfrak{g}$ of type $X_\ell^{(1)}$. In this case $\mathring{\mathfrak{g}}$ is of type X_ℓ. Recall that one has the "strange" formula of Freudenthal–de Vries:

$$(12.1.8) \qquad \frac{|\overline{\rho}|^2}{2h^\vee} = \frac{\dim \mathring{\mathfrak{g}}}{24}.$$

Later (see formula (12.3.7)) we will prove a much more general "very strange" formula. Recall also that M is the lattice spanned over $\mathbb{Z}$ by long roots of $\mathring{\mathfrak{g}}$. Setting $q = e(-\delta)$, we can rewrite (12.1.7) as follows:

$$(12.1.9) \qquad q^{\dim \mathring{\mathfrak{g}}/24} \prod_{n\geq 1}((1-q^n)^\ell \prod_{\alpha\in\mathring{\Delta}}(1-q^n e(\alpha))) = \sum_{\alpha\in M} \mathring{\chi}(h^\vee\alpha) q^{|\overline{\rho}+h^\vee\alpha|^2/2h^\vee}.$$

§**12.2.** The specialization of type $(1,0,\ldots,0)$ is called the *basic specialization*; we denote it by F. Note that

$$(12.2.1) \qquad F(e(-\alpha)) = q^{\langle\alpha,d\rangle}.$$

In particular, $F(e(\alpha)) = 1$ if $\alpha \in \mathring{\Delta}$. Hence, by Lemma 11.11 we obtain

$$(12.2.2) \qquad F(\mathring{\chi}(\lambda)) = \mathring{d}(\lambda), \text{ where } \mathring{d}(\lambda) = \prod_{\alpha\in\mathring{\Delta}_+^\vee} \langle\lambda+\overline{\rho},\alpha\rangle/\langle\overline{\rho},\alpha\rangle.$$

Introduce the Euler product

$$(12.2.3) \qquad \varphi(q) = \prod_{n=1}^{\infty}(1-q^n),$$

and the Dedekind η-function:

$$(12.2.4) \qquad \eta = q^{1/24}\varphi(q).$$

Applying F to both sides of (12.1.9) and using the equality $\dim\mathring{\mathfrak{g}} = \ell + |\mathring{\Delta}|$, we obtain Macdonald's η-function identities

$$(12.2.5) \qquad \eta^{\dim\mathring{\mathfrak{g}}} = \sum_{\alpha\in M} \mathring{d}(h^\vee\alpha) q^{|\overline{\rho}+h^\vee\alpha|^2/2h^\vee}.$$

§12.3. Here we derive a generalization of identity (12.2.5) using an arbitrary specialization of type s. We keep the notation of §8.2. Let $\mathfrak{g}$ be a simple finite-dimensional Lie algebra of type X_N, let μ be a diagram automorphism of $\mathfrak{g}$ of order r and $\mathfrak{g} = \bigoplus \mathfrak{g}_j$ the corresponding $\mathbb{Z}/r\mathbb{Z}$-gradation. Introduce elements E_i, F_i, H_i $(i = 0, \ldots, \ell)$ of $\mathfrak{g}$ as in §8.2.

Given a nonzero sequence of nonnegative integers $s = (s_0, \ldots, s_\ell)$, we set $m = r \sum_{i=0}^{\ell} a_i s_i$ (where a_i are given by Table Aff r). Let $\epsilon = \exp \frac{2\pi i}{m}$. Recall the definition of the automorphism $\sigma_{s;r}$ of type $(s; r)$ of $\mathfrak{g}$:

$$\sigma_{s;r}(E_j) = \epsilon^{s_j} E_j, \quad \sigma_{s;r}(F_j) = \epsilon^{-s_j} F_j, \quad \sigma_{s;r}(H_j) = H_j.$$

Let $\mathfrak{g} = \bigoplus_j \mathfrak{g}_{\bar{j}}(s; r)$ be the associated $\mathbb{Z}/m\mathbb{Z}$-gradation. Put

$$d_{\bar{j}}(s; r) = \dim \mathfrak{g}_{\bar{j}}(s; r) \quad (j \in \mathbb{Z}),$$

where $\bar{j} \in \mathbb{Z}/m\mathbb{Z}$ denote $j \bmod m$. By the last assertion of Theorem 8.7, we have

LEMMA 12.3. *Let $\mathfrak{g}(A)$ be the affine algebra of type $X_N^{(r)}$ and let $\mathfrak{g}(A) = \bigoplus_{j \in \mathbb{Z}} \mathfrak{g}_j(s)$ be its $\mathbb{Z}$-gradation of type s. Then*

$$\dim \mathfrak{g}_j(s) = d_{\bar{j}}(s; r).$$

□

Hence, the left-hand side of (11.11.5) is

$$R_{\text{left}} = \prod_{j \geq 1} (1 - q^j)^{d_{\bar{j}}(s;r)}. \tag{12.3.1}$$

Recall that the right-hand side of (11.11.5) is

$$R_{\text{right}} = \sum_{w \in W^s} \epsilon(w) D_s(w(\rho)) q^{\langle \rho - w(\rho), h^s \rangle}.$$

We rewrite R_{right} in terms of the Lie algebra $\mathring{\mathfrak{g}}$. Let Δ_s be the subset of $\mathring{\Delta}$ which consists of linear combinations of the roots from the set $\{\alpha_i | s_i = 0 \ (i = 0, \ldots, \ell)\}$. Let W_s be the subgroup of the Weyl group W generated by reflections in the roots from Δ_s. Let $\mathring{W}^s$ be the set of representatives of left cosets of $W_s T$ in W, so that $W = W_s T \mathring{W}^s$. We may choose $W^s = T\mathring{W}^s$. Then the set $\{t_\alpha w\}$, where $\alpha \in M, w \in \mathring{W}^s$, is a set of representatives of left cosets of W_s in W. Using (6.2.8) and (6.5.3) we obtain

$$(12.3.2) \qquad t_\alpha w(\rho) = h^\vee \Lambda_0 + w(\overline{\rho}) + h^\vee \alpha - (\frac{h^\vee}{2}(\alpha|\alpha) + (w(\overline{\rho})|\alpha))\delta.$$

Hence we have

$$\begin{aligned}&\langle \rho - t_\alpha w(\rho), h^s\rangle \\ &\quad = \langle \overline{\rho}, h^s\rangle - \langle w(\overline{\rho}), h^s\rangle - h^\vee \langle \alpha, h^s\rangle + \frac{m}{r}(\frac{h^\vee}{2}(\alpha|\alpha) + (w(\overline{\rho})|\alpha)).\end{aligned}$$

We define $\gamma_s \in \mathfrak{h}_0^*$ by

$$(12.3.3) \qquad (\gamma_s|\alpha_i) = rs_i/m \quad (i = 1, \ldots, \ell).$$

Then the last formula can be rewritten as follows:

$$(12.3.4) \quad \langle \rho - t_\alpha w(\rho), h^s\rangle = \frac{m}{2h^\vee r}(|h^\vee \alpha + w(\overline{\rho}) - h^\vee \gamma_s|^2 - |\overline{\rho} - h^\vee \gamma_s|^2).$$

Using (12.3.1 and 4) we deduce from (11.11.5) the following identity:

$$\begin{aligned}(12.3.5) \quad & q^{\frac{m}{2h^\vee r}|\overline{\rho} - h^\vee \gamma_s|^2} \prod_{j\geq 1}(1 - q^j)^{d_j(s;r)} \\ & = \sum_{w \in W_0^s} \epsilon(w) \sum_{\alpha \in M} D_s(w(\overline{\rho}) + h^\vee \alpha) q^{\frac{m}{2h^\vee r}|w(\overline{\rho}) + h^\vee(\alpha - \gamma_s)|^2}.\end{aligned}$$

Note that the η-function identity (12.2.5) is a special case of this identity for $s = (1, 0, \ldots, 0)$, $r = 1$.

As in that special case, one can express the first factor in the left-hand side of (12.3.5) entirely in terms of the $d_j(s;r)$. Namely, one has the following "very strange formula," which is a generalization of the "strange formula" (12.1.8):

$$(12.3.6) \qquad \frac{m}{2h^\vee r}|\overline{\rho} - h^\vee \gamma_s|^2 = \frac{m}{24}\dim \mathfrak{g} - \frac{1}{4m}\sum_{j=1}^{m-1} j(m-j)d_j(s;r).$$

The proof of this formula uses identity (12.3.5) and some elements of the theory of modular forms. It will be given in Chapter 13.

§12.4. Recall that the center of an affine algebra $\mathfrak{g}(A)$ is 1-dimensional and is spanned by the canonical central element (see §6.2):

$$K = \sum_{i=0}^{\ell} a_i^\vee \alpha_i^\vee .$$

It is clear that K operates on a $\mathfrak{g}(A)$-module $L(\Lambda)$ by the scalar operator $\langle \Lambda, K \rangle I_{L(\Lambda)}$. In particular, $\langle \Lambda, K \rangle = \langle \lambda, K \rangle$ for every $\lambda \in P(\Lambda)$. The number

$$k := \langle \Lambda, K \rangle = \sum_{i=0}^{\ell} a_i^\vee \langle \Lambda, \alpha_i^\vee \rangle \tag{12.4.1}$$

is called the *level* of $\Lambda \in \mathfrak{h}^*$, or of the module $L(\Lambda)$.

If $\Lambda \in P_+$, then the level of Λ is a nonnegative integer; it is zero if and only if all the labels $\langle \Lambda, \alpha_i^\vee \rangle$ of Λ are zero. Hence, by (9.10.1), an integrable $\mathfrak{g}(A)$-module $L(\Lambda)$ has level 0 if and only if $\dim L(\Lambda) = 1$; if $L(\Lambda)$ is integrable and $\dim L(\Lambda) \neq 1$, then the level of $L(\Lambda)$ is a positive integer.

Note that the level of ρ is equal to

$$\langle \rho, K \rangle = \sum_{i=0}^{\ell} a_i^\vee = h^\vee , \tag{12.4.2}$$

the dual Coxeter number.

Let Λ_i $(i = 0, \dots, \ell)$ be the *fundamental weights*:

$$\langle \Lambda_i, \alpha_j^\vee \rangle = \delta_{ij}, \ j = 0, \dots, \ell, \text{ and } \langle \Lambda_i, d \rangle = 0.$$

Note that

$$\Lambda_i = \overline{\Lambda}_i + a_i^\vee \Lambda_0, \tag{12.4.3}$$

where $\overline{\Lambda}_0 = 0$ and $\overline{\Lambda}_1, \dots, \overline{\Lambda}_\ell$ are the fundamental weights of $\mathring{\mathfrak{g}}$. Note also that

$$P_+ = \sum_{i=0}^{\ell} \mathbb{Z}_+ \Lambda_i + \mathbb{C}\delta. \tag{12.4.4}$$

It is clear that the level of Λ from P_+ is 1 if and only if $\Lambda \equiv \Lambda_i \mod \mathbb{C}\delta$ and i is such that $a_i^\vee = 1$; in particular, the level of Λ_0 is always 1. A glance at Table Aff gives that if A is symmetric or $r > 1$, then

$$\text{level}(\Lambda_i) = 1 \text{ if and only if } i \in (\operatorname{Aut} S(A)) \cdot 0. \tag{12.4.5}$$

Finally, the following observation, which follows from Propositions 3.12 b) and 5.8 b), is useful:

LEMMA 12.4. *If* $\text{Re}\langle\lambda, K\rangle > 0$, *then* $(W\cdot\lambda)\cap\{\mu\in\mathfrak{h}^* \mid \text{Re}\langle\mu, \alpha_i^\vee\rangle \geq 0$ *for all* $i\}$ *consists of a single element. In particular, if* $\lambda\in P$ *has a positive level, then the set* $P_+\cap W\cdot\lambda$ *consists of a single element.*

□

We let $P^k(\text{resp. } P_+^k) = \{\lambda\in P(\text{resp. } P_+) \mid \langle\lambda, K\rangle = k\}$.

§12.5. We collect here some facts about the weight system $P(\Lambda)$ of an integrable module $L(\Lambda)$ over an affine algebra $\mathfrak{g}(A)$, proved earlier in the general context of Kac–Moody algebras.

PROPOSITION 12.5. *Let* $L(\Lambda)$ *be an integrable module of positive level* k *over an affine algebra. Then*

a) $P(\Lambda) = W\cdot\{\lambda\in P_+ | \lambda\leq\Lambda\}$.

b) $P(\Lambda) = (\Lambda+Q)\cap$ *convex hull of* $W\cdot\Lambda$;

c) *If* $\lambda,\mu\in P(\Lambda)$ *and* μ *lies in the convex hull of* $W\cdot\lambda$, *then* $\text{mult}_{L(\Lambda)}\,\mu \geq \text{mult}_{L(\Lambda)}\,\lambda$.

d) $P(\Lambda)$ *lies in the paraboloid* $\{\lambda \in \mathfrak{h}_{\mathbb{R}}^* | |\overline{\lambda}|^2 + 2k(\lambda|\Lambda_0) \leq |\Lambda|^2$; $\langle\lambda, K\rangle = k\}$; *the intersection of* $P(\Lambda)$ *with the boundary of this paraboloid is* $W\cdot\Lambda$.

e) *For* $\lambda\in P(\Lambda)$ *the set of* $t\in\mathbb{Z}$ *such that* $\lambda - t\delta\in P(\Lambda)$, *is an interval* $[-p, +\infty)$ *with* $p\geq 0$, *and* $t\mapsto \text{mult}_{L(\Lambda)}(\lambda-t\delta)$ *is a nondecreasing function on this interval. Moreover, if* $x\in\mathfrak{g}_{-\delta}$, $x\neq 0$, *then the map* $x : L(\Lambda)_{\lambda-t\delta}\rightarrow L(\Lambda)_{\lambda-(t+1)\delta}$ *is injective.*

f) *Set* $\mathfrak{n}_-^{(\delta)} = \bigoplus_{n>0}\mathfrak{g}_{-n\delta}$; *then* $L(\Lambda)$ *is a free* $U(\mathfrak{n}_-^{(\delta)})$*-module.*

Proof. a) follows from Proposition 11.2 b), while b) and c) are special cases of Proposition 11.3 a) and b). d) follows from Proposition 11.4 a) and formula (6.2.7). e) follows from Corollary 11.9 b). f) is a special case of Proposition 11.9 c).

□

§12.6. We continue the study of the weight system $P(\Lambda)$ of an integrable module $L(\Lambda)$ of positive level k over an affine algebra.

A weight $\lambda\in P(\Lambda)$ is called *maximal* if $\lambda+\delta\notin P(\Lambda)$. Denote by $\max(\Lambda)$ the set of all maximal weights of $L(\Lambda)$. It is clear that $\max(\Lambda)$ is a W-invariant set (since $P(\Lambda)$ is W-invariant and δ is W-fixed) and hence, by Corollary 10.1, a maximal weight is W-equivalent to a unique dominant maximal weight. On the other hand, it follows from Proposition 12.5 e) that for every $\mu \in P(\Lambda)$ there exists a unique $\lambda \in \max(\Lambda)$ and a unique nonnegative integer n such that $\mu = \lambda - n\delta$, i.e., we have

$$(12.6.1) \qquad P(\Lambda) = \bigcup_{\lambda \in \max(\Lambda)} \{\lambda - n\delta | n \in \mathbb{Z}_+\} \quad \text{(disjoint union)}.$$

Here is a description of dominant maximal weights.

PROPOSITION 12.6. *The map $\lambda \mapsto \overline{\lambda}$ defines a bijection from $\max(\Lambda) \cap P_+$ onto $kC_{af} \cap (\overline{\Lambda} + \overline{Q})$. In particular, the set of dominant maximal weights of $L(\Lambda)$ is finite.*

Proof. Straightforward using Proposition 12.5. □

The following lemma describes explicitly the weight system of certain particularly important highest-weight modules.

LEMMA 12.6. *Let A be an affine matrix of type $X_N^{(r)}$, where $X = A$, D or E. Let $\Lambda \in P_+$ be of level 1. Then*

$$(12.6.2) \qquad \max(\Lambda) = W \cdot \Lambda = T \cdot \Lambda,$$

$$(12.6.3) \quad P(\Lambda) = \{\Lambda_0 + \tfrac{1}{2}|\Lambda|^2\delta + \alpha - (\tfrac{1}{2}|\alpha|^2 + s)\delta, \quad \text{where } \alpha \in \overline{\Lambda} + \overline{Q},\ s \in \mathbb{Z}_+\}.$$

Proof. Since level$(\Lambda) = 1$, Λ is a fundamental weight $\Lambda_i \mod \mathbb{C}\delta$. Using (12.4.5), one easily sees that i is a special vertex, and hence $W \cdot \Lambda_i = T \cdot \Lambda_i$ (see Remark 6.5). Hence (12.6.1 and 2) imply (12.6.3), using (6.5.2) and (6.2.6).

To prove (12.6.2) recall that $T = \{t_\alpha | \alpha \in \overline{Q}\}$ (see §6.5). Let $\lambda \in \max(\Lambda)$; we have $\lambda = \Lambda - \beta$, where $\beta \in Q$. Since $\beta = \overline{\beta} \mod \mathbb{C}\delta$, we have: $t_{\overline{\beta}}(\lambda) = \Lambda \mod \mathbb{C}\delta$ by (6.5.2). This proves (12.6.2). □

§12.7. Let $\Lambda \in P_+$. If follows from Proposition 11.10 and (11.10.1) that $\mathrm{ch}_{L(\Lambda)}$ converges absolutely to a holomorphic function in the region

$$Y = \{h \in \mathfrak{h} \mid \mathrm{Re}\langle \delta, h\rangle > 0\}.$$

In fact, Y is the region of convergence of $\mathrm{ch}_{L(\Lambda)}$ if level$(\Lambda) > 0$. Note also that for any highest-weight module V over an affine algebra, ch_V converges absolutely in the domain Y_0 (see Lemma 10.6 b).

For $\lambda \in \max(\Lambda)$ introduce the generating series

$$a_\lambda^\Lambda = \sum_{n=0}^{\infty} \mathrm{mult}_{L(\Lambda)}(\lambda - n\delta)e^{-n\delta elta}.$$

This series converges absolutely in the region Y since it is majorized by $e^{-\Lambda}\mathrm{ch}_{L(\Lambda)}$.

Since $W_\lambda \cap T = 1$ for $\lambda \in P(\Lambda)$ (see Proposition 6.6 c)) and $W(\delta) = \delta$, we deduce using (12.6.1):

$$\mathrm{ch}_{L(\Lambda)} = \sum_{\lambda \in \max(\Lambda)} e^\lambda a_\lambda^\Lambda = \sum_{\substack{\lambda \in \max(\Lambda) \\ \lambda \bmod T}} \left(\sum_{t\in T} e^{t(\lambda)}\right) a_\lambda^\Lambda. \tag{12.7.1}$$

This simple formula reduces the computation of the character of $L(\Lambda)$ to the computation of the functions a_λ^Λ. The following case is especially simple.

LEMMA 12.7. *Let A and Λ be as in Lemma 12.6. Then*

$$e^{-\frac{1}{2}|\Lambda|^2\delta}\,\mathrm{ch}_{L(\Lambda)} = a_{\Lambda_0}^{\Lambda_0} \sum_{\gamma \in \overline{Q} + \overline{\Lambda}} e^{\Lambda_0 + \gamma - \frac{1}{2}|\gamma|^2\delta}.$$

Proof. Use (12.6.3) and the fact that every a_λ^Λ is W-conjugate to $a_\Lambda^\Lambda = a_{\Lambda_0}^{\Lambda_0}$ due to (12.4.5) and Lemma 12.6.

□

The function $a_{\Lambda_0}^{\Lambda_0}$ will be calculated explicitly at the end of this chapter.

We proceed to rewrite character formulas (10.4.5) and (12.7.1) in terms of theta functions. For $\lambda \in \mathfrak{h}^*$ such that level$(\lambda) = k > 0$ set

$$\Theta_\lambda = e^{-\frac{|\lambda|^2}{2k}\delta} \sum_{t \in T} e^{t(\lambda)}, \quad A_\lambda = \sum_{w \in \mathring{W}} \varepsilon(w)\Theta_{w(\lambda)}. \tag{12.7.2}$$

Using (6.5.3) we obtain

(12.7.3) $$\Theta_\lambda = e^{k\Lambda_0} \sum_{\gamma \in M + k^{-1}\overline{\lambda}} e^{-\frac{1}{2}k|\gamma|^2\delta + k\gamma},$$

which is a classical theta function (see Chapter 13 for details). It is clear that this series converges absolutely on Y to a holomorphic function. Note also that Θ_λ depends only on $\lambda \mod kM + \mathbb{C}\delta$.

Using theta functions, we can rewrite the denominator identity (10.4.4) in yet another form. Recall that $\langle \rho, K\rangle = h^\vee$ (the dual Coxeter number) and $\rho = \overline{\rho} + h^\vee\Lambda_0$ (formula (6.2.8)). Using the decomposition $W = \mathring{W} \ltimes T$, we get

$$\sum_{w \in W} \epsilon(w) e^{w(\rho)-\rho} = e^{-\rho} \sum_{w \in \mathring{W}} \epsilon(w) \sum_{\alpha \in M} e^{t_\alpha w(\rho)} = e^{-\rho + \frac{|\rho|^2}{2h^\vee}\delta} A_\rho .$$

Hence, (10.4.4) can be rewritten as follows:

(12.7.4) $$A_\rho = \sum_{w \in \mathring{W}} \epsilon(w) \Theta_{h^\vee\Lambda_0 + w(\overline{\rho})} = e^{h^\vee\Lambda_0 + \overline{\rho} - \frac{|\overline{\rho}|^2}{2h^\vee}\delta} \prod_{\alpha \in \Delta_+} (1 - e^{-\alpha})^{\operatorname{mult}\alpha}.$$

Introduce the following number for $\Lambda \in P_+^k$:

(12.7.5) $$m_\Lambda = \frac{|\Lambda + \rho|^2}{2(k + h^\vee)} - \frac{|\rho|^2}{2h^\vee}.$$

For reasons which will become clear later we call this number the *modular anomaly of* Λ, and we introduce the *normalized character*

$$\chi_\Lambda = e^{-m_\Lambda\delta} \operatorname{ch}_{L(\Lambda)} .$$

For a weight $\lambda \in P(\Lambda)$ introduce the number

(12.7.6) $$m_{\Lambda,\lambda} = m_\Lambda - \frac{|\lambda|^2}{2k}.$$

It is clear that $m_{\Lambda,\lambda}$ is a rational number.

For $\lambda \in \mathfrak{h}^*$ set

$$c_\lambda^\Lambda = e^{-m_{\Lambda,\lambda}\delta} \sum_{n\in\mathbb{Z}} \text{mult}_{L(\Lambda)}(\lambda - n\delta)e^{-n\delta}. \tag{12.7.7}$$

Just as the series a_λ^Λ, this series converges absolutely to a holomorphic function on Y. Note that $c_\lambda^\Lambda = e^{-m_{\Lambda,\lambda}\delta}a_\lambda^\Lambda$ if $\lambda \in \max(\Lambda)$.

The function c_λ^Λ is called the *string function* of $\lambda \in \mathfrak{h}^*$. Note that

$$c_{w(\lambda)}^\Lambda = c_\lambda^\Lambda \text{ for } w \in W. \tag{12.7.8}$$

Since $W = \mathring{W} \ltimes T$, we use (6.5.2) to obtain

$$c_{w(\lambda)+k\gamma+a\delta}^\Lambda = c_\lambda^\Lambda \text{ for } \lambda \in \mathfrak{h}^*,\ w \in \mathring{W},\ \gamma \in M,\ a \in \mathbf{C}. \tag{12.7.9}$$

Note also that

$$c_\lambda^\Lambda = c_\lambda^{\Lambda+a\delta} \text{ for } a \in \mathbf{C}. \tag{12.7.10}$$

Using $W = \mathring{W} \ltimes T$ and (12.7.2) we can rewrite the character formula (10.4.5) as follows:

$$\chi_\Lambda = A_{\Lambda+\rho}/A_\rho. \tag{12.7.11}$$

On the other hand, we have by (12.7.1) and the definitions of Θ_λ and c_λ^Λ:

$$\chi_\Lambda = \sum_{\substack{\lambda\in P^k \bmod \mathbf{C}\delta \\ \lambda \bmod T}} c_\lambda^\Lambda \Theta_\lambda.$$

Using (12.7.9) we can rewrite this as follows:

$$\chi_\Lambda = \sum_{\lambda\in P^k \bmod (kM+\mathbf{C}\delta)} c_\lambda^\Lambda \Theta_\lambda. \tag{12.7.12}$$

Comparing (12.7.11 and 12) gives the theta function identity

$$A_{\Lambda+\rho} = A_\rho \sum_{\lambda\in P^k \bmod (kM+\mathbf{C}\delta)} c_\lambda^\Lambda \Theta_\lambda. \tag{12.7.13}$$

We shall use this important identity in the next chapter to study and compute the string functions.

§12.8. We turn now to the Sugawara construction. For the sake of simplicity, we consider only the case of a non-twisted affine algebra $\mathfrak{g}' = \mathfrak{g}'(X_\ell^{(1)})$, leaving the general case as an exercise (see Exercise 12.20). Recall that $\mathfrak{g}' = \tilde{\mathcal{L}}(\mathring{\mathfrak{g}}) = \mathbf{C}[t,t^{-1}] \otimes_{\mathbf{C}} \mathring{\mathfrak{g}} + \mathbf{C}K$ with commutation relations

$$[x^{(m)}, y^{(n)}] = [x,y]^{(m+n)} + m\delta_{m,-n}(x|y)K. \tag{12.8.1}$$

Here $\mathring{\mathfrak{g}}$ is a simple finite-dimensional Lie algebra of type X_ℓ, $(x|y)$ is the normalized (by (6.4.2)) invariant bilinear form on $\mathring{\mathfrak{g}}$, and $x^{(n)}$ stands for $t^n \otimes x$ ($n \in \mathbf{Z}$, $x \in \mathring{\mathfrak{g}}$).

Let $\{u_i\}$ and $\{u^i\}$ be dual bases of $\mathring{\mathfrak{g}}$, i.e. $(u_i \mid u^j) = \delta_{ij}$. Recall that

$$\mathring{\Omega} = \sum_i u_i u^i,$$

the Casimir operator of $\mathring{\mathfrak{g}}$, is independent of the choice of these dual bases. In particular, $\sum_i u_i u^i = \sum_i u^i u_i$. This proves the following useful formula:

$$\sum_i [u_i^{(m)}, u^{i(n)}] = m\delta_{m,-n}(\dim \mathring{\mathfrak{g}})K. \tag{12.8.2}$$

Furthermore, note that

$$\Omega = 2(K + h^\vee)d + \mathring{\Omega} + 2\sum_{n=1}^{\infty}\sum_i u_i^{(-n)} u^{i(n)} \tag{12.8.3}$$

is the Casimir operator of the affine algebra

$$\mathfrak{g} = \mathfrak{g}(X_\ell^{(1)}) = \mathfrak{g}' + \mathbf{C}d, \text{ where } [d, x^{(n)}] = nx^{(n)}.$$

Since Ω is independent of the choice of the dual bases $\{u_i\}$ and $\{u^i\}$, we can check this for a special choice of these bases. Consider the root space decomposition $\mathring{\mathfrak{g}} = \mathring{\mathfrak{h}} \oplus (\bigoplus_{\alpha \in \mathring{\Delta}} \mathbf{C}e_\alpha)$, such that $(e_\alpha \mid e_{-\alpha}) = 1$, and choose dual bases $h_1, \dots, h_\ell$ and $h^1, \dots, h^\ell$ of $\mathfrak{h}$. Then we have: $\mathring{\Omega} = \sum_i h_i h^i + \sum_{\alpha \in \mathring{\Delta}} e_{-\alpha} e_\alpha$. Using that $[e_\alpha, e_{-\alpha}] = \nu^{-1}(\alpha)$ (see Theorem 2.2e)), and that $2\mathring{\rho} = \sum_{\alpha \in \mathring{\Delta}_+} \alpha$ we obtain the form of $\mathring{\Omega}$ given in §2.5:

$$\mathring{\Omega} = \sum_{i=1}^{\ell} h_i h^i + 2\nu^{-1}(\mathring{\rho}) + 2\sum_{\alpha \in \mathring{\Delta}_+} e_{-\alpha} e_\alpha.$$

Hence (12.8.3) can be rewritten in the same form as Ω defined in §2.5, due to (6.2.5), (6.2.8) and §7.4 (see Exercise 7.16).

In order to perform calculations, it is convenient to introduce the *restricted completion* $U_c(\mathfrak{g}')$ of the universal enveloping algebra $U(\mathfrak{g}')$. Consider the category of all restricted $\mathfrak{g}'$-modules, i.e. modules V such that for any $v \in V$, $x^{(j)}(v) = 0$ for all $x \in \mathring{\mathfrak{g}}$ and all $j \gg 0$ (cf. §2.5). Consider all series $\sum_{j=1}^{\infty} u_j$ with $u_j \in U(\mathfrak{g}')$ such that for any restricted $\mathfrak{g}'$-module V and any $v \in V$, $u_j(v) = 0$ for all but finitely many u_j. We identify two such series if they represent the same operator in every restricted $\mathfrak{g}'$-module. We thus obtain an algebra $U_c(\mathfrak{g}')$ which contains $U(\mathfrak{g}')$ (since, by Exercise 11.22, $\bigoplus_{\Lambda \in P_+} L(\Lambda)$ is a faithful $U(\mathfrak{g}')$-module) and acts on every restricted $\mathfrak{g}'$-module. Note that the derivations d_n (see §7.3) extend from $U(\mathfrak{g}')$ to derivations of $U_c(\mathfrak{g}')$.

Now we introduce the *Sugawara operators* T_n $(n \in \mathbb{Z})$:

$$T_0 = \sum_i u_i u^i + 2\sum_{n=1}^{\infty}\sum_i u_i^{(-n)} u^{i(n)},$$
$$T_n = \sum_{m\in\mathbb{Z}}\sum_i u_i^{(-m)} u^{i(m+n)} \quad \text{if } n \neq 0. \tag{12.8.4}$$

Due to (12.8.2), these operators are contained in $U_c(\mathfrak{g}')$. Note also that these operators are independent of the choice of dual bases of $\mathring{\mathfrak{g}}$.

We have the following basic lemma.

LEMMA 12.8. a) *For $x \in \mathring{\mathfrak{g}}$ and $n, m \in \mathbb{Z}$ one has:*

$$[x^{(m)}, T_n] = 2(K + h^\vee) m x^{(m+n)}.$$

b) *Let V be a restricted $\mathfrak{g}$-module and let $v \in V$ be such that $\mathfrak{n}_+(v) = 0$ and $h(v) = \langle \lambda, h\rangle v$ for some $\lambda \in \mathfrak{h}^*$. Then*

$$T_0(v) = (\bar{\lambda} \mid \bar{\lambda} + 2\bar{\rho})v.$$

Proof. a) We perform the calculations in the semidirect product $\mathfrak{d} \ltimes U_c(\mathfrak{g}')$. First, note

$$T_0 = -2(K + h^\vee)d + \Omega. \tag{12.8.5}$$

Also, it is straightforward to check using (12.8.2) that

$$T_j = \frac{1}{j}[d_j, T_0] \quad \text{if } j \neq 0.$$

Now we have, using Theorem 2.6 and (12.8.5):

$$[x^{(m)}, T_0] = [x^{(m)}, 2(K + h^\vee)d_0] = 2(K + h^\vee)mx^{(m)}.$$

Finally, for $j \neq 0$, we have:

$$[x^{(m)}, T_j] = \frac{1}{j}[x^{(m)}, [d_j, T_0]] =$$

$$2(K + h^\vee)(\frac{1}{j}[mx^{(m+j)}, T_0] + \frac{1}{j}[d_j, mx^{(m)}]) = 2(K + h^\vee)mx^{(m+j)}.$$

b) follows immediately from (12.8.5) and (2.6.3).

□

Now we can calculate $[T_m, T_n]$. One should be careful about staying within the algebra $U_c(\mathfrak{g}')$. We may assume that $m > n$. Let first $m+n \neq 0$, $m \neq 0$, $n \neq 0$. Then we have:

$$\begin{aligned}
[T_m, T_n] &= \sum_{j\in\mathbb{Z}}\sum_i [u_i^{(-j)}u^{i(m+j)}, T_n] \\
&= \sum_{j\in\mathbb{Z}}\sum_i \left(u_i^{(-j)}[u^{i(m+j)}, T_n] + [u_i^{(-j)}, T_n]u^{i(m+j)}\right) \\
&= 2(K + h^\vee)\sum_{j\in\mathbb{Z}}\sum_i \left((m+j)u_i^{(-j)}u^{i(m+j+n)} - ju_i^{(-j+n)}u^{i(m+j)}\right)
\end{aligned}$$

by Lemma 12.8. Replacing j by $j + n$ in the second summand, we obtain:

$$[T_m, T_n] = 2(K + h^\vee)\sum_{j\in\mathbb{Z}}\sum_i (m - n)u_i^{(-j)}u^{i(m+n+j)} \tag{12.8.6}$$

Thus, we have, provided that $m + n \neq 0$, $m \neq 0$, $n \neq 0$:

$$[T_m, T_n] = 2(K + h^\vee)(m - n)T_{m+n}. \tag{12.8.7}$$

A similar calculation shows that (12.8.7) holds when $m + n \neq 0$ but m or $n = 0$.

Let now $m+n = 0$, $m > 0$. Then the right-hand side of (12.8.6) does not lie in $U_c(\mathfrak{g}')$. We proceed as follows. Since $\overset{\circ}{\Omega}$ is independent of the choice of dual bases we have $\sum_i u_i^{(-j)} u^{i(j+m)} = \sum_i u_i^{(j+m)} u^{i(-j)}$ when $m \neq 0$. Hence we can write (here and further we drop the sign of summation over i, but assume that it is present):

$$T_m = \sum_{j \geq 0} u_i^{(-j)} u^{i(j+m)} + \sum_{j>0} u_i^{(-j+m)} u^{i(j)}.$$

We have:

$$\begin{aligned}
[T_m, T_{-m}] &= 2(K + h^\vee) m \mathring{\Omega} \\
&\quad + 2(K + h^\vee) \sum_{j>0} ((j+m) u_i^{(-j)} u^{i(j)} - j u_i^{(-j-m)} u^{i(j+m)} \\
&\quad + j u_i^{(-j+m)} u^{i(j-m)} + (m-j) u_i^{(-j)} u^{i(j)}) \\
&= 2(K + h^\vee)(m T_0 + \sum_{j>0} j u_i^{(-j+m)} u^{i(j-m)} - \sum_{j>0} j u_i^{(-j-m)} u^{i(j+m)}).
\end{aligned}$$

Replacing j by $j+m$ in the first summation and j by $j-m$ in the second summation, we obtain:

$$\begin{aligned}
&[T_m, T_{-m}] \\
&\quad = 2(K + h^\vee)(m T_0 + \sum_{j > -m} (j+m) u_i^{(-j)} u^{i(j)} + \sum_{k>j} (m-j) u_i^{(-j)} u^{i(j)}) \\
&\quad = 2(K + h^\vee)(2m T_0 + \sum_{j=-m+1}^{-1} (j+m) u_i^{(-j)} u^{i(j)} - \sum_{j=1}^{m-1} (m-j) u_i^{(-j)} u^{i(j)}) \\
&\quad = 2(K + h^\vee)(2m T_0 + \sum_{j=0}^{m-1} j(m-j) [u^{i(j)}, u_i^{(-j)}]) \\
&\quad = 2(K + h^\vee)(2m T_0 + \sum_{j=0}^{m-1} j(m-j)(\dim \mathring{\mathfrak{g}}) K)
\end{aligned}$$

by (12.8.2). Since $\sum_{j=0}^{m-1} j(m-j) = (m^3 - m)/6$, combining with (12.8.7), we obtain the final formula:

(12.8.8) $$[T_m, T_n] = 2(K + h^\vee)((m-n) T_{m+n} + \delta_{m,-n} \frac{m^3 - m}{6} (\dim \mathring{\mathfrak{g}}) K).$$

We immediately obtain from (12.8.8) and Lemma 12.8 the following

COROLLARY 12.8. *Let V be a restricted $\mathfrak{g}'$-module such that K is a scalar operator kI, $k \neq -h^\vee$. Let*

(12.8.9) $$L_n = \frac{1}{2(k + h^\vee)} T_n, \quad n \in \mathbb{Z},$$

(12.8.10) $$c(k) = \frac{k (\dim \mathring{\mathfrak{g}})}{k + h^\vee},$$

(12.8.11) $$h_\Lambda = \frac{(\Lambda + 2\rho | \Lambda)}{2(k + h^\vee)} \quad \textit{if } V = L(\Lambda).$$

a) *Letting*

$$d_n \mapsto L_n, \quad c \mapsto c(k),$$

extends V *to a module over* $\mathfrak{g}' + \mathrm{Vir}$ *(the semidirect sum defined in §7.3). In particular,* V *extends to a module over* $\mathfrak{g}$ $(= \mathfrak{g}' + \mathbf{C}d)$ *by letting* $d \mapsto -L_0$.
b) *If* V *is the* $\mathfrak{g}$*-module* $L(\Lambda)$, *then* $L_0 = h_\Lambda I - d$.

□

The number $c(k)$ is called the *conformal anomaly* of the $\mathfrak{g}'$-module V, and the number h_Λ is called the *vacuum anomaly* of $L(\Lambda)$. Using the "strange" formula (12.1.8), we obtain from (12.7.5), (12.8.10) and (12.8.11) the following simple relation between the modular, conformal and vacuum anomalies for $\Lambda \in \mathfrak{h}^*$ of level k:

$$m_\Lambda = h_\Lambda - \tfrac{1}{24}c(k). \tag{12.8.12}$$

Another useful property of Sugawara operators is given by the following

PROPOSITION 12.8. *Let* V *be a restricted* $\mathfrak{g}'$*-module with a non-degenerate contravariant Hermitian form (see §11.5). Then the operators* T_n *and* T_{-n} *are adjoint with respect to this form.*

Proof. Let $\mathring{\mathfrak{k}}$ be the compact form of $\mathring{\mathfrak{g}}$, i.e., the fixed point set of ω_0. Then the bilinear form $(.|.)$ restricted to $\mathring{\mathfrak{k}}$ is negative definite (see Theorem 11.7a)). Hence we may choose a basis $\{v_j\}$ of $\mathring{\mathfrak{k}}$ such that $(v_i \mid v_j) = -\delta_{ij}$. Let $u_j = \sqrt{-1}\, v_j$; this is an orthonormal basis of $\mathring{\mathfrak{g}}$ such that $\omega_0(u_j) = -u_j$, hence $\omega_0(u_j(n)) = -u_j(-n)$. The proof of the proposition is now immediate.

□

Remark 12.8. Let $\mathring{\mathfrak{g}}_{\mathbf{R}}$ be a finite-dimensional vector space over $\mathbf{R}$ with a positive definite bilinear form $(.|.)$. Let $\mathring{\mathfrak{g}}$ be its complexification and extend $(.|.)$ to $\mathring{\mathfrak{g}}$ by bilinearity. Viewing $\mathring{\mathfrak{g}}$ as an abelian Lie algebra, we have: $\mathring{\mathfrak{g}} = \mathfrak{h}$, $\mathring{\Delta} = \emptyset$, $\mathring{Q} = 0$, $\mathring{Q}^\vee = 0$, $\mathring{P} = \mathfrak{h}^*$, $\mathring{W} = \{1\}$, $\mathring{\rho} = 0$. We let also $h^\vee = 0$. Then all results of this section (and their proofs) hold in this case as well. In particular, the conformal anomaly

$$c(k) = \dim \mathring{\mathfrak{g}}$$

is independent of k, the vacuum anomaly $h_\Lambda = \frac{(\Lambda|\Lambda)}{2k}$, where $k = \mathrm{level}(\Lambda) \neq 0$, and we let

$$m_\Lambda = h_\Lambda - \frac{1}{24}c(k).$$

Finally, for the $\mathfrak{g}$-module $L(\Lambda)$, $\Lambda \in \mathfrak{h}^*$, of level $k \neq 0$, $\mathrm{ch}_{L(\Lambda)} = e^{\Lambda}/\varphi(e^{-\delta})^{\dim \mathring{\mathfrak{g}}}$, where φ is defined by (12.2.3), hence

$$\chi_{\Lambda} := e^{-m_{\Lambda}\delta}\,\mathrm{ch}_{L(\Lambda)} = e^{-\frac{|\Lambda|^2}{2k}\delta+\Lambda}/\eta^{\dim \mathring{\mathfrak{g}}}, \tag{12.8.13}$$

where $\eta = e^{-\delta/24}\varphi(e^{-\delta})$. Also, $L(\Lambda)$ is unitarizable if and only if $\Lambda \in \mathfrak{h}^*_{\mathbb{R}}$ and k is a positive real number. This follows from (9.13.1).

§12.9. We shall extend the above construction to the case of a reductive finite-dimensional Lie algebra $\mathfrak{g}$ (it will be more convenient to use here this notation instead if $\mathring{\mathfrak{g}}$). We have the decomposition of $\mathfrak{g}$ into a direct sum of ideals:

$$\mathfrak{g} = \mathfrak{g}_{(0)} \oplus \mathfrak{g}_{(1)} \oplus \mathfrak{g}_{(2)} \oplus \cdots, \tag{12.9.1}$$

where $\mathfrak{g}_{(0)}$ is the center of $\mathfrak{g}$ and $\mathfrak{g}_{(i)}$ with $i \geq 1$ are simple. We fix on $\mathfrak{g}$ a non-degenerate invariant bilinear symmetric form $(\,.\,|\,.\,)$ so that (12.9.1) is an orthogonal decomposition. We shall assume that the restriction of $(\,.\,|\,.\,)$ to each $\mathfrak{g}_{(i)}$ with $i \geq 1$ is the normalized invariant form, and that $\mathfrak{g}_{(0)}$ and the form $(\,.\,|\,.\,)$ restricted to it are as described in Remark 12.8. We shall call such form a normalized invariant form on $\mathfrak{g}$.

We let

$$\widetilde{\mathcal{L}}(\mathfrak{g}) = \bigoplus_{i\geq 0} \widetilde{\mathcal{L}}(\mathfrak{g}_{(i)}), \quad \text{where } \widetilde{\mathcal{L}}(\mathfrak{g}_{(i)}) = \mathcal{L}(\mathfrak{g}_{(i)}) + \mathbb{C}K_i.$$

We also let (cf. §7.3):

$$\hat{\mathcal{L}}(\mathfrak{g}) = \widetilde{\mathcal{L}}(\mathfrak{g}) + \mathbb{C}d, \quad \text{where } d|_{\widetilde{\mathcal{L}}(\mathfrak{g}_{(i)})} = -t\frac{d}{dt},\ d(K_i) = 0.$$

The Lie algebras $\widetilde{\mathcal{L}}(\mathfrak{g})$ and $\hat{\mathcal{L}}(\mathfrak{g})$ are called affine algebras associated to the reductive Lie algebra $\mathfrak{g}$. The subalgebras $\widetilde{\mathcal{L}}(\mathfrak{g}_{(i)})$ (resp. $\widetilde{\mathcal{L}}(\mathfrak{g}_{(i)}) + \mathbb{C}d$) are called *components* of $\widetilde{\mathcal{L}}(\mathfrak{g})$ (resp. $\hat{\mathcal{L}}(\mathfrak{g})$).

Note that $\mathfrak{c} = \mathfrak{g}_{(0)} + \sum_{i\geq 1} \mathbb{C}K_i$ is the center of $\widetilde{\mathcal{L}}(\mathfrak{g})$ and $\hat{\mathcal{L}}(\mathfrak{g})$. As before, we identify $\mathfrak{g}$ with the subalgebra $1 \otimes \mathfrak{g}$. Let $\overline{\mathfrak{h}}$ be a Cartan subalgebra of $\mathfrak{g}$ and let $\mathfrak{g} = \overline{\mathfrak{n}}_- \oplus \overline{\mathfrak{h}} \oplus \overline{\mathfrak{n}}_+$ be a triangular decomposition of $\mathfrak{g}$. The subalgebra $\mathfrak{h} = \overline{\mathfrak{h}} + \mathfrak{c} + \mathbb{C}d$ is called the Cartan subalgebra of $\hat{\mathcal{L}}(\mathfrak{g})$. The triangular decomposition

$$\hat{\mathcal{L}}(\mathfrak{g}) = \mathfrak{n}_- \oplus \mathfrak{h} \oplus \mathfrak{n}_+$$

is defined in the same way as in §7.6. For $\lambda \in \mathfrak{h}^*$, we denote (as before) its restriction to $\overline{\mathfrak{h}}$ by $\overline{\lambda}$. As before, define $\delta \in \mathfrak{h}^*$ by

$$\delta|_{\overline{\mathfrak{h}}+\mathfrak{c}} = 0, \quad \langle \delta, d \rangle = 1.$$

Given $\Lambda \in \mathfrak{h}^*$, we denote (as before) by $L(\Lambda)$ the irreducible $\hat{\mathcal{L}}(\mathfrak{g})$-module which admits a non-zero vector v_Λ such that $\mathfrak{n}_+(v_\Lambda) = 0$ and $h(v_\Lambda) = \langle \Lambda, h \rangle v_\Lambda$ for $h \in \mathfrak{h}$. Using uniqueness of $L(\Lambda)$, we clearly have:

$$L(\Lambda) = \bigotimes_{i \geq 0} L(\Lambda_{(i)}), \tag{12.9.2}$$

where $\Lambda_{(i)}$ denote the restriction of Λ to $\mathfrak{h}_{(i)} := \mathfrak{h} \cap \hat{\mathcal{L}}(\mathfrak{g}_{(i)})$ and $L(\Lambda_{(i)})$ is the $\hat{\mathcal{L}}(\mathfrak{g}_{(i)})$-module with highest weight $\Lambda_{(i)}$.

We let k_i, the eigenvalue of K_i on $L(\Lambda)$, be the *i-th level* of Λ, and let $k = (k_0, k_1, \ldots,)$. Define $c(k) = \sum_i c(k_i)$, $h_\Lambda = \sum_i h_{\Lambda_{(i)}}$, $m_\Lambda = \sum_i m_{\Lambda_{(i)}}$. Due to (12.9.2), $\mathrm{ch}_{L(\Lambda)} = \prod_i \mathrm{ch}_{L(\Lambda_{(i)})}$ and

$$\chi_\Lambda := e^{-m_\Lambda \delta} \mathrm{ch}_{L(\Lambda)} = \prod_i \chi_{\Lambda_{(i)}}. \tag{12.9.3}$$

Let V be a restricted $\widetilde{\mathcal{L}}(\mathfrak{g})$-module such that k_i acts as $k_i I$ and $k_i \neq -h_i^\vee$ (where $h_i^\vee$ is the dual Coxeter number of $\widetilde{\mathcal{L}}(\mathfrak{g}_{(i)})$. Let $T_n^{(i)}$ be the Sugawara operators for $\widetilde{\mathcal{L}}(\mathfrak{g}_{(i)})$, and let $(n \in \mathbb{Z})$:

$$L_n^{(i)} = \frac{1}{2(k_i + h_i^\vee)} T_n^{(i)}, \; L_n^{\mathfrak{g}} = \sum_i L_n^{(i)}. \tag{12.9.4}$$

The operators $L_n^{\mathfrak{g}}$ are called the *Virasoro operators* for the $\mathfrak{g}$-module V. Then letting $d_n \mapsto L_n^{\mathfrak{g}}$, $c \mapsto c(k)$ extends V to a module over $\widetilde{\mathcal{L}}(\mathfrak{g}) + \mathrm{Vir}$. Note also the following useful formula (cf. (12.8.5)):

$$L_0^{\mathfrak{g}} = \sum_i \frac{\Omega_i}{2(k_i + h_i^\vee)} - d, \tag{12.9.5}$$

where Ω_i is the Casimir operator for $\hat{\mathcal{L}}(\mathfrak{g}_{(i)})$.

§12.10. In the remainder of this chapter, we let $\mathfrak{g}$ be a reductive finite-dimensional Lie algebra with a normalized invariant form $(\,.\,|\,.\,)$ and let $\dot{\mathfrak{g}}$ be a reductive subalgebra of $\mathfrak{g}$ such that $(\,.\,|\,.\,)|_{\dot{\mathfrak{g}}}$ is non-degenerate. Let $\mathfrak{g} = \bigoplus_{i\geq 0} \mathfrak{g}_{(i)}$ and $\dot{\mathfrak{g}} = \bigoplus_{i\geq 0} \dot{\mathfrak{g}}_{(i)}$ be the decompositions (12.9.1) of $\mathfrak{g}$ and $\dot{\mathfrak{g}}$. Let $(\,.\,|\,.\,)^{\cdot}$ be a normalized invariant form on $\dot{\mathfrak{g}}$, which coincides with $(\,.\,|\,.\,)$ on $\dot{\mathfrak{g}}_{(0)}$. Due to uniqueness of the invariant bilinear form on a simple Lie algebra we have for $x, y \in \dot{\mathfrak{g}}_{(s)}$, $s \geq 1$:

$$(x_{(r)}\,|\,y_{(r)}) = j_{sr}(x\,|\,y)^{\cdot},$$

where $x_{(r)}$ denotes the projection of x on $\mathfrak{g}_{(r)}$ and j_{sr} is a (positive) number independent of x and y; we let $j_{0r} = 1$. The numbers j_{sr} $(s, r \geq 0)$ are called *Dynkin indices*.

The inclusion homomorphism $\psi : \dot{\mathfrak{g}} \to \mathfrak{g}$ induces in an obvious way the inclusion homomorphism $\mathcal{L}(\dot{\mathfrak{g}}) \to \mathcal{L}(\mathfrak{g})$. This lifts uniquely to a homomorphism $\tilde{\psi} : \tilde{\mathcal{L}}(\dot{\mathfrak{g}}) \to \tilde{\mathcal{L}}(\mathfrak{g})$ by letting $\tilde{\psi}(\dot{K}_s) = \sum_r j_{sr} K_r$, which extends to a homomorphism $\hat{\psi} : \hat{\mathcal{L}}(\dot{\mathfrak{g}}) \to \hat{\mathcal{L}}(\mathfrak{g})$ by letting $\hat{\psi}(d) = d$. Here and further the overdot refers to an object associated to $\dot{\mathfrak{g}}$.

Let V be a restricted $\widetilde{\mathcal{L}}(\mathfrak{g})$-module such that K_i acts as $k_i I$, $k_i \neq -h_i^\vee$. Via $\tilde{\psi}$, this is a $\widetilde{\mathcal{L}}(\dot{\mathfrak{g}})$-module with $\dot{K}_i$ acting as $\dot{k}_i I$, where

$$\dot{k}_s = \sum_i j_{si} k_j. \tag{12.10.1}$$

We shall assume that $\dot{k}_i \neq -\dot{h}_i^\vee$. Let (see (12.9.4)):

$$L_n^{\mathfrak{g},\dot{\mathfrak{g}}} = L_n^{\mathfrak{g}} - L_n^{\dot{\mathfrak{g}}}.$$

PROPOSITION 12.10. a) *The operators $L_n^{\mathfrak{g},\dot{\mathfrak{g}}}$ commute with $\widetilde{\mathcal{L}}(\dot{\mathfrak{g}})$.*

b) *The map $d_n \mapsto L_n^{\mathfrak{g},\dot{\mathfrak{g}}}$, $c \to c(k) - \dot{c}(\dot{k})$ defines a representation of Vir on V.*

Proof. a) is immediate by Lemma 12.8. Furthermore, we have:

$$\begin{aligned}
[L_m^{\mathfrak{g},\dot{\mathfrak{g}}}, L_n^{\mathfrak{g},\dot{\mathfrak{g}}}] &= [L_m^{\mathfrak{g},\dot{\mathfrak{g}}}, L_n^{\mathfrak{g}}] \qquad (\text{since } L_n^{\dot{\mathfrak{g}}} \in U_c(\widetilde{\mathcal{L}}(\dot{\mathfrak{g}}))) \\
&= [L_m^{\mathfrak{g}}, L_n^{\mathfrak{g}}] - [L_m^{\dot{\mathfrak{g}}}, L_n^{\mathfrak{g}}] \\
&= [L_m^{\mathfrak{g}}, L_n^{\mathfrak{g}}] - [L_m^{\dot{\mathfrak{g}}}, L_n^{\mathfrak{g},\dot{\mathfrak{g}}} + L_n^{\dot{\mathfrak{g}}}] \\
&= [L_m^{\mathfrak{g}}, L_n^{\mathfrak{g}}] - [L_m^{\dot{\mathfrak{g}}}, L_n^{\dot{\mathfrak{g}}}] \quad (\text{since } L_m^{\dot{\mathfrak{g}}} \in U_c(\widetilde{\mathcal{L}}(\dot{\mathfrak{g}}))).
\end{aligned}$$

□

The Vir-module defined by Proposition 12.10 is called the *coset Vir-module*.

Choose Cartan subalgebras $\bar{\mathfrak{h}}$ and $\dot{\bar{\mathfrak{h}}}$ of $\mathfrak{g}$ and $\dot{\mathfrak{g}}$ such that $\dot{\bar{\mathfrak{h}}} \subset \bar{\mathfrak{h}}$. Choose a triangular decomposition $\mathfrak{g} = \bar{\mathfrak{n}}_- \oplus \bar{\mathfrak{h}} \oplus \bar{\mathfrak{n}}_+$; then we have the induced triangular decomposition $\dot{\mathfrak{g}} = \dot{\bar{\mathfrak{n}}}_- \oplus \dot{\bar{\mathfrak{h}}} \oplus \dot{\bar{\mathfrak{n}}}_+$, where $\dot{\bar{\mathfrak{n}}}_\pm = \bar{\mathfrak{n}}_\pm \cap \dot{\mathfrak{g}}$. We have the associated triangular decompositions: $\hat{\mathcal{L}}(\mathfrak{g}) = \mathfrak{n}_- + \mathfrak{h} + \mathfrak{n}_+$, $\hat{\mathcal{L}}(\dot{\mathfrak{g}}) = \dot{\mathfrak{n}}_- + \dot{\mathfrak{h}} + \dot{\mathfrak{n}}_+$, etc., and we have: $\varphi(\dot{\mathfrak{h}}) \subset \mathfrak{h}$, $\varphi(\dot{\mathfrak{n}}_+) \subset \mathfrak{n}_+$, etc.

Let $P_+ = \{\lambda \in \mathfrak{h}^* \mid \lambda|_{\mathfrak{h}\cap\hat{\mathcal{L}}(\mathfrak{g}_{(i)})} \subset P_{+(i)}$ for $i \geq 1$, and $\lambda|_{\mathfrak{h}\cap\mathfrak{g}_{(0)\mathbb{R}}}$ is real and $\langle\lambda, K_0\rangle > 0\}$ be the set of *dominant integral weights* for $\hat{\mathcal{L}}(\mathfrak{g})$. Let $P_+^k = \{\lambda \in P_+ \mid \lambda(K_i) = k_i \ (i = 0, 1, \ldots)\}$.

Fix $\Lambda \in P_+^k$. Using (12.9.2), Theorem 11.7 and §11.12, we see that the $\hat{\mathcal{L}}(\mathfrak{g})$-module $L(\Lambda)$ is unitarizable. It follows, by Proposition 11.8, that viewed as a $\hat{\mathcal{L}}(\dot{\mathfrak{g}})$-module, the module $L(\Lambda)$ decomposes into a direct sum of $\hat{\mathcal{L}}(\dot{\mathfrak{g}})$-modules $\dot{L}(\lambda)$ with $\lambda \in \dot{P}_+^{\dot{k}}$. Since the eigenspaces of d on $L(\Lambda)$ are finite-dimensional, it follows that the multiplicity of occurrence of $\dot{L}(\lambda)$ in this decomposition is finite. We denote this multiplicity by $\mathrm{mult}_\Lambda(\lambda; \dot{\mathfrak{g}})$.

§12.11. The following notion will play an important role in the sequel. The pair $M \in P_+^k$ and $\mu \in P(M)|_{\dot{\mathfrak{h}}} \cap \dot{P}_+^{\dot{k}}$ such that $h_M = \dot{h}_\mu$ is called a *vacuum pair of level* k (cf. Proposition 12.12b), below). Denote by R_k the set of all vacuum pairs of level k.

PROPOSITION 12.11. *Let* $(M; \mu) \in R_k$. *Then*

$$\mathrm{mult}_M(\mu; \dot{\mathfrak{g}}) = \sum_{\substack{\tilde{\mu} \in P(M) \\ \tilde{\mu}|_{\dot{\mathfrak{h}}} = \mu}} \mathrm{mult}_{L(M)}(\tilde{\mu}) > 0. \tag{12.11.1}$$

In particular, $(M; \mu) \in R_k$ *if and only if* $h_M = \dot{h}_\mu$ *and* $\mathrm{mult}_M(\mu; \dot{\mathfrak{g}}) \neq 0$.

Proof. Let $\tilde{\mu} \in P(M)$ be such that $\tilde{\mu}|_{\dot{\mathfrak{h}}} = \mu$. Let v be a non-zero vector from $L(M)_{\tilde{\mu}}$. We have to show that $\dot{\mathfrak{n}}_+(v) = 0$. In the contrary case, there exists $\beta \in \dot{Q}_+\backslash\{0\}$ such that $\mathrm{mult}_M(\mu + \beta; \dot{\mathfrak{g}}) > 0$. But then

$$\begin{aligned}\dot{h}_{\mu+\beta} - \dot{h}_\mu &= \sum_i \frac{|\mu_{(i)} + \beta_{(i)} + \dot{\rho}_{(i)}|^2 - |\mu_{(i)} + \dot{\rho}_{(i)}|^2}{2(\dot{k}_i + \dot{h}_i^\vee)} \\ &= \sum_i \frac{((\mu_{(i)} + \beta_{(i)}) + \mu_{(i)} + 2\dot{\rho}_i \mid \beta_{(i)})}{2(\dot{k}_i + \dot{h}_i^\vee)} > 0\end{aligned}$$

Hence $h_M < \dot{h}_{\mu+\beta}$, which contradicts Proposition 12.12b) below.

□

Remark 12.11. Let $M \in P_+$. Since the module $L(M)$ is completely reducible with respect to $\widetilde{\mathcal{L}}(\dot{\mathfrak{g}})$, we see from (12.9.5) that $L_0^{\mathfrak{g},\dot{\mathfrak{g}}}$ is diagonalizable on $L(M)$. It is clear from the proof of Proposition 12.12b) that its spectre is non-negative. Each of its eigenspaces is $\widetilde{\mathcal{L}}(\dot{\mathfrak{g}})$-invariant (by Proposition 12.10a)). Finally, its zero eigenspace (the vacuum space) decomposes into a direct sum of $\widetilde{\mathcal{L}}(\dot{\mathfrak{g}})$-modules $\dot{L}(\mu)$ such that $(M;\mu)$ is a vacuum pair, with multiplicities given by (12.11.1).

§12.12. For $\Lambda \in P_+^k$ and $\lambda \in \dot{\mathfrak{h}}^*$, let

$$b_\lambda^\Lambda(\dot{\mathfrak{g}}) = e^{-(m_\Lambda - \dot{m}_\lambda)\delta} \sum_{n\in\mathbb{Z}} \mathrm{mult}_\Lambda(\lambda - n\dot{\delta}; \dot{\mathfrak{g}}) e^{-n\dot{\delta}}.$$

This series converges absolutely to a holomorphic function on $\dot{Y}$, called a *branching function.*

Note that string functions are essentially special cases of branching functions:

$$c_\lambda^\Lambda = b_\lambda^\Lambda(\dot{\bar{\mathfrak{h}}})\eta^{-\ell}. \tag{12.12.1}$$

This follows from (12.8.13).

The branching functions have a simple representation theoretical meaning. To explain this, let

$$U(\Lambda,\lambda) = \{v \in L(\Lambda) | \dot{\mathfrak{n}}_+(v) = 0 \text{ and } h(v) = \langle\lambda, h\rangle v \text{ for } h \in \dot{\bar{\mathfrak{h}}}\}.$$

Note that, due to Proposition 12.10, the subspace $U(\Lambda,\lambda)$ is a coset Vir-submodule (with $c = c(k) - \dot{c}(\dot{k})$). Comparing Corollary 12.8b) with (12.8.12), we obtain the following interpretation of branching functions:

$$b_\lambda^\Lambda(\dot{\mathfrak{g}}) = \mathrm{tr}_{U(\Lambda,\lambda)}\, q^{L_0^{\mathfrak{g},\dot{\mathfrak{g}}} - \frac{1}{24}(c(k) - \dot{c}(\dot{k}))}, \tag{12.12.2}$$

where $q = e^{-\delta}$.

Thus, we have obtained the following decomposition of $L(\Lambda)$ with respect to $\tilde{\mathcal{L}}(\dot{\mathfrak{g}}) \oplus \mathrm{Vir}$ (direct sum of Lie algebras):

$$L(\Lambda) = \bigoplus_{\lambda\in\dot{P}_+^{\dot{k}} \bmod \mathbb{C}\dot{\delta}} \dot{L}(\lambda) \otimes U(\Lambda,\lambda), \tag{12.12.3}$$

which immediately implies an equation for normalized characters:

$$\chi_\Lambda = \sum_{\lambda\in\dot{P}_+^{\dot{k}} \bmod \mathbb{C}\dot{\delta}} \dot{\chi}_\lambda b_\lambda^\Lambda(\dot{\mathfrak{g}}). \tag{12.12.4}$$

Now we can prove the following important proposition.

PROPOSITION 12.12. a) *The module $L(\Lambda)$, $\Lambda \in P_+$, viewed as a coset Vir-module decomposes into an orthogonal direct sum of unitarizable irreducible highest-weight modules.*

b) *If $\Lambda \in P_+$ and $\mathrm{mult}_\Lambda(\lambda; \dot{\mathfrak{g}}) \neq 0$, then $h_\Lambda \geq \dot{h}_\lambda$.*

c) *If $k_0 \geq 0$ and $k_i \in \mathbf{Z}_+$ for $i > 0$, then $c(k) \geq \dot{c}(\dot{k})$.*

Proof. By Proposition 12.8, the coset Vir-module $L(\Lambda)$, $\Lambda \in P_+$ is unitary, hence all $U(\Lambda, \lambda)$ are unitary. Also, all eigenspaces of $L_0^{\mathfrak{g},\dot{\mathfrak{g}}}$ on $U(\Lambda, \lambda)$ are finite-dimensional and its spectrum is bounded below. Now a) follows from (12.2.3) and Proposition 11.12c).

Let $v \in L(\Lambda)$ be a highest-weight vector of a $\hat{\mathcal{L}}(\dot{\mathfrak{g}})$-submodule $\dot{L}(\lambda)$ of $L(\Lambda)$. Using Corollary 12.8b) we obtain:

$$L_0^{\mathfrak{g},\dot{\mathfrak{g}}}(v) = (h_\Lambda - \dot{h}_\lambda)v.$$

b) and c) follow now from Proposition 11.12.

□

Now, if $\dot{\mathfrak{g}}$ is semisimple, then the sum in (12.12.3) is clearly finite. This is, of course, not the case in general, as we have seen on the example $\dot{\mathfrak{g}} = \overline{\mathfrak{h}}$. We shall transform this sum to a finite one using the same trick as in §12.7. For this we shall assume that

$$\dot{\mathfrak{g}}_{(0)} \cap \overline{Q}^\vee \text{ spans } \dot{\mathfrak{g}}_{(0)} \text{ over } \mathbf{C}, \tag{12.12.5}$$

where $\overline{Q}^\vee \subset \overline{\mathfrak{h}}$ is the coroot lattice of $\mathfrak{g}$. Introduce the lattice $\dot{M}_0 = \dot{\nu}^{-1}(\dot{\mathfrak{g}}_{(0)} \cap \overline{Q}^\vee)$, which is a sublattice of the lattice M.

We let $\dot{\mathfrak{h}}' = \bigoplus_{i \geq 1} \dot{\mathfrak{h}}_{(i)}$. Then we have: $\dot{\mathfrak{h}} = \dot{\mathfrak{h}}_{(0)} + \dot{\mathfrak{h}}'$, $\dot{\mathfrak{h}}_{(0)} \cap \dot{\mathfrak{h}}' = \mathbf{C}d$, hence $\dot{\mathfrak{h}}^* = \dot{\mathfrak{h}}_{(0)} + \dot{\mathfrak{h}}'^*$ and $\dot{\mathfrak{h}}^*_{(0)} \cap \dot{\mathfrak{h}}'^* = \mathbf{C}\dot{\delta}$. Given $\lambda \in \dot{\mathfrak{h}}^*$, we have a decomposition

$$\lambda = \lambda_{(0)} + \lambda^{(1)}, \text{ where } \lambda_{(0)} \in \dot{\mathfrak{h}}^*_{(0)},\ \lambda^{(1)} \in \dot{\mathfrak{h}}'^*, \tag{12.12.6}$$

which is unique up to adding multiples of $\dot{\delta}$.

Due to (12.9.3) and (12.8.3) we have for $\lambda \in \dot{P}_+^k$:

$$\dot{\chi}_\lambda = \dot{\chi}_{\lambda^{(1)}}(e^{\lambda_{(0)} - \frac{|\lambda_{(0)}|^2}{2k}\dot{\delta}}/\eta^{\ell_0}), \text{ where } \ell_0 = \dim \dot{\mathfrak{g}}_{(0)}. \tag{12.12.7}$$

Here $\dot{\chi}_{\lambda^{(1)}} = \prod_{i \geq 1} \dot{\chi}_{\lambda_{(i)}}$ is the normalized character of the $\hat{\mathcal{L}}(\dot{\mathfrak{g}}')$-module $\dot{L}(\lambda^{(1)})$, where $\dot{\mathfrak{g}}'$ is the derived algebra of $\dot{\mathfrak{g}}$.

It is clear that for $\alpha \in \dot{M}_0$ we have:

$$t_\alpha(\dot{\chi}_{\lambda^{(1)}}) = \dot{\chi}_{\lambda^{(1)}} \text{ and } b^\Lambda_{t_\alpha(\lambda)} = b^\Lambda_\lambda.$$

Using this, we can rewrite (12.12.3) in the following form:

$$\chi_\Lambda = \sum_{\lambda \in \dot{P}^k_+ \bmod (\mathbb{C}\delta + \dot{k}_0 \dot{M}_0)} b^\Lambda_\lambda(\dot{\mathfrak{g}}) \dot{\chi}_{\lambda^{(1)}} (\dot{\Theta}^\circ_{\lambda^{(0)}} / \eta^{\ell_0}), \tag{12.12.8}$$

where $\dot{\Theta}^\circ_{\lambda^{(0)}} = e^{-\frac{|\lambda^{(0)}|^2}{2k_0}\dot{\delta}} \sum_{\alpha \in \dot{M}_0} e^{\dot{t}_\alpha(\lambda)}$ is the theta function associated to the lattice $\dot{M}_0$. It is clear that the sum on the right is finite. Equation (12.12.8) is a generalization of the theta function identity (12.7.3). It will be used in the next chapter to study branching functions.

§12.13. We apply here the developed machinery to calculate the functions a^Λ_Λ of level 1, for affine algebras of type $X^{(r)}_N$, where $X = A, D$ or E (cf. Lemma 12.7). For this we shall use the following well-known relation:

$$\dim \mathfrak{g}(X_N) = N(h^\vee + 1) \text{ if } X = A, D \text{ or } E. \tag{12.13.1}$$

Recall also that

$$a^\Lambda_\Lambda = 1 + a_1 q + \dots, \text{ where } q = e^{-\delta}. \tag{12.13.2}$$

Consider the coset Vir-submodule $U(\Lambda, \Lambda)$ of $L(\Lambda)$ (for the twisted case use Exercise 12.20). We have the following formula for its conformal central charge:

$$c(1) - \dot{c}(1) = \frac{\dim \mathfrak{g}(X_N)}{h^\vee + 1} - N = 0. \tag{12.13.3}$$

Hence, by Proposition 11.12b) and c), the coset Vir-module $U(\Lambda, \Lambda)$ is trivial. Hence, comparing (12.13.2) with (12.12.2), we obtain

$$b^\Lambda_\Lambda(\dot{\mathfrak{h}}) = 1.$$

Comparing this with (12.13.2) and (12.12.1) in the non-twisted case (resp.

Exercises 12.20 and 12.21 in the twisted case), and using Lemma 12.7, we deduce

PROPOSITION 12.13. *Let $\Lambda \in P^1_+$. Then for affine algebras of type $X_N^{(r)}$, where $X = A, D$ or E, we have:*

$$a^\Lambda_\Lambda = \prod_{n=1}^\infty (1 - e^{-n\delta})^{-\,\mathrm{mult}\, n\delta}, \tag{12.13.4}$$

$$\mathrm{mult}_{L(\Lambda)}\, \lambda = p^A((|\Lambda|^2 - |\lambda|^2)/2), \quad \text{where} \quad \sum_j p^A(j) q^j := \prod_{n\geq 1} (1 - q^n)^{-\,\mathrm{mult}\, n\delta} \tag{12.13.5}$$

$$e^{-\frac{1}{2}|\Lambda|^2\delta}\, \mathrm{ch}_{L(\Lambda)} = \sum_{\gamma \in \overline{Q} + \overline{\Lambda}} e^{\Lambda_0 + \gamma - \frac{1}{2}|\gamma|^2\delta} \Big/ \prod_{n\geq 1} (1 - e^{-n\delta})^{\mathrm{mult}\, n\delta}. \tag{12.13.6}$$

□

Remark 12.13. If A is of type $A_\ell^{(1)}$, $D_\ell^{(1)}$, $E_\ell^{(1)}$ or $A_{2\ell}^{(2)}$, then $\sum_j p^A(j) q^j = \varphi(q)^{-\ell}$, so that $p^A(j)$ is the number of partitions of j into positive integral parts of ℓ different colours; in particular, $p^{A_r^{(r)}}(j) = p(j)$ ($r = 1$ or 2) is the classical partition function.

Incidentally, using (12.13.3) and Proposition 11.12b), we obtain $h_\Lambda = \dot{h}_\Lambda$, hence the following

COROLLARY 12.13. *Under the hypothesis of Proposition 12.13 we have:*

$$(\Lambda \,|\, \Lambda) h^\vee = 2(\rho \,|\, \Lambda).$$

§12.14. Exercises.

12.1. Show that setting $q = e(-\delta)$, $z = e(-\alpha_0)$ in (12.1.4) for $\mathfrak{g}(A)$ of type $A_1^{(1)}$ and $A_2^{(2)}$,one gets the following classical triple and quintuple product identities (which are alternative forms of identities from Exercise 10.9):

$$\prod_{n=1}^\infty (1 - q^n)(1 - q^n z^{-1})(1 - q^{n-1} z) = \sum_{m \in \mathbb{Z}} (-1)^m q^{\frac{1}{2}m(m-1)} z^m;$$

$$\prod_{n=1}^\infty (1 - q^n)(1 - q^{n-1} z)(1 - q^n z^{-1})(1 - q^{2n-1} z^2)(1 - q^{2n-1} z^{-2})$$
$$= \sum_{m \in \mathbb{Z}} q^{\frac{1}{2}(3m^2 + m)} (z^{3m} - z^{3m-1}).$$

12.2. Let $\mathring{\mathfrak{g}}$ be a simple finite-dimensional Lie algebra of rank ℓ, let $\Phi(x,y)$ be the Killing form of $\mathring{\mathfrak{g}}$, let $\mathring{\mathfrak{h}}$ be a Cartan subalgebra, $\mathring{\Delta}$ the root system, M_P the lattice spanned by $\{\alpha/\Phi(\alpha,\alpha),\ \alpha\in\mathring{\Delta}\}$, $\mathring{\Delta}_+$ a set of positive roots, $\overline{\rho}$ their half-sum. Deduce from (12.2.3) the following identity:

$$\eta^{\dim \mathring{\mathfrak{g}}} = \sum_{\gamma\in M_P} \mathring{d}(\gamma) q^{\Phi(\gamma+\overline{\rho},\gamma+\overline{\rho})}.$$

12.3. In the notation of Exercise 12.2, let $\alpha_1,\dots,\alpha_\ell$ be the set of simple roots, h the Coxeter number of $\mathring{\mathfrak{g}}$, and let M_Q denote the lattice spanned by $\{h\alpha,\ \alpha\in\mathring{\Delta}\}$. Deduce the following identity from formula (12.1.7):

$$\prod_{i=1}^{\ell} \eta(q^{h\Phi(\alpha_i,\alpha_i)})^{h+1} = \sum_{\gamma\in M_Q} \mathring{d}(\gamma) q^{\Phi(\gamma+\overline{\rho},\gamma+\overline{\rho})}.$$

[Use the formula $\sum_{i=1}^{\ell} \Phi(\alpha_i,\alpha_i) \doteq \ell/h$].

12.4. Show that for automorphisms of $s\ell_2$ of type $(s,1;1)$, where $s = 0,1,2,3$, the identity (12.3.5) turns, respectively, into the following classical identities

$$\varphi(q)^3 = \sum_{n\in\mathbb{Z}} (4n+1) q^{2n^2+n} \qquad \text{(Jacobi)}$$

$$\varphi(q)^2/\varphi(q^2) = \sum_{n\in\mathbb{Z}} (-1)^n q^{n^2} \qquad \text{(Gauss)}$$

$$\varphi(q) = \sum_{n\in\mathbb{Z}} (-1)^n q^{(3n^2+n)/2} \qquad \text{(Euler)}$$

$$\varphi(q^2)^2/\varphi(q) = \sum_{n\in\mathbb{Z}} q^{2n^2+n} \qquad \text{(Gauss)}$$

12.5. Let W be the Weyl group of an affine algebra. Show that if $W = \mathring{W} W_1$, where W_1 is the set of representatives of minimal length of left cosets of $\mathring{W}$ in W, then $\overline{w(\rho)} - \overline{\rho} \in \overline{P}_+$ for $w \in W_1$. Show that $W_1 = \{w \in W \mid \mathring{\Delta}_+ \subset w(\Delta_+)\}$.
[Use Lemma 3.11 a).]

12.6. We keep the notation of Exercise 12.2 and identify $\mathring{\mathfrak{h}}$ with $\mathring{\mathfrak{h}}^*$ via the Killing form. Let $\overline{P}_+$ be the set of dominant integral weights and let $\mathring{L}(\lambda)$

denote an irreducible $\mathring{\mathfrak{g}}$-module with highest weight $\lambda \in \overline{P}_+$. Prove the following identity:

$$q^{\dim \mathring{\mathfrak{g}}/24} \prod_{n \geq 1} ((1-q^n)^\ell \prod_{\alpha \in \mathring{\Delta}} (1 - q^n e(\alpha)))$$

$$= \sum_{\lambda \in \bar{P}_+} \mathrm{tr}_{\mathring{L}(\lambda)} (\exp 4\pi i\, \overline{\rho}) \mathrm{ch} \mathring{L}(\lambda) q^{\Phi(\lambda+\overline{\rho}, \lambda+\overline{\rho})}.$$

[Consider the decomposition $W = \mathring{W} W_1$ from Exercise 12.5. We can write (cf. §12.1)

$$\sum_{w \in W} \epsilon(w) e(w(\rho) - \rho)/\mathring{R} = \sum_{w \in W_1} \epsilon(w) \,\mathrm{ch}\, \mathring{L}(w(\rho) - \overline{\rho}) e(\overline{\rho} - \rho)$$

$$= q^{-\Phi(\overline{\rho}, \overline{\rho})} \sum_{w \in W_1} \epsilon(w) \,\mathrm{ch}\, \mathring{L}(\lambda) q^{\Phi(\lambda+\overline{\rho}, \lambda+\overline{\rho})},$$

where $q = e(-\delta)$ and $\lambda = \overline{w(\rho)} - \overline{\rho} \in \bar{P}_+$ by Exercise 12.5. Note that $\overline{w(\rho)} = \mathrm{af}(w)(\overline{\rho})$ (see §6.6). Now we can use Exercise 10.19 to find that $\epsilon(w) = \mathrm{tr}_{\mathring{L}(\lambda)} (\exp 4\pi i \rho)$.]

12.7. We keep the notation of Exercise 12.6. Let $\alpha_1, \ldots, \alpha_\ell$ be simple roots of $\mathring{\mathfrak{g}}$ and let s be the number of short roots in this set. Let k be the ratio of square lengths of a long and a short roots, and let $\mathring{\Delta}_s, \mathring{\Delta}_\ell$ be the sets of all short and long roots. Let $\overline{\rho}^\vee$ be the half-sum of positive dual roots and h the Coxeter number of $\mathring{\mathfrak{g}}$. Set $a = h(h+1) \sum_{i=1}^{\ell} \Phi(\alpha_i, \alpha_i)$. Prove the following identity:

$$q^{a/24} \prod_{n \geq 1} ((1-q^n)^s (1-q^{kn})^{\ell - s} \prod_{\alpha \in \mathring{\Delta}_s} (1 - q e(\alpha)) \prod_{\alpha \in \mathring{\Delta}_\ell} (1 - q^k e(\alpha)))$$

$$= \sum_{\lambda \in \overline{P}_+} \mathrm{tr}_{\mathring{L}(\lambda)} (\exp 2\pi i \overline{\rho}^\vee / h) \,\mathrm{ch}\, \mathring{L}(\lambda) q^{\Phi(\lambda+\overline{\rho}, \lambda+\overline{\rho})}.$$

[The proof is similar to that of Exercise 12.6. Use Exercise 10.15 for $m = h$, and the hint from Exercise 12.3].

12.8. Show that the identities of Exercises 12.2 and 12.3 can be written as follows:

$$\eta^{\dim \mathring{\mathfrak{g}}} = \sum_{\lambda \in \overline{P}_+} \mathrm{tr}_{\mathring{L}(\lambda)} (\exp 4\pi i \overline{\rho}) \dim \mathring{L}(\lambda) q^{\Phi(\lambda+\overline{\rho}, \lambda+\overline{\rho})};$$

$$\prod_{i=1}^{\ell} \eta(q^{h\Phi(\alpha_i,\alpha_i)})^{h+1}$$
$$= \sum_{\lambda\in\overline{P}_+} \operatorname{tr}_{\mathring{L}(\lambda)} \exp(2\pi i\overline{\rho}^\vee/h) \dim \mathring{L}(\lambda) q^{\Phi(\lambda+\overline{\rho},\lambda+\overline{\rho})}.$$

In particular, the case $\mathring{\mathfrak{g}} = s\ell_2$ gives another form of Jacobi's identity:

$$\varphi(q)^3 = \sum_{n=0}^{\infty} (-1)^n (2n+1) q^{n(n+1)/2}.$$

12.9. We keep the notation of Exercises 12.2 and 3. Let $s=(s_0,\dots,s_\ell)$ be a set of nonnegative integers, not all of them zero. Set $m = s_0 + \sum_{i=1}^{\ell} a_i s_i$, where the a_i are defined by: $\sum_{i=1}^{\ell} a_i\alpha_i$ is the highest root. Define $h_s \in \mathfrak{h}$ by: $\langle \alpha_i, h_s\rangle = s_i/m$ for $i = 1,\dots,\ell$, and set $\sigma_s = \exp 2\pi i h_s$. Prove the following identity:

$$q^{\dim \mathring{\mathfrak{g}}/24} \prod_{n\geq 1} \det_{\mathring{\mathfrak{g}}} (1 - q^n \sigma_s) = \sum_{\alpha\in M} \mathring{\chi}_{g\alpha}(\sigma_s) q^{|\overline{\rho}+g\alpha|^2},$$

where $\mathring{\chi}_\lambda(\sigma_s) = (\sum_{w\in\mathring{W}} \epsilon(w) e^{w(\lambda+\overline{\rho})} / \sum_{w\in\mathring{W}} \epsilon(w) e^{w(\overline{\rho})})(-2\pi i h_s)$.

[Apply the following specialization to formula (12.1.7):

$$\Phi_s(e(-\alpha_0)) = q, \quad \Phi_s(e(-\alpha_t)) = e^{2\pi i s_t/m} \ (t = 1,\dots,\ell)].$$

12.10. Deduce from Exercise 12.6 another form of the identity of Exercise 12.9:

$$q^{\dim \mathring{\mathfrak{g}}/24} \prod_{n\geq 1} \det_{\mathring{\mathfrak{g}}} (1 - q^n \sigma_s)$$
$$= \sum_{\lambda\in\overline{P}_+} \operatorname{tr}_{\mathring{L}(\lambda)} (\exp 4\pi i\overline{\rho}) \operatorname{tr}_{\mathring{L}(\lambda)} (\sigma_s) q^{\Phi(\lambda+\overline{\rho},\lambda+\overline{\rho})}.$$

12.11. Let $\mathring{\mathfrak{g}}$ be a simple finite-dimensional Lie algebra, and let $\mathfrak{g}' = \widetilde{\mathcal{L}}(\mathring{\mathfrak{g}})$ (see §7.2). Given an irreducible finite-dimensional $\mathring{\mathfrak{g}}$-module $\mathring{L}(\lambda)$ and a non-zero complex number b, we can give a structure of a $\mathfrak{g}'$-module $\mathring{L}(\lambda; b)$ by letting $K \mapsto 0$, $t^j \otimes x \mapsto b^j x$ ($x \in \mathring{\mathfrak{g}}$). Show that all irreducible finite-dimensional $\mathfrak{g}'$-modules are of the form $\mathring{L}(\lambda_1; b_1) \otimes \cdots \otimes \mathring{L}(\lambda_n; b_n)$ where all the b_i are distinct. Show that all irreducible integrable $\mathbf{C}[t] \otimes \mathring{\mathfrak{g}}$-finite $\mathfrak{g}'$-modules are of the form $F \otimes L(\Lambda)$, where F is an irreducible finite-dimensional $\mathfrak{g}'$-module and $\Lambda \in P_+$.

12.12. We keep the notation of Exercise 12.11. Every integrable highest-weight $\mathfrak{g}'$-module of level k can be constructed as follows. Take a $\mathring{\mathfrak{g}}$-module $\mathring{L}(\lambda)$ with $\langle\lambda,\theta^\vee\rangle \leq k$, and consider it as a $\mathfrak{g}'_+ := \mathbb{C}K + (\mathbb{C}[t]\otimes\mathring{\mathfrak{g}})$-module by letting $K \mapsto kI$, $t^j \otimes \mathring{\mathfrak{g}} \mapsto 0$ if $j > 0$. Then the $\mathfrak{g}'$-module $U(\mathfrak{g}') \otimes_{U(\mathfrak{g}'_+)} \mathring{L}(\lambda)$ has a unique maximal submodule, this submodule is generated by the element $(F_0^{(-1)})^{k-\langle\lambda,\theta^\vee\rangle+1} \otimes v_\lambda$, and the quotient by this submodule is an integrable $\mathfrak{g}'$-module with highest weight $k\Lambda_0 + \lambda$.

12.13. Let $L(\Lambda)$ be an integrable $A_1^{(1)}$-module of positive level. Show that $\operatorname{mult}_{L(\Lambda)}(\Lambda - n\delta) \geq p(n)$, where $p(n)$ is the classical partition function.

12.14. Let $L(\Lambda)$ be an integrable $A_1^{(1)}$-module; set $s = \langle\Lambda,\alpha_0^\vee\rangle$, $r = \langle\Lambda,\alpha_1^\vee\rangle$. Show that all dominant maximal weights of $L(\Lambda)$ are either $\Lambda - j\alpha_0$, where $0 \leq j \leq [s/2]$, or $\Lambda - j\alpha_1$, where $0 \leq j \leq [r/2]$.

12.15. Prove that all the weights of the $A_1^{(1)}$-module $L(\Lambda_0)$ are of the form

$$\Lambda_0 - k^2\alpha_0 - (k^2 - k)\alpha_1 - s\delta, \text{ where } k \in \mathbb{Z}, s \in \mathbb{Z}_+.$$

12.16. Obtain the following decomposition of the tensor square of the $A_1^{(1)}$-module $L(\Lambda_0)$:

$$L(\Lambda_0) \otimes L(\Lambda_0) = \sum_{n\geq 0} a_n L(2\Lambda_0 - n\delta) + \sum_{n\geq 0} b_n L(2\Lambda_0 - \alpha_0 - n\delta),$$

where a_n and b_n can be determined from the equation

$$\sum_{n\geq 0} a_n q^{2n} + \sum_{n\geq 0} b_n q^{2n+1} = \prod_{n\geq 1}(1 + q^{2n-1}).$$

[Use the principal specialization to determine a_n and b_n].

12.17. Let $\mathfrak{g}(A)$ be an affine algebra, let $\Lambda \in P_+^k$ be such that $\langle\Lambda, d\rangle = 0$ and let $L(\Lambda) = \bigoplus_{j\in\mathbb{Z}_+} L(\Lambda)_j$ be the gradation of type $(1,0,\ldots,0)$ (the *basic gradation*). Then $L(\Lambda)_j$ is the eigenspace of d attached to the eigenvalue $-j$, and hence it is $\mathring{\mathfrak{g}}$-invariant. For $\lambda \in \mathring{P}_+$ put

$$\Phi_{\Lambda,\lambda}(q) = \sum_{n\geq 0}(\text{multiplicity of } \mathring{L}(\lambda) \text{ in } L(\Lambda)_{-n})q^n,$$

where $\mathring{L}(\lambda)$ denotes the integrable $\mathring{\mathfrak{g}}$-module with highest weight λ. Show that

$$\Phi_{\Lambda,\lambda}(q) = q^{-s_\Lambda} \sum_{w\in\mathring{W}} \epsilon(w) q^{\frac{1}{2k}|\lambda+\bar\rho-w(\bar\rho)|^2} c^{\Lambda}_{k\Lambda_0+\lambda+\bar\rho-w(\bar\rho)}.$$

Deduce that for A of type $X_N^{(r)}$, where $X = A$, D or E, one has

$$\Phi_{\Lambda_0,\lambda}(q) = q^{\frac{1}{2}|\lambda|^2} \prod_{\alpha \in \mathring{\Delta}_+} (1 - q^{(\lambda+\bar{\rho}|\alpha)}) / \prod_{j \geq 1} (1 - q^j)^{\operatorname{mult} j\delta} \text{ if } \lambda \in M,$$

and $\Phi_{\Lambda_0,\lambda}(q) = 0$ if $\lambda \notin M$. More generally, if $k = 1$ and Λ is the only maximal weight in $P(\Lambda) \cap P_+$, then

$$\Phi_{\Lambda,\lambda}(q) = a_\Lambda^\Lambda(q) q^{\frac{1}{2}(|\lambda|^2 - |\bar{\Lambda}|^2)} \prod_{\alpha \in \mathring{\Delta}_+} (1 - q^{(\lambda+\bar{\rho}|\alpha)}) \text{ if } \lambda \in \bar{\Lambda} + M,$$

and $\Phi_{\Lambda,\lambda}(q) = 0$ otherwise.

12.18. Show that the partition function $K(\alpha)$ for $A_1^{(1)}$ is given by the following formula:

$$K(k_0\alpha_0 + k_1\alpha_1) = \sum_{j \geq 0} (-1)^j p^{(3)}((j+1)k_0 - jk_1 - \tfrac{1}{2}j(j+1)),$$

where $p^{(3)}(j)$ is defined by:

$$\sum_{j \in \mathbb{Z}} p^{(3)}(j) q^j = \varphi(q)^{-3}.$$

[Show that for $\alpha = k_0\alpha_0 + k_1\alpha_1$ one has: $K(\alpha) + K(r_1(\alpha+\rho)-\rho) = p^{(3)}(k_0)$].

12.19. We keep the notation of Exercise 7.21 and Remark 3.8. Let (V, π) be an integrable highest-weight module of positive level k over an affine algebra $\mathfrak{g}(A)$, where A is the extended Cartan matrix of $\mathring{\mathfrak{g}}$. Show that the group G^π constructed in Remark 3.8 is a central extension φ: $G^\pi \to G$ of the group G (of Exercise 7.21) by $\mathbb{C}^\times$. Show that putting $T^\pi := \varphi^{-1}(f(\mathring{Q}^\vee)) \cap \widetilde{W}^\pi$, we get a central extension:

$$1 \to \{\pm I_V\} \to T^\pi \xrightarrow{\varphi} T \simeq \mathring{Q}^\vee \to 1.$$

Show that if $\tilde{\alpha}, \tilde{\beta} \in T^\pi$ are such that $\varphi(\tilde{\alpha}) = \alpha$, $\varphi(\tilde{\beta}) = \beta$, then:

$$\tilde{\alpha}\tilde{\beta}\tilde{\alpha}^{-1}\tilde{\beta}^{-1} = (-1)^{k(\alpha|\beta)} I_V.$$

12.20. Let $\mathfrak{g}$ be a simple finite-dimensional Lie algebra of type X_N and let $\mathfrak{g} = \oplus \mathfrak{g}_j(s;r)$ be its $\mathbb{Z}/m\mathbb{Z}$-gradation of type $(s;r)$. Consider the realization of type s of the affine algebra $\mathfrak{g}'(X_N^{(r)})$ given by Theorem 8.7:

$$\mathfrak{g}'(X_N^{(r)}) = \sum_{j \in \mathbb{Z}} t^j \otimes \mathfrak{g}_{\bar{j}}(s;r) \oplus \mathbb{C}K,$$

and use notation of this theorem. Choose a basis $u_{i;-j}$ of $\mathfrak{g}_{-j}(s;r)$ and the dual basis $u^{i;j}$ of $\mathfrak{g}_j(s;r)$, and define the following operator:

$$L_0^{\mathfrak{g};s} = \frac{1}{2r(k+h^\vee)} \sum_i \left(u_{i;0}^{(0)} u^{i;0(0)} + 2 \sum_{n>0} u_{i;-\bar{n}}^{(-n)} u^{i;\bar{n}(n)} \right) - H + rk\frac{(H|H)}{2} + \left(\frac{\dim \mathfrak{g}}{24} - \frac{|\rho|^2}{2h^\vee r} \right) \frac{k}{k+h^\vee}.$$

For $j \in \mathbb{Z}$, $j \neq 0$, let

$$L_j^{\mathfrak{g};s} = \frac{1}{mj}[d_{mj}, L_0^{\mathfrak{g};s}].$$

Show that $[x^{(n)}, L_j^{\mathfrak{g};s}] = nx^{(n+mj)}$, and that

$$[L_i^{\mathfrak{g};s}, L_j^{\mathfrak{g};s}] = (i-j)L_{i+j}^{\mathfrak{g};s} + \delta_{i,-j}\frac{i^3-i}{12} \dim \mathfrak{g}(X_N)\frac{k}{k+h^\vee}.$$

12.21. Under the hypothesis of Exercise 12.20, let $\dot{\mathfrak{g}}$ be a simple or abelian subalgebra of $\mathfrak{g}$, invariant with respect to the automorphism $\sigma_{s;r}$, and let $\dot{\sigma} = \dot{\sigma}_{\dot{s};\dot{r}}$ be the induced automorphism of $\dot{\mathfrak{g}}$, so that $\dot{r}\sum \dot{a}_i\dot{s}_i = m$ ($\dot{s}_i$ are not necessarily relatively prime). Let $L_j = L_j^{\mathfrak{g};s} - L_j^{\dot{\mathfrak{g}};\dot{s}}$. Show that the L_i commute with $\tilde{\mathcal{L}}(\dot{\mathfrak{g}}, \dot{\sigma}, m)$ and that they satisfy Virasoro commutation relations with central charge $c(k) - \dot{c}(k)$.

12.22. Applying Exercise 12.21 to $\dot{\mathfrak{g}} = \mathfrak{h}$, $k = 1$, derive the following formula:

$$\frac{\dim \mathfrak{g}(X_N)}{24} - \frac{|\rho|^2}{2h^\vee r} = \frac{(r+1)(h^\vee+1)(N-\ell)}{24r}.$$

12.23. Show that

$$b_\lambda^\Lambda(\dot{\mathfrak{h}}) = c_\lambda^\Lambda q^R \prod_{n\geq 1}(1-q^n)^{\operatorname{mult} n\delta},$$

where $R = \ell$ $(= |\bar{\rho}|^2/2h^\vee(h+1))$ if $r = 1$ and $R = |\bar{\rho}|^2/2h^\vee(h^\vee+1)$ if $r > 1$.

12.24. Show that for the $\hat{\mathcal{L}}(\mathfrak{g})$-module $L(\Lambda)$, $\Lambda \in P_+^k$, one has

$$\operatorname{tr}_{L(\Lambda)} q^{L_0^{\mathfrak{g}} - c(k)/24} = \frac{\sum_{\gamma\in M} d(\bar{\Lambda} + (k+h^\vee)\gamma) q^{|\bar{\Lambda}+\bar{\rho}+(k+h^\vee)\gamma|^2/2(k+h^\vee)}}{\eta^{\dim \mathfrak{g}}}.$$

§12.15. Bibliographical notes and comments.

Identities (12.1.4), (12.1.9) and (12.2.5) are due to Macdonald [1972]. His proof of (12.1.4), which is done in the framework of affine root systems ($= \Delta^{re}$), is quite lengthy and does not explain the "mysterious" factors corresponding to imaginary roots. (These factors were explained in Kac [1974] and Moody [1975]). These identities have been earlier obtained by Dyson [1972] in the classical case, but he did not notice the connection with root systems. Identities (12.3.5) were obtained by Kac [1978 A] and Lepowsky [1979].

The study of the series a_λ^Λ has been started by Feingold–Lepowsky [1978] and Kac [1978 A]. Lemma 12.6 for $r = 1$ is proved in Kac [1978 A]; its present proof is taken from Frenkel–Kac [1980]. The fact that the string functions c_λ^Λ are modular forms is pointed out in Kac [1980 B]. This observation was inspired by the "Monstrous moonshine" of Conway–Norton [1979].

The exposition of §12.4–12.7 closely follows Kac–Peterson [1984 A].

The construction of operators T_n goes back to Sugawara [1968]. The exposition of §12.8 uses Wakimoto [1986], Kac–Raina [1987] and Kac–Wakimoto [1988A]

Proposition 12.10 is due to Goddard–Kent–Olive [1985]. This coset construction was used in Goddard–Kent–Olive [1986] (and independently by Kac–Wakimoto [1986] and Tsuchiya–Kanie [1986 B]) to prove unitarizability of discrete series representations of the Virasoro algebra. This construction has been playing an important role in recent development of conformal field theory.

The exposition of §§12.9–12.12 follows Kac–Wakimoto [1988A], [1989A]. Results of §12.13 were obtained by Kac [1978A] and Kac–Peterson [1984A] by a more complicated method.

Exercises 12.2 and 12.3 are due to Macdonald [1972]. The form of Macdonald identities presented in Exercise 12.8 is due to Kostant [1976] (his proof is more complicated). Exercise 12.9 is due to Macdonald (unpublished). Exercises 12.13 and 12.16 are taken from Kac [1978 A]. Exercise 12.17 is taken from Kac [1980 B] and Kac–Peterson [1984 A]. Exercise 12.18 is taken from Kac–Peterson [1980], [1984 A]. A special case of Exercise 12.19 is treated in Frenkel–Kac [1980]; the general case may be deduced using the formula for the central extension via the tame symbol (described in Garland [1980]). Exercises 12.20–12.24 are taken from Kac–Wakimoto [1988A].

Chapter 13. Affine Algebras, Theta Functions, and Modular Forms

§13.0. We begin this chapter with an exposition of a theory of theta functions. Using the classical transformation properties of theta functions, we show that the linear span of normalized characters of given level is invariant under the action of $SL_2(\mathbf{Z})$. Using the theta function identities (12.7.13) and (12.12.8), we show that the string and branching functions are modular forms and find a transformation law for these forms. Furthermore, using the theory of modular forms, we prove the "very strange" formula (12.3.6), which in turn, is used, along with the Sugawara construction, to find an upper bound of orders of poles at all cusps for the string and branching functions. Furthermore, we study the "high-temperature limit" of characters and of string and branching functions. All this is applied to find explicit formulas for the weight multiplicities and branching rules for integrable highest-weight modules.

§13.1. We develop a theory of theta functions in the following general framework. (Keeping in mind applications to affine algebras, we use notation which is identical to that used in previous chapters.)

Let ℓ be a positive integer and let $\mathfrak{h}_{\mathbf{R}}$ be an $(\ell+2)$-dimensional vector space over $\mathbf{R}$ with a nondegenerate symmetric bilinear form $(.|.)$ of index $(\ell+1,1)$. We will identify $\mathfrak{h}_{\mathbf{R}}$ with $\mathfrak{h}_{\mathbf{R}}^*$ via this form. Fix a $\mathbf{Z}$-lattice M in $\mathfrak{h}_{\mathbf{R}}$ of rank ℓ, positive-definite and integral (i.e. $(a|b) \in \mathbf{Z}$ for all $a, b \in M$). Fix a vector $\delta \in \mathfrak{h}_{\mathbf{R}}$ such that $(\delta|\delta) = 0$ and $(\delta|M) = 0$.

Put $\mathring{\mathfrak{h}}_{\mathbf{R}} = \mathbf{R} \otimes_{\mathbf{Z}} M \subset \mathfrak{h}_{\mathbf{R}}$, $\mathfrak{h} = \mathbf{C} \otimes_{\mathbf{R}} \mathfrak{h}_{\mathbf{R}}$, $\mathring{\mathfrak{h}} = \mathbf{C} \otimes_{\mathbf{R}} \mathring{\mathfrak{h}}_{\mathbf{R}} \subset \mathfrak{h}$ and extend $(.|.)$ to $\mathfrak{h}$ by linearity. For $\lambda \in \mathfrak{h}$ we denote by $\bar{\lambda}$ the orthogonal projection of λ on $\mathring{\mathfrak{h}}$. Let

$$Y = \{v \in \mathfrak{h} \mid \mathrm{Re}(\delta|v) > 0\}.$$

For $\alpha \in \mathring{\mathfrak{h}}_{\mathbf{R}}$ let t_α denote the automorphism of $\mathfrak{h}$ defined by (cf. (6.5.2)):

$$t_\alpha(v) = v + (v|\delta)\alpha - \left((v|\alpha) + \tfrac{1}{2}(\alpha|\alpha)(v|\delta)\right)\delta.$$

Note that the automorphism t_α is characterized by the properties: a) $t_\alpha(\delta) = \delta$; b) $t_\alpha(v) = v + (v|\delta)\alpha \mod \mathbf{C}\delta$, and c) $(.|.)$ is t_α-invariant. It follows that $t_\alpha t_\beta = t_{\alpha+\beta}$.

For $\alpha \in \mathring{\mathfrak{h}}_{\mathbb{R}}$ let $p_\alpha(v) = v + 2\pi i\alpha, v \in \mathfrak{h}$. All the transformations p_α and t_β $(\alpha, \beta \in \mathring{\mathfrak{h}}_{\mathbb{R}})$ of $\mathfrak{h}$ generate a group N, called the Heisenberg group. More explicitly, $N = \mathring{\mathfrak{h}}_{\mathbb{R}} \times \mathring{\mathfrak{h}}_{\mathbb{R}} \times i\mathbb{R}$ with multiplication:

$$(\alpha, \beta, u)(\alpha', \beta', u') = (\alpha + \alpha', \beta + \beta', u + u' + \pi i((\alpha|\beta') - (\alpha'|\beta))).$$

The action of N on $\mathfrak{h}$ is given by

$$(\alpha, \beta, u)(v) = t_\beta(v) + 2\pi i\alpha + (u - \pi i(\alpha|\beta))\delta, \tag{13.1.1}$$

so that $(\alpha, 0, 0)(v) = p_\alpha(v)$, $(0, \beta, 0)(v) = t_\beta(v)$, $(0, 0, u)(v) = v + u\delta$, and Y is N-invariant.

Denote by $N_{\mathbb{Z}}$ the subgroup of N generated by all $(\alpha, 0, 0)$, $(0, \beta, 0)$ with $\alpha, \beta \in M$ and $(0, 0, u)$ with $u \in 2\pi i\mathbb{Z}$. Then

$$N_{\mathbb{Z}} = \{(\alpha, \beta, u) \in N \mid \alpha, \beta \in M, u + \pi i(\alpha|\beta) \in 2\pi i\mathbb{Z}\}.$$

§13.2. Fix a nonnegative integer k. A *theta function of degree* k is a holomorphic function F on the domain Y such that the following two conditions hold for all $v \in Y$:

$$F(n \cdot v) = F(v) \text{ for all } n \in N_{\mathbb{Z}}, \tag{T1}$$

$$F(v + a\delta) = e^{ka}F(v) \text{ for all } a \in \mathbb{C}. \tag{T2}$$

Let $\widetilde{Th}_k$ denote the space (over $\mathbb{C}$) of all theta functions of degree k. Then

$$\widetilde{Th} = \bigoplus_{k \geq 0} \widetilde{Th}_k$$

is a graded associative $\mathbb{C}$-algebra, called the *algebra of theta functions.*

In order to produce examples of theta functions let $M^* = \{\lambda \in \mathring{\mathfrak{h}}_{\mathbb{R}} \mid (\lambda|\alpha) \in \mathbb{Z} \text{ for all } \alpha \in M\}$ be the lattice dual to M; for a positive integer k put

$$P_k = \{\lambda \in \mathfrak{h} \mid (\lambda|\delta) = k \text{ and } \overline{\lambda} \in M^*\}.$$

Given $\lambda \in P_k$, we define the *classical theta function* of degree k with *characteristic* $\overline{\lambda}$ by the series

$$\Theta_\lambda = e^{-\frac{(\lambda|\lambda)}{2k}\delta} \sum_{\alpha \in M} e^{t_\alpha(\lambda)}. \tag{13.2.1}$$

Equivalently, we take $\lambda' \equiv \lambda \mod \mathbf{C}\delta$ such that $(\lambda'|\lambda') = 0$, and let

$$\Theta_\lambda = \sum_{\alpha \in M} e^{t_\alpha(\lambda')}.$$

Note that this function is exactly the one defined by (12.7.2) (which naturally arises in the theory of affine algebras). As in §12.7, we can rewrite Θ_λ in another form:

$$\Theta_\lambda = e^{k\Lambda_0} \sum_{\gamma \in M + k^{-1}\bar{\lambda}} q^{\frac{1}{2}k(\gamma|\gamma)} e^{k\gamma} \tag{13.2.2}$$

As before, q stands for $e^{-\delta}$, and $\Lambda_0 \in \mathfrak{h}_{\mathbb{R}}$ is the unique isotropic vector such that

$$(\Lambda_0|\delta) = 1 \qquad \text{and} \qquad (\Lambda_0|M) = 0.$$

It is clear that the series (13.2.2) converges absolutely on Y to a holomorphic function; one easily sees from (13.2.1) (or (13.2.2)) that Θ_λ satisfies (T1) and (T2), and hence is a theta function of degree k. Note also that

$$\Theta_{\lambda + k\alpha + a\delta} = \Theta_\lambda \quad \text{for } \alpha \in M,\ a \in \mathbf{C}. \tag{13.2.3}$$

Choose an orthonormal basis $v_1, \ldots, v_\ell$ of $\mathring{\mathfrak{h}}_{\mathbb{R}}$ and coordinatize $\mathfrak{h}$ by

$$v = 2\pi i \left(\sum_{s=1}^{\ell} z_s v_s - \tau\Lambda_0 + u\delta \right), \tag{13.2.4}$$

so that we shall often write (τ, z, u), where $z = \sum_{s=1}^{l} z_s v_s \in \mathring{\mathfrak{h}}$, $\tau, u \in \mathbf{C}$, in place of $v \in \mathfrak{h}$. Then $Y = \{(\tau, z, u) \mid z \in \mathring{\mathfrak{h}};\ \tau, u \in \mathbf{C}, \operatorname{Im}\tau > 0\}$, and we can rewrite Θ_λ in its classical form:

$$\Theta_\lambda(\tau, z, u) = e^{2\pi i k u} \sum_{\gamma \in M + k^{-1}\bar{\lambda}} e^{\pi i k \tau(\gamma|\gamma) + 2\pi i k(\gamma|z)} \tag{13.2.5}$$

(where $(\gamma|z) = \sum_{i=1}^{l} \gamma_i z_i$, $\gamma = \sum_{i=1}^{l} \gamma_i v_i$). Note also that in these coordinates we have: $q = e^{-\delta} = e^{2\pi i \tau}$.

Let $\mathcal{H} = \{\tau \in \mathbf{C} \mid \operatorname{Im}\tau > 0\}$ be the Poincaré upper half-plane. Note that a holomorphic function in $\tau \in \mathcal{H}$ lies in $\widetilde{Th}_0$. Conversely, if $F \in \widetilde{Th}_0$, then F is independent of u and for each fixed $\tau \in \mathcal{H}$, F is periodic in z

with periods in $M + \tau M$. Since $\mathring{\mathfrak{h}}/(M + \tau M)$ is compact, we deduce that F is a function in τ. So, we have proved

LEMMA 13.2. *$\widetilde{Th}_0$ is the algebra of holomorphic functions in $\tau \in \mathcal{H}$.*

□

The following multiplication formula is very useful for applications.

PROPOSITION 13.2. *Let Θ_λ and Θ_μ be classical theta functions of degree m and n, respectively. Then*

$$\Theta_\lambda \Theta_\mu = \sum_{\alpha \in M \text{mod } (m+n)M} \Theta_{\lambda+\mu+n\alpha} \psi_\alpha,$$

where

$$\psi_\alpha = \sum_{\gamma \in (m+n)M} q^{|n\overline{\lambda} - m\overline{\mu} - mn(\alpha+\gamma)|^2/2mn(m+n)}.$$

Proof. We may assume that $|\lambda|^2 = |\mu|^2 = 0$. Then

$$\Theta_\lambda = \sum_{\alpha \in M} e^{t_\alpha(\lambda)}, \quad \Theta_\mu = \sum_{\beta \in M} e^{t_\beta(\mu)}, \quad \text{and}$$

$$\begin{aligned}
\Theta_\lambda \Theta_\mu &= \sum_{\alpha,\gamma \in M} e^{t_\gamma(\lambda + t_\alpha(\mu))} \\
&= \sum_{\alpha \in M} q^{-|\lambda + t_\alpha(\mu)|^2/2(m+n)} \Theta_{\lambda + t_\alpha(\mu)} \\
&= \sum_{\alpha \in M \text{mod } (m+n)M} \Theta_{\lambda + t_\alpha(\mu)} \sum_{\gamma \in (m+n)M} q^{-|\lambda + t_{\alpha+\gamma}(\mu)|^2/2(m+n)} \\
&= \sum_{\alpha \in M \text{mod } (m+n)M} \Theta_{\lambda+\mu+n\alpha} \sum_{\gamma \in (m+n)M} q^{|n\lambda - m t_{\alpha+\gamma}(\mu)|^2/2mn(m+n)}.
\end{aligned}$$

□

§13.3. Let D be the Laplace operator associated with the form $(.|.)$. We have in coordinates (13.2.4):

$$D = \frac{1}{4\pi^2}\left(2\frac{\partial}{\partial u}\frac{\partial}{\partial \tau} - \sum_{s=1}^{\ell}\left(\frac{\partial}{\partial z_s}\right)^2\right). \tag{13.3.1}$$

Since $D(e^\lambda) = (\lambda|\lambda)e^\lambda$ we deduce from (13.2.1) that

$$D(\Theta_\lambda) = 0. \tag{13.3.2}$$

We put $Th_0 = \mathbf{C}, Th_k = \{F \in \widetilde{Th}_k \mid D(F) = 0\}$ for $k > 0$, $Th = \bigoplus_{k \geq 0} Th_k$.

Note that the subspace (over $\mathbf{C}$) Th of $\widetilde{Th}$ is not a subring (see Proposition 13.2).

PROPOSITION 13.3. *The set* $\{\Theta_\lambda \mid \lambda \in P_k \mod (kM + \mathbf{C}\delta)\}$ *is a* $\mathbf{C}$*-basis of* Th_k *(resp.* $\widetilde{Th}_0$*-basis of* $\widetilde{Th}_k$*) if* $k > 0$.

Proof. Let $F \in \widetilde{Th}_k$. Using $F(p_\alpha(v)) = F(v)$ for $\alpha \in M$ and (T2), we can, for a fixed τ, decompose F into a Fourier series:

$$F = e^{k\Lambda_0} \sum_{\gamma \in M^*} a_\tau(\gamma) e^\gamma.$$

By using (T1), we obtain that $a_\tau(\gamma)e^{-\pi i k^{-1}\tau(\gamma|\gamma)}$ depends only on $\gamma \mod kM$. It follows that

$$F = \sum_{\lambda \in P_k \mathrm{mod}\ kM + \mathbf{C}\delta} c_\lambda(\tau)\Theta_\lambda. \tag{13.3.3}$$

Furthermore, fix a positive real number a; then for $\alpha \in k^{-1}M^*$ we have

$$\Theta_\lambda(2\pi i\alpha + a\Lambda_0) = e^{2\pi i(\overline{\lambda}|\alpha)} \sum_{\gamma \in M + k^{-1}\overline{\lambda}} e^{-\frac{1}{2}k(\gamma|\gamma)a}.$$

Since the characters of the group $(k^{-1}M^*)/M$ are linearly independent, we deduce

$$\{\Theta_\lambda(a, z, 0) \mid \lambda \in P_k \mod kM + \mathbf{C}\delta\} \text{ is a linearly independent set over } \mathbf{C}, \text{ where the } \Theta_\lambda \text{ are viewed as functions in } z. \tag{13.3.4}$$

This completes the proof of the linear independence of the Θ_λ over $\mathbf{C}$ and over $\widetilde{Th}_0$.

Finally, if $D(F) = 0$, then, applying (13.3.2), we deduce from (13.3.3) that

$$0 = D(F) = \frac{ik}{\pi} \sum_\lambda (dc_\lambda/d\tau)\Theta_\lambda.$$

Using (13.3.4), we get $dc_\lambda/d\tau = 0$, hence the Θ_λ span Th_k over $\mathbb{C}$.

□

Example 13.3. Let $M = \mathbf{Z}\alpha$ be a 1-dimensional lattice with the bilinear form normalized by $(\alpha|\alpha) = 2$. Then $M^* = \frac{1}{2}M$ and the following classical theta functions form a basis of Th_m:

$$\Theta_{n,m}(\tau, z, u) := e^{2\pi imu} \sum_{k \in \mathbf{Z} + \frac{n}{2m}} e^{2\pi im(k^2\tau + kz)}, \quad n \in \mathbf{Z} \mathrm{mod}\ 2m\mathbf{Z}.$$

§13.4. We recall some elementary facts about the group $SL_2(\mathbf{R})$ and its discrete subgroups. The proofs may be found in the book Knopp [1970].

The group $SL_2(\mathbf{R})$ operates on $\mathcal{H}$ by

$$\begin{pmatrix} a & b \\ c & d \end{pmatrix} \cdot \tau = \frac{a\tau + b}{c\tau + d}.$$

For every positive integer n define the *principal congruence subgroup*

$$\Gamma(n) = \{\begin{pmatrix} a & b \\ c & d \end{pmatrix} \in SL_2(\mathbf{Z}) \mid a \equiv d \equiv 1 \mathrm{mod}\ n, \quad b \equiv c \equiv 0 \mathrm{mod}\ n\},$$

and the subgroup

$$\Gamma_0(n) = \{\begin{pmatrix} a & b \\ c & d \end{pmatrix} \in SL_2(\mathbf{Z}) \mid c \equiv 0 \mathrm{mod}\ n\}.$$

Another important subgroup is

$$\Gamma_\theta = \{\begin{pmatrix} a & b \\ c & d \end{pmatrix} \in SL_2(\mathbf{Z}) \mid ac \text{ and } bd \text{ are even}\}.$$

All these subgroups have finite index in $\Gamma(1) = SL_2(\mathbf{Z})$.

Put $S = \begin{pmatrix} 0 & -1 \\ 1 & 0 \end{pmatrix}$, $T = \begin{pmatrix} 1 & 1 \\ 0 & 1 \end{pmatrix}$. Then:

(13.4.1) S and T generate $\Gamma(1)$; S and T^2 generate Γ_θ,

(13.4.2) T, $({}^tT)^r$ and $-I$ generate $\Gamma_0(r)$ for $r = 2$ and 3.

Recall that the metaplectic group $Mp_2(\mathbf{R})$ is a double cover of $SL_2(\mathbf{R})$, defined as follows: $Mp_2(\mathbf{R}) = \{(A, j) \mid A = \begin{pmatrix} a & b \\ c & d \end{pmatrix} \in SL_2(\mathbf{R})$ and j is a holomorphic function in $\tau \in \mathcal{H}$ such that $j^2 = c\tau + d\}$, with multiplication $(A, j)(A_1, j_1) = (AA_1, j(A_1 \cdot \tau)j_1(\tau))$. $Mp_2(\mathbf{R})$ acts on $\mathcal{H}$ via the natural homomorphism $Mp_2(\mathbf{R}) \to SL_2(\mathbf{R})$. We put:

$$Mp_2(\mathbf{Z}) \text{ (resp. } Mp_2^\theta(\mathbf{Z})) = \{(A, j) \in Mp_2(\mathbf{R}) \mid A \in SL_2(\mathbf{Z}) \text{ (resp. } \in \Gamma_\theta)\}.$$

Furthermore, we introduce the following action of $Mp_2(\mathbf{R})$ on Y (actually, the $SL_2(\mathbf{R})$-action):

$$\begin{pmatrix} a & b \\ c & d \end{pmatrix} \cdot (\tau, z, u) = \left(\frac{a\tau + b}{c\tau + d}, \frac{z}{c\tau + d}, u - \frac{c(z|z)}{2(c\tau + d)}\right),$$

where $z \in \mathring{\mathfrak{h}}$, $\tau \in \mathcal{H}$, $u \in \mathbf{C}$.

It is clear that the groups N and $SL_2(\mathbf{R})$ act faithfully by holomorphic automorphisms of Y. One checks that $Mp_2(\mathbf{R})$ normalizes N; namely:

$$(13.4.3) \quad (\begin{pmatrix} a & b \\ c & d \end{pmatrix}, j)(\alpha, \beta, u)(\begin{pmatrix} a & b \\ c & d \end{pmatrix}, j)^{-1} = (a\alpha + b\beta, c\alpha + d\beta, u).$$

Hence, we have an action of the group $G := Mp_2(\mathbf{R}) \ltimes N$ on Y.

One checks directly the following

LEMMA 13.4. *The normalizer of $N_{\mathbf{Z}}$ in the subgroup $Mp_2(\mathbf{R})$ of G is $Mp_2(\mathbf{Z})$ if the lattice M is even (i.e., all $(\gamma|\gamma)$ are even for $\gamma \in M$), and is $Mp_2^{\theta}(\mathbf{Z})$ if M is odd (i.e., not even).*

□

Finally, we define a (right) action of G on holomorphic functions on Y as follows ($(A, j) \in Mp_2(\mathbf{R}), n \in N$):

$$F\,|_{(A,j)}\,(\tau, z, u) = j((\tau))^{-\ell} F(A \cdot (\tau, z, u)); \quad F\,|_n\,(v) = F(n(v)).$$

(At this point the use of $Mp_2(\mathbf{R})$ instead of $SL_2(\mathbf{R})$ is essential to have an action.) Obviously:

$$D(F)\,|_n = D(F\,|_n) \quad \text{for } n \in N. \tag{13.4.4}$$

We prove now the following important result:

PROPOSITION 13.4. a) *$\widetilde{Th}\,|_{(A,j)} = \widetilde{Th}$ if the lattice M is even and $A \in SL_2(\mathbf{Z})$ or if the lattice M is odd and $A \in \Gamma_\theta$.*

b) *$Th\,|_{(A,j)} = Th$ if the lattice M is even and $A \in SL_2(\mathbf{Z})$ or if the lattice M is odd and $A \in \Gamma_\theta$.*

Proof. a) follows immediately from Lemma 13.4 and (13.4.4).

Due to Proposition 13.3, in order to deduce b) from a), it suffices to check that

$$D(\Theta_\lambda\,|_{(T,1)}) = 0, \qquad D(\Theta_\lambda\,|_{(S,j)}) = 0.$$

The first equation is clear. The second is immediate from the following simple formula (recall that $j^2 = \tau$):

$$D(j^{-\ell} e^{2\pi i k(u - |\gamma - z|^2/2\tau)}) = 0, \quad \text{where } \gamma \in \mathring{\mathfrak{h}}.$$

□

We record also the following two simple transformation properties of classical theta functions of degree k, which follow directly from the definitions:

$$\Theta_\lambda|_{(\alpha,0,0)} = e^{2\pi i(\alpha|v)}\Theta_\lambda \quad \text{for } \alpha \in k^{-1}M^*; \tag{13.4.5}$$

$$\Theta_\lambda|_{(0,\alpha,0)} = \Theta_{\lambda - k\alpha} \quad \text{for } \alpha \in k^{-1}M^*. \tag{13.4.6}$$

We have the following corollary of Proposition 13.3 and formula (13.4.5):

COROLLARY 13.4. *The function Θ_λ (defined by (13.2.1)) is characterized among the holomorphic functions on Y by the properties* (T1), (T2), (13.3.2), *and* (13.4.5).

□

§13.5. Denote by $n = n(M)$ the least positive integer such that $nM^* \subset M$ and $n(\gamma|\gamma) \in 2\mathbf{Z}$ for all $\gamma \in M^*$.

Now we are in a position to prove the following transformation law which goes back to Jacobi.

THEOREM 13.5. *Let $\lambda \in P_k$. Then*

(13.5.1)
$$\Theta_\lambda(-\frac{1}{\tau}, \frac{z}{\tau}, u - \frac{(z|z)}{2\tau}) = (-i\tau)^{\frac{1}{2}\ell}|M^*/kM|^{-\frac{1}{2}} \times \sum_{\mu \in P_k \bmod (kM+\mathbf{C}\delta)} e^{-\frac{2\pi i}{k}(\overline{\lambda}|\overline{\mu})}\Theta_\mu(\tau, z, u);$$

(13.5.2)
$$\Theta_\lambda(\tau + 1, z, u) = e^{\pi i|\lambda|^2/k}\Theta_\lambda(\tau, z, u).$$

Furthermore, if $A \in \Gamma(kn)$ (resp. $\Gamma(kn) \cap \Gamma_\theta$) when M is even (resp. odd), then

(13.5.3)
$$\Theta_\lambda|_{(A,j)} = v(A, j; k)\Theta_\lambda,$$

where $v(A, j; k) \in \mathbf{C}$ and $|v(A, j; k)| = 1$.

Proof. Using that

$$\Theta_\lambda|_g = (\Theta_{k\Lambda_0}|_{(0,-k^{-1}\overline{\lambda},0)})|_g = (\Theta_{k\Lambda_0}|_g)|_{g^{-1}(0,-k^{-1}\overline{\lambda},0)g}$$

for $g \in G$, by (13.4.6) it suffices to prove the theorem for $\overline{\lambda} = 0$. Note also that when replacing $(.|.)$ by $k(.|.)$, the theta function $\Theta_\lambda(\tau, z, u)$ of degree k transforms to the theta function $\Theta_{k^{-1}\lambda}(\tau, z, ku)$ of degree 1. Hence we may (and shall) assume that $\lambda = \Lambda_0$.

By Proposition 13.4b), we may write for $A = \begin{pmatrix} a & b \\ c & d \end{pmatrix} \in SL_2(\mathbf{Z})$ (resp. $\in \Gamma_\theta$) if M is even (resp. odd):

(13.5.4)
$$\Theta_{\Lambda_0}|_{(A,j)} = \sum_{\mu \in M^* \bmod M} f(\mu)\Theta_{\Lambda_0+\mu}, \quad \text{where } f(\mu) \in \mathbf{C}.$$

Fix $\alpha \in M^*$. Since, by (13.4.5), $\Theta_{\Lambda_0}|_{(\alpha,0,0)} = \Theta_{\Lambda_0}$, we get by (13.4.3):

$$\Theta_{\Lambda_0}|_{(A,j)} = \Theta_{\Lambda_0}|_{(A,j)}\,|_{(A,j)^{-1}(\alpha,0,0)(A,j)} = \Theta_{\Lambda_0}|_{(A,j)}\,|_{(d\alpha,-c\alpha,0)}\,.$$

Hence, applying $(A,j)^{-1}(\alpha,0,0)(A,j)$ to both sides of (13.5.4), we get, thanks to (13.4.5 and 6):

$$\Theta_{\Lambda_0}|_{(A,j)} = \sum_{\mu\in M^* \bmod M} f(\mu)e^{\pi i(dc(\alpha|\alpha)+2d(\alpha|\mu))}\Theta_{\Lambda_0+\mu+c\alpha}.$$

If $A = S$, comparing this with (13.5.4), we get that $f(\mu+\alpha) = f(\mu)$ for all $\alpha, \mu \in M^*$ and hence:

$$\text{(13.5.5)}\qquad \Theta_{\Lambda_0}|_{(S,j)} = v(S,j)\sum_{\mu\in M^* \bmod M}\Theta_{\Lambda_0+\mu},\quad \text{where } v(S,j)\in \mathbf{C}.$$

If $A \in \Gamma(r)$, we get: $f(\mu) = f(\mu+c\alpha) = f(\mu)e^{2\pi i(\alpha|\mu)}$ for all $\alpha,\mu \in M^*$. Hence $f(\mu) = 0$ unless $\mu \in M$. This completes the proof of (13.5.3). The fact that $|v(A,j;m)| = 1$ follows from Corollary 13.5 below. (13.5.2) is obvious.

Furthermore, as explained in the beginning of the proof, (13.5.5) gives us that (13.5.1) holds up to a constant factor $v(S,j) \in \mathbf{C}$. In order to compute this constant, notice that $(S,j)^8 = I$ and that the rows of the matrix of the transformation (S,j) in the basis $\{\Theta_\lambda\}$ are pairwise orthogonal. It follows that this matrix is unitary. We deduce immediately that $|v(S,j)| = |M^*/kM|^{-\frac{1}{2}}$. Finally, using that $\Theta_\lambda(0,i,0) > 0$, we deduce from (13.5.5) that $v(S,j) = (-i)^{\frac{1}{2}\ell}|v(S,j)|$.

□

Since the matrix of the transformation (S,j) in the basis $\{\Theta_\lambda\}$ is unitary, and the matrix of the transformation (T,j) (resp. (T^2,j)) for M even (resp. odd), is (diagonal) unitary, using (13.4.1), we get the following useful result.

COROLLARY 13.5. *The matrix of a transformation from* $Mp_2(\mathbf{Z})$ *(resp.* $Mp_2^\theta(\mathbf{Z})$*) if M is even (resp. odd) in the basis* $\{\Theta_\lambda\}_{\lambda\in P_k \bmod kM+\mathbf{C}\delta}$ *is unitary.*

Example 13.5. Theorem 13.5 gives the following transformation law for theta functions $\Theta_{n,m}$ from Example 13.3:

$$\text{(13.5.6)}\qquad \begin{aligned}&\Theta_{n,m}\left(-\frac{1}{\tau},\frac{z}{\tau},u+\frac{z^2}{2\tau}\right)\\ &= (-i\tau)^{\frac{1}{2}}(2m)^{-\frac{1}{2}}\sum_{n'\in\mathbf{Z} \bmod 2m\mathbf{Z}} e^{-\frac{\pi i n n'}{m}}\Theta_{n',m}(\tau,z,u).\end{aligned}$$

§13.6. Here we give a brief account of some facts about modular forms which will be used in the sequel.

Fix a subgroup Γ of finite index in $\Gamma(1)$, a function $\chi : \Gamma \to \mathbf{C}^\times$ with $|\chi(A)| = 1$ for $A \in \Gamma$, and a real number k. Then a function $f : \mathcal{H} \to \mathbf{C}$ is called a ***modular form*** of ***weight*** k and ***multiplier system*** χ for Γ if f is holomorphic on $\mathcal{H}$ and $f\left(\frac{a\tau+b}{c\tau+d}\right) = \chi(A)(c\tau+d)^k f(\tau)$ for all $A = \left(\begin{smallmatrix} a & b \\ c & d \end{smallmatrix}\right) \in \Gamma$ and $\tau \in \mathcal{H}$.

Let f be such a modular form. Then, since Γ has a finite index in $\Gamma(1), T^s \in \Gamma$ for some positive integer s and hence $f(\tau + s) = e^{2\pi i C} f(\tau)$ for some $C \in \mathbf{R}$. Set

$$F(e^{2\pi i\tau/s}) = e^{-2\pi i C\tau/s} f(\tau).$$

Then F is a well-defined holomorphic function in $z = e^{2\pi i\tau/s}$ on the punctured disc $0 < |z| < 1$. Hence F has a Laurent expansion $F(z) = \sum_{n\in\mathbf{Z}} a_n z^n$ converging absolutely for $0 < |z| < 1$. Therefore, we have the Fourier expansion

$$f(\tau) = \sum_{n\in\mathbf{Z}} a_n e^{2\pi i(n+C)\tau/s} \quad \text{for } \tau \in \mathcal{H}.$$

We call f *meromorphic* at $i\infty$ if $a_n = 0$ for $n \ll 0$, ***holomorphic*** at $i\infty$ if $a_n \neq 0$ implies $n+C \geq 0$, ***vanishing*** at $i\infty$ if $a_n \neq 0$ implies $n+C > 0$. If f is holomorphic at $i\infty$, we say that the ***value*** of f at $i\infty$ is a_{-C} (interpreted as 0 if $C \notin \mathbf{Z}$). If $n_0 = \min\{n \mid a_n \neq 0\}$, we let $r = |(n_0 + C)/s|$ and say that it has *zero* (resp. *pole*) of order r at $i\infty$ if $n_0 + C \geq 0$ (resp. ≤ 0).

A *cusp* of a subgroup Γ of finite index in $SL_2(\mathbf{Z})$ is an orbit of Γ in $\mathbf{Q} \cup \{i\infty\}$, where $a/0$ is interpreted as $i\infty$ for $a \in \mathbf{Q}, a \neq 0$. Since $\Gamma(1)$ acts transitively on $\mathbf{Q} \cup \{i\infty\}$, the set of cusps of Γ is finite. Sometimes we speak of the cusp $\alpha \in \mathbf{Q} \cup \{i\infty\}$ of Γ; this means the orbit of α under Γ. For example, $\Gamma(1)$ has one cusp $i\infty$, Γ_θ has two cusps: $i\infty$ and -1, and $\Gamma_0(k)$ for prime k has two cusps: $i\infty$ and 0.

Let f be as above and consider a cusp α of Γ. Let $B = \left(\begin{smallmatrix} a & b \\ c & d \end{smallmatrix}\right) \in \Gamma(1)$ be such that $B(i\infty) = \alpha$. Then $f_0(\tau) := (c\tau + d)^{-k} f(B\tau)$ is a modular form of weight k and some multiplier system χ_0 for $B^{-1}\Gamma B$. We say f is *meromorphic*, ***holomorphic***, or ***has zero or pole of order*** r at α if f_0 is meromorphic, holomorphic, or has zero or pole of order r at $i\infty$. We say that f is ***R-singular*** if orders of poles of f at all cusps are $\leq R$.

A modular form of weight k and multiplier system χ for Γ is called a *meromorphic modular form*, a ***holomorphic modular form***, or a ***cusp form*** if it is meromorphic, holomorphic, or vanishes at all cusps of Γ, respectively.

A holomorphic modular form of weight 0 is a constant. This allows one to identify modular forms. In what follows, q will stand, for $e^{2\pi i\tau}$, as usual.

Using various specializations of classical theta functions, we can construct modular forms. In fact, given a holomorphic function F on Y and $\alpha, \beta \in \mathring{\mathfrak{h}}_{\mathbf{R}}$, we define a holomorphic function $F^{\alpha,\beta}(\tau)$ on $\mathcal{H}$ by

$$(13.6.1)\quad F^{\alpha,\beta}(\tau) := (F|_{(\alpha,\beta,0)})(\tau,0,0) = F(\tau, -\alpha+\tau\beta, -\tfrac{1}{2}(\beta|-\alpha+\tau\beta)).$$

For example,

$$(13.6.2)\qquad \Theta_\lambda^{\alpha,\beta}(\tau) = e^{\pi i k(\alpha|\beta)} \sum_{\gamma\in M+k^{-1}\overline{\lambda}-\beta} e^{2\pi i k(\alpha|\gamma)} q^{k|\gamma|^2/2}.$$

Furthermore, it is clear by (13.4.3) that

$$(13.6.3)\qquad \begin{aligned} &(F|_{(A,j)})^{\alpha,\beta}(\tau) = F^{a\alpha+b\beta, c\alpha+d\beta}(\tau)|_{(A,j)}, \\ &\text{where } \left(\begin{pmatrix} a & b \\ c & d\end{pmatrix}, j\right) \in Mp(2,\mathbf{R}). \end{aligned}$$

A special case of this is

$$(13.6.4)\qquad (F|_{(S,j)})^{\alpha,\beta}(\tau) = F^{-\beta,\alpha}(\tau)|_{(S,j)}.$$

Now it is easy to prove the following

PROPOSITION 13.6. *Given positive integers s and m, put*

$$\mathcal{F}_{m,s} = \{\Theta^{\alpha,\beta}(\tau) \mid \Theta \in Th_m,\ s\alpha\in M,\ s\beta\in M\}.$$

Then every function from $\mathcal{F}_{m,s}$ is a holomorphic modular form of weight $\frac{1}{2}\ell$ for $\Gamma(mn)\cap\Gamma(s)$ (resp. $\Gamma(mn)\cap\Gamma_\theta\cap\Gamma(s)$) if the lattice M is even (resp. odd).

Proof. Let $A = \left(\begin{smallmatrix} a & b \\ c & d\end{smallmatrix}\right) \in \Gamma(mn)$ (resp. $\Gamma(mn)\cap\Gamma_\theta$). Then by Theorem 13.5 and (13.6.3) we have

$$v(A)\Theta_\lambda^{\alpha,\beta}(\tau) = \Theta_\lambda^{a\alpha+b\beta, c\alpha+d\beta}\left(\frac{a\tau+b}{c\tau+d}\right)(c\tau+d)^{-\frac{\ell}{2}},$$

where $v(A)\in\mathbf{C}, |v(A)|=1$. On the other hand, by (13.6.2) we have

$$\Theta_\lambda^{\alpha,\beta}(\tau) = \pm\Theta_\lambda^{a\alpha+b\beta, c\alpha+d\beta}(\tau) \quad \text{if} \quad \begin{pmatrix} a & b \\ c & d\end{pmatrix} \in \Gamma(s).$$

Thus, every function from $\mathcal{F}_{m,s}$ satisfies the required transformation properties.

Furthermore, it is clear from (13.6.2), that $\mathcal{F}_{m,s}|_{(T^\epsilon,j)} = \mathcal{F}_{m,s}$ for $\epsilon = 1$ or 2, according to whether M is even or odd. Using this, as well as (13.6.4) and (13.4.1), we conclude that the linear span of $\mathcal{F}_{m,s}$ is invariant under $Mp_2(\mathbb{Z})$ if M is even. Since all functions from $\mathcal{F}_{m,s}$ are holomorphic at the cusp $i\infty$, we deduce, by Lemma 13.12 in §13.12, that all of them are holomorphic modular forms, provided that M is even. The general case is reduced to this one by the change of variables $\tau \to 2\tau$.

□

COROLLARY 13.6. *Let M be a $\mathbb{Z}$-lattice of rank ℓ and let $(.|.)$ be a positive-definite $\mathbb{Q}$-valued bilinear form on M. Let $\epsilon : M \to \mathbb{C}$ be constant on cosets of some sublattice of finite index and let $a \in \mathbb{Q} \otimes_{\mathbb{Z}} M$. Then*

$$f(\tau) = \sum_{\gamma \in M} \epsilon(\gamma) q^{|\gamma + a|^2}$$

is a holomorphic modular form of weight $\frac{1}{2}\ell$ for some $\Gamma(N)$ and some multiplier system.

Proof. Replacing M by a sublattice of finite index, we can assume that $\epsilon(\gamma) = 1$. Moreover, since replacing $(.|.)$ by $\frac{1}{k}(.|.)$ corresponds to replacing $\Gamma(N)$ by $\Gamma(kN)$, we can assume that M is an integral lattice. But then $f(\tau) = \Theta^{0,-a}_{\Lambda_0}$.

□

The most popular examples of holomorphic modular forms are these:

$$f_{n,m}(\tau) = \sum_{j \in \mathbb{Z}} q^{m(j+n/2m)^2}, \qquad g_{n,m}(\tau) = \sum_{j \in \mathbb{Z}} (-1)^j q^{m(j+n/2m)^2},$$

where $m, n \in \frac{1}{2}\mathbb{Z}$, $m > 0$. Since $f_{n,m}(\tau) = \Theta_{n,m}(\tau, 0, 0)$, we obtain from (13.5.6), provided that $n \in \mathbb{Z}$:

$$f_{n,m}(-\tau^{-1}) = \left(\frac{-i\tau}{2m}\right)^{1/2} \sum_{k \in \mathbb{Z} \bmod 2m\mathbb{Z}} e^{i\pi kn/m} f_{k,m}(\tau). \tag{13.6.5}$$

(In fact (13.5.6) implies (13.6.5) only for integral m; to get half-integral m, one needs to take $(\alpha|\alpha) = 1$ in Example 13.3.) Since $g_{n,m} = f_{n,4m} - f_{(n+2m),4m}$, we derive from (13.6.5), provided that $m, n \in \frac{1}{2} + \mathbb{Z}$:

$$g_{n,m}(-\tau^{-1}) = \left(\frac{-i\tau}{2m}\right)^{1/2} \sum_{\substack{k \in \mathbb{Z} \\ 0 \le k < m - \frac{1}{2}}} (2\cos\frac{(2k+1)n}{2m}) g_{k+\frac{1}{2},m}(\tau). \tag{13.6.6}$$

A special case of this is

$$g_{\frac{1}{2},\frac{3}{2}}(-\tau^{-1}) = (-i\tau)^{1/2} g_{\frac{1}{2},\frac{3}{2}}(\tau). \tag{13.6.7}$$

Now we consider some examples of modular forms defined by an infinite product, and among them the Dedekind η-function

$$\eta(\tau) = e^{\pi i\tau/12} \prod_{n\geq 1} \left(1 - e^{2\pi i n\tau}\right), \quad \tau \in \mathcal{H}.$$

Our starting point is the identity (12.7.4) for the affine algebra $A_1^{(1)}$. Using (12.1.5) and the information from Examples 12.1 and 13.3, it can be written as follows (we put $\alpha = \alpha_1$):

(13.6.8)

$$(\Theta_{1,2} - \Theta_{-1,2})(\tau, z, u)$$
$$= e^{2\pi i(\frac{1}{8}\tau - \frac{1}{2}(z|\alpha) - 2u)} \prod_{n\geq 1} (1 - q^n)(1 - q^{n-1}e^{2\pi i(z|\alpha)})(1 - q^n e^{-2\pi i(z|\alpha)}).$$

Note that we have for $m, r \in \mathbf{Z}$, $m > 0$:

$$g_{m/2-r,m/2}(\tau) = (\Theta_{1,2} - \Theta_{-1,2})^{0,\frac{1}{2m}r\alpha}(m\tau).$$

Hence, we obtain from (13.6.8) (recall that $(\alpha|\alpha) = 2$):

(13.6.9)

$$g_{m/2-r,m/2}(\tau) = q^{(2r-m)^2/8m} \prod_{j\geq 1} (1 - q^{mj})(1 - q^{mj-(m-r)})(1 - q^{mj-r})$$

A special case of this for $m = 3$, $r = 1$ is Euler's identity:

$$\eta(\tau) = \sum_{j\in\mathbf{Z}} (-1)^j q^{\frac{3}{2}(j+\frac{1}{6})^2} \quad (= g_{\frac{1}{2},\frac{3}{2}}(\tau)). \tag{13.6.10}$$

Comparing this with (13.6.7), we obtain the classical functional equation for the η-function:

$$\eta(-\frac{1}{\tau}) = (-i\tau)^{\frac{1}{2}} \eta(\tau). \tag{13.6.11}$$

Using (13.4.1) we deduce that $\eta(\tau)$ is a holomorphic cusp-form of weight $\frac{1}{2}$ for $\Gamma(1)$ and a multiplier system χ_η such that $\chi_\eta^{24} = 1$. Note that: $\chi_\eta(S) = e^{-\pi i/4}$, $\chi_\eta(T) = e^{\pi i/12}$.

§13.7. We shall apply the theory of theta functions to affine algebras. Let $\mathfrak{g}(A)$ be an affine algebra of type $X_N^{(r)}$ and let M be the lattice introduced in §6.5.

Since, by definition, $M = \mathbb{Z}(\mathring{W} \cdot \theta^\vee)$, and $(\theta^\vee|\theta^\vee) = 2a_0^{-1}$ (see §6.4), we see that the lattice M is always integral; moreover, it is even if and only if the affine algebra is not of type $A_{2\ell}^{(2)}$. Using the information about finite root systems given in §6.7, one easily computes for this lattice the constant $n = n(M)$ introduced in §13.5; it is found in the following table.

$\gamma(A)$	ℓ	$n = n(M)$
$A_\ell^{(1)}$	odd	$2(\ell+1)$
	even	$\ell+1$
$B_\ell^{(1)}, D_\ell^{(1)}, A_{2\ell-1}^{(2)}$	odd	8
	$2\|\ell$	4
	$4\|\ell$	2
$C_\ell^{(1)}, E_7^{(1)}, D_{\ell+1}^{(2)}$		4
$E_6^{(1)}, G_2^{(1)}, D_4^{(3)}$		3
$F_4^{(1)}, A_{2\ell}^{(2)}, E_6^{(2)}$		2
$E_8^{(1)}$		1

Fix a positive integer k. Using (6.1.1), we can write (cf. §10.1 and §12.4):

$$P^k = \{\lambda \in \mathfrak{h}^* \mid \text{level}(\lambda) = k,\ \langle\lambda, \alpha\rangle \in \mathbb{Z} \text{ for } \alpha \in \mathring{Q}^\vee\}.$$

Recall also the definition from §13.2:

$$P_k = \{\lambda \in \mathfrak{h}^* \mid \text{level}(\lambda) = k,\ (\lambda|\alpha) \in \mathbb{Z} \text{ for } \alpha \in M\}.$$

Using (6.5.8) and (6.5.9) we obtain:

(13.7.1) $$P_k \supset P^k,$$

(13.7.2) $$P_k = P^k \quad \text{if } r = 1 \text{ or } a_0 = 2.$$

LEMMA 13.7. *Let Θ_λ, $\lambda \in P_k$, be a classical theta function associated with the lattice M. If $A \in \Gamma(kn)$, then*

$$\Theta_\lambda|_{(A,j)} = v(A, j; k)\Theta_\lambda,$$

where $v(A,j;k) \in \mathbb{C}$, $|v(A,j;k)| = 1$.

Proof. Since $n = 2$ is the only case when M is odd, and since $\Gamma_\theta \subset \Gamma(2k)$, the lemma follows from Theorem 13.5.

□

Define the space of ***anti-invariant*** classical theta functions of degree k by:

$$(13.7.3) \qquad Th_k^- = \{F \in Th_k \mid F(w(v)) = \epsilon(w)F(v) \text{ for } w \in \overset{\circ}{W},\ v \in \mathfrak{h}\}.$$

Using Proposition 6.5 and (6.5.10), we see that the space Th_k^- consists of holomorphic functions F on Y, satisfying the following four properties:

(T1⁻) $F(w(\lambda)) = \epsilon(w)F(\lambda)$ for $w \in W$;

(T2⁻) $F(\lambda + 2\pi i\alpha) = F(\lambda)$ for $\alpha \in M$;

(T3⁻) $F(\lambda + a\delta) = e^{ka}F(\lambda)$ for $a \in \mathbf{C}$;

(T4⁻) $DF = 0$.

Given $\lambda \in P_k$, we introduce the anti-invariant classical theta function of degree k and characteristic $\overline{\lambda}$:

$$(13.7.4) \qquad A_\lambda = \sum_{w \in \overset{\circ}{W}} \epsilon(w)\Theta_{w(\lambda)}.$$

By (13.7.3), Proposition 6.5 and (6.5.10), we have:

$$(13.7.5) \qquad A_\lambda = e^{-\frac{(\lambda|\lambda)}{2k}\delta} \sum_{w \in W} \epsilon(w)e^{w(\lambda)}.$$

Let $P_{k+} = P_+ \cap P_k$, $P_{k++} = P_{++} \cap P_k$.

Proposition 13.7. a) *The set* $\{A_\lambda \mid \lambda \in P_{k++} \mod \mathbf{C}\delta\}$ *is a* $\mathbf{C}$*-basis of* Th_k^-.

b) *If* $\lambda \in P_{++}$, *then:*

$$A_\lambda|_{(S,j)} = (-i)^{\frac{1}{2}\ell}|M^*/kM|^{-\frac{1}{2}} \times \sum_{\mu \in P_{k++} \mathrm{mod}\ \mathbf{C}\delta} \sum_{w \in \overset{\circ}{W}} \epsilon(w)e^{-\frac{2\pi i}{k}(w(\overline{\lambda})|\overline{\mu})}A_\mu;$$

$$A_\lambda(\tau + 1, z, u) = e^{\pi i|\overline{\lambda}|^2/k}A_\lambda(\tau, z, u).$$

Proof. By Proposition 13.3, the set $\{A_\lambda \mid \lambda \in P_k\}$ spans Th_k^-. Thanks to Lemma 12.4 and Proposition 6.5 we have:

$$(13.7.6)\quad P_m \mod (mM + \mathbf{C}\delta) = \bigcup_{w \in \mathring{W}} w(P_{k+} \mathrm{mod}\ \mathbf{C}\delta) \quad \text{(disjoint union)}.$$

But we clearly have:

$$(13.7.7)\qquad A_\mu = 0 \text{ if } r_i(\mu) = \mu \text{ for some } i.$$

Hence, by (13.7.6), the set $\{A_\lambda \mid \lambda \in P_{k++} \mod \mathbf{C}\delta\}$ spans Th_k^-. These A_λ are linearly independent by Proposition 13.3 and (13.7.6), proving a).

By Theorem 13.5,

$$A_\lambda|_{(S,j)} = (-i)^{\frac{1}{2}\ell}|M^*/kM|^{-\frac{1}{2}} \times \sum_{\mu \in P_k \mathrm{mod}\ (kM + \mathbf{C}\delta)} \left(\sum_{w \in \mathring{W}} \epsilon(w) e^{-\frac{2\pi i}{k}(w(\overline{\lambda})|\overline{\mu})} \right) \Theta_\mu .$$

This together with (13.7.6 and 7) proves b).

□

§**13.8.** We turn now to the proof of the following crucial result:

LEMMA 13.8. *Let $\mathfrak{g}(A)$ be an affine algebra of type $X_\ell^{(1)}$ or $A_{2\ell}^{(2)}$. Then*

a) $Th_k^- = 0$ *if* $k < h^\vee$ *and* $Th_{h^\vee}^- = \mathbf{C}A_\rho$.

b) $A_\rho|_{(S,j)} = (-i)^{\frac{1}{2}\ell + |\mathring{\Delta}_+|} A_\rho$.

Proof. a) follows from (13.7.2) and Proposition 13.7a). By a) and Proposition 13.7b), we have:

$$A_\rho|_{(S,j)} = (-i)^{\frac{1}{2}\ell}|M^*/h^\vee M|^{-\frac{1}{2}} c A_\rho \text{ where } c = \sum_{w \in \mathring{W}} \epsilon(w) e^{-\frac{2\pi i}{h^\vee}(w(\overline{\rho})|\overline{\rho})}.$$

By Corollary 10.5, $|c| = |M^*/h^\vee M|^{\frac{1}{2}}$. On the other hand, identity (10.4.4), applied to $\mathring{\mathfrak{g}}$, gives

$$c = \prod_{\alpha \in \mathring{\Delta}_+} \left(-2i \sin \frac{\pi(\overline{\rho}|\alpha)}{h^\vee} \right).$$

It follows that $i^{|\mathring{\Delta}_+|} c$ is a positive real number, completing the proof.

□

The proof of Lemma 13.8 gives us the following useful identity

$$\prod_{\alpha\in\mathring{\Delta}_+} 2\sin\frac{\pi(\overline{\rho}|\alpha)}{h^\vee} = |M^*/h^\vee M|^{1/2}, \tag{13.8.1}$$

since,

$$(\overline{\rho}|\alpha) \le (\overline{\rho}|\theta) < h^\vee. \tag{13.8.2}$$

Recall the formula for the normalized character (see (12.7.11)) for $\Lambda \in P_+^k$:

$$\chi_\Lambda(\tau, z, u) := q^{m_\Lambda}\operatorname{ch}_{L(\Lambda)} = A_{\Lambda+\rho}/A_\rho.$$

Note that a special case of (12.2.2) for $\dot{\mathfrak{g}} = 0$ is

$$\chi_\Lambda(\tau, z, u) = e^{2\pi iku}\operatorname{tr}_{L(\Lambda)} e^{2\pi iz} q^{L_0^{\mathfrak{g}} - \frac{1}{24}c(k)} \quad \text{if} \quad \mathfrak{g}(A) = \hat{\mathcal{L}}(\mathfrak{g}). \tag{13.8.3}$$

Proposition 13.7b) together with (13.7.2) and Lemma 13.8.b) imply immediately the following important transformation law for normalized characters:

THEOREM 13.8. *Let $\mathfrak{g}(A)$ be an affine algebra of type $X_\ell^{(1)}$ or $A_{2\ell}^{(2)}$, and let $\Lambda \in P_+^k$. Then*

a) $\chi_\Lambda(-\frac{1}{\tau}, \frac{z}{\tau}, u - \frac{(z|z)}{2\tau}) = \sum_{\Lambda'\in P_+^k \bmod \mathbb{C}\delta} S_{\Lambda,\Lambda'}\chi_{\Lambda'}(\tau, z, u)$, where

$$S_{\Lambda,\Lambda'} = i^{|\mathring{\Delta}_+|}|M^*/(k+h^\vee)M|^{-\frac{1}{2}} \sum_{w\in\mathring{W}} \varepsilon(w) e^{-\frac{2\pi i(\overline{\Lambda}+\overline{\rho}|w(\overline{\Lambda}'+\overline{\rho}))}{k+h^\vee}};$$

$$\chi_\Lambda(\tau+1, z, u) = e^{2\pi i m_{\overline{\Lambda}}}\chi_\Lambda(\tau, z, u).$$

b) *The linear span of the $\{\chi_\Lambda\}_{\Lambda\in P_+^k \bmod \mathbb{C}\delta}$ is invariant under the following action of $SL_2(\mathbb{Z})$:*

$$\begin{pmatrix} a & b \\ c & d \end{pmatrix} \cdot f(\tau, z, u) = f(\frac{a\tau+b}{c\tau+d}, \frac{z}{c\tau+d}, u - \frac{c(z|z)}{2(c\tau+d)}). \tag{13.8.4}$$

□

The matrix

$$S_{(k)} := (S_{\Lambda,\Lambda'})_{\Lambda,\Lambda'\in P_+^k \bmod \mathbb{C}\delta}$$

has a number of remarkable properties. First, we obviously have

(13.8.5) $$ {}^tS_{(k)} = S_{(k)}. $$

Furthermore, Corollary 13.5 gives us

(13.8.6) $S_{(k)}$ is a unitary matrix,

and we have from (13.8.5) and (13.8.6):

(13.8.7) $S_{(k)}^{-1} = \overline{S}_{(k)}$ (complex conjugate).

Given $\Lambda \in P_+^k$, we let

$$ {}^t\Lambda = k\Lambda_0 + {}^t\,\overline{\Lambda} + \langle \Lambda, d\rangle\delta, $$

where ${}^t\overline{\Lambda}$ is the highest weight of the $\mathring{\mathfrak{g}}$-module contragredient to $\mathring{L}(\overline{\Lambda})$. Then

(13.8.8) $$ S_{(k)}^2 = (\delta_{\Lambda, {}^tM})_{\Lambda, M \in P_+^k \bmod \mathbb{C}\delta}. $$

This is immediate from $S^2 \cdot f(\tau, z, u) = f(\tau, -z, u)$. Finally, letting for $\lambda \in \overline{P}_+$

$$ \mathring{\chi}_\lambda(e^y) = \operatorname{tr}_{\mathring{L}(\lambda)} e^y, \quad y \in \overline{\mathfrak{h}}, $$

and using the Weyl character formula, we can rewrite the formula for $S_{\Lambda,\Lambda'}$ as follows:

(13.8.9) $$ S_{\Lambda,\Lambda'} = a(\Lambda)\mathring{\chi}_{\overline{\Lambda}'}\left(e^{-2\pi i \frac{\nu^{-1}(\overline{\Lambda}+\overline{\rho})}{k+h^\vee}}\right), $$

where

(13.8.10) $$ a(\Lambda)\ (= S_{\Lambda, k\Lambda_0}) = |M^*/(k+h^\vee)M|^{-\frac{1}{2}} \prod_{\alpha \in \mathring{\Delta}_+} 2\sin\frac{\pi(\overline{\Lambda}+\overline{\rho}|\alpha)}{k+h^\vee}. $$

Remark 13.8. Due to (13.8.2), $a(\Lambda)$ is a positive number since $(\overline{\Lambda}|\alpha) \le k$ if $\alpha \in \mathring{\Delta}_+$.

Incidentally, (13.8.6) gives the following generalization of (13.8.1):

(13.8.11) $$ \sum_{\Lambda \in P_+^k \bmod \mathbb{C}\delta} a(\Lambda)^2 = 1. $$

Example 13.8. If $\mathfrak{g}(A)$ is of type $A_1^{(1)}$, then

$$ S_{(k-j)\Lambda_0 + j\Lambda_1, (k-j')\Lambda_0 + j'\Lambda_1} = \left(\frac{2}{k+2}\right)^{1/2} \sin\frac{\pi(j+1)(j'+1)}{k+2}. $$

§13.9. Unfortunately $P_k \neq P^k$ in the rest of the cases, i.e., when $a_0 = 1$ and $r = 2$ or 3. As a result, Lemma 13.8 and Theorem 13.8 fail in these cases. We shall indicate now how to handle them.

For A of type $X_N^{(r)} = A_{2\ell-1}^{(2)}, D_{\ell+1}^{(2)}, E_6^{(2)}$, and $D_4^{(3)}$ we let A' be of type $X_{N'}^{(r')} = D_{\ell+1}^{(2)}, A_{2\ell-1}^{(2)}, E_6^{(2)}$, and $D_4^{(3)}$, respectively. Let $\overline{\Lambda}'_j = (a_j/a_j^\vee)\overline{\Lambda}_j$ (resp. $= (a_{\ell+1-j}/a_{\ell+1-j}^\vee)\overline{\Lambda}_{\ell+1-j}$) if $A = A_{2\ell-1}^{(2)}$ or $D_{\ell+1}^{(2)}$ (resp. $= E_6^{(2)}$ or $D_4^{(3)}$). Let $\Lambda'_i = \overline{\Lambda}'_i + a'^\vee_i\Lambda_0$ $(i = 1, \ldots, \ell)$, $\Lambda'_0 = \Lambda_0$, $\rho' = \sum_{i=0}^{\ell}\Lambda'_i$, $k' = k$, $P' = \sum_{i=0}^{\ell}\mathbf{Z}\Lambda'_i + \mathbf{C}\delta$, $M' = \mathring{Q}^\vee$, $W' = W$. Then P' is the set of integral weights for $\mathfrak{g}(A')$, etc., and the analogue of (13.7.2) is

$$P_k = P'^k \tag{13.9.1}$$

We denote by Θ'_λ, A'_λ, χ'_λ the corresponding functions for $\mathfrak{g}(A')$. Then the formula (13.5.1) and Proposition 13.7b) can be rewritten as follows:

(13.9.2)
$$\Theta_\lambda(-\frac{1}{\tau}, \frac{z}{\tau}, u - \frac{(z|z)}{2\tau}) = (-i\tau)^{\ell/2}|M^*/kM|^{-\frac{1}{2}} \times \sum_{\mu \in P'^k \bmod kM'} e^{-\frac{2\pi i}{k}(\overline{\lambda}|\overline{\mu})}\Theta'_\lambda(\frac{\tau}{r}, \frac{z}{r}, u),$$

(13.9.3)
$$A_\lambda(-\frac{1}{\tau}, \frac{z}{\tau}, u - \frac{(z|z)}{2\tau}) = (-i\tau)^{\ell/2}|M^*/kM|^{-\frac{1}{2}} \times \sum_{\mu \in P'^k_{++} \bmod \mathbf{C}\delta} \sum_{w \in \mathring{W}} \varepsilon(w) e^{-\frac{2\pi i}{k}(w(\overline{\lambda})|\overline{\mu})} A_\mu(\frac{\tau}{r}, \frac{z}{r}, u).$$

In the same way as Lemma 13.8b), we deduce

$$A_\rho(-\frac{1}{\tau}, \frac{z}{\tau}, u - \frac{(z|z)}{2\tau}) = (-i\tau)^{\ell/2}(-i)^{|\mathring{\Delta}_+|}|M'/M|^{-\frac{1}{2}}A'_{\rho'}(\frac{\tau}{r}, \frac{z}{r}, u), \tag{13.9.4}$$

and the following analogue of (13.8.1):

$$\prod_{\alpha \in \mathring{\Delta}_+^\vee} 2\sin\frac{\pi(\overline{\rho}|\alpha)}{h^\vee} = |M^*/h^\vee M'|^{1/2} \quad \text{if} \quad a_0 r = 2 \text{ or } 3. \tag{13.9.5}$$

From (13.9.3) and (13.9.4) we deduce the following analogue of Theorem 13.8.

THEOREM 13.9. *Let $\mathfrak{g}(A)$ be an affine algebra of type $X_N^{(r)}$ with $r > 1$, $a_0 = 1$, and let $\Lambda \in P_+^k$. Then:*

a) $\chi_\Lambda(-\frac{1}{\tau}, \frac{z}{\tau}, u - \frac{(z|z)}{2\tau}) = \sum_{\Lambda' \in P'^k_+ \bmod \mathbb{C}\delta} S_{\Lambda,\Lambda'} \chi'_{\Lambda'}(\frac{\tau}{r}, \frac{z}{r}, u)$, *where*

$$S_{\Lambda,\Lambda'} = i^{|\mathring{\Delta}_+|} |M^*/(k+h^\vee)M|^{-\frac{1}{2}} |M'/M|^{\frac{1}{2}} \sum_{w \in \mathring{W}} \varepsilon(w) e^{-\frac{2\pi i}{k+h^\vee}(w(\overline{\Lambda}+\overline{\rho})|\overline{\Lambda}'+\overline{\rho}')};$$

$$\chi_\Lambda(\tau+1, z, u) = e^{2\pi i m_{\overline{\Lambda}}} \chi_\Lambda(\tau, z, u).$$

b) *The linear span of the $\{\chi_\Lambda\}_{\Lambda \in P_+^k \bmod \mathbb{C}\delta}$ is invariant with respect to $\Gamma_0(r)$ under the action (13.8.4).*

Proof. Since ${}^tT = ST^{-1}S^{-1}$, using (13.4.2) we see that a) implies b). □

In what follows, in order to state all cases in a uniform fashion, in the cases $r = 1$ or $a_0 = 2$ we let $A' = A$, $\Lambda_i = \Lambda'_i$, $P = P'$, $M = M'$, $\rho = \rho'$, etc., and let $r' = 1$.

§13.10. Now we can prove a transformation law for the string functions for general affine algebras, and branching functions in the nontwisted case. If $\mathfrak{g}$ is semisimple, we let for $\widehat{\mathcal{L}}(\mathfrak{g})$:

$$S_{\Lambda,\Lambda'} = \prod_{j \geq 1} S_{\Lambda_{(j)},\Lambda'_{(j)}}.$$

If $\dot{\mathfrak{g}}$ is a reductive subalgebra of $\mathfrak{g}$ satisfying conditions of Theorem 13.10b) below, we let

$$\dot{S}_{\lambda_{(0)},\lambda'_{(0)}} = |\dot{M}_0^*/\dot{k}_0\dot{M}_0|^{-\frac{1}{2}} e^{-2\pi i(\overline{\lambda}_{(0)}|\overline{\lambda}'_{(0)})^{\cdot}/\dot{k}_0},$$

$$\dot{S}_{\lambda,\lambda'} = \prod_{j \geq 0} \dot{S}_{\lambda_{(j)},\lambda'_{(j)}}.$$

THEOREM 13.10. a) *Let $\mathfrak{g}(A)$ be an affine algebra of rank $\ell + 1$ and let $\overline{\Lambda} \in P_+^k$. Then*

$$c_\lambda^\Lambda(-\frac{1}{\tau}) = |M^*/kM|^{-1/2}(-i\tau)^{-\ell/2} \sum_{\substack{\Lambda' \in P'^k_+ \bmod \mathbb{C}\delta \\ \lambda' \in P'^k \bmod (kM' + \mathbb{C}\delta)}} S_{\Lambda,\Lambda'} e^{2\pi i(\overline{\lambda}|\overline{\lambda}')/k} c_{\lambda'}^{\prime\Lambda'}(\frac{\tau}{r}).$$

b) *Let $\mathfrak{g}$ be a semisimple finite-dimensional Lie algebra with a Cartan subalgebra $\overline{\mathfrak{h}}$ and the coroot lattice $\overline{Q}^\vee \subset \overline{\mathfrak{h}}$, and let $\dot{\mathfrak{g}}$ be a reductive subalgebra of $\mathfrak{g}$ such that its center $\dot{\mathfrak{g}}_{(0)}$ is spanned over $\mathbf{C}$ by the lattice $\dot{M}_0 := \overline{Q}^\vee \cap \dot{\mathfrak{g}}_{(0)}$. Let $\Lambda \in P_+^k$, where $k = (k_1, k_2, \ldots)$, $k_i > 0$. Then:*

$$(13.10.1) \qquad b_\lambda^\Lambda(-\frac{1}{\tau}; \dot{\mathfrak{g}}) = \sum_{\substack{\Lambda' \in P_+^k \bmod \mathbf{C}\delta \\ \lambda' \in \dot{P}_+^{\dot{k}} \bmod (\mathbf{C}\delta + \dot{k}_0 \dot{M}_0)}} S_{\Lambda,\Lambda'} \overline{\dot{S}}_{\lambda,\lambda'} b_{\lambda'}^{\Lambda'}(\tau; \dot{\mathfrak{g}}).$$

Proof. We shall give a proof of b). The proof of a) is the same (in the nontwisted case, a) is a special case of b)).

Consider the column vectors

$$\overrightarrow{\chi}(\tau, z, u) = (\chi_\Lambda)_{\Lambda \in P_+^k \bmod \mathbf{C}\delta}, \qquad \dot{\overrightarrow{\chi}}(\tau, z, u) = (\dot{\chi}_\lambda)_{\lambda \in \dot{P}_+^{\dot{k}} \bmod (\mathbf{C}\delta + \dot{k}_0 \dot{M}_0)},$$

where we let $\dot{\chi}_\lambda = \dot{\chi}_{\lambda^{(1)}}(\dot{\Theta}^0_{\lambda_{(0)}}/\eta(\tau)^{\ell_0})$ (see (12.12.8)), and consider the matrix of branching functions:

$$B(\tau) = (b_\lambda^\Lambda(\tau; \dot{\mathfrak{g}}))_{\substack{\Lambda \in P_+^k \bmod \mathbf{C}\delta \\ \lambda \in \dot{P}_+^{\dot{k}} \bmod (\mathbf{C}\dot{\delta} + \dot{k}_0 \dot{M}_0)}}$$

Then equation (12.12.8) can be written in a matrix form:

$$(13.10.2) \qquad \overrightarrow{\chi}(\tau, z, u) = B(\tau) \dot{\overrightarrow{\chi}}(\tau, z, u), \quad z \in \dot{\mathfrak{h}}.$$

Apply the transformation S to both sides of (13.10.2), using Theorem 13.8a) and (13.5.1):

$$S_{(k)} \overrightarrow{\chi}(\tau, z, u) = B(-\frac{1}{\tau}) \dot{S}_{(\dot{k})} \dot{\overrightarrow{\chi}}(\tau, z, u).$$

Plugging the expression for $\overrightarrow{\chi}$ given by (13.10.2) in this equation, we get:

$$S_{(k)} B(\tau) \dot{\overrightarrow{\chi}}(\tau, z, u) = B(-\frac{1}{\tau}) \dot{S}_{(\dot{k})} \dot{\overrightarrow{\chi}}(\tau, z, u).$$

Since the function $\{\dot{\chi}_\lambda\}_{\lambda \in \dot{P}_+^{\dot{k}} \bmod (\mathbf{C}\delta + \dot{k}_0 \dot{M}_0)}$ are linearly independent (see Propositions 13.7a) and 13.3), we deduce: $S_{(k)} B(\tau) = B(-\frac{1}{\tau}) \dot{S}_{(\dot{k})}$, which, due to (13.8.5) is equivalent to

$$(13.10.3) \qquad B(-\frac{1}{\tau}) = S_{(k)} B(\tau) \overline{\dot{S}}_{(\dot{k})},$$

proving b). □

Remark 13.10. Conditions on $\mathfrak{g} \supset \dot{\mathfrak{g}}$ in Theorem 13.10b), simply mean that $\dot{\mathfrak{g}}$ is a Lie algebra of a closed reductive subgroup of the semisimple Lie group corresponding to $\mathfrak{g}$.

Since obviously

$$c_\lambda^\Lambda(\tau+1) = e^{2\pi i m_{\Lambda,\lambda}} c_\lambda^\Lambda(\tau); \tag{13.10.4}$$
$$b_\lambda^\Lambda(\tau+1;\dot{\mathfrak{g}}) = e^{2\pi i (m_\Lambda - \dot{m}_\lambda)} b_\lambda^\Lambda(\tau;\dot{\mathfrak{g}}),$$

Theorem 13.10 implies

COROLLARY 13.10. a) *If $\mathfrak{g}(A)$ is an affine algebra and $k \in \mathbb{Z}_+$, then the linear span of all the string functions c_λ^Λ with $\Lambda \in P_+^k$ is invariant under the action of $\Gamma_0(r')$ under the (projective) action*

$$\begin{pmatrix} a & b \\ c & d \end{pmatrix} \cdot f(\tau) = (c\tau + d)^{\ell/2} f(\frac{a\tau+b}{c\tau+d}). \tag{13.10.5}$$

b) *Under the assumption of Theorem 13.10b), given $k \in \mathbb{Z}_+$, the linear span of all the branching functions $b_\lambda^\Lambda(\dot{\mathfrak{g}})$ with $\Lambda \in P_+^k \mod \mathbb{C}\delta$ is $SL_2(\mathbb{Z})$-invariant under the action*

$$\begin{pmatrix} a & b \\ c & d \end{pmatrix} \cdot f(\tau) = f(\frac{a\tau+b}{c\tau+d}). \tag{13.10.6}$$

Proof. a) for $r' = 1$ and b) follow from Theorem 13.10, (13.10.4), and (13.4.1). a) for $r' > 1$ is proved by the same argument as that used in the proof of Theorem 13.9. □

§13.11. In this section $\mathfrak{g}(A)$ is an affine algebra of type $X_N^{(r)}$ and rank $\ell+1$. We give here the proof of the "very strange" formula. Given $\alpha, \beta \in \mathring{\mathfrak{h}}_{\mathbb{R}}$, we define the associated specialization $f^{\alpha,\beta}$ of A_ρ as follows. Put:

$$\Delta^{\alpha,\beta} = \{\gamma \in \Delta \mid (\gamma|\Lambda_0 + \beta) = 0 \text{ and } (\gamma|\alpha) \in \mathbb{Z}\}.$$

It is easy to see that $\Delta^{\alpha,\beta}$ is a finite reduced root system and that

$$\Delta_+^{\alpha,\beta} := \{\gamma \in \Delta^{\alpha\beta} \mid \bar{\gamma} \in \mathring{\Delta}_+ \cup \tfrac{1}{2}\mathring{\Delta}_+\}$$

is a set of positive roots. Recalling the definition (13.6.1), we put

$$f^{\alpha,\beta}(\tau) := \Big(A_\rho \prod_{\gamma \in \Delta_+^{\alpha,\beta}} (1 - e^{-\gamma})^{-1}\Big)^{\alpha,\beta}(\tau).$$

LEMMA 13.11.1. *Let* $\alpha, \beta \in \sum_{j=1}^{\ell} \mathbf{Q}\alpha_j$. *Then* $f^{\alpha,\beta}(\tau)$ *is a modular form of weight* $\frac{1}{2}(\ell + |\Delta_+^{\alpha,\beta}|)$ *for* $\Gamma(n)$, *some* n.

Proof. This is similar to that of Proposition 13.6, using the formula

$$f^{\alpha,\beta}(\tau) = \lim_{\substack{\alpha' \to \alpha \\ \beta' \to \beta}} \frac{A_\rho(\tau, -\alpha + \tau\beta, \frac{1}{2}(\beta|\alpha - \tau\beta))}{\prod_{\gamma \in \Delta_+^{\alpha,\beta}} 2\pi i(-\gamma|(\alpha - \alpha') + \tau(\beta - \beta'))}, \tag{13.11.1}$$

where α', β' are such that $\Delta^{\alpha',\beta'} = \emptyset$.

□

We need one more fact.

LEMMA 13.11.2. *Let* $b_1, b_2, \ldots$ *be a periodic sequence of integers with period* m, *such that* $b_j = b_{m-j}$ *for* $j = 1, \ldots, m-1$. *Set* $b = \sum_{j=1}^{m} b_j$. *For* $c \in \mathbf{C}$ *put*

$$f_c(\tau) = q^c \prod_{j=1}^{\infty} (1 - q^j)^{b_j}.$$

Then $f_c(\tau)$ *is a modular form (for* $\Gamma(n)$, *some* n*) if and only if:*

$$c = \frac{bm}{24} - \frac{1}{4m} \sum_{j=1}^{m-1} j(m-j)b_j. \tag{13.11.2}$$

Proof. Recall that the functions $g_{n,r}(\tau)$ given by (13.6.9) are modular forms. It is easy to see that if c is given by formula (13.11.2), then $f_c(\tau)$ can be represented as a finite product of functions of the form $g_{m,r}(\tau)$ $(1 \le r \le m-1)$ and a power of $\eta(m\tau)$; hence, $f_c(\tau)$ is a modular form. Conversely, if $f_{c'}(\tau)$ is a modular form, then $q^{c-c'}$ is also a modular form; it follows that $c = c'$.

□

Now we are in a position to prove the "very strange" formula (12.3.7).

Proof of (12.3.7). Note that f^{0,γ_s} is nothing other than the specialization of type s of $A_\rho \prod_{\substack{\gamma \in \Delta_+ \\ (\gamma|\Lambda_0 + \gamma_s) = 0}} (1 - e^{-\gamma})^{-1}$. Hence setting $e^{2\pi i\tau} = q_1^{m/r}$, we obtain (using (12.3.1)):

$$f^{0,\gamma_s}(\tau) = q_1^{\frac{m}{2rh^\vee}|\bar{\rho} - h^\vee \gamma_s|^2} \prod_{j \ge 1} (1 - q_1^j)^{d_j(s;r)}. \tag{13.11.3}$$

But f^{0,γ_*} is a modular form by Lemma 13.11.1. Using (13.11.3), we apply Lemma 13.11.2 to complete the proof.

□

We have the following important corollary of the "very strange" formula:

COROLLARY 13.11. a) *For an arbitrary affine algebra of type* $X_N^{(r)}$ *one has:*

$$\frac{|\rho|^2}{2h^\vee} = \frac{\dim \mathfrak{g}(X'_{N'})}{24a_0} \tag{13.11.4}$$

b) *If* $r > 1$ *and* $a_0 = 1$, *then*

$$\frac{|\rho^\vee|^2}{2h^\vee} = \frac{|\rho'|^2}{2h^\vee} = \frac{\dim \mathfrak{g}(X_N)}{24r}. \tag{13.11.5}$$

Proof. The classical "strange" formula (13.11.4) for $r = 1$ is a special case of the "very strange" formula for $s = 0$, since $|\overline{\rho}|^2 = |\rho|^2$. The formula (13.11.5) for $|\rho^\vee|^2/2h^\vee$ follows from this since ρ for A is the same as $\rho^\vee$ for ${}^tA$ and $h^\vee$ for $X_N^{(r)}$ is independent of r. The remaining formulae are checked, case after case, by making use of the "very strange" formula for $s = 0$.

□

The following theorem also may be deduced from the "very strange" formula, but we shall give another proof, which is simpler. In order to state the theorem, we need to introduce the following subset of the set $\{0, 1, \ldots, \ell\}$:

$$J = \{j \mid a_j \text{ (resp. } a_j^\vee) = 1 \quad \text{if } r = 1 \text{ (resp. } r > 1)\} \\ (= (\operatorname{Aut} S(A)) \cdot 0).$$

THEOREM 13.11. *Let* $\Lambda \in P_+^k$. *Then*

$$2k(\Lambda|\rho) \geq h^\vee(\Lambda|\Lambda), \tag{13.11.6}$$

and the equality holds if and only if $\Lambda = k\Lambda_j \mod \mathbb{C}\delta$ *with* $j \in J$.

Proof. Inequality (13.11.6) is equivalent to

$$|\frac{\rho}{h^\vee} - \frac{\Lambda}{k}|^2 \leq |\frac{\rho}{h^\vee}|^2,$$

which follows from

$$(13.11.7)\qquad |\frac{\overline{\rho}}{h^\vee} - z|^2 \le |\frac{\overline{\rho}}{h^\vee}|^2 \quad \text{if } z \in C_{\rm af}.$$

To prove this inequality note that

$$0 \le (\alpha|z) \le 1 \quad \text{if } z \in C_{\rm af} \text{ for all } \alpha \in \mathring{\Delta}_+.$$

Hence, in the case $r = 1$ we have, using Corollary 8.7:

$$0 \ge \sum_{\alpha \in \mathring{\Delta}_+} ((\alpha|z)^2 - (\alpha|z)) = h^\vee(z|z) - 2(\overline{\rho}|z),$$

which is an equivalent form of (13.11.7). We have an equality if and only if $(\alpha|z) = 0$ or 1 for all $\alpha \in \mathring{\Delta}_+$, which means $z = \overline{\Lambda}_s$ with $a_s = 1$, proving the theorem in the case $r = 1$.

In the case $a_0 = 1$ and $r > 1$ we have, using Corollary 8.7 (note that $\phi(z,z) = 2r \sum_{\alpha \in \mathring{\Delta}_{+s}} (\alpha|z)^2 + 2 \sum_{\alpha \in \mathring{\Delta}_{+\ell}} (\alpha|z)^2$):

$$0 \ge r \sum_{\alpha \in \mathring{\Delta}_{+s}} ((\alpha|z)^2 - (\alpha|z)) + \sum_{\alpha \in \mathring{\Delta}_{+\ell}} ((\alpha|z)^2 - r(\alpha|z))$$
$$= rh^\vee(z|z) - 2r(\overline{\rho}|z),$$

which is (13.11.7). We have an equality if and only if $\langle z, \alpha^\vee \rangle = 0$ or 1 for all $\alpha \in \mathring{\Delta}_+$, which means $z = \overline{\Lambda}_s$ with $a_s^\vee = 1$, proving the theorem in this case as well. In the remaining case, $A_{2\ell}^{(2)}$, one checks directly that (13.11.7) holds with equality only for $z = 0$ (or uses a similar argument with $\mathring{\Pi}$ replaced by $\Pi \backslash \{\alpha_\ell\}$).

□

PROPOSITION 13.11. *Let $\Lambda \in P_+^k$, $k > 0$, and let $\lambda \in P(\Lambda)$. Then*

$$(13.11.8)\qquad m_{\Lambda,\lambda} \ge -\frac{|\rho|^2}{2h^\vee}\frac{k}{k+h^\vee}$$

with equality if and only if $\Lambda = k\Lambda_i \mod \mathbb{C}\delta$ and $a_i = 1$ (resp. $a_i^\vee = 1$) in the case $r = 1$ (resp. $r > 1$), and $\lambda = w(\Lambda)$, $w \in W$.

Proof. By definition (12.7.6) of $m_{\Lambda,\lambda}$, (13.11.8) means

$$\frac{|\Lambda + \rho|^2 - |\rho|^2}{k + h^\vee} \ge \frac{|\lambda|^2}{k}.$$

By Proposition 11.4, it suffices to prove this for $\lambda = \Lambda$, in which case this inequality is equivalent to (13.11.6). The cases of equality follow from Proposition 11.4 and Theorem 13.11.

□

Note that the inequality (13.11.8) in the nontwisted case is a special case of Proposition 12.12b).

§13.12. We shall now find levels and upper bounds of orders of poles at all cusps for the string and branching functions. For this we need the following

LEMMA 13.12. a) *Let V be a space of modular forms of weight k which is invariant with respect to the (projective) action (13.10.5) with $\ell = -2k$. Suppose that all functions from V have poles of order $\leq R$ at $i\infty$. Then orders of poles of these functions at all cusps are $\leq R$ (i.e., all functions from V are R-singular).*
b) *Let $f(\tau)$ be a modular form of integral (resp. half-integral) weight for some principal congruence subgroup and some multiplier system. Suppose that $f(\tau + n) = f(\tau)$ for some $n \in \mathbb{Z}$, $n > 0$. Then $f(\tau)$ is a modular form for $\Gamma(n)$ with a trivial multiplier system (resp. with a multiplier system with values in ± 1).*

Proof. a) is clear since $SL_2(\mathbb{Z})$ acts transitively on the set of cusps. For b) see, e.g., Schoenberg [1973].

THEOREM 13.12. a) *Let $\mathfrak{g}(A)$ be an affine algebra of type $X_N^{(r)}$ and rank $\ell + 1$ and let $\Lambda \in P_+^k$, $k > 0$. Let*

$$s = \operatorname{lcm}\{k, h^\vee, k + h^\vee, n\}.$$

Then all string functions $c_\lambda^\Lambda(\tau)$ are modular forms of weight $-\frac{1}{2}\ell$ and a multiplier system with values in $(\pm 1)^\ell$ for the group $\Gamma(s)$. Furthermore, all of them are R-singular, where

$$R = \frac{\dim \mathfrak{g}(X'_{N'})}{24 a_0} \, \frac{k}{k + h^\vee}.$$

b) *Under the assumptions and notation of Theorem 13.10, let*

$$s = \operatorname{lcm}\{k_i + h_i^\vee, n_i, \dot{k}_i + \dot{h}_i^\vee \ (i \geq 1),\ \dot{k}_0, \dot{n}_0 := n(\dot{M}_0), \dot{n}_i \ (i \geq 1), 24\}.$$

Then all branching functions $b_\lambda^\Lambda(\tau;\dot{\mathfrak{g}})$ are modular forms of weight 0 with a trivial multiplier system for the group $\Gamma(s)$. Furthermore, all of them are $(c(k)-\dot{c}(\dot{k}))/24$-singular.

Proof. To prove a), we use identity (12.7.13). Due to Lemma 13.7, this identity implies that the function $F := \sum_{\lambda \in P_+^k \bmod (kM+\mathbb{C}\delta)} c_\lambda^\Lambda(\tau)\Theta_\lambda$ satisfies the transformation property $F(A\cdot(\tau,z,u)) = C(A)F(\tau,z,u)$, where $C(A) \in \mathbb{C}$ and $|C(A)| = 1$, for $A \in \Gamma(nh^\vee) \cap \Gamma(n(k+h^\vee))$. Since the summation in F is taken over a subset of $P_k \bmod (kM+\mathbb{C}\delta)$ (see 13.7.1), applying Proposition 13.3 and again Lemma 13.7. we see that c_λ^Λ are modular forms of weight $-\frac{1}{2}\ell$ for the group $\Gamma(nh^\vee) \cap \Gamma(n(k+h^\vee)) \cap \Gamma(nk)$, and apply Lemma 13.12a).

To complete the proof of a), note that $m_{\Lambda,\lambda} \geq -R$, by Corollary 13.11 and Proposition 13.11, hence the order of a pole at $i\infty$ for all c_λ^Λ is $\leq R$. If $r = 1$ or $a_0 = 2$, we derive that all c_λ^Λ are R-singular, due to Corollary 13.10a) and Lemma 13.12b). In the rest of the cases the proof is similar using Theorem 13.10a), Corollary 13.11 b), and the fact that $\Gamma_0(2)$ and $\Gamma_0(3)$ have only two cusps, $i\infty$ and 0.

The same argument as above, using (12.12.8) in place of (12.7.13), proves that $b_\lambda^\Lambda(\tau;\dot{\mathfrak{g}})$ are modular forms of weight 0 for $\Gamma(s)$ with a trivial multiplier system. All of them are $(c(k)-\dot{c}(\dot{k}))/24$-singular, since due to (12.8.12), Proposition 12.12b) implies that

$$m_\Lambda - \dot{m}_\lambda \geq -\frac{1}{24}(c(k)-\dot{c}(\dot{k})).$$

□

Note that Theorem 13.12 tells us that there exists an effective integer b, depending only on $\mathfrak{g}$, $\dot{\mathfrak{g}}$, and k, such that the first b coefficients of a branching function determine the whole function.

§13.13. In this section we study the asymptotic behavior of characters as $\beta := -i\tau \downarrow 0$, of an affine algebra of type $X_N^{(r)}$ and rank $\ell+1$. We shall write $f(\beta) \sim g(\beta)$ if $\lim_{\beta\downarrow 0} f(\beta)/g(\beta) = 1$.

First of all, we have from the definitions:

$$\Theta_\lambda(-(i\beta)^{-1}, z, -i\beta(z|z)/2) \sim 1 \text{ or } 0 \tag{13.13.1}$$

depending on whether $\lambda \in \Lambda_0 + kM + \mathbb{C}\delta$ or not, where $k =$ level(λ);

$$\eta(-1/i\beta)^{-1} \sim e^{\pi/12\beta}. \tag{13.13.2}$$

Now, $S^{-1}(i\beta, -i\beta z, 0) = (-(i\beta)^{-1}, z, -i\beta(z|z)/2)$, hence, applying $(S, \tau^{1/2})$ to both sides of (13.5.1), we obtain:

(13.13.3)

$$\Theta_\lambda(i\beta, -i\beta z, 0) = |M^*/kM|^{-1/2}\beta^{-\ell/2} \times \sum_{\mu \in P_k \bmod (kM+\mathbb{C}\delta)} e^{\frac{-2\pi i}{k}(\overline{\lambda}/\overline{\mu})}\Theta_\mu(-(i\beta)^{-1}, z, -i\beta(z|z)/2).$$

Hence, by (13.13.1) we have:

$$\Theta_\lambda(i\beta, -i\beta z, 0) \sim |M^*/kM|^{-1/2}\beta^{-\ell/2}. \tag{13.13.4}$$

Similarly, from (13.6.11) and (13.13.2) we deduce:

$$\eta(i\beta)^{-1} \sim \beta^{1/2}e^{\pi/12\beta}. \tag{13.13.5}$$

Let $R_+ = \mathring{\Delta}_+$ if $r = 1$ or $a_0 = 2$, and $R_+ = \mathring{\Delta}_+^\vee$ otherwise. The asymptotics of the A_λ is given by the following

PROPOSITION 13.13. *Let $\lambda \in P_{k++}$. Then*

$$A_\lambda(i\beta, -i\beta z, 0) \sim b(\lambda, z)\beta^{-\ell/2}e^{-\frac{\pi h^\vee \dim \mathfrak{g}(X_N)}{12\beta k r}}, \tag{13.13.6}$$

where

(13.13.7)

$$b(\lambda, z) = |M^*/kM|^{-1/2}b(\rho, z)\prod_{\alpha \in R_+} 2\sin\pi\frac{(\overline{\lambda}|\alpha)}{k},$$

(13.13.8)

$$b(\rho, z) = |M'/M|^{-1/2}\prod_{\alpha \in R_+} 2\sin\pi(z|\alpha).$$

Proof. We have from (13.13.3):

$$A_\lambda(i\beta, -i\beta z, 0) = |M^*/kM|^{-1/2}\beta^{-\ell/2} \times \sum_{w \in \mathring{W}} \varepsilon(w) \sum_{\substack{\mu \in P_k \bmod (kM+\mathbb{C}\delta) \\ \mu \text{ regular}}} e^{-2\pi i(w(\overline{\lambda})|\overline{\mu})/k}\Theta_\mu(-(i\beta)^{-1}, z, -i\beta(z|z)/2).$$

We may take μ to be regular (i.e., such that $\mathring{W}_\mu = 1$), since nonregular μ give a zero contribution (cf. (13.7.7)). Plugging in the explicit expression (13.2.5) for Θ_λ, we get:

$$A_\lambda(i\beta, -i\beta z, 0) = |M^*/kM|^{-1/2}\beta^{-\ell/2}e^{\pi k\beta(z|z)} \tag{13.13.9}$$
$$\times \sum_{w \in \mathring{W}} \varepsilon(w) \sum_{\substack{\mu \in P_k \bmod (kM+\mathbb{C}\delta) \\ \mu \text{ regular}}} \sum_{\gamma \in M+\frac{\overline{\mu}}{k}} e^{-\frac{2\pi i}{k}(w(\overline{\lambda})|\overline{\mu}) - \frac{\pi k|\gamma|^2}{\beta} + 2\pi i k(\gamma|z)}.$$

In order to complete the proof we need the following:

LEMMA 13.13. *Let $\mu \in M^k$ be regular, and let $\xi = \rho$ if $r = 1$ or $a_0 = 2$, and $\xi = \rho^\vee$ otherwise. Then we have:*

$$|\mu|^2 \geq |\overline{\xi}|^2 \text{ with equality if and only if } \mu = \sigma(\overline{\rho}), \sigma \in \mathring{W}.$$

Proof. We consider the case when $r = 1$ or $a_0 = 2$; then $\overline{\xi} = \overline{\rho}$. (In the rest of the cases $\xi = \rho^\vee$ and the proof is similar.) In this case $M = \mathring{Q}^\vee$ (see (6.5.8)), hence $M^* = \sum_{i=1}^{\ell} \mathbf{Z}\overline{\Lambda}_i$. Let $\sigma \in \mathring{W}$ be such that $\sigma^{-1}(\mu) \in \overline{P}_{++}$. Then $\sigma^{-1}(\mu) = \sum_{i=1}^{\ell} b_i\overline{\Lambda}_i = \sum_{i=1}^{\ell} b_i'\alpha_i^\vee$, where $b_i, b_i' > 0$, $b_i \in \mathbf{Z}$. Hence we have:

$$\begin{aligned} |\mu|^2 - |\overline{\rho}|^2 &= |\sigma^{-1}(\mu)|^2 - |\overline{\rho}|^2 = (\sigma^{-1}(\mu) - \overline{\rho}|\sigma^{-1}(\mu) + \overline{\rho}) \\ &= \sum_{i=1}^{\ell}(b_i - 1)(\overline{\Lambda}_i|\sigma^{-1}(\mu) + \overline{\rho}). \end{aligned}$$

This completes the proof since $\sigma^{-1}(\mu) + \overline{\rho}$ is a linear combination of the $\alpha_i^\vee$ $(i = 1, \ldots, \ell)$ with positive coefficients.

□

End of the proof of Proposition 13.13. By Lemma 13.13, it is clear from (13.13.9) that the asymptotic behavior of $A_\lambda(i\beta, -i\beta z, 0)$ as $\beta \downarrow 0$ is determined by the terms with $\gamma = \sigma(\overline{\xi})/k$, $\overline{\mu} = \sigma(\overline{\xi})$, $\sigma \in \mathring{W}$. Thus, we have as $\beta \downarrow 0$:

$$\begin{aligned} A_\lambda(i\beta, -i\beta z, 0) &\sim |M^*/kM|^{-1/2}\beta^{-\ell/2}e^{-\pi|\overline{\xi}|^2/k\beta} \\ &\quad \times \sum_{w \in \mathring{W}} \varepsilon(w) \sum_{\sigma \in \mathring{W}} e^{-\frac{2\pi i}{k}(\sigma w(\lambda)|\overline{\xi})}e^{-2\pi i(\sigma(\overline{\xi})|z)} \\ &= |M^*/kM|^{-1/2}\beta^{-e/2}e^{-\pi|\overline{\xi}|^2/k\beta} \\ &\quad \times \left(\sum_{w \in \mathring{W}} \varepsilon(w)e^{-\frac{2\pi i}{k}(w(\overline{\xi})|\lambda)}\right)\left(\sum_{\sigma \in \mathring{W}} \varepsilon(w)e^{2\pi i(\sigma(\overline{\xi})|z)}\right). \end{aligned}$$

Hence, from the Weyl denominator identity and the "strange formulas" (13.11.4) and (13.11.5) we obtain asymptotics (13.13.6). Formula (13.13.8) is equivalent to identity (13.8.1) in the cases $r = 1$ or $a_0 = 2$. In the remaining cases it is equivalent to the identity (13.9.5).

□

An immediate corollary of Proposition 13.13 is the asymptotics of characters. For $\Lambda \in P_+^k$ one has:

$$(13.13.10) \qquad \chi_\Lambda(i\beta, -i\beta z, 0) \sim a(\Lambda) e^{\pi c(k)/12r\beta},$$

where (cf. (13.8.10))

$$(13.13.11) \quad a(\Lambda) = |M^*/(k+h^\vee)M'|^{-1/2} \prod_{\alpha \in R_+} 2 \sin \pi \frac{(\overline{\Lambda} + \overline{\rho}|\alpha)}{k+h^\vee},$$

$$(13.13.12) \quad c(k) = \frac{k \dim \mathfrak{g}(X_N)}{k+h^\vee}.$$

Remark 13.13. Let $z \in \mathring{\mathfrak{h}}_{\mathbb{R}}$ and consider the "Hamiltonian" $H = -d - z$. Then

$$(13.13.13) \qquad \mathrm{ch}_{L(\Lambda)}(i\beta, -i\beta z, 0) = \mathrm{tr}_{L(\Lambda)}\, e^{-2\pi\beta H},$$

hence (13.13.10) has the following interpretation:

$$(13.13.14) \qquad \mathrm{tr}_{L(\Lambda)}\, e^{-2\beta \mathrm{H}} \sim a(\Lambda) e^{\pi c(k)/12r\beta} \quad \text{as} \quad \beta \downarrow 0.$$

In analogy with statistical mechanics, $\beta = \frac{1}{T}$, where T is the temperature. Thus (13.13.13) is the "partition function" and (13.13.14) is its "high temperature limit." It is remarkable that this limit is independent of z.

§13.14. We are now in a position to study the asymptotic behavior of string and branching functions.

Consider the set R_k of vacuum pairs of level k (see §12.11). If $(\Lambda; \lambda) \in R_k$, then clearly $(\Lambda + a\delta; \lambda + a\dot{\delta} + \dot{k}_0\gamma) \in R_k$ for any $a \in \mathbb{C}$ and $\gamma \in \dot{M}_0$. Such pairs are called equivalent. Denote the set of all equivalence classes by $\widetilde{R}_k$. This is a finite set. Here and further, in the case of branching functions we keep the assumptions and notation of Theorem 13.10b); in the case of string functions we have: $\dot{k}_0 = k$ and $\dot{M}_0 = M$.

In general, the set $\widetilde{R}_k$ is quite complicated, but in the case of string functions the answer is very simple. Namely, recalling the definition of J in §13.11, due to Proposition 13.11 and (12.4.5) we have:

(13.14.1) $\quad \widetilde{R}_k = \{(k\Lambda_j; k\Lambda_j) \mid j \in J\}$ in the case of string functions.

THEOREM 13.14. a) *Under the assumptions of Theorem* 13.10a) *one has:*

$$c_\lambda^\Lambda(i\beta) \sim |J|\, |\overline{P}/M|^{-1} |M^*/M|^{1/2} a(\Lambda) (\beta/k)^{\ell/2} e^{\pi c(k)/12r\beta}.$$

b) *Under the assumptions of Theorem* 13.10b), *let*

$$d(\Lambda,\lambda) = \sum_{(\Lambda';\lambda')\in\widetilde{R}_k} S_{\Lambda,\Lambda'}\overline{\dot{S}}_{\lambda,\lambda'}\,\mathrm{mult}_{\Lambda'}(\lambda';\dot{\mathfrak{g}}).$$

Then one of the two possibilities (i) *or* (ii) *holds:*
(i) $d(\Lambda,\lambda)=0$ *and* $\lim\limits_{\beta\downarrow 0} b_\lambda^\Lambda(i\beta;\dot{\mathfrak{g}})/e^{\pi(c(k)-\dot{c}(k))/12\beta}=0$,
(ii) $d(\Lambda,\lambda)$ *is a positive real number and*

$$b_\lambda^\Lambda(i\beta;\dot{\mathfrak{g}}) \sim d(\Lambda,\lambda)e^{\pi(c(k)-\dot{c}(k))/12\beta}.$$

Proof. We first prove b). By the definition of the branching function in §12.12 and since it is a convergent series for $\operatorname{Im}\tau>0$, we have:

$$b_{\lambda'}^{\Lambda'}\left(-\frac{1}{i\beta};\dot{\mathfrak{g}}\right) \sim e^{-2\pi(m_{\Lambda'}-\dot{m}_{\lambda'})/\beta}\,\mathrm{mult}_{\Lambda'}(\lambda';\dot{\mathfrak{g}}) \qquad (13.14.2)$$

if $\mathrm{mult}_{\Lambda'}(\lambda';\dot{\mathfrak{g}})\neq 0$, but $\mathrm{mult}_{\Lambda'}(\lambda'+n\dot{\delta};\mathfrak{g})=0$ for $n>0$. Replacing τ by $-\frac{1}{\tau}$ in (13.10.1), and using Proposition 12.12b), we see that the leading term of $b_\lambda^\Lambda(\tau;\dot{\mathfrak{g}})$ as $\tau\to 0$, is given by the summands corresponding to the pairs $(\Lambda';\lambda')$ from $\widetilde{R}_k$. Due to (12.8.12) we thus obtain:

$$\lim_{\beta\downarrow 0} b_\lambda^\Lambda(i\beta;\dot{\mathfrak{g}})/e^{\pi(c(k)-\dot{c}(k))/12\beta} = d(\Lambda,\lambda).$$

Since the left-hand side is a positive real number as soon as β is a positive real number, this completes the proof of b).

We could prove a) in a similar way by explicitly evaluating $d(\Lambda,\lambda)$. We shall give, however, another, more illuminating proof. Note that if $\lambda\in P(\Lambda)$, then, by Proposition 12.5, there exists $p\in\mathbb{Z}_+$ such that $\Lambda-p\delta\in[W(\lambda)]$, where [ ] stands for convex hull. Then $\Lambda-(n+p)\delta\in[W(\lambda-n\delta)]$ for all $n\in\mathbb{Z}_+$, and since also $\lambda-n\delta\in[W(\Lambda-n\delta)]$ for all $n\in\mathbb{Z}_+$, we obtain by Proposition 12.5:

$$\mathrm{mult}_\Lambda(\Lambda-n\delta)\le\mathrm{mult}_\Lambda(\lambda-n\delta)\le\mathrm{mult}_\Lambda(\Lambda-(n+p)\delta)$$

for all $n\in\mathbb{Z}_+$. This implies for $\beta>0$:

$$c_\lambda^\Lambda(i\beta)\sim c_\Lambda^\Lambda(i\beta)\quad \text{if } \lambda\in P(\Lambda). \qquad (13.14.3)$$

Considering the theta function identity (12.7.13) asymptotically as $\beta\downarrow 0$ we obtain, using (13.14.3), (13.13.4), and (13.13.10):

$$a(\Lambda)e^{\pi c(k)/12r\beta}\sim b_\Lambda|M^*/kM|^{-1/2}\beta^{-\ell/2}c_\lambda^\Lambda(i\beta), \qquad (13.14.4)$$

where $b_\Lambda = \#\{\lambda \in P_1^k \mod (kM + \mathbb{C}\delta) \mid \lambda \equiv \Lambda \mod Q\}$. The proof of a) now follows from (13.14.4) and the following simple formula:

$$b_\Lambda = a_0 |\mathring{Q}^{\vee *}/kM| / |\mathring{Q}^{\vee *}/\mathring{Q}|. \tag{13.14.5}$$

□

It is clear that

$$d(k\Lambda_0, \dot{k}\dot{\Lambda}_0) = \sum_{(\Lambda',\lambda')\in\widetilde{R}_k} a(\Lambda')\dot{a}(\lambda') \operatorname{mult}_{\Lambda'}(\lambda'; \dot{\mathfrak{g}}) > 0$$

since $(k\Lambda_0, \dot{k}\dot{\Lambda}_0) \in \widetilde{R}_k$. On the other hand no example when $\dot{L}(\lambda) \subset L(\Lambda)$, but $d(\Lambda, \lambda) = 0$, is known. It is natural to conjecture that no such example exists. In fact, I would like to propose an even stronger conjecture:

CONJECTURE 13.14. *If* $(\Lambda'; \lambda') \in R_k$ *and* $\operatorname{mult}_\Lambda(\lambda; \dot{\mathfrak{g}}) \neq 0$, *then* $S_{\Lambda,\Lambda'}\overline{\dot{S}}_{\lambda,\lambda'} \geq 0$.

Let $(\Lambda; \lambda)$ be a vacuum pair; then we have by (13.14.2):

$$b_\lambda^\Lambda(-\frac{1}{i\beta}; \dot{\mathfrak{g}}) \sim e^{\pi(c(k)-\dot{c}(\dot{k}))/12\beta} \operatorname{mult}_\Lambda(\lambda; \dot{\mathfrak{g}}).$$

We deduce from this and Theorem 13.14 as $\beta \downarrow 0$, the following result.

PROPOSITION 13.14. *The vector* $(\operatorname{mult}_\Lambda(\lambda; \dot{\mathfrak{g}}))_{(\Lambda;\lambda)\in\widetilde{R}_k}$ *is an eigenvector with eigenvalue* 1 *of the matrix* $(S_{\Lambda,\Lambda'}\overline{\dot{S}}_{\lambda,\lambda'})_{(\Lambda;\lambda),(\Lambda';\lambda')\in\widetilde{R}_k}$.

□

Note that if Conjecture 13.14 is true, then, by the Frobenius–Perron theory Proposition 13.14 determines uniquely the vector, since $\operatorname{mult}_{k\Lambda_0}(\dot{k}\dot{\Lambda}_0; \dot{\mathfrak{g}}) = 1$.

We conclude this chapter with a very special case of the developed theory, which is important to the string theory. Let $\mathfrak{g} \supset \dot{\mathfrak{g}}$ be as in Theorem 13.10b). One checks directly that $c(k) - \dot{c}(\dot{k})$ is a strictly increasing function of k when $k \in \mathbb{N}$. Hence, due to Proposition 12.12c), the equality $c(k) = \dot{c}(\dot{k})$ is possible only if $k = 1$. In this case $\dot{\mathfrak{g}}$ is called a *conformal subalgebra* of $\mathfrak{g}$. It follows from (13.13.10) that a $\widehat{\mathcal{L}}(\mathfrak{g})$-module $L(\Lambda)$, $\Lambda \in P_+^k$, $k > 0$, viewed as a $\widehat{\mathcal{L}}(\dot{\mathfrak{g}})$-module, decomposes into a finite sum:

$$L(\Lambda) = \bigoplus_{\lambda\in\dot{P}_+^{\dot{k}}} \operatorname{mult}_\Lambda(\lambda; \dot{\mathfrak{g}}) \dot{L}(\lambda) \tag{13.14.6}$$

if and only if $c(k) = \dot{c}(\dot{k})$ and $\dot{\mathfrak{g}}$ is semisimple (then necessarily $k = 1$). Looking asymptotically at characters as $\beta \downarrow 0$ in (13.14.6), we deduce from (13.13.10):

$$a(\Lambda) = \sum_{\lambda \in \dot{P}_+^{\dot{k}}} \dot{a}(\lambda) \operatorname{mult}_\Lambda(\lambda; \dot{\mathfrak{g}}). \tag{13.14.7}$$

Thus, the (generally irrational) number $a(\Lambda)$ plays a role of a dimension. It is called the *asymptotic dimension* of $L(\Lambda)$.

Finally, notice that, due to Proposition 11.12b), the coset Vir-module is trivial if $c(k) = \dot{c}(\dot{k})$, hence

$$\operatorname{mult}_\Lambda(\lambda; \dot{\mathfrak{g}}) \neq 0 \Rightarrow (\Lambda; \lambda) \in R_k, \quad \text{if } c(k) = \dot{c}(\dot{k}), \tag{13.14.8}$$

and we may use Proposition 13.14 to compute these multiplicities.

§13.15. Exercises.

13.1. Show that $\theta_{(\ell)}(\tau) = \sum_{n \in \mathbb{Z}^\ell} e^{\pi i n^2 \tau}$ (where $n^2 = \sum n_i^2$) is a holomorphic modular form of weight $\frac{1}{2}\ell$ for Γ_θ and a multiplier system v_ℓ such that $v_\ell(T^2) = 1$, $v_\ell(S) = e^{-\pi i \ell/4}$. Using Gauss identity (Exercise 12.4) show that $\theta_{(1)}(\tau) = \frac{\eta^2((\tau+1)/2)}{\eta(\tau+1)}$.

13.2. Let

$$f_1(\tau) = q^{-1/48} \prod_{j \in \mathbb{Z}_+} (1 + q^{j+1/2}) = e^{-\pi i/24} \frac{\eta(\frac{1}{2}(\tau+1))}{\eta(\tau)} = \frac{\eta(\tau)^2}{\eta(\tau/2)\eta(2\tau)},$$

$$f_2(\tau) = q^{-1/48} \prod_{j \in \mathbb{Z}_+} (1 - q^{j+1/2}) = \frac{\eta(\tau/2)}{\eta(\tau)},$$

$$f_3(\tau) = q^{1/24} \prod_{j \in \mathbb{Z}_+} (1 + q^{j+1}) = \frac{\eta(2\tau)}{\eta(\tau)}.$$

Show that these are modular forms of weight 0 for $\Gamma(48)$ with a trivial multiplier system. Check that $f_1(-1/\tau) = f_1(\tau)$, $f_2(-1/\tau) = \sqrt{2} f_3(\tau)$, and deduce that the linear span of f_1, f_2, and f_3 is $SL_2(\mathbb{Z})$-invariant under the action (13.10.6).

13.3. Let $d_1, d_2, \ldots$ be an m-periodic sequence of integers. Put

$$n_k = \sum_{j|k} d_j \mu\left(\frac{k}{j}\right) \quad (k = 1, 2, \ldots), \tag{13.15.1}$$

where μ is the classical Möbius function ($\mu(1) = 1$, $\mu(n) = (-1)^s$ if n is a product of s distinct primes, $\mu(n) = 0$ otherwise). Suppose that

$$(t,m) = (s,m) \text{ implies } d_t = d_s. \tag{13.15.2}$$

Show that then $n_k = 0$ unless $k|m$. Prove a converse statement.
[If there exists a prime p such that $k = p^t k_1$, p does not divide k_1, p^t does not divide m, then

$$n_k = \sum_{a|k_1} \pm(d_{p^t a} - d_{p^{t-1}a}) = 0.]$$

13.4. Let $\mathfrak{g}$ be a simple finite-dimensional Lie algebra of rank ℓ, Φ the Killing form on $\mathfrak{g}$, $\mathfrak{h}$ a Cartan subalgebra, Δ the root system, Δ_+ a subset of positive roots, and ρ their half-sum, $\Pi = \{\alpha_1, \ldots \alpha_\ell\}$ the set of simple roots, $-\alpha_0 = \sum_{i=1}^{\ell} a_i\alpha_i$ the highest root, W the Weyl group. Let M be the lattice spanned by long roots. Let $s = (s_0, s_1, \ldots, s_\ell)$ be a sequence of nonnegative integers; put $m = s_0 + \sum_{i=1}^{\ell} a_i s_i$. Define $\lambda_s \in \mathfrak{h}^*$ by $\Phi(\lambda_s, \alpha_i) = s_i/2m$ $(i = 1, \ldots, \ell)$. Let $\Delta_{s+} = \Delta_+ \cap \mathbb{Z}\{\alpha_i \mid s_i = 0\ (i = 0, \ldots, \ell)\}$, let W_s be the subgroup of W generated by reflections in $\alpha \in \Delta_{s+}$, and let ρ_s be the half-sum of the elements from Δ_{s+}. Put $D_s(\lambda) = \prod_{\alpha\in\Delta_{s+}} (\lambda|\alpha)/(\rho_s|\alpha)$ for $\lambda \in \mathfrak{h}^*$. Let e_j, f_j $(j = 1, \ldots, \ell)$ be the Chevalley generators of $\mathfrak{g}$. Define the automorphism σ_s of $\mathfrak{g}$ by

$$\sigma_s(e_j) = e^{2\pi i s_j/m},\ \sigma_s(f_j) = e^{-2\pi i s_j/m} f_j \quad (j = 1, \ldots, \ell),$$

let $\mathfrak{g} = \bigoplus_j \mathfrak{g}_j$ be the corresponding $\mathbb{Z}/m\mathbb{Z}$-gradation and put $d_j(s) = \dim \mathfrak{g}_j$. Show that (cf. (12.3.6)):

$$\begin{aligned} &q^{\|\rho-\lambda_s\|^2} \prod_{j\geq 1}(1-q^j)^{d_j} \\ &\quad = \sum_{w\in W_s\backslash W} \epsilon(w) \sum_{\alpha\in\frac{1}{2}M} D_s(w(\rho)+\alpha) q^{m\|w(\rho)+\alpha-\lambda_s\|^2} \end{aligned} \tag{13.15.3}$$

and prove the "very strange" formula:

$$\|\rho - \lambda_s\|^2 = \frac{1}{24}\dim\mathfrak{g} - \frac{1}{4m^2}\sum_{j=1}^{m-1} j(m-j)d_j(s). \tag{13.15.4}$$

Here and further $\|\lambda\|^2$ stands for $\Phi(\lambda, \lambda)$.

13.5. In the notation of Exercises 13.3 and 13.4, show that

$$\det_{\mathfrak{g}/\mathfrak{h}}(1 - e^{4\pi i\rho}) = dh^{\vee\ell},$$

where d is the determinant of the Cartan matrix of $\mathfrak{g}$ and $h^\vee = \|\alpha_0\|^{-2}$.

13.6. In the notations of Exercise 13.4, assume that the sequence $d_j = d_{j \bmod m}(s)$ satisfies condition (13.15.2), i.e., that the automorphism σ_s is quasirational (see Exercise 8.12). Define n_k by (13.15.1). Deduce from Exercises 13.3 and 13.4 that

$$\prod_{k|m} \eta(k\tau)^{n_k} = \text{right-hand side of (13.15.3)}.$$

Show that

$$m\|\rho - \lambda_s\|^2 = \frac{1}{24}\sum_{k|m} kn_k.$$

13.7. Let σ_s be a quasirational automorphism of $\mathfrak{g}$ and let n_k be defined by (13.15.1). Show that

$$\det_{\mathfrak{g}}(1 - q\sigma) = \prod_{j|m}(1 - q^j)^{n_{m/j}}.$$

Deduce from that another type of η-function identities, using Exercise 12.10 and the "strange" formula.

In Exercises 13.8–13.12, $\mathfrak{g}$ is an affine algebra with a symmetric Cartan matrix, so that $\mathring{\mathfrak{g}}$ is of type A_ℓ, D_ℓ, or E_ℓ.

13.8. Let $\mathring{\mathfrak{g}} = \bigoplus_{j\in\mathbb{Z}/h\mathbb{Z}} \mathfrak{g}_j(\mathbb{1}; 1)$ be the gradation corresponding to the automorphism of type $(\mathbb{1}; 1)$, let $d_j = \dim \mathfrak{g}_j(\mathbb{1}; 1)$ and let n_k be defined by (13.15.1). Show that $n_1 = \ell + 1$; $n_h = -1$; $n_i = 0$ if i does not divide h; $n_i = -n_j = \pm 1$ or 0 if $ij = h$ and $i \neq 1$.

[Use Lemma 14.2d).]

13.9. Show that the application of the specialization $\Phi_{\mathbb{1}}$ in Exercise12.9 gives the following identity:

$$\sum_{\gamma\in\mathring{Q}} q^{\|h\gamma+\bar{\rho}\|^2 - \|\bar{\rho}\|^2} = \varphi(q^h)^\ell \prod_{j|h} \varphi(q^j)^{n_{h/j}}.$$

13.10. Using (14.2.8), deduce from (0.10.1) that

$$F_{\mathbb{1}}(e(-\Lambda_0)L(\Lambda_0)) = \dim_q L(\Lambda_0) = \varphi(q)^\ell \prod_{j|h} \varphi(q^j)^{-n_j}.$$

13.11. Apply the principal specialization to the formula given by Lemma 12.7 to deduce that

$$F_{\mathbb{1}}(e(-\Lambda_0)L(\Lambda_0)) = \left(\sum_{j=0}^{\infty} \mathrm{mult}(\Lambda_0 - j\delta)q^j\right) \sum_{\gamma \in \mathring{Q}} q^{\|h\gamma+\bar{\rho}\|^2 - \|\bar{\rho}\|^2}.$$

13.12. Deduce from Exercises 13.8–13.11 the formula:

$$\sum_{j \geq 0} \mathrm{mult}_{L(\Lambda_0)}(\Lambda_0 - j\delta)q^j = \varphi(q)^{-\ell}.$$

Using the same method, give an alternative proof of (12.13.4).

13.13. Consider an affine algebra of type $X_N^{(r)}$. Let

$$G(\tau) = \exp \frac{\pi i \tau \dim \mathfrak{g}(X'_{N'})}{12(a+1)} \prod_{n \geq 1} (1 - e^{2\pi i n \tau})^{\mathrm{mult}\, n\delta},$$

where $a = h$ if $r = 1$ and $a = h^\vee$ if $r > 1$. Check the following table:

$\mathfrak{g}(A)$	$G(\tau)$
$X_\ell^{(1)}$ or $A_{2\ell}^{(2)}$	$\eta(\tau)^\ell$
$A_{2\ell-1}^{(2)}$	$\eta(\tau)^{\ell-1}\eta(2\tau)$
$D_{\ell+1}^{(2)}$	$\eta(\tau)\eta(2\tau)^{\ell-1}$
$E_6^{(2)}$	$\eta(\tau)^2\eta(2\tau)^2$
$D_4^{(3)}$	$\eta(\tau)\eta(3\tau)$

13.14. Let $\Lambda \in P_+$ be such that $\langle \Lambda, d \rangle = 0$. Then the eigenspace decomposition of $L(\Lambda)$ with respect to $a_0^{-1}d$ is of the form

$$L(\Lambda) = \bigoplus_{j \in \mathbb{Z}_+} L(\Lambda)_{-j}, \tag{13.15.5}$$

where $L(\Lambda)_{-j}$ is the eigenspace with the eigenvalue $-j$. Show that $\dim L(\Lambda)_{-j} < \infty$ and that (13.15.5) is the basic gradation of $L(\Lambda)$, i.e., the gradation of type $(1, 0, \ldots, 0)$. Provided that A is a symmetric affine matrix, show that

$$q^{-\ell/24} \sum_{j \geq 0} (\dim L(\Lambda_0)_{-j}) q^j = \eta(\tau)^{-\ell} \sum_{\gamma \in M} q^{\frac{1}{2}|\gamma|^2}.$$

In particular, in the case $\mathring{\mathfrak{g}} = E_8$, the right-hand side is $(qj(\tau))^{\frac{1}{3}}$, where $j(\tau)$ is the celebrated modular invariant, the generator of the field of modular forms of weight 0 for $SL_2(\mathbb{Z})$.

13.15. Generalize Exercise 13.14 to the case of twisted affine algebras.

13.16. Show that for $B_\ell^{(1)}$, all $\Lambda \in P_+ \mod \mathbf{C}\delta$ of level 1 are Λ_0, Λ_1, and Λ_ℓ. Show that, up to W-equivalence, all the string functions of level 1 are:

$$c_{\Lambda_0}^{\Lambda_0} = c_{\Lambda_1}^{\Lambda_1}, \quad c_{\Lambda_1}^{\Lambda_0} = c_{\Lambda_0}^{\Lambda_1}, \quad \text{and} \quad c_{\Lambda_\ell}^{\Lambda_\ell}.$$

Show that (cf. Exercise 13.2):

$$c_{\Lambda_\ell}^{\Lambda_\ell} = \eta(\tau)^{-\ell-1}\eta(2\tau) = \eta(\tau)^{-\ell} f_3(\tau), \tag{13.15.6}$$

$$c_{\Lambda_0}^{\Lambda_0} - c_{\Lambda_1}^{\Lambda_0} = \eta(\tau)^{-\ell-1}\eta(\tfrac{1}{2}\tau) = \eta(\tau)^{-\ell} f_2(\tau), \tag{13.15.7}$$

$$c_{\Lambda_0}^{\Lambda_0} + c_{\Lambda_1}^{\Lambda_0} = \eta(\tfrac{1}{2}\tau)^{-1}\eta(\tau)^{2-\ell}\eta(2\tau)^{-1} = \eta(\tau)^{-\ell} f_1(\tau). \tag{13.15.8}$$

[$A(\tau) := \eta(\tau)^{\ell+1}\eta(2\tau)^{-1}c_{\Lambda_\ell}^{\Lambda_\ell}$ is holomorphic at cusps $i\infty$ and 0 of $\Gamma_0(2)$ and hence is a constant. This implies (13.15.6). Formula (13.15.7) is deduced from (13.15.6) by replacing τ by $-\frac{1}{\tau}$ and using Theorem 13.10a). Formula (13.15.8) follows from (13.15.7) by replacing τ by $\tau + 1$.]

13.17. Show that for the hyperbolic Kac–Moody algebra $\mathfrak{g}(A)$, where A is from Exercise 3.8, one has:

$$\operatorname{mult}(k\alpha_1 + k\alpha_2 + \alpha_3) = p(k).$$

[Use Exercise 11.7 and Exercise 13.12 for $\mathfrak{g}$ of type $A_1^{(1)}$.]

13.18. Let $\lambda \in P_{++}^k$. Using Proposition 13.2, show that

$$A_\lambda \Theta_{\Lambda_0} = \sum_{\mu \in P_{++}^{k+1} \bmod \mathbf{C}\delta} \varphi_{\lambda,\mu}^{(k)} A_\mu,$$

where

$$\varphi_{\lambda,\mu}^{(k)}(\tau) = \sum_{w \in \mathring{W}} \varepsilon(w) \sum_{\gamma \in M} q^{\frac{1}{2}k(k+1)|\gamma + (k+1)^{-1}w(\mu) - k^{-1}\lambda|^2}.$$

13.19. Deduce from Exercise 13.18 that for the affine algebra of type $X_N^{(r)}$, $X = A$, D, or E, one has:

$$\chi_\Lambda \chi_{\Lambda_0} = \sum_{\mu \in P_{++}^{k+1} \bmod \mathbf{C}\delta} \left(\varphi_{\lambda,\mu}^{(k)}(\tau)/G(\tau)\right) \chi_\mu,$$

where $G(\tau)$ is defined in Exercise 13.13, so that we have the following formula for the branching functions for the tensor product decomposition of the $\mathfrak{g}(X_N^{(r)})$-module $L(\Lambda) \otimes L(\Lambda_0)$:

$$b_\mu^{\Lambda,\Lambda_0}(\tau) = \varphi_{\lambda,\mu}^{(k)}(\tau)/G(\tau).$$

13.20. Let $\dot{\mathfrak{g}}$ be a simple finite-dimensional Lie algebra and let $\dot{\mathfrak{g}} \subset \mathfrak{g} := \dot{\mathfrak{g}} \oplus \dot{\mathfrak{g}}$ be the diagonal imbedding. Consider the $\widetilde{\mathcal{L}}(\mathfrak{g})$-module $L(\Lambda') \otimes L(\Lambda'')$, where $L(\Lambda')$ and $L(\Lambda'')$ are integrable $\widetilde{\mathcal{L}}(\dot{\mathfrak{g}})$-modules of level k' and k'', respectively. Show that the coset Vir-module has conformal anomaly

$$c(k', k'') = (\dim \dot{\mathfrak{g}}) \left(\frac{k'}{k' + h^\vee} + \frac{k''}{k'' + h^\vee} - \frac{k' + k''}{k' + k'' + h^\vee} \right),$$

where $h^\vee$ is the dual Coxeter number of $\widetilde{\mathcal{L}}(\dot{\mathfrak{g}})$, and that

$$L_0^{\mathfrak{g},\dot{\mathfrak{g}}} = \frac{1}{2} \left(\frac{(\Lambda'|\Lambda' + 2\rho)}{k' + h^\vee} + \frac{(\Lambda''|\Lambda'' + 2\rho)}{k'' + h^\vee} - \frac{\Omega}{k' + k'' + h^\vee} \right),$$

where Ω is the Casimir operator for $\widehat{\mathcal{L}}(\dot{\mathfrak{g}})$.

13.21. Deduce the following decomposition as

$$\mathfrak{g}'\left(A_1^{(1)}\right) \oplus \text{ Vir-modules } (j, k \in \mathbb{Z}_+,\ j \le k):$$

$$L((k-j)\Lambda_0 + j\Lambda_1) \otimes L(\Lambda_0) = \sum_{\substack{0 \le s \le k+1 \\ s \equiv j \bmod 2}} L((k+1-s)\Lambda_0 + s\Lambda_1) \otimes \widetilde{L}(c(k), h_{j+1,s+1}^{(k)})$$

where

$$c^{(k)} = 1 - \frac{6}{(k+2)(k+3)}, \qquad h_{r,s}^{(k)} = \frac{((k+3)r - (k+2)s)^2 - 1}{4(k+2)(k+3)},$$

and $\widetilde{L}(c^{(k)}, h_{r,s}^{(k)})$ is a positive energy Vir-module such that $c = c^{(k)}$, the lowest eigenvalue of L_0 is equal to $h_{r,s}^{(k)}$ and has multiplicity 1, and $\operatorname{tr}_{\widetilde{L}(c^{(k)}, h_{r,s}^{(k)})} q^{d_0 - c^{(k)}/24} = \varphi_{r,s}^{(k)}$, where

$$\varphi_{r,s}^{(k)} = \frac{1}{\eta} \left(f_{r(k+3)-s(k+2),(k+2)(k+3)} - f_{r(k+3)+s(k+2),(k+2)(k+3)} \right).$$

13.22. Deduce that the Vir-module $L(c^{(k)}, h_{r,s}^{(k)})$ with $k \in \mathbb{Z}_+$ and $1 \le s \le r \le k+1$ are unitarizable, and that, letting

$$\chi_{r,s}^{(k)} = \operatorname{tr}_{L(c^{(k)}, h_{r,s}^{(k)})} q^{d_0 - c^{(k)}/24},$$

one has for these modules the following coefficient-wise inequality of q-expansions:

$$\chi_{r,s}^{(k)} \le \varphi_{r,s}^{(k)}. \tag{13.15.9}$$

13.23. Show that (see Exercise 13.2):

$$f_1 = \varphi_{1,1}^{(1)} + \varphi_{2,1}^{(1)}, \qquad f_2 = \varphi_{1,1}^{(1)} - \varphi_{2,1}^{(1)}, \qquad f_3 = \varphi_{2,2}^{(1)}.$$

13.24. Let $\delta = 0$ or $1/2$ and let Cl_δ denote the associative algebra on generators ψ_j, $j \in \mathbb{Z} + \delta$, and the following defining relations:

$$\psi_i \psi_j + \psi_j \psi_i = \delta_{ij}.$$

Let V_δ denote the irreducible Cl_δ-module which admits a nonzero vector $|0\rangle$ such that

$$\psi_j |0\rangle = 0 \quad \text{for } j > 0.$$

Let

$$L_n = -\frac{1}{2} \sum_{j \in \delta + \mathbb{Z}} j \psi_j \psi_{n-j} \quad \text{for } n \ne 0,$$

$$L_0 = \frac{1}{8}\left(\frac{1}{2} - \delta\right) + \sum_{j \in \delta + \mathbb{Z}_+} j \psi_{-j} \psi_j.$$

Show that $[\psi_m, L_n] = (m + n/2)\psi_{m+n}$ and deduce that the map $d_n \mapsto L_n$ defines a representation of Vir in V_δ with $c = \frac{1}{2}$.

13.25. Show that the elements

$$\psi_{j_s} \cdots \psi_{j_2} \psi_{j_1} |0\rangle \quad \text{with } 0 \ge j_1 > j_2 > \cdots > j_s$$

form a basis of V_δ. Define an Hermitian form on V_δ by declaring these elements to be orthogonal and to have square length 1 (resp. 2), if $j_1 < 0$ (resp. $j_1 = 0$). Show that ψ_j and ψ_{-j} are adjoint operators and deduce that V_δ is a unitarizable Vir-module. Show that

$$\mathrm{tr}_{V_0}\, q^{L_0 - 1/48} = 2 f_3, \qquad \mathrm{tr}_{V_{1/2}}\, q^{L_0 - 1/48} = f_1.$$

13.26. Let V_δ^+ (resp. V_δ^-) denote the linear span of elements of even (resp. odd) degree. Taking for granted that for $h \ne 0$, $\frac{1}{2}$, or $\frac{1}{16}$, the Vir-module $L(\frac{1}{2}, h)$ is not unitarizable (see Kac–Raina [1987] for a proof), show that $V_{\frac{1}{2}}^+ \simeq L(\frac{1}{2}, 0)$, $V_{\frac{1}{2}}^- \simeq L(\frac{1}{2}, \frac{1}{2})$, $V_0^+ \simeq V_0^- \simeq L(\frac{1}{2}, \frac{1}{16})$ (cf. Exercise 13.22). Deduce that $\chi_{r,s}^{(1)} = \varphi_{r,s}^{(1)}$ for $1 \le s \le r \le 2$. (It follows from Feigin–Fuchs [1984 A, B] that this is true for all Vir-modules from Exercise 13.22.)

13.27. Deduce from Theorem 13.10b) the following transformation formula ($1 \le b \le a \le k+1$):

$$\varphi_{a,b}^{(k)}(-\frac{1}{\tau}) = (8/(k+2)(k+3))^{\frac{1}{2}} \times \sum_{1 \le s \le r \le k+1} (-1)^{(a+b)(r+s)} \sin\frac{\pi a r}{k+2} \sin\frac{\pi b s}{k+3}\, \varphi_{r,s}^{(k)}(\tau).$$

13.28. Deduce the following decomposition of the $\mathfrak{g}(E_8^{(1)})$-module (viewed as a $\mathfrak{g}(E_8^{(1)}) \oplus$ Vir-module):

$$L(\Lambda_0) \otimes L(\Lambda_0) = L(2\Lambda_0) \otimes L(\frac{1}{2}, 0) + L(\Lambda_7) \otimes L(\frac{1}{2}, \frac{1}{2}) + L(\Lambda_1) \otimes L(\frac{1}{2}, \frac{1}{16}).$$

13.29. Show that $a(\sigma \cdot \Lambda) = a(\Lambda)$ for any $\sigma \in \operatorname{Aut} S(A)$. Deduce that for A of type $X_N^{(r)}$ with $X = A$, D, or E one has (see §13.14):

$$a(\Lambda) = |J|^{-\frac{1}{2}} \quad \text{if } \Lambda \in P_+^1.$$

13.30. Let $\mathfrak{g}$ be a simple simply-laced finite-dimensional Lie algebra of rank ℓ and let $\dot{\mathfrak{g}}$ be its subalgebra generated by elements E_i, F_i, $i \in \{0, 1, \dots, \ell\} \setminus \{j\}$ (see §8.2). Show that a level 1 $\widehat{\mathcal{L}}(\mathfrak{g})$-module $L(\Lambda)$ viewed as a $\widehat{\mathcal{L}}(\dot{\mathfrak{g}})$-module decomposes into a direct sum of a_j irreducible summands as follows:

$$L(\Lambda) = \bigoplus_{\lambda \in \dot{P}_+^1 \cap P(\Lambda)} \dot{L}(\lambda).$$

13.31. Recall that we have a surjective map $\psi : M^* \to \operatorname{Aut} S(X_\ell^{(1)})$. Show that

$$e^{2\pi i (\overline{\Lambda}|\alpha)} S_{\Lambda, M} = S_{\Lambda, \psi(\alpha)(M)}.$$

Deduce that the matrix $S_{(1)}$ in the simply-laced case is the character table of the group M^*/M, divided by $|J|^{1/2}$.

13.32. Show that for the diagonal embedding $\dot{\mathfrak{g}} \subset \mathfrak{g} := \dot{\mathfrak{g}} \oplus \dot{\mathfrak{g}}$ one has

$$\widetilde{R}_{k',k''} = \{k'\Lambda_j, k''\Lambda_j; (k'+k'')\Lambda_j \mid j \in J\},$$

and therefore Conjecture 13.14 holds in this case. Deduce, using Exercise 13.31, that

$$b_\Lambda^{\Lambda',\Lambda''}(i\beta) \sim |M^*/\overline{Q}| a(\Lambda') a(\Lambda'') a(\Lambda) e^{\pi(\dot{c}(k') - \dot{c}(k'') - \dot{c}(k'+k''))/12\beta}.$$

13.33. Show that for $0 < z < 1$ one has:

$$\prod_{n\geq 1}(1-e^{-2\pi n\beta-2\pi iz})(1-e^{-2\pi(n-1)\beta+2\pi i}) \sim -ie^{\pi iz}e^{-\pi(z^2-z+1/6)/\beta}.$$

Deduce that for $z \in \operatorname{Int} C_{af}$ and $r = 1$ one has:

$$A_\rho(i\beta,-z,0) \sim \beta^{-\ell/2}(-i)^{|\mathring{\Delta}_+|}e^{-\pi h^\vee|z-\overline{\rho}/h^\vee|^2/\beta}.$$

Deduce that for $\lambda, \mu \in P_+^k$ one has:

$$\chi_\lambda(i\beta, -\frac{\overline{\mu}+\overline{\rho}}{k+h^\vee}, 0) \sim S_{\lambda,\mu}e^{\frac{\pi h^\vee}{\beta}|\frac{\overline{\mu}+\overline{\rho}}{k+h^\vee}-\frac{\overline{\rho}}{h^\vee}|^2}.$$

13.34. Let I be a finite set with a distinguished element 0 and let V be a finite-dimensional vector space with a basis $\{\chi_\lambda\}$, indexed by $\lambda \in I$. Let S be an invertible operator on V such that all the entries of the 0-th row of its matrix $(S_{\lambda,\mu})$ are nonzero. Show that there exists unique operators ϕ'_λ and ϕ''_λ, indexed by $\lambda \in I$, on V satisfying the following three properties:
(i) ϕ'_λ are diagonal in the basis $\{\chi_\lambda\}$,
(ii) $\phi''_\lambda(\chi_0) = \chi_\lambda$,
(iii) $\phi''_\lambda = S\phi'_\lambda S^{-1}$.
Show that (both for $'$ and $''$): $\phi_\lambda\phi_\mu = \sum_\nu N^\nu_{\lambda\mu}\phi_\nu$ (the numbers $N^\nu_{\lambda\mu}$ are called *fusion coefficients*), where

$$\frac{S_{\lambda\epsilon}}{S_{0\epsilon}}\frac{S_{\mu\epsilon}}{S_{0\epsilon}} = \sum_\nu N^\nu_{\lambda\mu}\frac{S_{\nu\epsilon}}{S_{0\epsilon}}. \tag{13.15.10}$$

Equivalently:

$$N^\nu_{\lambda\mu} = \sum_\epsilon \frac{S_{\lambda\epsilon}S_{\mu\epsilon}(S^{-1})_{\nu\epsilon}}{S_{0\epsilon}},$$

13.35. Let $\mathfrak{g}$ be a simple finite-dimensional Lie algebra, let $\overline{\mathfrak{h}}$ be its Cartan subalgebra, let $\overline{\Delta} \subset \overline{\mathfrak{h}}^*$ be the root system, $\overline{\Delta}_+$ a subset of positive roots, $\overline{\rho}$ their half-sum, θ the highest root, $h^\vee$ the dual Coxeter number, let $\overline{Q}^\vee$ be the coroot lattice, $M = \nu(\overline{Q}^\vee)$ for the normalized invariant form $(.|.)$, let $\overline{P}_+ \subset \overline{\mathfrak{h}}^*$ be the set of dominant weights, let $\overline{P}^k_+ = \{\lambda \in \overline{P}_+ \mid (\lambda|\theta^\vee) \leq k\}$. Given $\lambda \in \overline{P}_+$, either $r_\alpha(\lambda+\overline{\rho}) = \lambda+\overline{\rho} \mod (k+h^\vee)M$ for some $\alpha \in \overline{\Delta}$ and we let $\varepsilon(\lambda) = 0$, or else there exists a unique $w \in \overline{W}$ and unique $\tilde{\lambda} \in \overline{P}^k_+$ such that $w(\lambda+\overline{\rho}) \equiv \tilde{\lambda}+\overline{\rho} \mod (k+h^\vee)M$ and we let $\varepsilon(\lambda) = \varepsilon(w)$. Now let in Exercise 13.34, $I := \overline{P}^k_+$ and $S_{\lambda\mu} := S_{\lambda+k\Lambda_0,\mu+k\Lambda_0}$. Show that the fusion coefficients, as defined by (13.15.10), are given by the following formula:

$$N^\nu_{\lambda\mu} = \sum_{\substack{\tau\in P^{2k}_+ \\ \tilde{\tau}=\nu}} \varepsilon(\tau)\,\mathrm{mult}_{\lambda\otimes\mu}\,\nu,$$

where $\mathrm{mult}_{\lambda\otimes\mu}\,\nu$ stands for the multiplicity of $\overline{L}(\nu)$ in $\overline{L}(\lambda)\otimes\overline{L}(\mu)$. Show that $N_{\lambda\mu\nu} := N^{*\nu}_{\lambda\mu}$ is symmetric in all three indices.

13.36. Show that for $\mathfrak{g} = s\ell_2$ one has ($i, j, s \in \mathbb{Z}_+ = \overline{P}_+$):

$$N_{ij}^s = \begin{cases} 1 & \text{if } |i-j| \leq s \leq i+j,\ i+j+s \in 2\mathbb{Z} \text{ and } \leq 2k, \\ 0 & \text{otherwise.} \end{cases}$$

13.37. (Open problem). Show that for the Kac–Moody algebra from Exercise 3.8 one has

$$\operatorname{mult} \alpha \leq p(1 - (\alpha|\alpha)),$$

where $(\alpha|\alpha)$ is defined by $(\alpha_i|\alpha_j) = \frac{1}{2}a_{ij}$ and $p(n)$ is the classical partition function (cf. Table H_3 in §11.15).

13.38. (Open problem). If $\lambda \in \dot{P}_+ \cap P(\Lambda)|_{\dot{\mathfrak{h}}}$, then $b_\lambda^\Lambda(\dot{\mathfrak{g}}) \neq 0$.

13.39. (Open problem). If $L(\Lambda)$ is an irreducible highest-weight module over an affine algebra, then for $z \in \operatorname{Int}(C_{\mathrm{af}})$ one has:

$$\operatorname{tr}_{L(\Lambda)} e^{2\pi\beta(d+z)} \sim D\beta^w e^{\pi G/12\beta}$$

for some positive real numbers D and G (called the *asymptotic dimension* and *growth* of $L(\Lambda)$, respectively) and a nonnegative real number w (called the *weight* of $L(\Lambda)$).

§13.16. Bibliographical notes and comments.

The theory of theta functions is an extensive subject which has its origin in the works of Jacobi and Riemann. The treatment of the part of the theory presented in §§13.1–13.5 is fairly nonstandard; it is based on the ideas of Kac–Peterson [1984 A]. A presentation of other topics of theta function theory may be found in Mumford [1983], [1984].

The exposition of §§13.7–13.14 mainly follows Kac–Peterson [1984 A] and Kac–Wakimoto [1988 A]. Some of the results of these sections have been previously known. Thus, Lemma 13.8b (excluding the case $A_{2\ell}^{(2)}$) is due to Looijenga [1976], who used it to give a theta-function proof of the Macdonald identities. An independent theta-function proof of the Macdonald identities was also found by Bernstein–Schvartzman [1978]. Van Asch [1976] was the first to use properties of modular functions to prove the Macdonald specialized identities.

Modular invariance of characters and branching functions, especially the matrices $S_{(k)}$ (computed by Kac–Peterson [1984 A]), have been playing a fundamental role in recent developments of conformal field theory, see Cardy [1986], Gepner–Witten [1986], Capelli–Itzykson–Zuber [1987], Verlinde [1988], and many others. A special case of the "very strange" formula, reproduced in Exercise 13.6, is proved in Kac [1978 A] by the same method as (12.3.6). Important cases of the "very strange" formula were found earlier by Macdonald [1972]. Exercises 13.8–13.12 are due to Kac [1978 A]. For $A_1^{(1)}$ Exercise 13.12 was previously obtained by Feingold–Lepowsky [1978]. Exercise 13.16 is taken from Kac–Peterson [1984 A]. Exercise 13.17 is proved by Feingold–Frenkel [1983] by a more complicated method.

Exercise 13.14 is taken from Kac [1980 B]. It is intimately related to the work of Conway–Norton [1979], who suggested that there must exist a "natural" graded module over the Monster group, such that the corresponding generating function is $qj(\tau)$. In Kac [1980 E] such a module was constructed for the centralizer of an involution of the Monster, by analogy with the Frenkel–Kac [1980] construction. Frenkel–Lepowsky–Meurman [1984] and [1989] (see also Borcherds [1986]) managed to extend a modification of this construction to the whole Monster, recovering the construction of Griess [1982] of the action of the Monster on a 196883-dimensional commutative nonassociative algebra. It would be fair to say that though the relation of the representation theory of affine algebras to the theory of modular forms is quite clear, the similar relation for the Monster group remains mysterious.

An explicit expression for all string functions is known only for the simplest affine Lie algebra, of type $A_1^{(1)}$, due to Kac–Peterson [1980], [1984 A]. After a lengthy calculation, they turn out to be certain peculiar "indefinite" modular forms discovered by Hecke around 1925. Another expression for these string functions was found by Distler–Qiu [1989] and Nemeschansky [1989 A] by using the free field construction described by Exercise 9.18. The open problem posed in Exercise 13.19 of previous editions has been solved by Kac–Wakimoto [1988 A]. Exercise 13.18 is taken from Kac–Wakimoto [1988 B]. Exercises 13.21 and 13.22 were found independently by Goddard–Kent–Olive [1986] and Kac–Wakimoto [1986]. The inequality (13.15.9) is actually an equality, which follows from Feigin–Fuchs [1984 A, B]. Due to Friedan–Qiu–Shenker [1985] the Vir-module $L(c,h)$ with $c < 1$ that is not listed in Exercise 13.22 is not unitarizable. The unitarizable Vir-modules $L(c,h)$ with $c \geq 1$ are those with $h \geq 0$ (Kac [1982 B]). Exercises 13.24–13.26 are well-known. Exercise 13.27 is taken from Cappelli–Itzykson–Zuber [1987 A], and Exercise 13.28 from Kac–Wakimoto [1986]. Exercises 13.29–13.32 are taken from Kac–Wakimoto [1988 A] and Exercise 13.33 from Kac–Wakimoto [1989 A]. Exercise 13.34 is a formalization of the work of Verlinde [1988].

As shown by Kac–Wakimoto [1988 B], [1989] the class of modules $L(\lambda)$ with modular invariance properties is much larger than the class of integrable modules. They probably correspond to some non-unitary CFT, and are related to the quantum gravity theory of Knizhnik–Polyakov–Zamolodchikov [1988]. Highest weights λ of all these modules lie in the "positive" half-space $\{\lambda \in \mathfrak{h}^* \mid \text{Re}\langle\lambda, K\rangle > -h^\vee\}$ (see the list in Kac–Wakimoto [1989 A]), and on the critical hyperplane $\{\lambda \in \mathfrak{h}^* \mid \langle\lambda, K\rangle = -h^\vee\}$. Several authors studied the $L(\lambda)$ corresponding to λ on the critical hyperplane: Wakimoto [1986], Hayashi [1988], Ku [1988 B], Malikov [1989], Feigin–Frenkel [1989 A] (these modules are modular invariant if and only if $\langle\lambda + \rho, \alpha^\vee\rangle \notin \mathbb{N}$ for all $\alpha \in \Delta_+^{re}$. Finally, at least a conjectural connection between the modules from the "negative" half-space of an affine Kac–Moody algebra and the modules over the corresponding quantized finite type Kac–Moody algebra at q a root of unity was found by Lusztig [1989 C].

Chapter 14. The Principal and Homogeneous Vertex Operator Constructions of the Basic Representation. Boson–Fermion Correspondence. Application to the Soliton Equations.

§**14.0.** The highest-weight module $L(\Lambda_0)$ over an affine algebra $\mathfrak{g}(A)$ is called the ***basic representation*** of $\mathfrak{g}(A)$. In this chapter we construct the basic representation explicitly in terms of certain (infinite-order) differential operators in infinitely many indeterminates, called the vertex operators. In a similar fashion, we construct representations of affine algebras of infinite rank. In the case of $A = A_\infty$ this construction is essentially the boson–fermion correspondence in the 2-dimensional quantum field theory.

These realizations are applied to construct the so-called soliton solutions of hierarchies of partial differential equations, the celebrated KdV and KP equations among them.

§**14.1.** Let $\mathfrak{g}(A)$ be an affine algebra of type $X_N^{(r)}$ and rank $\ell+1$ (from Table Aff r). Recall that $\mathfrak{g}(A) = \mathfrak{g}'(A) + \mathbf{C}d$, where $\mathfrak{g}'(A)$, the derived subalgebra, is generated by the Chevalley generators e_i, f_i ($i = 0, \ldots, \ell$), and d is the scaling element. Recall that the center of $\mathfrak{g}'(A)$ coincides with that of $\mathfrak{g}(A)$ and is spanned by the canonical central element K (see Chapter 6). Let $\overline{\mathfrak{g}}(A) = \mathfrak{g}'(A)/\mathbf{C}K$, so that we have the exact sequence

$$0 \to \mathbf{C}K \to \mathfrak{g}'(A) \xrightarrow{\pi} \overline{\mathfrak{g}}(A) \to 0. \tag{14.1.1}$$

Recall that relations $\deg e_i = -\deg f_i = 1$ ($i = 0, \ldots, \ell$) define the principal gradation $\mathfrak{g}'(A) = \bigoplus_{j\in\mathbb{Z}} \mathfrak{g}_j(\mathbb{1})$; it induces the principal gradation $\overline{\mathfrak{g}}(A) = \bigoplus_{j\in\mathbb{Z}} \overline{\mathfrak{g}}_j(\mathbb{1})$ so that $\dim \overline{\mathfrak{g}}_j(\mathbb{1}) = \dim \mathfrak{g}_j(\mathbb{1})$ for $j \neq 0$.

The element

$$\overline{e} = \sum_{i=0}^{\ell} \pi(e_i) \in \overline{\mathfrak{g}}_1(\mathbb{1})$$

is called the *cyclic element* of $\overline{\mathfrak{g}}(A)$. Let $\overline{\mathfrak{s}} = \{x \in \overline{\mathfrak{g}} \mid [x, \overline{e}] = 0\}$ be the centralizer of $\overline{e}$ in $\overline{\mathfrak{g}}(A)$. It is clear that $\overline{\mathfrak{s}}$ is graded with respect to the principal gradation of $\overline{\mathfrak{g}}(A)$:

$$\overline{\mathfrak{s}} = \bigoplus_{j \in \mathbb{Z}} \overline{\mathfrak{s}}_j.$$

The subalgebra $\mathfrak{s} = \pi^{-1}(\overline{\mathfrak{s}})$ is called the *principal subalgebra* of $\mathfrak{g}'(A)$ (or $\mathfrak{g}(A)$). It is graded with respect to the principal gradation of $\mathfrak{g}'(A)$:

$$\mathfrak{s} = \bigoplus_{j \in \mathbb{Z}} \mathfrak{s}_j.$$

Note that

$$\dim \mathfrak{s}_j = \dim \overline{\mathfrak{s}}_j \text{ for } j \neq 0, \tag{14.1.2}$$

$$\mathfrak{s}_0 = \mathbb{C}K. \tag{14.1.3}$$

The last relation is clear by Proposition 1.6.

The nonzero integers of the set which contains j with multiplicity $\dim \mathfrak{s}_j$ are called *exponents* of the affine algebra $\mathfrak{g}(A)$. We will compute the exponents below.

§14.2. We study the principal subalgebra $\mathfrak{s}$ by making use of an explicit construction of $\overline{\mathfrak{g}}(A)$, discussed in Chapter 8.

Let $\mathfrak{g}$ be a simple finite-dimensional Lie algebra of type X_N and let μ be a diagram automorphism of $\mathfrak{g}$ of order r $(= 1, 2, \text{ or } 3)$.

Let E_i, F_i, H_i $(i = 0, \ldots, \ell)$ be the elements of $\mathfrak{g}$ introduced in §8.2. Using the results of §6.2, we see that

$$\sum_{i=0}^{\ell} a_i^\vee H_i = 0. \tag{14.2.1}$$

Here $a_i^\vee$ are the labels of the diagram of the transpose of the affine matrix of type $X_N^{(r)}$ in Tables Aff. Recall that the elements E_i $(i = 0, \ldots, \ell)$ generate the Lie algebra $\mathfrak{g}$.

Let a_i be the labels of the diagram of the affine matrix $X_N^{(r)}$. The integer $h^{(r)} = r \sum_{i=0}^{\ell} a_i$ is called the *r-th Coxeter number* of $\mathfrak{g}$. By Theorem 8.6 a), the relations

$$\deg E_i = -\deg F_i = 1, \quad \deg H_i = 0 \quad (i = 0, \ldots, \ell),$$

define a $\mathbb{Z}/h^{(r)}\mathbb{Z}$-gradation

$$\mathfrak{g} = \bigoplus_j \mathfrak{g}_j(\mathbb{1}\,; r), \tag{14.2.2}$$

called the ***r*-principal gradation** of $\mathfrak{g}$.

The element

$$E = \sum_{i=0}^{\ell} E_i$$

is called the ***r*-cyclic element** of $\mathfrak{g}$. Denote by $S^{(r)}$ the centralizer of E in $\mathfrak{g}$. It is graded with respect to the r-principal gradation:

$$S^{(r)} = \bigoplus_{j \in \mathbb{Z}/h^{(r)}\mathbb{Z}} S_j^{(r)}.$$

PROPOSITION 14.2. a) $\dim \mathfrak{g}_j(\mathbb{1}\,; r) = \ell + \dim S_j^{(r)}$ $(j \in \mathbb{Z}/h^{(r)}\mathbb{Z})$.

b) *$S^{(r)}$ is a Cartan subalgebra of $\mathfrak{g}$.*

c) *The subspaces $S_i^{(r)}$ and $S_j^{(r)}$ are orthogonal (resp. nondegenerately paired) with respect to a nondegenerate invariant bilinear form on $\mathfrak{g}$ if $i + j \not\equiv 0 \mod h^{(r)}$ (resp. $i + j \equiv 0 \mod h^{(r)}$).*

Proof. Using automorphisms of $\mathfrak{g}$ of the form $E_i \mapsto \lambda_i E_i$, $F_i \mapsto \lambda_i^{-1} F_i$, the r-cyclic element E is conjugate to a multiple of an arbitrary element of the form $\sum_{i=0}^{\ell} c_i E_i$, where all $c_i \neq 0$. Therefore, it is sufficient to prove the lemma for one of the elements of this form, say E'. We take $E' = \sum_{i=0}^{\ell} \sqrt{a_i^\vee} E_i$. Put $F' = -\omega_0(E')$, where ω_0 is the compact involution of $\mathfrak{g}$; then $F' = \sum_{i=0}^{\ell} \sqrt{a_i^\vee} F_i$, and we have

$$[E', F'] = \sum_{i=0}^{\ell} a_i^\vee H_i = 0$$

by (14.2.1). Hence $\operatorname{ad} E'$ commutes with its adjoint operator $\operatorname{ad} F'$, with respect to the (positive-definite) Hermitian form $(.|.)_0$ (defined in §2.7). Therefore, E is a semisimple element, the orthogonal complement $B = \bigoplus_{j \in \mathbb{Z}/h^{(r)}\mathbb{Z}} B_j$ to $S^{(r)}$ in $\mathfrak{g}$ is $\operatorname{ad} E$-invariant, and the restriction of $\operatorname{ad} E$ to B is invertible. Since $[E, B_j] \subset B_{j+\bar{1}}$, we conclude that $\dim B_j = \ell$ $(= \dim B_0)$ for all $j \in \mathbb{Z}/h^{(r)}\mathbb{Z}$, proving a).

From a) we deduce

$$\dim \mathfrak{g} = \dim S^{(r)} + h^{(r)}\ell. \tag{14.2.3}$$

But one easily verifies by inspection that

$$\dim \mathfrak{g} - h^{(r)}\ell = N. \tag{14.2.4}$$

Comparing (14.2.3 and 4) gives

$$\dim S^{(r)} = N = \operatorname{rank} \mathfrak{g}. \tag{14.2.5}$$

Since $S^{(r)}$ is the centralizer of the semisimple element E, (14.2.5) proves b).

The first part of c) follows from Lemma 8.1; the second part now follows from the fact that the restriction of a nondegenerate invariant bilinear form to any Cartan subalgebra is nondegenerate.

□

The nondecreasing sequence of integers $m_1^{(r)} \leq m_2^{(r)} \leq \cdots$ from the interval $[1, h^{(r)} - 1]$, in which j appears with multiplicity $\dim S_j^{(r)}$, is called the set of r-*exponents* of $\mathfrak{g}$. They have the following properties.

LEMMA 14.2. a) *The number of r-exponents is equal to* $N = \operatorname{rank} \mathfrak{g}$, *and we have:*

$$1 = m_1^{(r)} < m_2^{(r)} \leq \cdots < m_N^{(r)} = h^{(r)} - 1.$$

b) $m_j^{(r)} + m_{N-j+1}^{(r)} = h^{(r)}$.

c) *If* i *and* j *have the same greatest common divisor with* $h^{(r)}$, *then* i *and* j *have the same multiplicity among the exponents.*

d) $\dim \mathfrak{g}_j(\mathbb{1}; r) = \ell +$ (*multiplicity of* j *among r-exponents*).

e) *Let* $r = 1$, *so that* $N = \ell$. *Denote by* c_j *the number of roots of* $\mathfrak{g}$ *of height* j. *Then*

$$\dim \mathfrak{g}_j(\mathbb{1}; 1) = c_j + c_{h^{(1)}-j} \quad \text{for } 1 \leq j \leq h^{(1)} - 1; \tag{14.2.6}$$

$$c_j + c_{h^{(1)}+1-j} = \ell \quad \text{for } 1 \leq j \leq h^{(1)}; \tag{14.2.7}$$

$$\dim \mathfrak{g}_j(2, 1, \ldots, 1; 1) = \ell \quad \text{for all } j. \tag{14.2.8}$$

f) *Let* $\mathfrak{g}$ *be of type* $A_{2\ell}$. *Then*

$$\dim \mathfrak{g}_j(2, 1, \ldots, 1; 2) = \ell \quad \text{for all } j. \tag{14.2.9}$$

Proof. Proposition 14.2 b) implies that the number of exponents equals N. It is clear that 1 appears among the exponents with multiplicity 1. Now a) and b) follow by Proposition 14.2 c). Part d) follows from Proposition 14.2 a) and the definitions.

The proof of the rest of the statements uses the following:

SUBLEMMA 14.2. *Automorphisms σ of type $(1,1,\dots,1;r)$ or $(2,1,\dots,1;1)$ are rational, i.e., σ^s is conjugate to σ if the order of σ and s are relatively prime. In particular, they are quasirational (see Exercise 8.12 for the definition).*

Proof. Let σ be of type $(1,\dots,1;r)$. Then the order of σ is $h^{(r)}$ and the dimension of its fixed point set is ℓ. The element σ^s, where s and $h^{(r)}$ are relatively prime, has the same properties. But Theorem 8.6 shows that there is a unique, up to conjugation, automorphism with these properties. The proof for σ of type $(2,1,\dots,1;1)$ is similar, using (12.4.3). □

The statement c) of Lemma 14.2 follows from d) and the sublemma. In order to prove e) consider the automorphism $\sigma_{h^{(1)}} = \sigma_{\mathbb{1},1}$ (resp. $\sigma_{h^{(1)}+1} = \sigma_{(2,1,\dots,1;1)}$) of $\mathfrak{g}$ of order $h^{(1)}$ (resp. $= h^{(1)}+1$). We have for $m = h^{(1)}$ (resp. $= h^{(1)}+1$):

$$\sigma_m(E_j) = \left(\exp\frac{2\pi i}{m}\right)E_j, \quad \sigma_m(F_j) = \left(\exp -\frac{2\pi i}{m}\right)F_j. \tag{14.2.10}$$

This implies (14.2.6) and

$$\dim \mathfrak{g}_j(2,1,\dots,1;1) = c_j + c_{h^{(1)}+1-j} \text{ for } 1 \le j \le h^{(1)}. \tag{14.2.11}$$

If $h^{(1)}+1$ is a prime number, then, by the sublemma, all the eigenspaces of $\sigma_{h^{(1)}+1}$ have the same dimension ℓ; this together with (14.2.11) proves (14.2.7 and 8) for an exceptional Lie algebra $\mathfrak{g}$, since for $\mathfrak{g}$ of type G_2, F_4, E_6, E_7, E_8, respectively, $h^{(1)}+1 = 7, 13, 13, 19, 31$ is a prime number. If $\mathfrak{g}$ is of classical type A_ℓ, B_ℓ, C_ℓ, or D_ℓ, one checks (14.2.7) directly; using (14.2.11) this gives (14.2.8) as well, proving e). Finally, f) is checked directly. □

Using Lemma 14.2 one easily computes the r-exponents. For instance, if j and $h^{(r)}$ are relatively prime, then j is a r-exponent of multiplicity 1 by Lemma 14.2 a) and c) (this takes care of all exceptional Lie algebras except E_6, $r = 1$, in which case one should check that 2 is not an exponent). A complete list of r-exponents is given in Table E_0.

Remark 14.2. The numbers $m_j^{(1)}$ $(j = 1, \dots \ell)$ are the ordinary exponents, and $h^{(1)}$ is the ordinary Coxeter number of a simple Lie algebra $\mathfrak{g}$ of type X_ℓ. Indeed, comparing Lemma 14.2 d) with (14.2.6) gives

$$c_j + c_{h^{(1)}-j} = \ell + \text{ (multiplicity of } j \text{ among exponents)},$$

which is one of the definitions of the exponents and the Coxeter number of $\mathfrak{g}$. Note also that all r-exponents have simple multiplicity with one exception: $\mathfrak{g} = D_\ell$, where ℓ is even, and $r = 1$; in such case $\ell - 1$ has multiplicity 2.

TABLE E_0

$\mathfrak{g}$	r	$h^{(r)}$	$m_1^{(r)}, m_2^{(r)}, \dots, m_N^{(r)}$
A_ℓ	1	$\ell + 1$	$1, 2, 3, \dots, \ell$
B_ℓ	1	2ℓ	$1, 3, 5, \dots, 2\ell - 1$
C_ℓ	1	2ℓ	$1, 3, 5, \dots, 2\ell - 1$
D_ℓ	1	$2\ell - 2$	$1, 3, 5, \dots, 2\ell - 3, \ell - 1$
E_6	1	12	$1, 4, 5, 7, 8, 11$
E_7	1	18	$1, 5, 7, 9, 11, 13, 17$
E_8	1	30	$1, 7, 11, 13, 17, 19, 23, 29$
F_4	1	12	$1, 5, 7, 11$
G_2	1	6	$1, 5$
$A_{2\ell}$	2	$4\ell + 2$	$1, 3, 5, \dots, 2\ell - 1, 2\ell + 3, \dots, 4\ell + 1$
$A_{2\ell-1}$	2	$4\ell - 2$	$1, 3, 5, \dots, 4\ell - 3$
$D_{\ell+1}$	2	$2\ell + 2$	$1, 3, 5, \dots, 2\ell + 1$
D_4	3	12	$1, 5, 7, 11$
E_6	2	18	$1, 5, 7, 11, 13, 17$

§14.3. Now we return to the affine algebra $\mathfrak{g}'(A)$ of type $X_N^{(r)}$. By Theorem 8.7, we have

$$\overline{\mathfrak{g}}(A) \simeq \bigoplus_{j \in \mathbb{Z}} (t^j \otimes \mathfrak{g}_{j \bmod h^{(r)}}(\mathbb{1}; r)), \tag{14.3.1}$$

where the isomorphism is defined by

$$\pi(e_i) \mapsto t \otimes E_i, \quad \pi(f_i) \mapsto t^{-1} \otimes F_i \quad (i = 0, \dots, \ell). \tag{14.3.2}$$

Moreover, (14.3.1) is the principal gradation of $\overline{\mathfrak{g}}(A)$. Hence

$$\overline{e} = t \otimes E \text{ is the cyclic element of } \overline{\mathfrak{g}}(A); \tag{14.3.3}$$

$$\overline{\mathfrak{s}}_j = t^j \otimes S^{(r)}_{j \bmod h^{(r)}} \quad (j \in \mathbb{Z}). \tag{14.3.4}$$

Finally, notice that

$$h^{(r)} = rh,$$

where h is the Coxeter number of $\mathfrak{g}'(A)$.

The following proposition now follows from Proposition 14.2.

PROPOSITION 14.3. a) $\dim \mathfrak{g}_j(\mathbb{1}) = \ell + \dim \mathfrak{s}_j$ *for* $j \in \mathbb{Z}\backslash\{0\}$.
b) $\bar{\mathfrak{s}}$ *is a commutative subalgebra of* $\bar{\mathfrak{g}}(A)$.
c) *The set of exponents of* $\mathfrak{g}'(A)$ *is*

$$\{m_s^{(r)} + jh^{(r)}, \text{ where } s = 1, \dots, N; j \in \mathbb{Z}\}.$$

□

From Proposition 14.3 a) and Proposition 10.8 we deduce:

COROLLARY 14.3. *Dual affine algebras have the same exponents.*

□

Using Table E_0 one immediately gets the list of exponents of all affine algebras as the set of all integers satisfying the conditions given by the following table:

TABLE *E*

$A_\ell^{(1)}$	$i \not\equiv 0 \mod (\ell+1)$
$B_\ell^{(1)}$	$i \equiv 1 \mod 2$
$C_\ell^{(1)}$	$i \equiv 1 \mod 2$
$D_\ell^{(1)}$	$i \equiv 1 \mod 2, i \equiv \ell - 1 \mod (2\ell - 2)$
$E_6^{(1)}$	$i \equiv \pm1, \pm4, \pm5 \mod 12$
$E_7^{(1)}$	$i \equiv \pm1, \pm5, \pm7, 9 \mod 18$
$E_8^{(1)}$	$i \equiv \pm1, \pm7, \pm11, \pm13 \mod 30$
$F_4^{(1)}$	$i \equiv \pm1 \mod 6$
$G_2^{(1)}$	$i \equiv \pm1 \mod 6$
$A_{2\ell}^{(2)}$	$i \equiv 1 \mod 2$ such that $i \not\equiv 0 \mod (2\ell+1)$
$A_{2\ell-1}^{(2)}$	$i \equiv 1 \mod 2$
$D_{\ell+1}^{(2)}$	$i \equiv 1 \mod 2$
$D_4^{(3)}$	$i \equiv \pm1 \mod 6$
$E_6^{(2)}$	$i \equiv \pm1 \mod 6$

Note that the multiplicities of exponents are 1, except for the case $A = D_\ell^{(1)}$, ℓ even, when the multiplicity of $(\ell-1)(2s+1)(s \in \mathbb{Z})$ is 2.

Finally, recall that by Theorem 8.7, $\mathfrak{g}'(A)$ can be constructed as the central extension of $\bar{\mathfrak{g}}(A)$ by $\mathbb{C}K$:

$$\mathfrak{g}'(A) = \bigoplus_{j \in \mathbb{Z}} (t^j \otimes \mathfrak{g}_{j \bmod h^{(r)}}(\mathbb{1}; r)) \oplus \mathbb{C}K, \tag{14.3.5}$$

with the bracket
(14.3.6)

$$[P_1(t)\otimes g_1, P_2(t)\otimes g_2] = P_1(t)P_2(t)\,[g_1, g_2]\oplus\frac{1}{hr}\left(\operatorname{Res}\frac{dP_1(t)}{dt}P_2(t)\right)(g_1|g_2)K,$$

where $(.|.)$ is the normalized invariant bilinear form on $\mathfrak{g}$ and h is the Coxeter number of $\mathfrak{g}(A)$.

§14.4. Recalling the definition of the infinite-dimensional Heisenberg algebra (§9.13), note the following simple lemma:

LEMMA 14.4. *The principal subalgebra $\mathfrak{s}$ of $\mathfrak{g}'(A)$ is isomorphic to the infinite Heisenberg algebra.*

Proof. Proposition 14.3 b) implies $[\mathfrak{s}_i, \mathfrak{s}_j] \subset \mathbb{C}K$ for all $i, j \in \mathbb{Z}$. By Proposition 14.2 c) and formula (14.3.6), we have $[\mathfrak{s}_i, \mathfrak{s}_j] = 0$ for $i \neq -j$ and $[a, \mathfrak{s}_{-i}] = \mathbb{C}K$ for every $a \in \mathfrak{s}_i$, $a \neq 0$, proving the proposition.

□

Now we are in a position to make the first step in the "principal construction" of the basic representation.

PROPOSITION 14.4. *Let $\mathfrak{g}'(A)$ be an affine algebra, where A is of type $X_N^{(r)}$, with $X = A$, D, or E, and let $\mathfrak{s}$ be the principal subalgebra of $\mathfrak{g}'(A)$. Then the basic $\mathfrak{g}'(A)$-module $L(\Lambda_0)$, considered as an $\mathfrak{s}$-module, remains irreducible.*

Proof. Let $L(\Lambda_0) = \bigoplus_{j\geq 0} L(\Lambda_0)_j$ be the principal gradation of $L(\Lambda_0)$ and let $\dim_q L(\Lambda_0) = \sum_{j\geq 0} \dim L(\Lambda_0)_j q^j$ be the q-dimension of $L(\Lambda_0)$ (see §10.10). Then by Proposition 10.10 we have

$$\dim_q L(\Lambda_0) = \prod_{j\geq 1}(1-q^j)^{\dim^t \mathfrak{g}_j(2,1\dots,1)-\dim \mathfrak{g}_j(\mathbb{1})}. \tag{14.4.1}$$

But the algebras $\mathfrak{g}({}^tA)$ under consideration are precisely the algebras from Table Aff 1 or $A_{2\ell}^{(2)}$. Hence, thanks to (14.2.8 and 9), we have:

$$\dim^t \mathfrak{g}_j(2,1,\dots,1) = \ell \text{ for all } j \neq 0. \tag{14.4.2}$$

Furthermore, by Proposition 14.2 and formula (14.3.4), we have

$$\dim \mathfrak{g}_j(\mathbb{1}) = \ell + \dim \mathfrak{s}_j. \tag{14.4.3}$$

Substituting (14.4.2 and 3) into (14.4.1), we obtain

$$\dim_q L(\Lambda_0) = \prod_{j\geq 1}(1-q^j)^{-\dim \mathfrak{s}_j}. \tag{14.4.4}$$

On the other hand, the $\mathbb{Z}$-gradation $\mathfrak{s} = \bigoplus_{j\in\mathbb{Z}} \mathfrak{s}_j$ induces a $\mathbb{Z}$-gradation $U(\mathfrak{s}) = \bigoplus_{j\in\mathbb{Z}} U(\mathfrak{s})_j$; set $U_j = U(\mathfrak{s})_{-j}(v_0)$, where $v_0 \in L(\Lambda_0)_{\Lambda_0}$ is a nonzero vector. Then

$$U_j \subset L(\Lambda_0)_j. \tag{14.4.5}$$

But by Lemma 9.13 a), the $\mathfrak{s}$-module $U = \bigoplus_{j\geq 0} U_j$ is isomorphic to the canonical commutation relations representation; hence we have

$$\sum_{j\geq 0}(\dim U_j)q^j = \prod_{j\geq 1}(1-q^j)^{-\dim \mathfrak{s}_j}. \tag{14.4.6}$$

Comparing (14.4.4) with (14.4.6) gives equality in (14.4.5) for all j. Hence the $\mathfrak{s}$-module $L(\Lambda_0) = U$ is irreducible.

□

§14.5. Proposition 14.4 allows us to identify the restriction of the basic representation to the principal subalgebra with the canonical commutation relations representation. In order to extend this to the whole Lie algebra $\mathfrak{g}(A)$ we need a lemma about differential operators.

Let $R = \mathbb{C}[x_1, x_2, \ldots]$ be the algebra of polynomials in infinitely many indeterminates x_i and let $\widehat{R} = \mathbb{C}[[x_1, x_2, \ldots]]$ be the algebra of formal power series (i.e., algebra of all linear combinations of (finite) monomials in the x_i or, in other words, the formal completion of R (see §1.5)). A *differential operator* on R is a (generally infinite) sum of the form

$$\sum_{r\geq 0}\ \sum_{1\leq i_1\leq\cdots\leq i_r} P_{i_1\ldots i_r}\frac{\partial}{\partial x_{i_1}}\cdots\frac{\partial}{\partial x_{i_r}}, \text{ where } P_{i_1\ldots i_r}\in\widehat{R}.$$

A differential operator defines a linear map $D: R \to \widehat{R}$ (in fact, every such linear map can be realized uniquely as a differential operator). A particular case is the multiplication by $P \in R$. Another example is the operator T_λ defined by

$$(T_\lambda f)(x_1, x_2, \ldots) = f(x_1+\lambda_1, x_2+\lambda_2, \ldots).$$

Note that

$$T_\lambda = \exp \sum_{i\geq 1} \lambda_i \frac{\partial}{\partial x_i} \tag{14.5.1}$$

by Taylor's formula.

LEMMA 14.5. *Let $D : R \to \widehat{R}$ be a linear map.*

a) *If* $[x_i, D] = \lambda_i D,\ i = 1, 2, \ldots,$ *then* $D = D(1)\exp(-\sum_i \lambda_i \frac{\partial}{\partial x_i})$.

b) *If* $[\frac{\partial}{\partial x_i}, D] = \mu_i D,\ i = 1, 2, \ldots,$ *then* $D(1) = c\exp\sum_i \mu_i x_i,\ c \in \mathbb{C}$.

c) *If* $[x_i, D] = \lambda_i D$ *and* $\left[\frac{\partial}{\partial x_i}, D\right] = \mu_i D,\ i = 1, 2, \ldots,$ *then* $D = c\exp\left(\sum_i \mu_i x_i\right)\exp\left(-\sum_i \lambda_i \frac{\partial}{\partial x_i}\right)$ *for some* $c \in \mathbb{C}$.

Proof. a) We replace D by DT_λ; then our statement is equivalent to the following: if $[x_i, D] = 0,\ i = 1, 2, \ldots,$ then $D = D(1)$. But $D(f) = D(1)f$ for all $f \in R$ by an easy induction on the total degree of f.

b) We replace D by $\exp(-\sum \mu_i x_i)D$; then our statement is equivalent to the following: if $\left[\frac{\partial}{\partial x_i}, D\right] = 0,\ i = 1, 2, \ldots,$ then $D(1) = \text{const}$. This is obvious: $\left[\frac{\partial}{\partial x_i}, D\right] = 0$ implies $\frac{\partial}{\partial x_i}(D(1)) = 0,\ i = 1, 2, \ldots,$ and hence $D(1) = \text{const}$.

Part c) follows from a) and b).

□

The operators of the form $\left(\exp\sum_j \lambda_j x_j\right)\left(\exp - \sum_j \mu_j \frac{\partial}{\partial x_j}\right)$ are called *vertex operators*. These operators were discovered by physicists in the string theory.

§14.6. In this section we shall construct, in terms of differential operators in infinitely many indeterminates, the basic representation $L(\Lambda_0)$ of an affine Lie algebra $\mathfrak{g}'(A)$ with the Cartan matrix A of type $X_N^{(r)}$, where either $r = 1$ and A is symmetric or $r > 1$. This construction is called the *principal realization* of the basic representation.

Let $\mathfrak{g}$ be a simple finite-dimensional Lie algebra of type X_N, let $\mathfrak{g} = \bigoplus_{j \in \mathbb{Z}/h^{(r)}\mathbb{Z}} \mathfrak{g}_j(\mathbb{1}; r)$ be its r-principal gradation, $E \in \mathfrak{g}_{\bar{1}}(\mathbb{1}, r)$ the r-cyclic element, $S^{(r)} = \bigoplus_j S_j^{(r)}$ the centralizer of E in $\mathfrak{g}$ (see §14.2). According to Proposition 14.2 b), $S^{(r)}$ is a Cartan subalgebra of $\mathfrak{g}$ (graded by the r-principal gradation). Let $\Delta \subset S^{(r)*}$ be the set of roots of $\mathfrak{g}$ with respect to $S^{(r)}$. For $\beta \in \Delta$ pick a nonzero root vector $A_\beta \in \mathfrak{g}$ and decompose it with respect to the principal gradation:

$$A_\beta = \sum_j A_{\beta,j} \quad (\text{summation over } j \in \mathbb{Z}/h^{(r)}\mathbb{Z}).$$

Since $S^{(r)} \cap \mathfrak{g}_{\overline{0}}(\mathbb{1};r) = 0$, the subspace $\mathfrak{g}_{\overline{0}}(\mathbb{1};r)$ $\left(= \sum_{j=1}^{\ell} \mathbb{C}H_j\right)$ is the linear span of the $A_{\beta,0}$, $\beta \in \Delta$. Therefore, one can choose ℓ root vectors $A_{\beta_1},\ldots,A_{\beta_\ell}$ with respect to $S^{(r)}$, corresponding to some roots $\beta_1,\ldots,\beta_\ell \in (S^{(r)})^*$, such that their projections on $\mathfrak{g}_{\overline{0}}(\mathbb{1};r)$ form a basis of this space.

Since $\operatorname{ad} E$ is a nondegenerate operator on $\mathfrak{g}/S^{(r)}$, which shifts the induced gradation by $\overline{1}$, we obtain that the images of the elements $A_{\beta_i,j}$ in $\mathfrak{g}/S^{(r)}$ are linearly independent. For all β and j we have

$$[E, A_{\beta,j}] = \langle\beta, E\rangle A_{\beta,j+\overline{1}} \text{ with } \langle\beta,E\rangle \neq 0.$$

Let T_j $(j = 1,\ldots,N)$ be a basis of $S^{(r)}$ such that $T_j \in S^{(r)}_{m_j^{(r)}}$ and $(T_i|T_{N+1-j}) = h\delta_{ij}$ for all $i,j = 1,\ldots,N$ (see Proposition 14.2 c)). We set

$$\lambda_{\beta j} = \langle\beta, T_j\rangle. \tag{14.6.1}$$

Then, clearly, we have

$$[T_k, A_{\beta,j}] = \lambda_{\beta k} A_{\beta,j+m_k^{(r)}} \tag{14.6.2}$$

(here $j + m_k^{(r)}$ is viewed as an element of $\mathbb{Z}/h^{(r)}\mathbb{Z}$).

Note that by (14.2.4), all the elements $A_{\beta_i,j}$ and T_k form a basis of $\mathfrak{g}$, since they are linearly independent. Note also that $(A_{\beta,j}|T_k) = 0$.

We use the "principal" realization of the Lie algebra $\mathfrak{g}'(A)$ defined by (14.3.5 and 6). The map

$$e_i \mapsto t\otimes E_i, \quad f_i \mapsto t^{-1}\otimes F_i, \quad \alpha_i^\vee \mapsto H_i + (a_i/a_i^\vee h)K \tag{14.6.3}$$

provides an isomorphism with this realization.

Let $E_+ = \{b_1, b_2, \ldots\}$ be the sequence of positive exponents of $\mathfrak{g}(A)$ arranged in nondecreasing order. We introduce the following basis of the principal subalgebra $\mathfrak{s}$:

$$K, \; p_i = t^{b_i}\otimes T_{i'}, \quad q_i = b_i^{-1}t^{-b_i}\otimes T_{N+1-i'},$$

where $i = 1,2,\ldots$. Here and further on, i' is defined to be the element of $\{1,\ldots,N\}$ congruent to i mod N. By (14.3.6) we have

$$[p_i, q_j] = \delta_{ij}K \text{ for all } i,j = 1,2,\ldots. \tag{14.6.4}$$

The degrees of these elements in the principal gradation are

$$\deg p_i = b_i = -\deg q_i. \tag{14.6.5}$$

Assume now that $\mathfrak{g}(A)$ is an affine algebra of one of the types listed at the beginning of the section and let $L(\Lambda_0)$ be its basic module. Since $L(\Lambda_0)$ has level 1, the element K is represented by the identity operator. Due to Proposition 14.4 and Corollary 9.13 we can identify $L(\Lambda_0)$ with the space $R = \mathbb{C}[x_1, x_2, \ldots]$ so that K operates as an identity, p_i as $\frac{\partial}{\partial x_i}$, and q_i as multiplication by x_i $(i = 1, 2, \ldots)$. From (14.6.5) we see that the relations

$$\deg x_i = b_i \quad (i = 1, 2, \ldots), \tag{14.6.6}$$

(together with $\deg PQ = \deg P + \deg Q$) define the principal gradation of $L(\Lambda_0)$.

In order to extend the realization of the basic representation from $\mathfrak{s}$ to the whole Lie algebra $\mathfrak{g}'(A)$, we extend, in an obvious way, the identification of $L(\Lambda_0)$ with R to that of $\hat{L}(\Lambda_0)$ with $\widehat{R}$, where $\hat{L}(\Lambda_0)$ denotes the formal completion of $L(\Lambda_0)$ by its principal gradation (i.e., $\hat{L}(\Lambda_0) = \prod_j L(\Lambda_0)_j$). Let $\prod_\alpha \mathfrak{g}_\alpha$ be the formal completion of $\mathfrak{g}(A)$. This space is not a Lie algebra; however, the adjoint action of $\mathfrak{g}(A)$ can be extended to it in an obvious way. We introduce the following elements of this completion, depending on a parameter $z \in \mathbb{C}^\times$:

$$X^\beta(z) = \sum_{j \in \mathbb{Z}} z^{-j} \left(t^j \otimes A_{\beta,j}\right), \quad \beta \in \Delta.$$

Then $X^\beta(z)$ maps $L(\Lambda_0)$ into $\hat{L}(\Lambda_0)$.

LEMMA 14.6. *The operator $X^\beta(z) : R \to \widehat{R}$ acts as the following vertex operator:*

$$\begin{aligned} \Gamma^\beta(z) = \langle \Lambda_0, A_{\beta,0} \rangle &\left(\exp \sum_{j=1}^{\infty} \lambda_{\beta j'} z^{b_j} x_j \right) \\ \times &\left(\exp - \sum_{j=1}^{\infty} \lambda_{\beta, N+1-j'} b_j^{-1} z^{-b_j} \frac{\partial}{\partial x_j} \right). \end{aligned} \tag{14.6.7}$$

Proof. Using (14.6.2) and (14.3.6), we write:

$$\begin{aligned} [p_s, X^\beta(z)] &= [t^{b_s} \otimes T_{s'}, \sum_{j \in \mathbb{Z}} z^{-j} \left(t^j \otimes A_{\beta,j}\right)] \\ &= \sum_{j \in \mathbb{Z}} z^{-j} t^{j+b_s} \otimes [T_{s'},\ A_{\beta,j}] \\ &= \lambda_{\beta s'} z^{b_s} \sum_{j \in \mathbb{Z}} z^{-j-b_s} t^{j+b_s} \otimes A_{\beta, j+b_s} \\ &= \lambda_{\beta s'} z^{b_s} X^\beta(z). \end{aligned}$$

Similarly,

$$[q_s, X^\beta(z)] = \lambda_{\beta,N+1-s'} b_s^{-1} z^{-b_s} X^\beta(z).$$

Now the lemma follows from Lemma 14.5.

□

We expand the vertex operator $\Gamma^\beta(z)$ defined by (14.6.7) in powers of z:

$$\Gamma^\beta(z) = \sum_{j\in\mathbb{Z}} \Gamma_{\beta,j} z^j.$$

Then $\Gamma_{\beta,j}$ are infinite-order differential operators, which map $R = L(\Lambda_0)$ into itself. Since all the elements $A_{\beta_i,j}$ and T_s form a basis of $\mathfrak{g}$, we can reformulate Lemma 14.6 as follows.

THEOREM 14.6. *Let $\mathfrak{g}'(A)$ be an affine algebra such that either A is symmetric (from Table Aff 1) or A is from Tables Aff 2 or 3. Set $R = \mathbb{C}[x_1, x_2, \ldots]$. Then the identity operator, the operators x_j, $\frac{\partial}{\partial x_j}$ ($j = 1, 2, \ldots$), and $\Gamma_{\beta_i,j}$ ($i = 1, \ldots, \ell; j \in \mathbb{Z}$) form a basis of a Lie subalgebra of the algebra of differential operators preserving R. This subalgebra is isomorphic to $\mathfrak{g}'(A)$, and the representation of it on R is equivalent to the basic representation of $\mathfrak{g}'(A)$.*

The realization of the basic representation given by Theorem 11.6 is called the *principal vertex operator construction*.

§14.7. We write down the explicit formulas for the principal vertex operator construction of the affine algebra $\mathfrak{g}'(A)$ of type $A_{n-1}^{(1)}$ ($n \geq 2$). In this case, $\mathfrak{g} = s\ell_n(\mathbb{C})$, and $(x|y) = \operatorname{tr} xy$. Let E_{ij} ($i, j = 1, \ldots, n$) denote the $n \times n$ matrix which is 1 in the i, j-entry and 0 everywhere else. We take:

$$E_0 = E_{n1}, \qquad E_i = E_{i,i+1} \quad (i = 1, \ldots, n-1),$$

$$F_0 = E_{1n}, \qquad F_i = E_{i+1,i} \quad (i = 1, \ldots, n-1),$$

$$H_0 = E_{nn} - E_{11}, \qquad H_i = E_{ii} - E_{i+1,i+1} \quad (i = 1, \ldots, n-1).$$

The 1-principal $\mathbb{Z}/n\mathbb{Z}$-gradation of $\mathfrak{g}$ is given by setting $\deg E_{ij} = j - i$ for $i \neq j$ and $\deg D = 0$ for a traceless diagonal matrix D. We set

$$E = \sum_{i=0}^{n-1} E_i = \begin{pmatrix} 0 & 1 & 0 & \ldots & 0 \\ 0 & 0 & 1 & \ldots & 0 \\ \ldots & \ldots & \ldots & \ldots & \ldots \\ 0 & 0 & 0 & \ldots & 1 \\ 1 & 0 & 0 & \ldots & 0 \end{pmatrix}$$

and let S be the centralizer of E in $\mathfrak{g}$. Then S is a Cartan subalgebra of $\mathfrak{g}$, with basis

$$T_j = E^j \quad (j = 1, \ldots, n-1).$$

We have $\deg T_j = j \mod n$, and

$$(T_i | T_{n-j}) = n\delta_{ij} \text{ for } i, j = 1, \ldots, n-1.$$

For two distinct n-th roots of unity, ε and η, define $n \times n$ matrices

$$A_{(\varepsilon,\eta)} = (\varepsilon^i \eta^{-j})_{i,j=1}^n.$$

Then $A_{(\varepsilon,\eta)} \in \mathfrak{g}$ are root vectors with respect to S:

$$[T_i, A_{(\varepsilon,\eta)}] = (\varepsilon^i - \eta^i)\, A_{(\varepsilon,\eta)} \text{ for } i = 1, \ldots, n-1.$$

Let $A_{(\varepsilon,\eta),j}$ be the homogeneous components of the $A_{(\varepsilon,\eta)}$. The T_k together with the homogeneous components of the $A_{(\varepsilon,1)}$ form a basis of $\mathfrak{g}$. In the notation of (14.6.1), we have

$$\lambda_{(\varepsilon,\eta),j} = \varepsilon^j - \eta^j \text{ for } j = 1, \ldots, n-1.$$

We use the principal realization of $\mathfrak{g}'(A)$:

$$\mathfrak{g}'(A) = \bigoplus_{j \in \mathbb{Z}} (t^j \mathfrak{g}_{j \mod n}(\mathbb{1}; 1)) \oplus \mathbb{C}K,$$

with the bracket given by

$$[a(t) \oplus \lambda K, b(t) \oplus \mu K] = a(t)b(t) - b(t)a(t) \oplus \frac{1}{n}\left(\operatorname{Res} \operatorname{tr} \frac{da(t)}{dt} b(t)\right) K.$$

We also have $h = n$ and $E_+ = \{j \in \mathbb{Z}_+ \mid j \not\equiv 0 \mod n\}$. For $m = 1, 2, \ldots;$ $m \not\equiv 0 \mod n$, we set

$$p_m = t^m T_{m'}, \quad q_m = m^{-1} t^{-m} T_{n-m'}.$$

We can use this simpler indexing than in the general case above, because no exponent has multiplicity greater than 1. We have

$$[p_\ell, q_m] = \delta_{\ell m} K$$

for all $\ell, m > 0$ not divisible by n.

We identify the space of the basic representation with the space of polynomials

$$R = \mathbb{C}[x_m; m \in \mathbb{Z}_+, m \not\equiv 0 \bmod n].$$

Then 1 is a highest-weight vector, and putting $\deg x_m = m$ defines the principal gradation of R. For $i = 1, \dots, n-1$, we set

$$X^{(\varepsilon,\eta)}(z) = \sum_{j\in\mathbb{Z}} z^{-j} t^j A_{(\varepsilon,\eta),j}.$$

Then K, p_m, q_m ($m \in \mathbb{Z}_+, \dots, m \not\equiv 0 \bmod n$) and the homogeneous components $t^j A_{(\varepsilon,1),j}$ of $X^{(\varepsilon,1)}(z)$ form a basis of $\mathfrak{g}'(A)$. By Lemma 14.6 and Theorem 14.6, the basic representation σ of $\mathfrak{g}'(A)$ on R is given by:

$$\sigma(K) = 1,$$

$$\sigma(p_m) = \frac{\partial}{\partial x_m}, \quad \sigma(q_m) = x_m \quad (m \in \mathbb{Z}_+, m \not\equiv 0 \bmod n),$$

$$\sigma(t^j A_{(\varepsilon,\eta),j}) = \text{ the coefficient of } z^{-j} \text{ in}$$

$$\frac{\varepsilon}{\varepsilon-\eta} \exp\left(\sum_{m=1}^{\infty} z^m (\varepsilon^m - \eta^m) x_m\right) \exp\left(\sum_{m=1}^{\infty} \frac{z^{-m}}{m} (\eta^{-m} - \varepsilon^{-m}) \frac{\partial}{\partial x_m}\right).$$

Indeed, to compute $\langle \Lambda_0, A_{(\varepsilon,\eta),0} \rangle$ in (14.6.7), we observe that

$$\alpha_0^\vee = (E_{nn} - E_{11}) + \tfrac{1}{n} K, \quad \alpha_i^\vee = (E_{ii} - E_{i+1,i+1}) + \tfrac{1}{n} K \quad (i = 1, \dots, n-1).$$

Hence $A_{(\varepsilon,\eta),0} = \sum_{j=1}^{n-1} ((\varepsilon/\eta) + (\varepsilon/\eta)^2 + \dots + (\varepsilon/\eta)^j) \alpha_j^\vee - \frac{\varepsilon/\eta}{1-\varepsilon/\eta} K$. Since $\langle \Lambda_0, \alpha_i^\vee \rangle = \delta_{0,i}$ and $\langle \Lambda_0, K \rangle = 1$, we obtain $\langle \Lambda_0, A_{(\varepsilon,\eta),0} \rangle = \frac{\varepsilon}{\varepsilon-\eta}$.

§14.8. We turn now to the homogeneous vertex operator construction. This construction is based on considering the homogeneous Heisenberg subalgebra $\mathfrak{t}$, see §8.4 (in place of the homogeneous Heisenberg subalgebra $\mathfrak{s}$ in the principal vertex operator construction). It is not true that $L(\Lambda_0)$ is irreducible as a $\mathfrak{t}$-module. One can still prove an analogue of Proposition 14.4 (see Exercise 14.6) and proceed along the lines of the proof in the principal case. This way allows one to "discover" the construction. A shorter way to prove that the construction "works" is to apply some vertex operator calculus (which is more familiar to physicists). Here we shall use the latter method, referring the reader to Exercises 14.5–14.7 for the former. For the sake of simplicity, we shall not consider the twisted case.

Let $\mathfrak{g} = \mathfrak{h} \oplus (\bigoplus_{\alpha\in\Delta} \mathbf{C}E_\alpha)$ be a simple finite-dimensional Lie algebra of type A_ℓ , D_ℓ, or E_ℓ, with commutation relations defined by (7.8.5). Let $(.|.)$ be the normalized invariant form on $\mathfrak{g}$ (defined by (7.8.6)). Let

$$\hat{\mathcal{L}}(\mathfrak{g}) = \mathbf{C}[t,t^{-1}] \otimes_{\mathbf{C}} \mathfrak{g} + \mathbf{C}K + \mathbf{C}d$$

be the associated affine algebra of type $A_\ell^{(1)}$, $D_\ell^{(1)}$, and $E_\ell^{(1)}$, respectively.

Consider the complex commutative associative algebra

$$V = S(\bigoplus_{j<0}(t^j \otimes \mathfrak{h})) \otimes_{\mathbf{C}} \mathbf{C}[Q], \tag{14.8.1}$$

where S stands for the symmetric algebra and $\mathbf{C}[Q]$ stands for the group algebra of the root lattice $Q \subset \mathfrak{h}$ of $\mathfrak{g}(X_\ell)$. Let $\alpha \mapsto e^\alpha$ denote the inclusion $Q \subset \mathbf{C}[Q]$. As before, $u^{(n)}$ will stand for $t^n \otimes u$ ($n \in \mathbf{Z}$, $u \in \mathfrak{g}$). For $n > 0$, $u \in \mathfrak{h}$, denote by $u(-n)$ the operator on V of multiplication by $u^{(-n)}$. For $n \geq 0$, $u \in \mathfrak{h}$, denote by $u(n)$ the derivation of the algebra V defined by the formula

$$u(n)(v^{(-m)} \otimes e^\alpha) = n\delta_{n,-m}(u|v) \otimes e^\alpha + \delta_{n,0}(\alpha|u)v^{(-m)} \otimes e^\alpha. \tag{14.8.2}$$

Choosing dual bases u_i and u^i of $\mathfrak{h}$, define the operator D_0 on V by the formula

$$D_0 = \sum_{i=1}^{\ell}\left(\frac{1}{2}u_i(0)u^i(0) + \sum_{n\geq 1} u_i(-n)u^i(n)\right). \tag{14.8.3}$$

Furthermore, for $\alpha \in Q$, define the sign operator c_α:

$$c_\alpha(f \otimes e^\beta) = \varepsilon(\alpha,\beta) f \otimes e^\beta. \tag{14.8.4}$$

Finally, for $\alpha \in \Delta \subset Q$ introduce the *vertex operator*

$$\Gamma_\alpha(z) = \left(\exp \sum_{j\geq 1} \frac{z^j}{j}\alpha(-j)\right)\left(\exp - \sum_{j\geq 1}\frac{z^{-j}}{j}\alpha(j)\right) e^\alpha z^{\alpha(0)} c_\alpha. \tag{14.8.5}$$

Here z is viewed as an indeterminate. Expanding in powers of z:

$$\Gamma_\alpha(z) = \sum_{j\in\mathbf{Z}} \Gamma_\alpha^{(j)} z^{-j-1},$$

we obtain a sequence of operators $\Gamma_\alpha^{(j)}$ on V. Now we can state the result.

THEOREM 14.8. *The map* $\sigma : \hat{\mathcal{L}}(\mathfrak{g}) \to \operatorname{End} V$ *given by*

$$\begin{aligned} K &\mapsto 1, \\ u^{(n)} &\mapsto u(n), \quad \text{for } u \in \mathfrak{h},\ n \in \mathbf{Z}, \\ E_\alpha^{(n)} &\mapsto \Gamma_\alpha^{(n)}, \quad \text{for } \gamma \in \Delta,\ n \in \mathbf{Z}, \\ d &\mapsto -D_0 \end{aligned}$$

defines the basic representation of the affine algebra $\hat{\mathcal{L}}(\mathfrak{g})$ *on* V.

Proof. First, we check that this map is a representation. Since $E_\alpha(z) \mapsto \Gamma_\alpha(z)$, according to (7.7.3) and (7.7.4) we need to check the following relations (all other relations trivially hold):

$$[D_0, \Gamma_\alpha(z)] = (z\tfrac{d}{dz} + 1)\Gamma_\alpha(z), \tag{14.8.6}$$

$$[\Gamma_\alpha(z_1), \Gamma_{-\alpha}(z_2)] = -\alpha(z_1)\delta(z_1 - z_2) + \delta'_{z_1}(z_1 - z_2), \tag{14.8.7}$$

$$[\Gamma_\alpha(z_1), \Gamma_\beta(z_2)] = \varepsilon(\alpha,\beta)\Gamma_{\alpha+\beta}(z_1)\delta(z_1 - z_2) \text{ if } \alpha + \beta \in \Delta. \tag{14.8.8}$$

Relation (14.8.6) follows from relations

$$[D_0, u(n)] = -nu(n), \qquad [D_0, e^\gamma] = -\frac{1}{2}(\gamma|\gamma)e^\gamma, \tag{14.8.9}$$

which are obvious. In order to check (14.8.7 and 8), we let for $\gamma \in \Delta$:

$$\Gamma_\gamma^\pm(z) = \exp \sum_{j \ge 1} \frac{\gamma(\pm j)}{\mp j} z^{\mp j}, \qquad \Gamma_\gamma^0(z) = e^\gamma z^{\gamma(0)} c_\gamma.$$

Then we have:

$$\Gamma_\alpha^-(z_1)\Gamma_\beta^+(z_2) = \Gamma_\beta^+(z_2)\Gamma_\alpha^-(z_1)\left(1 - \frac{z_2}{z_1}\right)^{(\alpha|\beta)}, \tag{14.8.10}$$

$$\Gamma_\alpha^0(z_1)\Gamma_\beta^0(z_2) = e^{\alpha+\beta} z_1^{\alpha(0)} z_2^{\beta(0)} z_1^{(\alpha|\beta)} c_\alpha c_\beta \varepsilon(\alpha,\beta). \tag{14.8.11}$$

Here and further on, by $(1 - z_2/z_1)^m$, $m \in \mathbf{Z}$, we mean its power series expansion in z_2/z_1. The second of these formulas is trivial and the first one, viewed as an identity of formal power series in $z_1^{\pm 1}$ and $z_2^{\pm 1}$, follows from the following two facts:

$$e^A e^B = e^B e^A e^{[A,B]} \text{ for two operators } A \text{ and } B \text{ such that } [A,B] \text{ commutes with } A \text{ and } B; \tag{14.8.12}$$

$$\exp\Big(-\sum_{j \ge 1} x^j/j\Big) = 1 - x. \tag{14.8.13}$$

The following formula is immediate by (14.8.10 and 11):

(14.8.14)
$$\Gamma_\alpha(z_1)\Gamma_\beta(z_2) = \left(1 - \frac{z_2}{z_1}\right)^{(\alpha|\beta)} z_1^{(\alpha|\beta)} \varepsilon(\alpha,\beta)$$

$$\times \left(\exp \sum_{j\geq 1} \frac{1}{j}(\alpha(-j)z_1^j + \beta(-j)z_2^j)\right)\left(\exp - \sum_{j\geq 1} \frac{1}{j}(\alpha(j)z_1^{-j} + \beta(j)z_2^{-j})\right)$$

$$\times\, e^{\alpha+\beta} z_1^{\alpha(0)} z_2^{\beta(0)} c_\alpha c_\beta.$$

Using (14.8.14) we obtain for $\alpha \in \Delta$:

(14.8.15)
$$[\Gamma_\alpha(z_1), \Gamma_{-\alpha}(z_2)] = \delta'_{z_1}(z_1 - z_2)$$

$$\times \left(\exp \sum_{j\geq 1} \frac{\alpha(-j)}{j}(z_1^j - z_2^j)\right)\left(\exp - \sum_{j\geq 1} \frac{\alpha(j)}{j}(z_1^{-j} - z_2^{-j})\right)\left(\frac{z_1}{z_2}\right)^{\alpha(0)}.$$

Here we have used that

(14.8.16)
$$\delta(z_1 - z_2) = z_2^{-1}\left(1 - \frac{z_1}{z_2}\right)^{-1} + z_1^{-1}\left(1 - \frac{z_2}{z_1}\right)^{-1},$$

and therefore

(14.8.17)
$$\delta'_{z_1}(z_1 - z_2) = z_2^{-2}\left(1 - \frac{z_1}{z_2}\right)^{-2} - z_1^{-2}\left(1 - \frac{z_2}{z_1}\right)^{-2}.$$

Applying (7.7.5) to (14.8.15) proves (14.8.7). The proof of (14.8.8) is similar. Thus, σ is a representation of $\hat{\mathcal{L}}(\mathfrak{g})$ on V.

Furthermore, it follows from (14.8.9) that

$$\operatorname{tr}_V q^{D_0} = \sum_{\gamma\in Q} q^{(\gamma|\gamma)/2} \Big/ \varphi(q)^\ell.$$

Comparing this with (12.13.6), we see that σ is the basic representation with the highest-weight vector $1 \otimes 1$.

□

§14.9. The basic representation of the infinite rank affine algebra of type A_∞ (and B_∞) can be constructed using one of the two approaches discussed in §§14.6 and 14.8. In this section and the next, we shall discuss yet another approach, which is of fundamental importance for mathematical physics. This is usually referred to as the boson-fermion correspondence.

An infinite expression of the form

$$\underline{i_1} \wedge \underline{i_2} \wedge \underline{i_3} \wedge \cdots,$$

where $i_1, i_2, \ldots$ are integers such that

$$i_1 > i_2 > i_3 > \cdots, \text{ and } i_n = i_{n-1} - 1 \text{ for } n \gg 0,$$

is called a *semi-infinite monomial.* Let F be the complex vector space with a basis consisting of all semi-infinite monomials, and let $H(.,.)$ denote the Hermitian form on F for which this basis is orthonormal.

Define the *charge* decomposition

$$F = \bigoplus_{m \in \mathbb{Z}} F^{(m)}$$

by letting

$$|m\rangle = \underline{m} \wedge \underline{m-1} \wedge \underline{m-2} \wedge \cdots$$

denote the *vacuum vector* of charge m and $F^{(m)}$ denote the linear span of all semi-infinite monomials of *charge* m, that is, those which differ from $|m\rangle$ only at a finite number of places.

Given a semi-infinite monomial $\varphi = \underline{i_1} \wedge \underline{i_2} \wedge \cdots$ of charge m, we associate with it a partition $\lambda^\varphi = \{\lambda_1 \geq \lambda_2 \geq \cdots \geq 0\}$ by letting $\lambda_1 = i_1 - m$, $\lambda_2 = i_2 - (m-1), \ldots$. This is clearly a bijective correspondence between the set of all semi-infinite monomials of given charge m and the set of all partitions Par (i.e., the set of all finite nonincreasing sequences of non-negative integers). We define the *energy* of φ to be equal to the size of the partition λ^φ, $|\lambda^\varphi| := \sum_i \lambda_i$. Let $F_j^{(m)}$ denote the linear span of all semi-infinite monomials of charge m and energy j.

We have the energy decomposition:

$$F^{(m)} = \sum_{j \in \mathbb{Z}} F_j^{(m)}. \tag{14.9.1}$$

It follows from the above discussion that

$$\dim F_j^{(m)} = p(j),$$

hence we have

$$\dim_q F^{(m)} := \sum_{j \in \mathbb{Z}} \dim F_j^{(m)} q^j = 1/\varphi(q). \tag{14.9.2}$$

For $j \in \mathbb{Z}$, introduce the *wedging* and *contracting* operators ψ_j and ψ_j^* on F by the following formulas:

$$\psi_j(\underline{i}_1 \wedge \underline{i}_2 \wedge \cdots) = \begin{cases} 0 \text{ if } j = i_s \text{ for some } s, \\ (-1)^s \underline{i}_1 \wedge \cdots \underline{i}_s \wedge \underline{j} \wedge \underline{i}_{s+1} \wedge \cdots \quad \text{if } i_s > j > i_{s+1}. \end{cases}$$

$$\psi_j^*(\underline{i}_1 \wedge \underline{i}_2 \wedge \cdots) = \begin{cases} 0 \text{ if } j \neq i_s \text{ for all } s, \\ (-1)^{s+1} \underline{i}_1 \wedge \underline{i}_2 \wedge \cdots \underline{i}_{s-1} \wedge \underline{i}_{s+1} \wedge \cdots \quad \text{if } j = i_s. \end{cases}$$

It is straightforward to check that the operators ψ_j and ψ_j^* are adjoint with respect to the Hermitian form $H(.,.)$, and that the following relations hold:

$$\psi_i \psi_j^* + \psi_j^* \psi_i = \delta_{ij}, \quad \psi_i \psi_j + \psi_j \psi_i = 0, \quad \psi_i^* \psi_j^* + \psi_j^* \psi_i^* = 0. \tag{14.9.3}$$

Thus, the operators ψ_j and ψ_j^* generate a Clifford algebra Cl. It is clear that the Cl-module F is irreducible and that

$$\psi_j|0\rangle = 0 \quad \text{for} \quad j \leq 0, \qquad \psi_j^*|0\rangle = 0 \quad \text{for} \quad j > 0.$$

Remark 14.9. Recall that given a vector space V with a symmetric bilinear form $(.|.)$, the associated *Clifford algebra* $\mathrm{Cl}\, V$ is defined as follows:

$$\mathrm{Cl}\, V = T(V)/J,$$

where $T(V)$ is the tensor algebra over V and J is its 2-sided ideal generated by elements of the form $x \otimes y - (x|y)$ $(x, y \in V)$. Given a maximal isotropic subspace U of V, the algebra $\mathrm{Cl}\, V$ has a unique irreducible module F_U, called the *spin module*, which admits a nonzero vector $|0\rangle$ such that $U|0\rangle = 0$. The algebra Cl is the Clifford algebra associated with the space $V = \sum_i \mathbb{C}\psi_i + \sum_i \mathbb{C}\psi_i^*$ with the symmetric bilinear form $(\psi_i|\psi_j^*) = \delta_{ij}$, all other $= 0$, and F is its spin module associated with the subspace $U = \sum_{i \leq 0} \mathbb{C}\psi_i + \sum_{i>0} \mathbb{C}\psi_i^*$.

All our further calculations are based on the following commutation relations, which follow from (14.9.3):

$$[\psi_i \psi_j^*, \psi_k] = \delta_{kj} \psi_i; \qquad [\psi_i \psi_j^*, \psi_k^*] = -\delta_{ki} \psi_j^*. \tag{14.9.4}$$

The embedding $r : gl_\infty \to \mathrm{Cl}$ defined by

$$r(E_{ij}) = \psi_i \psi_j^*$$

defines a representation r of gl_∞ on F, producing thereby a representation r_m of gl_∞ on $F^{(m)}$ for each $m \in \mathbb{Z}$. Note that the representation r is unitarizable for the antilinear involution $a \mapsto -{}^t\overline{a}$ of gl_∞, in the sense that $r({}^t\overline{a})$ and $r(a)$ are adjoint operators with respect to the Hermitian form H for all $a \in gl_\infty$.

It is easy to check that gl_∞ acts by "derivations" on F, i.e., for $a = (a_{ij}) \in gl_\infty$ one has:

$$r(a)(\underline{i}_1 \wedge \underline{i}_2 \wedge \cdots) = a \cdot \underline{i}_1 \wedge \underline{i}_2 \wedge \cdots + \underline{i}_1 \wedge a \cdot \underline{i}_2 \wedge \cdots + \cdots , \tag{14.9.5}$$

where $a \cdot \underline{j} = \sum_i a_{ij} \underline{i}$. Here (and further on) we use the usual rules of the exterior algebra to express the right-hand side of (14.9.5) in terms of semi-infinite monomials. Thus F can be viewed as an infinite generalization of the usual exterior algebra. For this reason r is usually called the *infinite wedge representation.*

gl_∞ is the Lie algebra of the group

$$GL_\infty = \{a = (a_{ij})_{i,j \in \mathbb{Z}} \mid a \text{ is invertible and all but a finite number of } a_{ij} - \delta_{ij} \text{ are } 0\}.$$

The corresponding action R of GL_∞ on F is given by

$$R(g)(\underline{i}_1 \wedge \underline{i}_2 \wedge \cdots) = g \cdot \underline{i}_1 \wedge g \cdot \underline{i}_2 \wedge \cdots , \tag{14.9.6}$$

that is (14.9.5 and 6) are related by

$$\exp r(a) = R(\exp a), \quad a \in gl_\infty. \tag{14.9.7}$$

Using the standard exterior algebra calculus, we get the following formula for the representation R_m of $g \in GL_\infty$ on $F^{(m)}$:

$$R_m(g)(\underline{i}_1 \wedge \underline{i}_2 \wedge \cdots) = \sum_{j_1 > j_2 > \cdots} \left(\det g^{i_1, i_2, \dots}_{j_1, j_2, \dots}\right) \underline{j}_1 \wedge \underline{j}_2 \wedge \cdots , \tag{14.9.8}$$

where $g^{i_1, i_2, \dots}_{j_1, j_2, \dots}$ denotes the matrix located at the intersection of the rows $j_1, j_2, \dots$ and columns $i_1, i_2, \dots$ of g.

It is clear that the representations r_m are irreducible. Moreover, we have:

$$r_m(E_{ij})|m\rangle = 0 \quad \text{if } i < j, \text{ or } i = j > m,$$
$$r_m(E_{jj})|m\rangle = |m\rangle \quad \text{if } j \geq m.$$

Thus, as an $s\ell_\infty$-module, $F^{(m)}$ is isomorphic to $L(\Lambda_m)$, where Λ_m is the fundamental weight, defined as usual by $\langle \Lambda_m, \alpha_j^\vee \rangle = \delta_{mj}$ (see §7.11 for notation concerning gl_∞, $\overline{a}_\infty$, etc.).

The representation r_0 of gl_∞ on $F^{(0)}$ is called the *basic representation*.

The gl_∞-module $F^{(m)}$ does not extend to $\overline{a}_\infty$ since for example

$$r_0(\text{diag}(\ldots, \lambda_i, \lambda_{i+1}, \ldots))|0\rangle = (\lambda_0 + \lambda_{-1} + \cdots)|0\rangle,$$

which is generally divergent. In order to avoid this anomaly, we modify the representation r by letting

$$\hat{r}(E_{ij}) = \begin{cases} \psi_i \psi_j^* & \text{if } i \neq j \text{ or } i + j > 0, \\ \psi_j^* \psi_i & \text{if } i + j \leq 0. \end{cases}$$

It is easy to see that this extends by linearity to a representation $\hat{r}$ of the completed infinite rank affine algebra a_∞ on F with $K = 1$ (see §7.11 for the definitions). We obtain thereby for each $m \in \mathbb{Z}$ an irreducible representation $\hat{r}_m$ of a_∞ on $F^{(m)}$.

As in §14.1, we call

$$\overline{e} = \sum_i \pi(e_i) \in \overline{x}_\infty$$

the *cyclic element* of $\overline{x}_\infty$, and denote by $\overline{s}$ its centralizer in $\overline{x}_\infty$. Here $\pi : x_\infty \to \overline{x}_\infty$ is the canonical homomorphism. The subalgebra $s = \pi^{-1}(\overline{s})$ is called the *principal subalgebra* of x_∞.

In the case of a_∞, we have

$$\overline{e} = \sum_{i \in \mathbb{Z}} E_{i,i+1},$$

hence the elements $\overline{e}^j = \sum_{i \in \mathbb{Z}} E_{i,i+j}$ $(j \in \mathbb{Z})$ form a basis of $\overline{s}$. One immediately finds using (7.12.1):

$$\psi(\overline{e}^m, \overline{e}^n) = m\delta_{m,-n}.$$

Hence $s = \sum_{m \in \mathbb{Z}} \mathbb{C}s_m + \mathbb{C}K$, where $\pi(s_m) = \overline{e}^m$ and $[s, K] = 0$, with commutation relations

$$[s_m, s_n] = m\delta_{m,-n}K, \tag{14.9.9}$$

i.e., s is the oscillator algebra. Comparing (14.9.2) with (14.9.9), we obtain, in the same way as in §14.4, the following:

PROPOSITION 14.9. *Viewed as an $\mathfrak{s}$-module, the $\mathfrak{a}_\infty$-modules $F^{(m)}$ are irreducible.*

§14.10. The operators ψ_j and ψ_j^* are called *free fermions*. The bosonization consists of introducing *free bosons*:

$$\alpha_n = \hat{r}(s_n), \quad n \in \mathbf{Z}.$$

Explicitly:

$$\alpha_n = \sum_{j \in \mathbf{Z}} \psi_j \psi_{j+n}^* \quad \text{if } n \in \mathbf{Z} \backslash \{0\},$$

$$\alpha_0 = \sum_{j>0} \psi_j \psi_j^* - \sum_{j \le 0} \psi_j^* \psi_j.$$

By (14.9.9) we have:

(14.10.1) $$[\alpha_m, \alpha_n] = m\delta_{m,-n}.$$

Note also that

(14.10.2) $$\alpha_0 \big|_{F^{(m)}} = mI,$$

and that α_m and α_{-m} are adjoint operators.

Physicists would call F the fermionic Fock space. We introduce now the bosonic Fock space

$$B = \mathbf{C}[x_1, x_2, \dots; q, q^{-1}],$$

which is a polynomial algebra on indeterminates $x_1, x_2, \dots$ and q, q^{-1}. Define a representation r^B of the oscillator algebra $\mathfrak{s}$ on B as follows:

$$r^B(s_m) = \frac{\partial}{\partial x_m}, \quad r^B(s_{-m}) = m x_m \quad \text{if } m > 0;$$

$$r^B(s_0) = q\frac{\partial}{\partial q}, \quad r^B(K) = 1.$$

Then by Proposition 14.9 and the uniqueness of the canonical commutation relations (Corollary 9.13), there exists a (unique) isomorphism of $\mathfrak{s}$-modules

$$\sigma : F \xrightarrow{\sim} B,$$

such that $\sigma(|m\rangle) = q^m$. Note that $\sigma(F^{(m)}) = B^{(m)} := q^m \mathbf{C}[x_1, x_2, \dots]$.

The map σ transports the Hermitian form $H(.,.)$ on F to a Hermitian form $H_B(.,.)$ on B, and the energy decomposition (14.9.1) of F to that of B. Explicitly they are given by the following proposition.

PROPOSITION 14.10.

a) $H_B(q^m P(x), q^n Q(x)) = \delta_{mn} \overline{P}\left(\frac{\partial}{\partial x_1}, \frac{1}{2}\frac{\partial}{\partial x_2}, \dots\right) Q(x)\big|_{x=0}$.

b) *Energy of* $q^m x_1^{j_1} x_2^{j_2} \dots = j_1 + 2j_2 + \cdots$.

Proof. a) follows from (9.13.4). b) follows from the simple observation that the operator $\psi_i \psi_j^*$ increases energy by $i - j$ (thus the energy decomposition is nothing else but the principal gradation).

□

Using σ, we can also transport the operator of multiplication by q from B to F, obtaining the (unique) operator on F, which we again denote by q, such that

$$q|m\rangle = |m+1\rangle, \quad q\psi_i = \psi_{i+1} q \quad (m, i \in \mathbb{Z}).$$

(The second equation follows from the first one and the equation $q\alpha_i = \alpha_i q$, $i \in \mathbb{Z} \setminus \{0\}$.) Note that q and q^{-1} are adjoint operators.

The fermionization consists of reconstructing fermions ψ_i and ψ_i^* in terms of bosons α_i. In order to do that introduce (as before) the generating series (fermionic fields):

$$\psi(z) = \sum_{j \in \mathbb{Z}} z^j \psi_j, \qquad \psi^*(z) = \sum_{j \in \mathbb{Z}} z^{-j} \psi_j^*.$$

Introduce also the following operators:

$$\Gamma_+(z) = \exp \sum_{n \geq 1} \frac{z^{-n}}{n} \alpha_n, \qquad \Gamma_-(z) = \exp \sum_{n \geq 1} \frac{z^n}{n} \alpha_{-n}.$$

As before, we view z as a formal parameter, so that $\Gamma_\pm(z)$ are viewed as generating series of operators $\Gamma_n^\pm$ on F:

$$\Gamma_\pm(z) = \sum_{n \in \mathbb{Z}_+} \Gamma_n^\pm z^{\mp n}.$$

Since $\Gamma_0^\pm = 1$, we may consider the generating series $\Gamma_\pm(z)^{-1}$. Note that $\Gamma_+(z)$ and $\Gamma_-(z)$ are adjoint operators in the sense that Γ_n^+ and Γ_{-n}^- are adjoint operators on F. Note also that in the bosonic picture we have:

$$\Gamma_+(z) = \exp \sum_{n \geq 1} \frac{z^{-n}}{n} \frac{\partial}{\partial x_n}, \qquad \Gamma_-(z) = \exp \sum_{n \geq 1} z^n x_n. \tag{14.10.3}$$

Finally, with a partition $\lambda \in \text{Par}$, we associate the *Schur polynomial* $S_\lambda(x)$ as usual. First, define the elementary Schur polynomials $S_m(x)$ by the generating series

$$\sum_{m\in\mathbb{Z}} S_m(x)z^m = \exp\sum_{n\geq 1} z^n x_n \quad (= \prod_{n\geq 1} e^{z^n x_n}). \tag{14.10.4}$$

We have:

$$S_m(x) = 0 \text{ for } m < 0, \qquad S_0(x) = 1, \qquad \text{and}$$
$$S_m(x) = \sum_{k_1+2k_2+\cdots=m} \frac{x_1^{k_1}}{k_1!}\frac{x_2^{k_2}}{k_2!}\cdots \quad \text{for } m > 0.$$

Given $\lambda = \{\lambda_1 \geq \lambda_2 \geq \cdots\} \in \text{Par}$, let

$$S_\lambda(x) = \det(S_{\lambda_i+j-i}(x))_{1\leq i,j\leq|\lambda|}.$$

Here and further on, x stands for $(x_1, x_2, \ldots)$.

Now we are in a position to state the boson–fermion correspondence:

THEOREM 14.10. a) *One has:*

$$\psi(z) = z^{\alpha_0} q\Gamma_-(z)\Gamma_+(z)^{-1},$$
$$\psi^*(z) = q^{-1}z^{-\alpha_0}\Gamma_-(z)^{-1}\Gamma_+(z).$$

b) *If $\varphi \in F^{(m)}$ is a semi-infinite monomial, then*

$$\sigma(\varphi) = q^m S_{\lambda^\varphi}(x).$$

Proof. In order to prove a) we directly check the following relations of operators on F:

$$[\alpha_j, \psi(z)] = z^j\psi(z), \qquad [\alpha_j, \psi^*(z)] = -z^j\psi^*(z).$$

The first of these two equations transports to B as follows:

$$[\tfrac{\partial}{\partial x_j}, \sigma\psi(z)\sigma^{-1}] = z^j(\sigma\psi(z)\sigma^{-1}),$$
$$[x_j, \sigma\psi(z)\sigma^{-1}] = \frac{z^{-j}}{j}(\sigma\psi(z)\sigma^{-1}).$$

Hence, by Lemma 14.5, the operator

$$\sigma\psi(z)\sigma^{-1} : B^{(m)} \to B^{(m+1)}$$

is of the form

$$\sigma\psi(z)\sigma^{-1} = C_m(z)q\Gamma(z), \quad \text{where}$$
$$\Gamma(z) = \left(\exp\sum_{j\geq 1} z^j x_j\right)\left(\exp - \sum_{j\geq 1}\frac{z^{-j}}{j}\frac{\partial}{\partial x_j}\right).$$

But the coefficient of $|m+1\rangle$ in $\psi(z)|m\rangle$ is z^{m+1}, hence $C_m(z) = z^{m+1}$. This proves the first of the two formulas in a). The proof of the second formula is similar.

In order to prove b) we let $g(y) = \exp \sum_{j\geq 1} y_j \bar{e}^j$, where $y_1, y_2, \ldots$ are some complex numbers, and compute

$$\sigma(R_m(g(y))\underline{i}_1 \wedge \underline{i}_2 \wedge \cdots) = R_m^B(g(y))P(x), \tag{14.10.5}$$

where $P(x) = \sigma(\underline{i}_1 \wedge \underline{i}_2 \wedge \cdots)$. We shall obtain the result by comparing the coefficient of the vacuum $|m\rangle$ (= vacuum expectation value), which we denote by $F(y)$, on both sides of (14.10.5).

We must first settle a minor technical problem since $g(y)$ does not lie in GL_∞. It does lie, however, in the group

$$\widetilde{GL}_\infty = \{a = (a_{ij})_{i,j\in\mathbb{Z}} \mid a \text{ is invertible and all but a finite number of } a_{ij} - \delta_{ij} \text{ with } i \geq j \text{ are } 0\}.$$

The corresponding Lie algebra is

$$\widetilde{gl}_\infty := \{(a_{ij})_{i,j\in\mathbb{Z}} \mid \text{ all but a finite number of } a_{ij} \text{ with } i\geq j \text{ are } 0\}.$$

The representations R and r extend from GL_∞ (resp. gl_∞) to $\widetilde{GL}_\infty$ (resp. $\widetilde{gl}_\infty$), and formulas (14.9.7) and (14.9.8) still hold.

Since $R_m^B(g(y)) = \exp \sum_{j\geq 1} y_j \frac{\partial}{\partial x_j}$, we obtain:

$$F(y) = \left(\exp\sum_{j\geq 1} y_j \frac{\partial}{\partial x_j}\right)P(x)\Big|_{x=0} = P(x+y)\Big|_{x=0}.$$

Thus, from the calculation in the bosonic picture, we have

$$F(y) = P(y). \tag{14.10.6}$$

On the other hand, by definition:

$$g(y) = \exp\sum_{j\geq 1} \bar{e}^j y_j = \sum_{m\geq 0} S_m(y)\bar{e}^m.$$

The latter expression is a matrix a with matrix entries $a_{mn} = S_{n-m}(y)$. Now we can read off the coefficient of $|m\rangle$ in the expansion of $\sigma_m(R(a)(\underline{i}_1 \wedge \underline{i}_2 \wedge \cdots))$ from (14.9.8), which gives $\det a^{i_1,i_2,i_3,\ldots}_{m,m-1,m-2,\ldots}$. This is easily seen to be $S_{i_1-m,i_2-(m-1),\ldots}(y)$. Comparing this with (14.10.6) completes the proof.

□

As a consequence of (14.9.8) and Theorem 14.10 b), we obtain the following formula for the action of GL_∞ on $B^{(m)}$:

$$R^B_m(a)S_\lambda = \sum_{\mu \in \mathrm{Par}} \left(\det a^{\lambda_1+m,\lambda_2+m-1,\ldots}_{\mu_1+m,\mu_2+m-1,\ldots}\right) S_\mu. \tag{14.10.7}$$

Another corollary of Theorem 14.10 is a vertex operator construction of the gl_∞- and a_∞-module $L(\Lambda_m)$:

COROLLARY 14.10. *Let r_m (resp. $\hat{r}_m$) denote the representation of gl_∞ (resp. a_∞) on $L(\Lambda_m)$. Then $L(\Lambda_m)$ can be identified with the space $\mathbb{C}[x_1, x_2, \ldots]$ such that*

$$\textstyle\sum_{i,j\in\mathbb{Z}} z_1^i z_2^{-j} r_m(E_{ij}) = \frac{(z_1/z_2)^m}{1-z_2/z_1}\Gamma(z_1,z_2), \tag{14.10.8}$$

$$\textstyle\sum_{i,j\in\mathbb{Z}} z_1^i z_2^{-j} \hat{r}_m(E_{ij}) = \frac{1}{1-z_2/z_1}((z_1/z_2)^m\Gamma(z_1,z_2) - 1), \tag{14.10.9}$$

where

$$\Gamma(z_1,z_2) = \left(\exp \textstyle\sum_{j\geq 1}(z_1^j - z_2^j)x_j\right)\left(\exp - \textstyle\sum_{j\geq 1} \frac{z_1^{-j}-z_2^{-j}}{j}\frac{\partial}{\partial x_j}\right). \tag{14.10.10}$$

Proof. Note that we have on $F^{(m)}$:

$$\sum_{i,j\in\mathbb{Z}} z_1^i z_2^{-j} r_m(E_{ij}) = \psi(z_1)\psi^*(z_2). \tag{14.10.11}$$

Note also that, by (14.8.12 and 13):

$$\Gamma_+(z_1)^{-1}\Gamma_-(z_2)^{-1} = \Gamma_-(z_2)^{-1}\Gamma_+(z_2)^{-1}(1-z_2/z_1)^{-1}. \tag{14.10.12}$$

Applying Theorem 14.10 b) to (14.10.11) and using (14.10.12) and (14.10.2 and 3) gives (14.10.8).

In the case of $\hat{r}_m$ we observe from (7.12.1) that we must simply subtract $\sum_{i\leq 0}(z_1/z_2)^i = (1-z_2/z_1)^{-1}$ from the right-hand side of (14.10.8) to get the right-hand side of (14.10.9).

□

Remark 14.10. It follows from the proof of Theorem 10.12 b) that the boson–fermion correspondence $\sigma : F \to B$ can be constructed explicitly as follows. Let $H(x) = \sum_{j\in\mathbf{N}} x_j\alpha_j$ and let $\varphi_m(a)$ denote the coefficient of $|m\rangle$ in $a \in F$ in the semi-infinite monomial basis. Then

$$\sigma(v) = \sum_{m\in\mathbf{Z}} \varphi_m(e^{H(x)}v)q^m.$$

§14.11. We proceed to explain how one uses the representation theory developed in previous sections to construct solutions of soliton equations. The starting point is the following trivial observation:

Remark 14.11. Consider a representation of a group G on a vector space V and a vector $v_0 \in V$. Let Ω_1 be an operator on V commuting with the action of G and suppose that v_0 satisfies the equation

$$\Omega_1 v = \lambda v, \quad \lambda \in \mathbf{C}. \tag{14.11.1}$$

Then any element of the orbit $G \cdot v_0$ satisfies this equation.

In this section we shall apply this observation to the group GL_∞ acting diagonally on $F \otimes F$, $v_0 = |0\rangle \otimes |0\rangle$ and

$$\Omega_1 = \sum_{j\in\mathbf{Z}} \psi_j \otimes \psi_j^*.$$

It is clear that $\Omega_1 v_0 = 0$; using (14.9.4) one checks immediately that Ω_1 commutes with gl_∞ and hence with GL_∞. Thus, due to Remark 14.11, any element τ of the orbit $GL_\infty|0\rangle$ satisfies the equation

$$\sum_{j\in\mathbf{Z}} \psi_j(\tau) \otimes \psi_j^*(\tau) = 0. \tag{14.11.2}$$

LEMMA 14.11. *The orbit $GL_\infty|0\rangle$ is the set of all nonzero solutions $\tau \in F^{(0)}$ of equation* (14.11.2).

Proof. Write $\tau = \tau_0 + \sum_{j=1}^{N} c_j\tau_j$ as a linear combination of semi-infinite monomials τ_i, among which τ_0 has the greatest energy. If among the τ_i with $i \geq 1$ there exists one, say τ_2, of the form

$$r_0(E_{ij})\tau_0 \text{ with } i < j, \tag{14.11.3}$$

we can kill off the term $c_2\tau_2$ by replacing τ with $R_0(\exp -c_2E_{ij})\tau$. Repeating this procedure a finite number of times we arrive at an element of the form $\tau_0 + \varphi$, where none of the semi-infinite monomials appearing in φ is equal to τ_0 or is of the form (14.11.3). Since $\tau_0 + \varphi$ satisfies (14.11.2), it follows that $\varphi = 0$. Since $\tau_0 \in GL_\infty|0\rangle$, we obtain that $\tau \in GL_\infty|0\rangle$.

□

Using the boson–fermion correspondence $\sigma : F \to B$, we may view (14.11.2) as an equation on $\tau \in B^{(0)} = \mathbb{C}[x_1, x_2, \ldots]$. We shall rewrite it in a more explicit form using Theorem 14.10 a). By definition, (14.11.2) can be rewritten as

$$z^0\text{-term of } \psi(z)\tau \otimes \psi^*(z)\tau = 0. \tag{14.11.4}$$

The isomorphism $\sigma : F^{(0)} \xrightarrow{\sim} B^{(0)} = \mathbb{C}[x_1, x_2, \cdots]$ extends to an isomorphism

$$\sigma \otimes \sigma : F^{(0)} \otimes F^{(0)} \xrightarrow{\sim} \mathbb{C}[x'_1, x'_2, \ldots ; x''_1, x''_2, \ldots].$$

We can transform (14.11.4) to the bosonic picture using Theorem 14.10 a) and (14.10.3), obtaining

(14.11.5)

$$\mathrm{Res}_{z=0}\left(\exp\sum_{j\geq 1} z^j(x'_j - x''_j)\right)\left(\exp -\sum_{j\geq 1}\frac{z^{-j}}{j}\left(\frac{\partial}{\partial x'_j} - \frac{\partial}{\partial x''_j}\right)\right)\tau(x')\tau(x'')=0.$$

Introducing new variables

$$x_j = \frac{1}{2}(x'_j + x''_j), \quad y_j = \frac{1}{2}(x'_j - x''_j), \tag{14.11.6}$$

so that

$$\frac{\partial}{\partial x_j} = \frac{\partial}{\partial x'_j} + \frac{\partial}{\partial x''_j}, \qquad \frac{\partial}{\partial y_j} = \frac{\partial}{\partial x'_j} - \frac{\partial}{\partial x''_j},$$

the latter equation becomes:

$$\mathrm{Res}_{z=0}\left(\exp 2\sum_{j\geq 1} z^j y_j\right)\left(\exp -\sum_{j\geq 1}\frac{z^{-j}}{j}\frac{\partial}{\partial y_j}\right)\tau(x+y)\tau(x-y) = 0.$$

After expanding exponentials with the help of the generating series (14.10.4) for elementary Schur polynomials and taking the z^{-1}- term, this equation becomes:

$$\sum_{j\geq 0} S_j(2y)S_{j+1}(-\tilde{\partial}_y)\tau(x+y)\tau(x-y) = 0, \tag{14.11.7}$$

where $\tilde{\partial}_y$ stands for $\left(\frac{\partial}{\partial y_1}, \frac{1}{2}\frac{\partial}{\partial y_2}, \frac{1}{3}\frac{\partial}{\partial y_3}, \dots\right)$.

Given a polynomial $P(x_1, x_2, \dots)$ depending on a finite number of the x_j, and two C^∞-functions $f(x)$ and $g(x)$, we denote by $P(D_1, D_2, \dots)f \cdot g$ the expression

$$P\left(\frac{\partial}{\partial u_1}, \frac{\partial}{\partial u_2}, \dots\right) f(x_1 + u_1, x_2 + u_2, \dots) g(x_1 - u_1, x_2 - u_2, \dots)\big|_{u=0}.$$

The equation $P(D)f \cdot g = 0$ is called a *Hirota bilinear equation*. For example, if $P = x_1^n$, then from Leibniz' formula we obtain:

$$D_1^n f \cdot g = \sum_{k=0}^{n} (-1)^k \binom{n}{k} \frac{\partial^k f}{\partial x_1^k} \frac{\partial^{n-k} g}{\partial x_1^{n-k}}.$$

Note that $Pf \cdot f \equiv 0$ if and only if $P(x) = -P(-x)$. This is called a trivial Hirota bilinear equation.

Now we rewrite equation (14.11.7) in a Hirota bilinear form by using the following trick based on Taylor's formula:

$$\begin{aligned} P(\tilde{\partial}_y)\tau(x+y)\tau(x-y) &= P(\tilde{\partial}_u)\tau(x+y+u)\tau(x-y-u)\big|_{u=0} \\ &= P(\tilde{\partial}_u)\left(\exp \sum_{j \geq 1} y_j \frac{\partial}{\partial u_j}\right)\tau(x+u)\tau(x-u)\big|_{u=0}. \end{aligned} \tag{14.11.8}$$

We obtain:

$$\sum_{j \geq 0} S_j(2y) S_{j+1}(-\tilde{D})\left(\exp \sum_{s \geq 1} y_s D_s\right)\tau \cdot \tau = 0. \tag{14.11.9}$$

Here $\tilde{D}$ stands for $(D_1, \frac{1}{2}D_2, \frac{1}{3}D_3, \dots)$.

If we expand (14.11.9) to a multiple Taylor series in the variables y_1, $y_2, \dots$, then each coefficient of the series must vanish, giving us thereby a nonlinear partial differential equation in a Hirota bilinear form. This system of equations is called the *KP hierarchy*.

For example, the coefficient of y_r is the following Hirota bilinear equation ($r \geq 1$):

$$(2S_{r+1}(-\tilde{D}) - D_1 D_r)\tau \cdot \tau = 0. \tag{14.11.10$_r$}$$

Using the explicit formula for $S_m(x)$ given in §14.10, we see that the Hirota bilinear equation $(14.11.10)_r$ is trivial for $r = 1$ and 2, whereas $(14.11.10)_3$ becomes, after dropping odd monomials and multiplying by 12:

$$(D_1^4 + 3D_2^2 - 4D_1 D_3)\tau \cdot \tau = 0. \tag{14.11.11}$$

This is the Kadomtzev–Petviashvili (KP) equation in the form of Hirota. Namely, putting

(14.11.12) $$u = 2\frac{\partial^2}{\partial x_1^2}\log\tau; \quad x_1 = x,\ x_2 = y,\ x_3 = t,$$

the equation (14.11.11) takes its classical form:

(14.11.13) $$\frac{3}{4}\frac{\partial^2 u}{\partial y^2} = \frac{\partial}{\partial x}\left(\frac{\partial u}{\partial t} - \frac{3}{2}u\frac{\partial u}{\partial x} - \frac{1}{4}\frac{\partial^3 u}{\partial x^3}\right).$$

In order to construct polynomial solutions of the KP hierarchy, note that any semi-infinite monomial of charge 0 lies in the orbit $GL_\infty|0\rangle$, and hence satisfies (14.11.2). Using Theorem 14.10 b), we obtain the following remarkable result:

PROPOSITION 14.11.1. *All Schur polynomials $S_\lambda(x)$ are solutions of the KP hierarchy.*

□

Another type of solution, the so-called soliton solutions, are constructed by using the vertex operator $\Gamma(z_1, z_2)$ (see (14.10.10)). Let $u_1, \ldots, u_N$, $v_1, \ldots, v_N$ be some indeterminates. In what follows, by the expressions $(u_j - u_i)^k$, $(u_j - v_i)^k$, etc., with $i < j$, we mean the formal power series expansion $u_j^k(1 - u_i/u_j)^k$, $u_j^k(1 - v_i/u_j)^k$, etc.

Let $\tau(x_1, x_2, \ldots)$ be a formal power series. Then using (14.8.13) we derive by induction on N the following formula:

(14.11.14)
$$\Gamma(u_N, v_N)\ldots\Gamma(u_1, v_1)\tau(x_1, x_2, \ldots)$$
$$= \prod_{1\le i<j\le N}\frac{(u_j - u_i)(v_j - v_i)}{(u_j - v_i)(v_j - u_i)}\left(\exp\sum_k\sum_{j=1}^N(u_j^k - v_j^k)x_k\right)$$
$$\times\,\tau\left(\ldots, x_k - \sum_{j=1}^N\frac{1}{k}(u_j^{-k} - v_j^{-k}), \ldots\right).$$

In particular, we have: $\Gamma(u, v)^2\tau(x) = 0$, so that

(14.11.15) $$\exp a\Gamma(u, v) = 1 + a\Gamma(u, v), \quad a \in \mathbf{C}.$$

Given complex numbers $a_1, \ldots, a_N$; $u_1, \ldots, u_N$; and $v_1, \ldots, v_N$ such that $u_i \ne v_j$ if $i \ne j$, we let

$$\tau_{a_1,\ldots,a_N;u_1,\ldots,u_N;v_1,\ldots,v_N}(x) = (1 + a_N\Gamma(u_N, v_N))\ldots(1 + a_1\Gamma(u_1, v_1))\cdot 1.$$

Since 1 is a solution of the KP hierarchy and $\Gamma(u, v)$ lies in a completion of gl_∞ by Corollary 14.10, we obtain:

PROPOSITION 14.11.2. *The function*

$$\begin{aligned}
&\tau_{a_1,\dots,a_N;u_1,\dots,u_N;v_1,\dots,v_N}(x) \\
&= \sum_{\substack{0\le r\le N \\ 1\le j_1<j_2<\dots j_r\le N}} \prod_{\nu=1}^{r} a_{j_\nu} \prod_{1\le\nu<\mu\le r} \frac{(u_{j_\nu}-u_{j_\mu})(v_{j_\nu}-v_{j_\mu})}{(u_{j_\nu}-v_{j_\mu})(v_{j_\nu}-u_{j_\mu})} \\
&\times \exp \sum_{k\ge 1}\sum_{\nu=1}^{r}(u_{j_\nu}^k - v_{j_\nu}^k)x_k.
\end{aligned}$$

is the solution of the KP *hierarchy.*

□

The τ-function $\tau_{a_1,\dots,a_N;u_1,\dots,u_N;v_1,\dots,v_N}(x)$ is called an *N-soliton solution* of the KP hierarchy. Of course, from the point of view of the single equation (14.12.4), the indeterminates $x_4, x_5, \dots$ are arbitrary parameters; an N-soliton solution τ, or rather $u(x) = 2\frac{\partial^2 \log \tau}{\partial x_1^2}$, describes the interaction of N waves during time x_3, in shallow water in plane coordinates x_1 and x_2.

For example, the 1-soliton solution

$$u(x,y,t) = (2\log \tau_{1;a;b}(x,y,t))_{xx}$$

of the classical KP equation (14.11.13) takes the following form:

$$u(x,y,t) = \frac{1}{2}(a-b)^2[\cosh\frac{1}{2}((a-b)x + (a^2-b^2)y + (a^3-b^3)t + \text{const})]^{-2}.$$

§14.12. We shall examine here what Remark 14.11 gives us when applied to an arbitrary principal vertex operator construction. We start with some general remarks concerning an arbitrary symmetrizable Kac–Moody algebra $\mathfrak{g}(A)$. Let $L(\Lambda)$ be an integrable highest-weight module over $\mathfrak{g}(A)$ and let v_Λ be a highest-weight vector. Recall that $L(\Lambda)$ carries a (unique) positive-definite Hermitian contravariant form H such that $H(v_\Lambda, v_\Lambda) = 1$ (see Theorem 11.7 b)). This form determines a positive-definite Hermitian form on the tensor product $L(\Lambda)\otimes L(\Lambda)$ by letting $H(u\otimes v, u'\otimes v') = H(u,u')H(v,v')$. Then we have an orthogonal direct sum of modules

(14.12.1) $$L(\Lambda)\otimes L(\Lambda) = L_{\text{high}} \oplus L_{\text{low}},$$

where L_{high}, the ***highest component***, is the $\mathfrak{g}(A)$-submodule generated by the vector $v_\Lambda \otimes v_\Lambda$, and L_{low} is its orthocomplement. Notice that L_{high} is isomorphic to $L(2\Lambda)$ by the complete reducibility theorem (Corollary 10.7).

Let G be the group of automorphisms of $L(\Lambda)$ generated by 1-parameter groups $\exp te_i$ and $\exp tf_i$, $i = 1, 2, \ldots, n$, where e_i, f_i are the Chevalley generators. Denote by $\mathcal{V}_\Lambda$ the G-orbit of v_Λ and by $\widehat{\mathcal{V}}_\Lambda$ its formal completion (see §1.5).

Finally, let Ω be the generalized Casimir operator (§2.5), and let Ω_2 be the operator on $L(\Lambda) \otimes L(\Lambda)$ introduced in §2.8, which commutes with $\mathfrak{g}(A)$. We can now state the result.

PROPOSITION 14.12. a) *If* $v \in \mathcal{V}_\Lambda$, *then*

$$v \otimes v \in L_{\text{high}}; \tag{14.12.2}$$

$$\Omega(v \otimes v) = 4(\Lambda + \rho|\Lambda) v \otimes v; \tag{14.12.3}$$

$$\Omega_2(v \otimes v) = (\Lambda|\Lambda) v \otimes v. \tag{14.12.4}$$

b) *The conditions* (14.12.2)–(14.12.4) *on* $v \in L(\Lambda)$ *are equivalent. (In fact, any nonzero vector* $v \in L(\Lambda)$ *satisfying* (14.12.3) *lies in* $\mathcal{V}_\Lambda$*; see Peterson–Kac [1983].).*

Proof. If $v = v_\Lambda$, then (14.12.2 and 3) are obvious, and (14.12.4) is clear if we choose

$$\Omega_2 = \sum_{\alpha \in \Delta \cup \{0\}} \sum_i e_\alpha^{(i)} \otimes e_{-\alpha}^{(i)}, \tag{14.12.5}$$

where $e_\alpha^{(i)}$ are bases of $\mathfrak{g}_\alpha$, $\alpha \in \Delta \cup \{0\}$ such that $(e_\alpha^{(i)}|e_{-\alpha}^{(j)}) = \delta_{ij}$. This implies a).

It is straightforward to check that equations (14.12.3 and 4) on $v \in L(\Lambda)$ are equivalent. The equivalence of (14.12.2 and 3) follows from the following lemma (since L_{low} is a direct sum of the $L(\Lambda')$ with $\Lambda' < 2\Lambda$):

LEMMA 14.12. *Let* $\Lambda, \Lambda' \in P_+$. *If* $\Lambda > \Lambda'$, *then*

$$(\Lambda + 2\rho|\Lambda) - (\Lambda' + 2\rho|\Lambda') > 0.$$

Proof. $|\Lambda + \rho|^2 - |\Lambda' + \rho|^2 = (\Lambda + \Lambda' + 2\rho|\Lambda - \Lambda') > 0.$

□

Now let $L(\Lambda_0) = \mathbb{C}[x_j; j \in E_+]$ be the principal vertex operator construction of the basic representation of an affine algebra. Then $L(\Lambda_0) \otimes L(\Lambda_0)$ can be thought of as the space of polynomials on two sets of variables: $\mathbb{C}[x_j', x_j''; j \in E_+]$. We introduce new variables x_j and y_j, $j \in E_+$, by

(14.11.6), so that $L(\Lambda_0) \otimes L(\Lambda_0) = \mathbb{C}[x_j, y_j; j \in E_+]$. Then the principal subalgebra $\mathfrak{s}$ acts on $L(\Lambda_0) \otimes L(\Lambda_0)$ as follows:

$$p_j \mapsto \frac{\partial}{\partial x_j}, \quad q_j \mapsto 2x_j, \quad K \mapsto 2.$$

Note that by (9.13.4) and Proposition 14.4, we have the following formula for the contravariant Hermitian form on $L(\Lambda_0)$:

$$H(P,Q) = \overline{P}(\ldots, \frac{1}{j}\frac{\partial}{\partial x_j}, \ldots)Q(\ldots, x_j, \ldots)(0), \quad j \in E_+. \tag{14.12.6}$$

Now we are in a position to rewrite equation (14.12.2) in terms of the Hirota bilinear equations.

LEMMA 14.12. *$\tau \otimes \tau \in L_{\text{high}}$ if and only if*

$$P(\ldots, \frac{1}{2j}D_j, \ldots)\tau \cdot \tau = 0$$

for all $P(y) \in \mathbb{C}[y_j; j \in E_+] \cap L_{\text{low}} \subset L(\Lambda_0) \otimes L(\Lambda_0)$.

Proof. Applying Lemma 9.13 b) to the $\mathfrak{s}$-module L_{low}, we have

$$L_{\text{low}} = (L_{\text{low}} \cap \mathbb{C}[y_j; j \in E_+]) \otimes \mathbb{C}[x_j; j \in E_+]. \tag{14.12.7}$$

Hence, $\tau \otimes \tau \in L_{\text{high}}$ if and only if

$$H\left(\sum_i Q_i(x)P_i(y),\ \tau(x')\tau(x'')\right) = 0, \tag{14.12.8}$$

where $Q_i \in \mathbb{C}[x_j; j \in E_+]$ are arbitrary and $P_i \in \mathbb{C}[y_j; j \in E_+] \cap L_{\text{low}}$. Applying (14.12.6), condition (14.12.8) is equivalent to:

$$\sum_i \overline{Q}_i\left(\frac{\partial}{\partial x_1}, \frac{1}{2j}\frac{\partial}{\partial x_j}, \ldots\right)\overline{P}_i\left(\frac{\partial}{\partial y_1}, \frac{1}{2j}\frac{\partial}{\partial y_j}, \ldots\right)\tau(x+y)\tau(x-y)\big|_{y=0,x=0} = 0,$$

where P and Q are the same as above. Since this holds for arbitrary Q_i, and since the subspace $\mathbb{C}[y] \cap L_{\text{low}}$ is invariant under complex conjugation, we get the result.

□

The elements from the space

$$Hir := \mathbb{C}[y_j; j \in E_+] \cap L_{\text{low}}$$

are called *Hirota polynomials*, and the elements from the formal completion $\widehat{V}_{\Lambda_0}$ in $\mathbb{C}[[x_j; j \in E_+]]$ of the orbit of 1 are called τ*-functions*. Thus, in order to derive applications of Proposition 14.12 to PDE, we have to be able to construct explicitly two things: a) even Hirota polynomials; b) τ-functions.

Denote by $H_k \subset \mathbb{C}[y_j; j \in E_+]$ the space of Hirota polynomials of principal degree k, so that $Hir = \bigoplus_k H_k$ (recall that the principal degree is defined by $\deg y_j = j$). By formulas (14.12.7) and (14.12.1), we have

$$\begin{aligned}\dim_q Hir &:= \sum_{k=0}^{\infty} (\dim H_k) q^k \\ &= \left((\dim_q L(\Lambda_0))^2 - \dim_q L(2\Lambda_0)\right) \prod_{k \in E_+} (1-q^k).\end{aligned} \tag{14.12.9}$$

(Recall that $\dim_q L(\Lambda)$ can be computed by Proposition 10.10.)

Let us apply formula (14.12.9) to the infinite rank affine algebra of type A_∞. Formula (10.10.1) can be written in this case in the following nice form:

$$\dim_q L(\Lambda_{s_1} + \cdots + \Lambda_{s_n}) = \prod_{1 \le i < j \le n} (1 - q^{s_i - s_j + j - i}) / \varphi(q)^n, \tag{14.12.10}$$

where $s_1 \ge s_2 \ge \cdots \ge s_n$ are arbitrary integers. In particular, we have:

$$\dim_q L(\Lambda_0) = \varphi(q)^{-1}, \qquad \dim_q L(2\Lambda_0) = (1-q)\varphi(q)^{-2}.$$

Hence we obtain: $\dim_q Hir = q\varphi(q)^{-1}$, i.e.,

$$\dim H_k = p(k-1). \tag{14.12.11}$$

Denote by $P_{\lambda_1, \lambda_2, \ldots}(D)$ the coefficient of the monomial $y_1^{\lambda_1} y_2^{\lambda_2} \ldots$ in (14.11.9). It is easy to see that

$$P_{\lambda_1, \lambda_2, \ldots}(y) \in Hir_{\lambda_1 + \lambda_2 + \cdots + 1}. \tag{14.12.12}$$

On the other hand, note that

$$\Omega_2 = \Omega_1^* \Omega_1.$$

Hence equation (14.11.2) implies (14.12.4). Conversely, if v satisfies (14.12.4), then $H(\Omega_1^* \Omega_1 (v \otimes v), (v \otimes v)) = 0$, hence $H(\Omega_1(v \otimes v), \Omega_1(v \otimes v)) = 0$, hence v satisfies (14.11.2). It follows that the polynomials $P_{\lambda_1, \lambda_2, \ldots}(y)$ span Hir, by Proposition 14.12 b). Comparing (14.12.11) and (14.12.12), we arrive at the following result.

THEOREM 14.12. *The Hirota bilinear equations $P_{\lambda_1,\lambda_2,\ldots}(D)\tau \cdot \tau = 0$ form a basis of the* KP *hierarchy of Hirota bilinear equations.*

□

Note that for $j \leq 3$ (resp. $j = 4$) the number of trivial Hirota equations is $p(j-1)$ (resp. $p(3)-1$). Hence (14.11.11) is the nontrivial Hirota equation of lowest principal degree.

§14.13. The most celebrated example, the Korteweg–de Vries (KdV) equation, occurs in the context of the principal vertex operator construction of the basic representation of the affine algebra $\mathfrak{g}$ of type $A_1^{(1)}$. Recall the basic realization of this algebra:

$$\mathfrak{g} = s\ell_2(\mathbf{C}[t,t^{-1}]) + \mathbf{C}K + \mathbf{C}d,$$
$$[a(t), b(t)] = a(t)b(t) - b(t)a(t) + (\mathrm{Res}_{t=0}\, \frac{da}{dt} b)K,$$
$$[\mathfrak{g}, K] = 0, \qquad [d, a(t)] = t\frac{da(t)}{dt}.$$

Recall the principal vertex operator construction of the $\mathfrak{g}$-module $L(\Lambda_0)$ (cf. §14.6 and Exercise 14.12):

$$L(\Lambda_0) = \mathbf{C}[x_1, x_3, x_5, \ldots];$$
$$H_j \mapsto \frac{\partial}{\partial x_j}, \quad H_{-j} \mapsto jx_j, \quad j \in \mathbf{N}^{\mathrm{odd}};$$
$$K \mapsto 1, \qquad 2d - \frac{1}{2}A_0 \mapsto - \sum_{j\in \mathbf{N}^{\mathrm{odd}}} jx_j \frac{\partial}{\partial x_j},$$
$$A(z) := \sum_{j\in\mathbf{Z}} z^{-j}A_j \mapsto \frac{1}{2}(\Gamma(z) - 1),$$

where

$$H_{2j+1} = t^j \begin{pmatrix} 0 & 1 \\ t & 0 \end{pmatrix}, \qquad A_{2j} = t^j \begin{pmatrix} -1 & 0 \\ 0 & 1 \end{pmatrix}, \qquad A_{2j+1} = t^j \begin{pmatrix} 0 & 1 \\ -t & 0 \end{pmatrix},$$

and

$$\Gamma(z) = \left(\exp 2 \sum_{j\in\mathbf{N}^{\mathrm{odd}}} z^j x_j\right)\left(\exp -2 \sum_{j\in\mathbf{N}^{\mathrm{odd}}} \frac{z^{-j}}{j}\frac{\partial}{\partial x_j}\right).$$

We choose the following dual bases of $\mathfrak{g}$:

$$\left\{\frac{1}{\sqrt{2}}H_{2j+1},\ \frac{1}{\sqrt{2}}A_j,\ K,\ d\right\} \quad \text{and}$$
$$\left\{\frac{1}{\sqrt{2}}H_{-2j-1},\ \frac{1}{\sqrt{2}}A_{-j},\ d,\ K\right\}, \quad j \in \mathbf{Z},$$

and take the corresponding operator Ω_2 on $L(\Lambda_0) \otimes L(\Lambda_0)$. Then calculations similar to those in §14.11 show that equation (14.12.4) is equivalent to the following hierarchy of Hirota bilinear equations:

(14.13.1)

$$\left(\sum_{n\in\mathbb{N}} S_n(4y_1, 0, 4y_3, 0, \ldots) S_n(-\frac{2}{1}D_1, 0, -\frac{2}{3}D_3, 0, \ldots) - 8 \sum_{j\in\mathbb{N}^{\mathrm{odd}}} j y_j D_j\right) \times \left(\exp \sum_{j\in\mathbb{N}^{\mathrm{odd}}} y_j D_j\right)\tau\cdot\tau = 0.$$

On the other hand, formula (14.12.9) gives in this case (cf. Exercise 14.3):

$$\dim_q Hir = \prod_{j\geq 1}(1-q^{2j-1})^{-1} - \prod_{j\geq 1}(1-q^{4j-2})^{-1}. \tag{14.13.2}$$

It is clear that all Hirota equations of odd principal degree are trivial. Also, by (14.13.2), we have $\dim H_2 = 0$ and $\dim H_4 = 1$. Looking at the coefficient of $y_1 y_2$ in (14.13.1), we see that the unique Hirota bilinear equation of (the lowest) principal degree 4 is

$$(D_1^4 - 4D_1D_3)\tau\cdot\tau = 0. \tag{14.13.3}$$

The same change of functions and variables in the corresponding Hirota bilinear equation as in (14.11.2) (except that there is no x_2), gives the Korteweg–de Vries equation:

$$\frac{\partial u}{\partial t} = \frac{3}{2}u\frac{\partial u}{\partial x} + \frac{1}{4}\frac{\partial^3 u}{\partial x^3}. \tag{14.13.4}$$

The N-soliton solutions for the *KdV hierarchy* (14.13.1) are constructed in the same way as those for the KP hierarchy. The answer is as follows:

$$\tau_{a_1,\ldots,a_N;u_1,\ldots,u_N}(x_1, x_3, \ldots) = \sum_{\substack{0\leq r\leq N\\ 1\leq j_1<j_2<\cdots j_r\leq N}} \prod_{\nu=1}^{r} a_{j_\nu} \prod_{1\leq\nu<\mu\leq r} \frac{(u_{j_\nu}-u_{j_\mu})^2}{(u_{j_\nu}+u_{j_\mu})^2} \exp 2\sum_{\substack{k\geq 1\\ k \text{ odd}}}\sum_{\nu=1}^{r} u_{j_\nu}^k x_k.$$

The function $2\frac{\partial^2}{\partial x_1^2}\log\tau_{a_1,\ldots,a_N;u_1,\ldots,u_N}(x)$ describes the interaction of N waves during time x_3 in a narrow channel in coordinate x_1.

In particular, the 1-soliton solution $u(x,t) = 2(\log\tau_{1;a}(x,t))_{xx}$ describes a solitary wave:

$$u(x,t) = 2a[\cosh(ax + a^3t + \text{const})]^{-2}.$$

Actually, all these results for the KdV hierarchy can be deduced from the corresponding results for the KP hierarchy using the so-called reduction procedure. This amounts to putting $v_i = -u_i$. The representation-theoretical meaning of this procedure is explained in Exercises 14.11 and 14.12.

Another celebrated example, the nonlinear Schrödinger (NLS) equation, is related to the homogeneous vertex operator construction of the same basic representation of the same affine algebra $\mathfrak{g}$ of type $A_1^{(1)}$. We let

$$\alpha = \begin{pmatrix} 1 & 0 \\ 0 & -1 \end{pmatrix}, \qquad e = \begin{pmatrix} 0 & 1 \\ 0 & 0 \end{pmatrix}, \qquad f = \begin{pmatrix} 0 & 0 \\ 1 & 0 \end{pmatrix}$$

and choose the following dual bases of $\mathfrak{g}$:

$$\{t^n\alpha,\ t^n e,\ t^n f,\ K,\ d\}, \text{ and } \{\tfrac{1}{2}t^{-n}\alpha,\ t^{-n}f,\ t^{-n}e,\ d,\ K\}.$$

We have: $Q = \mathbb{Z}\alpha$, $(\alpha|\alpha) = 2$, $\varepsilon(m\alpha, n\alpha) = (-1)^{mn}$, so that $c_{\pm\alpha}(f \otimes e^{n\alpha}) = (-1)^n$. Letting $q = e^\alpha$, we identify $\mathbb{C}[Q]$ with $\mathbb{C}[q, q^{-1}]$. Thus, the homogeneous vertex operator construction can be described as follows:

$$L(\Lambda_0) = \mathbb{C}[x_1, x_2, \ldots; q, q^{-1}];$$

$$\alpha^{(n)} \mapsto 2\frac{\partial}{\partial x_n} \quad \text{and} \quad \alpha^{(-n)} \mapsto nx_n \quad \text{for} \quad n > 0, \qquad \alpha^{(0)} \mapsto 2q\frac{\partial}{\partial q};$$

$$K \mapsto 1, \qquad d \mapsto -\left(q\frac{\partial}{\partial q}\right)^2 - \sum_{n \geq 1} nx_n \frac{\partial}{\partial x_n};$$

$$E(z) := \sum_{n \in \mathbb{Z}} E^{(n)} z^{-n-1} \mapsto \Gamma_+(z), \qquad F(z) := \sum_{n \in \mathbb{Z}} F^{(n)} z^{-n-1} \mapsto \Gamma_-(z),$$

where

$$\Gamma_\pm(z) = \left(\exp \pm \sum_{j \geq 1} z^j x_j\right)\left(\exp \mp 2 \sum_{j \geq 1} \frac{z^{-j}}{j}\frac{\partial}{\partial x_j}\right) q^{\pm 1} z^{\pm 2q\frac{\partial}{\partial q}} c_{\pm\alpha}$$

(note that $z^{\pm 2q\frac{\partial}{\partial q}}(q^n) = z^{\pm 2n} q^n$).

Taking the operator Ω_2 on $L(\Lambda_0) \otimes L(\Lambda_0)$ corresponding to the above choice of dual bases and performing calculations similar to those in §14.11, we find that equation (14.12.4) is equivalent to the following hierarchy of

Hirota bilinear equations on $\tau = \sum_{s\in\mathbb{Z}} \tau_s(x)q^s$ $(m, n \in \mathbb{Z})$:

(14.13.5)

$$\Big((m-n)^2 + 2\sum_{j\geq 1} jy_jD_j\Big)\Big(\exp\sum_{s\geq 1} y_sD_s\Big)\tau_m\cdot\tau_n$$
$$+(-1)^{m-n}\sum_{j\geq 1} S_j(2y)S_{j+2m-2n-2}(-2\tilde{D})\Big(\exp\sum_{s\geq 1} y_sD_s\Big)\tau_{m-1}\cdot\tau_{n+1}$$
$$+(-1)^{m-n}\sum_{j\geq 1} S_j(-2y)S_{j-2m+2n-2}(2\tilde{D})\Big(\exp\sum_{s\geq 1} y_sD_s\Big)\tau_{m+1}\cdot\tau_{n-1} = 0.$$

Note that this hierarchy is invariant with respect to the translation $m \mapsto m+r$, $n \mapsto n+r$, $r \in \mathbb{Z}$. We look at the coefficient of y_1^s and restrict ourselves to the case $m-n = 0, -1,$ or 1. For $s = 0$ or 1 we then get trivial equations, whereas for $s = 2$ we obtain the following equations respectively on functions $\tau_n(x)$, $\tau_{n-1}(x)$, and $\tau_{n+1}(x)$:

$$\begin{cases} D_1^2\tau_n\cdot\tau_n + 2\tau_{n-1}\tau_{n+1} = 0, \\ (D_1^2 + D_2)\tau_n\cdot\tau_{n+1} = 0, \\ (D_1^2 + D_2)\tau_{n-1}\cdot\tau_n = 0. \end{cases} \tag{14.13.6}$$

Letting $x = x_1$, $t = x_2$, $q(x,t) = \tau_1/\tau_0$, $q^*(x,t) = \tau_{-1}/\tau_0$, $u(x,t) = \log\tau_0$, we transform equations (14.13.6) into the following equations respectively:

$$u_{xx} = -qq^*, \qquad -q_t + q_{xx} + 2qu_{xx} = 0, \qquad q_t^* + q_{xx}^* + 2q^*u_{xx} = 0.$$

Excluding u, we arrive at the NLS system of PDE's on functions q and q^*:

$$\begin{cases} q_t = q_{xx} - 2q^2q^*, \\ q_t^* = -q_{xx}^* + 2qq^{*2}. \end{cases} \tag{14.13.7}$$

Note that imposing the constraint

$$q^*(x, it) = \pm\overline{q(x,it)} := g(x,t),$$

we get the classical nonlinear Schrödinger equation on the function g:

$$ig_t = -g_{xx} \pm 2|g|^2g. \tag{14.13.8}$$

For any sequence of signs, say, $++-\cdots$, and numbers $a_1, a_2, \ldots \in \mathbb{C}$, $z_1, z_2, \ldots \in \mathbb{C}^\times$ such that $|z_1| < |z_2| < \cdots$ we can construct a solution of the *NLS hierarchy* (14.13.5):

$$\tau^{++-\cdots}_{a_1,a_2,\ldots;z_1,z_2,\ldots}(x,q)$$
$$= \ldots(1 + a_3\Gamma_-(z_3))(1 + a_2\Gamma_+(z_2))(1 + a_1\Gamma_+(z_1))1\otimes 1.$$

The proof that this is a solution is the same as in §14.11 for the KP hierarchy.

§14.14. Exercises.

14.1. Show that Proposition 14.4, Lemma 14.6, and Theorem 14.6 hold if Λ_0 is replaced by $\Lambda \in P_+^1$.

14.2. Let $\mathfrak{g}$ be a simple finite-dimensional Lie algebra of rank ℓ, let h be its Coxeter number, and E_0 be the set of exponents. Given a positive integer j, let c_j denote the number of roots of $\mathfrak{g}$ of height j and let M_j denote the number of times j appears in E_0. Check their following properties:

$$c_j = c_{j-1} - M_{j-1}, \qquad M_j = M_{h-j}.$$

Deduce by induction on j the following formulas (cf. Lemma 14.2 e)):

$$\begin{aligned} c_j + c_{h-j} &= \ell + M_j, \\ c_j + c_{h+1-j} &= \ell, \\ c_j + c_{h+2-j} &= \ell - M_{j-1}. \end{aligned}$$

14.3. Let A be an affine type matrix such that ${}^tA = X_\ell^{(1)}$ and let $\mathfrak{g}$ be a simple finite-dimensional Lie algebra of type X_ℓ. Let $L(\Lambda)$ be an integrable $\mathfrak{g}(A)$-module of level 1. Deduce from Exercise 14.2 the following formulas for q-dimensions (cf. (14.4.4)):

$$\begin{aligned} \dim_q L(\Lambda) &= \prod_{j\in E_0} \prod_{n\in\mathbb{Z}_+} (1 - q^{j+nh})^{-1}, \\ \dim_q L(2\Lambda) &= \prod_{j\in E_0} \prod_{n\in\mathbb{Z}_+} (1 - q^{j+nh})^{-1}(1 - q^{j+1+n(h+2)})^{-1}. \end{aligned}$$

Using these formulas, rewrite (14.12.9) in a more explicit form:

$$\dim_q Hir = \prod_{j\in E_0} \prod_{n\in\mathbb{Z}_+} (1 - q^{j+nh})^{-1} - \prod_{j\in E_0} \prod_{n\in\mathbb{Z}_+} (1 - q^{j+1+n(h+2)})^{-1}.$$

14.4. Show that the principal vertex operator construction of the basic $\mathfrak{g}'(A)$-module $L(\Lambda_0)$ given by Theorem 14.6 extends to the semidirect product $\mathfrak{g}'(A) \rtimes Vir$ by letting

$$d_0 \mapsto \frac{1}{h^{(r)}} \sum_{j\in E_+} \alpha_{-j}\alpha_j + \frac{1}{4h^{(r)2}} \sum_{j\in E_0} j(h^{(r)} - j), \quad d_n \mapsto \frac{1}{2h^{(r)}} \sum_{j\in E} \alpha_{nh^{(r)}-j}\alpha_j,$$

where, for $j \in E_+$, we let $\alpha_j = \frac{\partial}{\partial x_j}$, $\alpha_{-j} = jx_j$. The conformal central charge is then equal to N.

In Exercises 14.5–14.7 we discuss the homogeneous vertex operator construction of the basic representation σ.

14.5. Let T^σ be the group introduced in Exercise 12.19 (acting on the space $L(\Lambda_0)$), and let $\epsilon : \mathring{Q}^\vee \times \mathring{Q}^\vee \to \{\pm 1\}$ be an asymmetry function. Show that $T^\sigma \xrightarrow{\sim} \mathring{Q}^\vee \times \{\pm 1\}$ with multiplication:

$$(\alpha, a)(\beta, b) = \epsilon(\alpha, \beta)(\alpha + \beta, ab), \quad \alpha, \beta \in \mathring{Q}^\vee,\ a, b \in \{\pm 1\}.$$

14.6. Let $\mathfrak{g}'(A)$ be an affine algebra, where either $r = 1$ and A is symmetric, or $r = 2$ or 3. Let $\tilde{\mathfrak{t}} = \mathbf{C}K \oplus (\bigoplus_{j \in \mathbf{Z}} \mathfrak{g}_{j\delta})$. Using Proposition 12.13, show that there are no nontrivial subspaces in $L(\Lambda_0)$, invariant with respect to all operators from $\sigma(\tilde{\mathfrak{t}})$ and T^σ.

14.7. Assume now that A is a symmetric affine matrix, so that $\mathring{A}$ is of type A_ℓ, D_ℓ, or E_ℓ. Then $\mathring{Q}^\vee$ can be identified with the root lattice $\mathring{Q}$ of $\mathfrak{g}(\mathring{A})$ via identifying their bases $\alpha_i^\vee \leftrightarrow \alpha_i$, preserving the bilinear form $(.|.)$; let $\mathring{\Delta} \subset \mathring{Q}$ be the root system of $\mathfrak{g}(\mathring{A})$. Recall that $\mathring{\mathfrak{g}} := \mathfrak{g}(\mathring{A}) = \mathring{\mathfrak{h}} \oplus (\bigoplus_{\alpha \in \mathring{\Delta}} \mathbf{C}E_\alpha)$ with commutation relations (7.8.5). Choose $t_\gamma^\sigma \in T^\sigma$ with image $\gamma \in \mathring{Q}$, and such that $t_\beta^\sigma t_\gamma^\sigma = \epsilon(\beta, \gamma) t_{\beta+\gamma}^\sigma$, $\beta, \gamma \in \mathring{Q}$. Recall that $\mathfrak{g}'(A) = \mathbf{C}[t, t^{-1}] \otimes_{\mathbf{C}} \mathring{\mathfrak{g}} \oplus \mathbf{C}K$ and $\tilde{\mathfrak{t}} = \mathbf{C}K \oplus (\bigoplus_{j \in \mathbf{Z}} (t^j \otimes \mathring{\mathfrak{h}}))$. Put $\mathfrak{t}_\pm = \bigoplus_{j>0} (t^{\pm j} \otimes \mathring{\mathfrak{h}})$, so that $\mathfrak{t} = \mathfrak{t}_- \oplus \mathbf{C}K \oplus \mathfrak{t}_+$ is the homogeneous Heisenberg subalgebra. Using Exercise 14.6, show that $L(\Lambda_0)$ considered as $(T^\sigma, \tilde{\mathfrak{t}})$-module, can be identified with the space $V := \mathbf{C}[\mathring{Q}] \otimes_{\mathbf{C}} S(\mathfrak{t}_-)$ with the following action of T^σ and $\tilde{\mathfrak{t}}$:

$$t_\beta^\sigma (e^\gamma \otimes P) = \epsilon(\beta, \gamma) e^{\beta+\gamma} \otimes P,$$

$$\sigma(t^k \otimes h) = h(k), \quad \sigma(K) = I.$$

Check that

$$\begin{cases} [t^k \otimes h, \widehat{E}_\gamma(z)] = \langle \gamma, h \rangle z^k \widehat{E}_\gamma(z), & h \in \mathfrak{h},\ k \in \mathbf{Z}; \\ t_\alpha^{\mathrm{ad}} \cdot \widehat{E}_\gamma(z) = z^{-(\alpha, \gamma)} \widehat{E}_\gamma(z), & \alpha \in \mathring{Q}^\vee. \end{cases}$$

Deduce the homogeneous vertex operator construction of the basic representation of $\mathfrak{g}'(A)$. Let $\{u_i\}$ and $\{u^i\}$ be dual bases of $\mathring{\mathfrak{h}}$. Introduce the following operators on the space V: D_0 by formula (14.8.3), and

$$D_m = \frac{1}{2} \sum_{j \in \mathbf{Z}} \sum_{i=1}^{\ell} u_i(-j) u_i(j + m) \text{ for } m \in \mathbf{Z} \backslash \{0\}. \qquad (14.14.1)$$

Show that the map $d_i \mapsto D_i$, $c \mapsto \ell I$ extends the $\mathfrak{g}'(A)$-module V to a $\mathfrak{g}'(A) \rtimes Vir$-module.

14.8. Let Q be an arbitrary even lattice and let $\mathfrak{h}$ be the complex span of Q. Introduce the space V by formula (14.8.1) and operators $u(n)$, for $u \in \mathfrak{h}$ and $n \in \mathbf{Z}$ by (14.8.2). Let D_n be operators on V defined by formulas (14.8.3) and (14.14.1). Show that they form a representation of the Virasoro algebra in V with conformal central charge $c = \dim \mathfrak{h}$. Let $\varepsilon : Q \times Q \to \{\pm 1\}$ be an asymmetry function and let c_γ, $\gamma \in Q$, be the corresponding sign operators on V. For $\gamma \in Q$ define the vertex operator $\Gamma_\gamma(z)$ by formula (14.8.5). Show that this is a primary field of conformal weight $\frac{1}{2}(\gamma|\gamma)$. Prove the following formula:
(14.14.2)

$$\Gamma_{\gamma_N}(z_N)\ldots\Gamma_{\gamma_1}(z_1)(1\otimes 1) = \prod_{1\le i<j\le N} \varepsilon(\gamma_i,\gamma_j)(z_i - z_j)^{(\gamma_i|\gamma_j)}$$

$$\times \left(\prod_{i=1}^{N} \exp \sum_{j\in\mathbf{N}} \frac{z_i^j}{j}\gamma_i(-j)\right) \exp \sum_{i=1}^{N} \gamma_i,$$

where by $(z_i - z_j)^m$ we mean the power series expansion of $(-z_j)^m \times (1 - z_i/z_j)^m$.

14.9. In the proof of Theorem 14.8 deduce that σ is a basic representation from (14.14.2) (avoiding thereby the use of the character formula).

14.10. Equation (14.12.4) is equivalent to the system of equations of the form $P = 0$, where $P \in S^2(L^*(\Lambda))$; show that these P generate the ideal of functions from $S(L^*(\Lambda))$ vanishing on $\mathcal{V}_\Lambda$.

[Let c_M denote the eigenvalue of the Casimir operator Ω on $L(M)$. Show that $(\Omega - c_{s\Lambda})v^s = \frac{1}{2}s(s-1)((\Omega - c_{2\Lambda})v^2)v^{s-2}$. Now use Lemma 14.12.]

14.11. Fix $n > 0$. Let $\overline{v}_1, \ldots, \overline{v}_n$ be the standard basis of $\mathbf{C}^n$. We identify $\mathbf{C}[t,t^{-1}] \otimes \mathbf{C}^n$ with $\mathbf{C}^\infty = \bigoplus_{j\in\mathbf{Z}} \mathbf{C}v_j$ by setting $v_{nk+i} = t^{-k} \otimes \overline{v}_i$. This gives us an embedding $gl_n(\mathbf{C}[t,t^{-1}]) \to \overline{a}_\infty$. Show that the subalgebra thus obtained consists of all matrices $(c_{ij})_{i,j\in\mathbf{Z}} \in \overline{a}_\infty$ such that $c_{i+n,j+n} = c_{ij}$. Show that the restriction of the central extension $a_\infty \to \overline{a}_\infty$ to this subalgebra is isomorphic to $\widetilde{\mathcal{L}}(gl_n) := gl_n(\mathbf{C}[t,t^{-1}]) \oplus \mathbf{C}K$ with bracket

$$[A(t), B(t)] = A(t)B(t) - B(t)A(t) \oplus \left(\operatorname{Res}\operatorname{tr} \frac{dA(t)}{dt} B(t)\right) K.$$

Show that the fundamental a_∞-module $L(\Lambda_s)$ remains irreducible when restricted to $\widetilde{\mathcal{L}}(gl_n)$.

14.12. Consider the vertex operator construction of the fundamental a_∞-module $L(\Lambda_s)$, $s = 0, \ldots, n-1$, on the space $\mathbf{C}[x_j; j = 1, 2, \ldots]$. Show that the subspace $\mathbb{C}[x_j; j \not\equiv 0 \bmod n]$ is invariant with respect to the affine algebra $\widetilde{\mathcal{L}}(s\ell_n) := \{A(t) + \mathbb{C}K \mid A(t) \in \widetilde{\mathcal{L}}(gl_n),\ \mathrm{tr}\ A(t) = 0\}$ of type $A_{n-1}^{(1)}$, **viewed as a subalgebra of a_∞. Show that we obtain the principal vertex operator construction of the $\widetilde{\mathcal{L}}(s\ell_n)$-module $L(\Lambda_s)$, given by the following formulas:**

$$\left(E_{n1}^{(1)} + \sum_{i=1}^{n-1} E_{i,i+1}^{(0)}\right)^j \mapsto \frac{\partial}{\partial x_j}\ (\text{resp. } -jx_{-j})$$

$$\text{if } j > 0\ (\text{resp.} < 0), j \not\equiv 0 \mod n;$$

$$\sum_{k=0}^{n-1}\sum_{m\in\mathbb{Z}} z^{-k-nm}\left(\sum_{j=k+1}^{n} \varepsilon^{-j} E_{j-k,j}^{(m)} + \sum_{j=1}^{k} \varepsilon^{-j} E_{j-k+n,j}^{(m+1)}\right)$$

$$\mapsto \frac{\epsilon^{s+1}}{\epsilon - 1}\Gamma(\epsilon z, z) - \frac{\epsilon}{\epsilon - 1},$$

where ε runs over all nth roots of 1 different from 1.

Deduce that a τ-function for the KP hierarchy is a τ-function for the $A_{n-1}^{(1)}$-hierarchy if and only if τ is independent of x_j with $j \equiv 0 \mod n$. Deduce that the set of all solutions of equation (14.2.4) in $L(\Lambda_s)$ is $\mathcal{V}_{\Lambda_0} \cup \{0\}$.

14.13. **Consider the Clifford algebra Cl_B on generators ϕ_i, $i \in \mathbb{Z}$, with defining relations**

$$\phi_i\phi_j + \phi_j\phi_i = (-1)^i\delta_{i,-j},$$

and consider it s irreducible module V with the vacuum vector $|0\rangle$ subject to conditions

$$\phi_i|0\rangle = 0 \text{ for } i < 0.$$

Introduce the *neutral fermionic field*

$$\phi(z) = \sum_{i\in\mathbb{Z}} \phi_i z^i,$$

and the operator q on V defined by

$$q|0\rangle = \phi_0|0\rangle, \qquad q\phi_i = \phi_i q.$$

We have: $q^2 = \frac{1}{2}$. Introduce the *neutral bosons* λ_n, $n \in \mathbb{Z}^{\mathrm{odd}}$, by

$$\lambda_n = \frac{1}{2}\sum_{j\geq 1}(-1)^{j+1}\phi_j\phi_{-j-n},$$

and show that

$$[\lambda_m, \lambda_n] = \frac{1}{2}m\delta_{m,-n}.$$

Let $\sigma = \sigma_B : V \xrightarrow{\sim} \mathbf{C}[x_1, x_3, x_5, \dots ; q]/(q^2 - \frac{1}{2})$ be an isomorphism of vector spaces such that

$$\sigma|0\rangle = 1, \sigma\phi_0|0\rangle = q;$$
$$\sigma\lambda_m\sigma^{-1} = \frac{\partial}{\partial x_m}, \quad \sigma\lambda_{-m}\sigma^{-1} = \frac{1}{2}mx_m \quad \text{for } m \in \mathbf{N}^{\text{odd}}.$$

Show that

$$\sigma\phi(z)\sigma^{-1} = q\left(\exp \sum_{j\in\mathbf{N}_{\text{odd}}} x_j z^j\right)\left(\exp -2\sum_{j\in\mathbf{N}^{\text{odd}}} \frac{z^{-j}}{j}\frac{\partial}{\partial x_j}\right).$$

The map σ is called the *boson-fermion correspondence of type B*. Explicitly:

$$\sigma(v) = \varphi_0(e^{H_B(x)}v) + \varphi_1(e^{H_B(x)}v)q,$$

where $H_B(x) = \sum_{n\in\mathbf{N}^{\text{odd}}} x_n\lambda_n$ and φ_0 (resp. φ_1) (a) is the coefficient of $|0\rangle$ (resp. $q|0\rangle$) in a.

14.14. Let $F_{ij} = (-1)^j E_{ij} - (-1)^i E_{-j,-i}$, $i, j \in \mathbf{Z}$, $(i,j) \neq (0,0)$, be the standard basis of so_∞. Show that the map

$$\rho(F_{ij}) = \phi_i\phi_{-j}$$

defines an irreducible so_∞-module on the space V_0 (resp. V_1) of even (resp. odd) elements of V, which is isomorphic to $L(\Lambda_0)$. We have: $\sigma = \sigma_0 \oplus \sigma_1$: $V_0 \oplus V_1 \xrightarrow{\sim} \mathbf{C}[x_1, x_3, \dots] \oplus q\mathbf{C}[x_1, x_3, \dots]$, where σ is the boson-fermion correspondence of type B. Show that the map

$$\hat{\rho}(F_{ij}) = \begin{cases} \phi_i\phi_{-j} & \text{if } i \neq j \text{ or } i = j > 0, \\ \phi_i\phi_{-j} - \frac{1}{2}I & \text{if } i = j < 0, \end{cases}$$
$$\hat{\rho}(K) = 1$$

extends by linearity to the b_∞-module $L(\Lambda_0)$. Show that the map σ_0 : $V_0 \xrightarrow{\sim} \mathbf{C}[x_1, x_3, \dots]$ gives us an equivalence of the representation $\hat{\rho}$ and the following vertex operator construction of b_∞:

$$\sum_{i,j\in\mathbf{Z}} z_1^i z_2^{-j} F_{ij} \mapsto \frac{1}{2}\frac{1 - z_2/z_1}{1 + z_2/z_1}(\Gamma_B(z_1, z_2) - 1),$$

where $\Gamma_B(z_1, z_2) = \left(\exp \sum_{j\in\mathbf{N}^{\text{odd}}} (z_1^j + z_2^j)x_j\right)\left(\exp -2\sum_{j\in\mathbf{N}^{\text{odd}}} \frac{z_1^{-j} - z_2^{-j}}{j}\frac{\partial}{\partial x_j}\right)$.

14.15. Show that the operator

$$\Omega_1^B = \sum_{j\in\mathbb{Z}} (-1)^j \phi_j \otimes \phi_{-j}$$

on $V \otimes V$ commutes with b_∞. Show that the equation

$$\Omega_1^B(\tau\otimes\tau) = \phi_0(\tau)\otimes\phi_0(\tau), \quad \tau \in V_0,$$

transferred from V_0 to $\mathbb{C}[x_1, x_3, \dots]$ via σ_0 gives the so-called *BKP hierarchy* of Hirota bilinear equations:

$$\sum_{j\geq 1} S_j(2y_1, 0, 2y_3, \dots) S_j(-\frac{2}{1}D_1, 0, -\frac{2}{3}D_3, \dots)\Big(\exp \sum_{s\in\mathbb{N}^{\text{odd}}} y_s D_s\Big)\tau\cdot\tau = 0.$$

Show that the equation of the lowest principal degree of this hierarchy is:

$$(D_1^6 - 5D_1^3D_3 - 5D_3^2 + 9D_1D_5)\tau\cdot\tau = 0.$$

Construct soliton solutions of the BKP hierarchy.

14.16. Given $n > 0$, show that the subalgebra $\{(c_{ij})_{i,j\in\mathbb{Z}} \oplus \mathbb{C}K \in b_\infty$ such that $c_{i+n,j+n} = c_{ij}\}$ is isomorphic to an affine algebra $\mathfrak{g}'(A)$, where A is of type $A_{2\ell}^{(2)}$ (resp. $D_{\ell+1}^{(2)}$) if $n = 2\ell+1$ (resp. $n = 2\ell$). Consider the realization of the basic representation of b_∞ on the space $\mathbb{C}[x_j; j$ positive and odd$]$. Show that the subspace $\mathbb{C}[x_j; j \not\equiv 0 \mod n]$ is invariant with respect to $\mathfrak{g}'(A)$, giving the principal vertex operator construction of its basic representation. Show that the corresponding vertex operators are (ϵ is an nth root of unity, $\epsilon \neq 1$):

$$\Big(\exp \sum_{j\in\mathbb{N}^{\text{odd}}} (1-\epsilon^j) z^j x_j\Big)\Big(\exp -2 \sum_{j\in\mathbb{N}^{\text{odd}}} (1-\epsilon^{-j})\frac{z^{-j}}{j}\frac{\partial}{\partial x_j}\Big).$$

14.17. Show that the energy decomposition (principal gradation) of the module $F^{(s)}$ is: $\deg(\underline{i}_1 \wedge \underline{i}_2 \wedge \cdots) =$ (sum of the $j > s$ that occur in the set $\{i_1, i_2, \dots\}$) $-$ (sum of the $j \leq s$ that do not occur in this set). Equating q-dimensions, deduce Euler's identity:

$$\varphi(q)^{-1} = 1 + \sum_{k\geq 1} q^{k^2}/(1-q)^2 \dots (1-q^k)^2.$$

Let H be the energy operator on F, i.e. $H|_{F_j^{(m)}} = jI$, and let α_0 be the charge operator (see §14.10). Computing in bosonic and fermionic pictures $\operatorname{tr}_F q^H z^{\alpha_0}$ derive once more the Jacobi triple product identity:

$$\prod_{n\in\mathbb{N}} (1+zq^n)(1+z^{-1}q^{n-1}) = \sum_{j\in\mathbb{Z}} z^j q^{j(j+1)/2}/\varphi(q).$$

14.18. Using the boson–fermion correspondence, show that

$$S_\lambda\left(\tilde{\partial}\right) S_\mu(x)\big|_{x=0} = \delta_{\lambda\mu}.$$

14.19. Let x_∞, where $x = b$, c, or d, be the completed infinite rank affine algebra of type X_∞ (see §7.11). We call

$$\overline{e} := \sum_i \pi(e_i) \in \overline{x}_\infty$$

the *cyclic element* of $\overline{x}_\infty$, and let $\overline{\mathfrak{s}}$ be the centralizer of $\overline{e}$ in $\overline{x}_\infty$. Then $\overline{\mathfrak{s}}$ is graded with respect to the principal gradation: $\overline{\mathfrak{s}} = \bigoplus_{j\in\mathbb{Z}} \overline{\mathfrak{s}}_j$. The subalgebra $\mathfrak{s} = \pi^{-1}(\overline{\mathfrak{s}})$ is called the *principal subalgebra* of x_∞. Show that $\dim \mathfrak{s}_j = \dim \overline{\mathfrak{s}}_j = 1$ for j odd, $= 0$ for j nonzero even and $\mathfrak{s}_0 = \mathbb{C}K$. Show that $\overline{e} = \sum_{j\in\mathbb{Z}} E_{i,i+1}$ for $\overline{b}_\infty$ or $\overline{c}_\infty$; $\overline{e} = E_{0,2} - E_{-1,1} + \sum_{i\geq 1} E_{i,i+1} - \sum_{i\leq -1} E_{i,i+1}$ for $\overline{d}_\infty$; and that $\{\overline{e}^j\}_{j\in\mathbb{Z}^{\mathrm{odd}}}$ form a basis of $\overline{\mathfrak{s}}$ in the cases $\overline{b}_\infty$ and $\overline{c}_\infty$; $\{\overline{e}^j\}_{j\in\mathbb{N}^{\mathrm{odd}}} \cup \{({}^t\overline{e})^j\}_{j\in\mathbb{N}^{\mathrm{odd}}}$ form a basis of $\overline{\mathfrak{s}}$ in the case d_∞. For b_∞, c_∞, or d_∞ put $r = \frac{1}{2}$, 1, or $\frac{1}{2}$, respectively. Introduce the following elements of $\overline{x}_\infty$:

$$p_j = \overline{e}^j, \quad q_j = \frac{1}{rj}\overline{e}^{-j} \quad (j \in \mathbb{N}^{\mathrm{odd}}).$$

Then we have the following commutation relations in x_∞:

$$[p_i, q_j] = \delta_{ij} K \quad (i, j \in \mathbb{N}^{\mathrm{odd}}).$$

Show that in the cases b_∞ and d_∞,

$$\dim_q L(\Lambda_0) = \prod_{j\in\mathbb{N}^{\mathrm{odd}}} (1 - q^j)^{-1},$$

and deduce that $L(\Lambda_0)$ remains irreducible when restricted to $\mathfrak{s}$.

14.20. Show that for the fundamental c_∞-module $L(\Lambda_s)$ $(s = 0, 1, \dots)$ one has

$$\dim_q L(\Lambda_s) = (1 - q^{2s+2})/\varphi(q).$$

Denote by $\rho_s : F^{(-s-2)} \to F^{(-s)}$ $(s \in \mathbb{Z}_+)$ the map of the exterior multiplication by $\sum_{j\geq 0} \underline{-j} \wedge \underline{j+1}$. Show that ρ_s is an injective homomorphism of c_∞-modules. Deduce that the c_∞-modules $L(\Lambda_s)$ and $F^{(-s)}/\rho_s(F^{(-s-2)})$ are isomorphic.

14.21. Define a subgroup P of GL_∞ by:

$$P = \{(a_{i,j})_{i,j\in\mathbf{Z}} \mid a_{ij} = 0 \text{ for } j \leq 0 \text{ and } i > 0\}.$$

This is the normalizer of the weight space of $L(\Lambda_0)_{\Lambda_0}$. The group Σ_∞ of all finite permutations of $\mathbf{Z}$ has a natural embedding in GL_∞ (by $\sigma(v_j) = v_{\sigma(j)}$). Denote by Σ_- (resp. Σ_+) the subgroup of Σ_∞ of permutations which fix all $j \leq 0$ (resp. $j > 0$). Let U_+ (resp. U_-) denote the subgroup of GL_∞ of all upper- (resp. lower-) triangular matrices with ones on the diagonal. For a right coset $w \in \Sigma_\infty/\Sigma_+ \times \Sigma_-$ we put $P_w^\pm = U_\pm \cap (\cap_{u\in w} uU_-u^{-1})$; P_w^+ is a finite-dimensional subgroup of U_+. Show that $G = \bigcup_w P_w^\pm wP$ is a disjoint union, where w runs over $\Sigma_\infty/\Sigma_+ \times \Sigma_-$, and that presentation on the right is unique.

14.22. Deduce from Exercise 14.21 that we have a disjoint union:

$$SL_n(\mathbf{C}[t,t^{-1}]) = \bigcup SL_n(\mathbf{C}[t^{\pm 1}])\,\mathrm{diag}(t^{k_1},\ldots,t^{k_n})SL_n(\mathbf{C}[t]),$$

where $k_1 \leq \cdots \leq k_n$ are integers and $\sum k_i = 0$. (These kinds of decompositions are called *Bruhat* and *Birkhoff decompositions* for $+$ and $-$, respectively.)

14.23. Consider the upper-triangular matrix $g(z) \in \widetilde{GL}_\infty$ introduced in §14.10, and let $C \in GL_\infty$, $w \in \Sigma_\infty$. Show that, in notation of §14.10, for $v = Cw|0\rangle$ we have:

$$F_v(z) = \det(g(z)C)^{i_0,\ldots,i_m}_{0,\ldots,m}.$$

Deduce that in the principal realization of the basic a_∞-module in the space $\mathbf{C}[x_1, x_2, \ldots]$, the orbit $GL_\infty \cdot 1$ together with 0 contains all polynomials of the form:

$$\det(g(x)C)^{i_0,\ldots,i_m}_{0,\ldots,m}, \tag{14.14.3}$$

where $C \in GL_\infty$. Show that, moreover, every polynomial from $GL_\infty \cdot 1$ can be uniquely represented in the form (14.14.3) such that $i_0 < \cdots < i_m$ and $C \in P_w$, where $w \mid 0\rangle = \pm \underline{i_0} \wedge \cdots \wedge \underline{i_m} \wedge |-m-1\rangle$ and m is sufficiently large. Thus, we obtain all polynomial solutions of the KP hierarchy. Using the principal subalgebra, show that if $P(x_1, x_2, \ldots)$ lies in $GL_\infty \cdot 1$, then its "translate" $P(x_1 + c_1, x_2 + c_2, \ldots)$, where $c_i \in \mathbf{C}$, also lies in $GL_\infty \cdot 1$.

14.25. Show that one has a natural bijection between the set $\Sigma_\infty/\Sigma_- \times \Sigma_+$ and the set of all partitions. Namely, given $w \in \Sigma_\infty$, let m be the greatest integer, such that $w(m) = m$; arrange all $w(j)$ with $j \leq m$ in decreasing order: $j_m > j_{m+1} > \cdots$, and put $b_0 = j_m - m$, $b_1 = j_{m-1} - (m-1), \ldots$. Then $b_0 \geq b_1 \geq \cdots$ is the associated partition.

14.26. Given the two partitions b and b', define the *skew Schur polynomial* $S_{b/b'}(x) = S_{b'}(\tilde{\partial}_x)S_b(x)$; in particular $S_{b/\{0\}}(x) = S_b(x)$ is a Schur polynomial. Show that $S_{b/b'}(z)$ is, up to a sign, the coefficient of $w'(|0\rangle)$ in $g(z)w(|0\rangle)$ where b and b' are the partitions associated with w and w'.

14.27. Let V be the n-dimensional vector space ($n = 1, 2, \ldots, \infty$) over $\mathbb{C}$ and let $\Lambda V = \bigoplus_{k \in \mathbb{Z}_+} \Lambda^k V$ be the Grassmann algebra over V. Denote by $\mathcal{V}_k$ the set of all decomposable elements of $\Lambda^k V$ (i.e., elements of the form $v_1 \wedge \cdots \wedge v_k$, where $v_i \in V$). Denote by $\beta_k : \mathcal{V}_k \backslash \{0\} \to \mathrm{Gr}_{n,k}$ the canonical map onto the Grassmannian $\mathrm{Gr}_{n,k}$ of k-dimensional subspaces of V defined by $\beta_k(v_1 \wedge \cdots \wedge v_k) = \sum_j \mathbb{C} v_j$. An element $v \in V$ defines a wedging operator on ΛV:

$$v(v_1 \wedge v_2 \wedge \cdots) = v \wedge v_1 \wedge v_2 \wedge \cdots,$$

and an element $f \in V^*$ defines a contracting operator:

$$f(v_1 \wedge v_2 \wedge \cdots) = \langle f, v_1 \rangle v_2 \wedge v_3 \wedge \cdots - \langle f, v_2 \rangle v_1 \wedge v_3 \wedge \cdots + \cdots.$$

Show that the wedging (resp. contracting) operators anticommute and that $vf + fv = \langle f, v \rangle$. Deduce that the wedging and contracting operators generate a Clifford algebra on $V \oplus V^*$ with the bilinear form

$$(v \oplus f | v' \oplus f') = \langle v, f' \rangle + \langle f, v' \rangle.$$

Choose bases $\{e_i\}$ of V and $\{e_i^*\}$ of V^* such that $\langle e_i^*, e_j \rangle = \delta_{ij}$, and let $\Omega_1 = \sum_{j=1}^{n} e_j \otimes e_j^*$ be an operator on $\Lambda V \otimes \Lambda V$. Show that Ω_1 commutes with the action of $GL(V)$ on ΛV and $\tau \in \Lambda^k V$ is a decomposable element if and only if

$$\Omega_1(\tau \otimes \tau) = 0. \tag{14.14.4}$$

Show that written in the basis $e_{i_1} \wedge \cdots \wedge e_{i_k}$ ($i_1 < i_2 < \cdots < i_k$) of $\Lambda^k V$, equation (14.14.4) gives the classical Plücker relations. Deduce from Exercise 14.10 that Plücker relations generate the ideal of relations of $\mathcal{V}_k$. Show that the equation

$$\Omega_1(\tau \otimes \tau') = 0, \quad \tau \in \Lambda^k V,\ \tau' \in \Lambda^r V,\ k \geq r, \tag{14.14.5}$$

is equivalent to $\beta_k(\tau) \supset \beta_r(\tau')$. Show that equations given by (14.14.5) generate the ideal of relations of the variety (called the *affine flag variety*)

$$\mathcal{V} := \{(\tau_1, \ldots, \tau_n) \in \mathcal{V}_1 \times \cdots \times \mathcal{V}_n \mid \beta_k(\tau_k) \supset \beta_r(\tau_r) \text{ for } k > r\}.$$

14.28. Put $x_j = \frac{1}{j}(\epsilon_1^j + \cdots + \epsilon_N^j)$. Show that $S_k(x_1, x_2, \ldots)$ is equal to the trace of the matrix $\mathrm{diag}(\epsilon_1, \ldots, \epsilon_N)$ in the $GL_N(\mathbb{C})$-module $S^k(\mathbb{C}^N)$ (and hence, by Schur, $S_b(x)$ is the trace of this matrix in the $GL_N(\mathbb{C})$-module corresponding to the partition b; see, e.g., Macdonald [1979]).

14.29. We identify $\mathbf{C}^\infty$ with $\mathbf{C}[t,t^{-1}]\otimes\mathbf{C}^n$ as in Exercise 14.11. This gives us an embedding of the group $T := \{\mathrm{diag}(t^{k_1},\dots,t^{k_n}) \mid k_i \in \mathbf{Z}, \sum k_i = 0\}$ into the group of all permutations of $\mathbf{Z}$. For $g \in T$ there exists a permutation $w \in \Sigma_\infty$ such that $g = wg'$, where g' is a permutation of $\mathbf{Z}$ which leaves invariant the set of nonpositive integers. Let b be the partition associated with w and put $s_g = S_b$; show that, up to a sign, the polynomial s_g is independent of the choice of w. Let $\mathfrak{g}$ be an affine algebra of type $A_{n-1}^{(1)}$; we identify the basic $\mathfrak{g}$-module with the subspace $\mathbf{C}[x_j; j \not\equiv 0 \mod n]$ of $\mathbf{C}[x_1,x_2,\dots]$. Let G be the associated group and let $\mathcal{V} = G\cdot 1$ be the orbit of the highest-weight vector. Show that $s_g \in \mathcal{V}$ ($g \in T$), i.e., s_g is a polynomial solution of the $A_{n-1}^{(1)}$-hierarchy. Let $P \subset G$ be the preimage of $SL_n(\mathbf{C}[t])$ under the canonical homomorphism $G \to SL_n(\mathbf{C}[t,t^{-1}])$. Using the Bruhat decomposition (see Exercise 14.21), show that

$$\mathcal{V} = \bigcup_{g\in T} P\cdot s_g \quad \text{(a disjoint union).}$$

14.30. We keep the notation of Exercise 14.29. Let $n = 2$; for $k \in \mathbf{Z}$ we denote by s_k the polynomial s_g with $g = \mathrm{diag}(t^k, t^{-k})$. Show that $s_k = S_{\{2k-1,\dots,1\}}$ if $k > 0$ and $s_k = S_{\{-2k,-2k-1,\dots,1\}}$ if $k \le 0$. Show that $\dim(\mathfrak{s}_+(s_k) + \mathbf{C}s_k) = k+1 = \dim P\cdot s_k$. Deduce that all polynomial solutions of the KdV hierarchy are of the form

$$c_0 S_{\{k,k-1,\dots,1\}}(x_1+c_1, x_2+c_2,\dots), \text{ where } c_i \in \mathbf{C}. \tag{14.14.6}$$

[The projectivisation of the set $P\cdot s_k$ is isomorphic to $\mathbf{C}^k$. We have an injective map $f : \mathbf{C}^k \to \mathbf{C}^k$ defined by $f((c_1,\dots,c_k)) = s_k(x_1+c_1, x_2+c_2,\dots)$ and such that $\dim f(\mathbf{C}^k) = k$. Deduce that f is an isomorphism.]

14.31. Show that the polynomial $S_{\{2,1\}} = \frac{1}{3}x_1^3 - x_3$ is a solution of the $A_{n-1}^{(1)}$-hierarchy for $n \ge 4$, but that not all polynomials of the set $P\cdot S_{\{2,1\}}$ are of the form (14.14.6).

14.32. Define the *infinite Grassmannian* Gr as follows. Consider the infinite-dimensional space $\mathbf{C}^\infty = \bigoplus_{j\in\mathbf{Z}} \mathbf{C}v_j$ and its subspaces $\mathbf{C}_k^\infty = \bigoplus_{j\le k} \mathbf{C}v_j$. Denote by Gr (resp. Gr_n) the set of all subspaces U of $\mathbf{C}^\infty$ such that $U \supset \mathbf{C}_k^\infty$ (resp. $U \supset \mathbf{C}_k^\infty$ and $\dim U/\mathbf{C}_k^\infty = n-k$) for $k \ll 0$. Then $\mathrm{Gr} = \bigcup_{n\in\mathbf{Z}} \mathrm{Gr}_n$ (disjoint union) and GL_∞ acts transitively on each Gr_n. Let

$$\mathcal{V}_n = \{\tau \in F^{(n)} \mid \Omega_1(\tau\otimes\tau) = 0\} \quad (= GL_\infty|n\rangle \cup \{0\})$$

(see §14.11), and denote by $\beta_n : V_n \backslash \{0\} \to \mathrm{Gr}_n$ the map defined by $\beta_n(\underline{i}_1 \wedge \underline{i}_2 \wedge \cdots) = \sum_j \mathbf{C} v_{i_j}$, which is GL_∞-equivariant, surjective, with fibers $\mathbf{C}^\times$. Prove infinite-dimensional analogues of all facts stated in Exercise 14.27.

14.33. Show that the equation

$$\Omega_1(\tau \otimes \tau') = 0, \quad \tau \in F^{(k+n)}, \ \tau' \in F^{(k)}, \ n \geq 0,$$

when transported to $\mathcal{B}$, is equivalent to the following hierarchy of Hirota bilinear equations (called the n-th MKP hierarchy):

$$\sum_{j \geq 0} S_j(2y) S_{j+n+1}(-\widetilde{D}) \left(\exp \sum_{r \geq 1} y_r D_r \right) \tau \cdot \tau' = 0. \tag{14.14.7}$$

Deduce from Exercise 14.32 that the set of all polynomial solutions considered up to a constant factor of the KP hierarchy is naturally parametrized by Gr; and that of the union of all n-th MKP hierarchies, by the *infinite flag variety*

$$\mathcal{F} = \{\cdots \supset U_1 \supset U_0 \supset U_{-1} \supset \cdots \mid U_j \in \mathrm{Gr}_j\}.$$

14.34. Denote by $P_{\lambda_1, \lambda_2, \ldots; n}(D)$ the coefficient of the monomial $y_1^{\lambda_1} y_2^{\lambda_2} \ldots$ in (14.14.7). Let $L^{(n)}_{\text{high}}$ denote the highest component of the gl_∞-module $B^{(k+n)} \otimes B^{(k)}$; $L^{(n)}_{\text{low}}$, its orthocomplement; $Hir^{(n)} = L^{(n)}_{\text{low}} \cap \mathbf{C}[y]$ and $Hir^{(n)} = \bigoplus_j Hir^{(n)}_j$, the principal gradation (cf. §14.12). Show that the polynomials $P_{\lambda_1, \lambda_2, \ldots; n}(y)$ form a basis of the space $H^{(n)}_{|\lambda|+n+1}$, so that $\dim H^{(n)}_j = p(j - n - 1)$. Show that $2P_{0;1}(x) = x_1^2 - x_2$, and hence $(D_1^2 - 2D_2)\tau' \cdot \tau = 0$ is the equation of lowest degree of the first MKP. Show that using the change (14.11.12) of variables and functions together with $v(x, y, t) = \log(\tau'/\tau)$, we obtain:

$$u = v_y - v_x^2 - v_{xx},$$

called the Miura transformation.

14.35. Show that a Hirota polynomial $P(y)$ is odd if and only if $P(y) \in \Lambda^2(L(\Lambda_0))$, and that any odd polynomial is a Hirota polynomial.
[Use the fact that $2y_j = x'_j - x''_j$.]

14.36. Show that for the operators L_n introduced in Exercise 9.17 one has (cf. Exercise 7.24):

$$[L_n, \psi_k] = (-n/2 - k + i\lambda n)\psi_{k-n}.$$

14.37. Show that for the KP hierarchy, the formal completion of the space Hir is invariant with respect to the vertex operator

$$Z(u,v) = \left(\exp \sum_k (u^k - v^k) y_k\right)\left(\exp -\frac{1}{2}\sum_k \frac{1}{k}(u^{-k} - v^{-k})\frac{\partial}{\partial y_k}\right) + \left(\exp - \sum_k (u^k - v^k) y_k\right)\left(\exp \frac{1}{2}\sum_k \frac{1}{k}(u^{-k} - v^{-k})\frac{\partial}{\partial y_k}\right).$$

14.38. Put $a_{-k} = 2^{-\frac{1}{2}} k y_k$ and $a_k = 2^{\frac{1}{2}} \frac{\partial}{\partial y_k}$ for $k > 0$, and let $L_{n;0}$ denote the operators on the space $V = \mathbf{C}[y_1, y_2, \ldots]$ introduced in Exercise 9.17 with $\lambda = \mu = 0$. Let

$$Z(u) = \lim_{v \to u} (1 - \frac{v}{u})^{-2} (Z(u,v) - 2),$$

where $Z(u,v)$ is the vertex operator introduced in Exercise 14.37. Show that $Z(u) = \sum_j u^{-j} L_{j;0}$, and therefore, the subspace Hir of V of Hirota polynomials for the KP hierarchy and its orthocomplement $Hir^{\perp}$ are invariant with respect to the Virasoro operators $L_{j;0}, j \in \mathbf{Z}$.

14.39. Using the boson-fermion correspondence, transport the action of the operators $L_{k;0}$ defined in Exercise 14.38 to $F^{(0)}$, and show that the elements of this space killed by all $L_{k;0}$ with $k > 0$ are linear combinations of the elements $\underline{n} \wedge \underline{n-1} \wedge \cdots \wedge \underline{1} \wedge \mid -n \rangle$, where $n = 0, 1, 2, \ldots$. For $n = 0, 1, 2, \ldots$, let $Q_n(y)$ denote the Schur polynomial $S_{\{n,\ldots,n\}}(y)$, where n is repeated n times. Deduce that if a polynomial $P(y)$ is killed by all $L_{k;0}$ with $k > 0$, then $P(y)$ is a linear combination of the $Q_n(y)$, $n = 0, 1, 2, \ldots$.

14.40. Deduce from Exercise 14.39 that the polynomials $Q_n(\tilde{y})$, $n = 1, 2, \ldots$, are Hirota polynomials for the KP hierarchy. (For $n = 2$ we thus recover the polynomial in (14.11.11).)

14.41. Show that $Hir^{\perp}$ is irreducible under Vir and that Hir decomposes into a direct sum of irreducible Vir-modules generated by (highest-weight) vectors $Q_n(y)$, $n = 1, 2, \ldots$.

14.42. Deduce from Exercise 14.39 that the highest-weight vectors of the a_∞-module $L(\Lambda_0) \otimes L(\Lambda_0)$ are scalar multiples of the polynomials $Q_n(y)$, $n = 0, 1, 2, \ldots$.

14.43. Let $L(\Lambda)$ be an integrable highest-weight module over a_∞ ($\Lambda = \sum_i k_i\Lambda_i$, where k_i are nonnegative integers and all but a finite number of them are zero). Let $\lambda = \{\lambda_i\}_{i\in\mathbb{Z}}$ be such that $k_i = \lambda_i - \lambda_{i-1}$ and $\lambda_i = 0$ for sufficiently big i, and let h_i be the hook lengths of the (infinite) Young diagram of λ. Show that $\dim_q L(\Lambda) = \prod_i (1-q^{h_i})^{-1}$.

In Exercises 14.44–14.51 we discuss a pseudodifferential operator approach to the KP and KdV hierarchies.

14.44. A (formal) pseudodifferential operator is an expression of the form

$$P(x,\partial) = \sum_{j\in\mathbb{Z}} a_j(x)\partial^j,$$

where $a_j(x)$ are formal power series in x and $a_j(x) = 0$ for $j \gg 0$. The multiplication of pseudodifferential operators is defined by letting

$$\partial^j \cdot b(x) = \sum_{k\geq 0} \binom{j}{k} \frac{d^k b}{dx^k}\partial^{j-k}.$$

Show that the pseudodifferential operators form thereby an associative algebra Ψ. Show that Ψ carries an anti-involution * determined by conditions

$$b(x)^* = b(x), \qquad \partial^* = -\partial.$$

14.45. Denote by U_+ (resp. U_-) the space of expressions of the form (formal oscillating functions):

$$\sum_{j\in\mathbb{Z}} a_j(x)z^j e^{zx} \quad (\text{resp. } \sum_{j\in\mathbb{Z}} a_j(x)z^j e^{-zx}),$$

where $a_j(x)$ are formal power series in x and $a_j(x) = 0$ for $j \gg 0$. Show that one has a representation of the algebra Ψ on the space U_+ (resp. U_-) determined by the condition $(b(x)\partial^j)e^{zx} = b(x)z^i e^{zx}$ (similarly for U_-). For $P = \sum_{j\in\mathbb{Z}} a_j\partial^j \in \Psi$ let

$$P_+ = \sum_{j\geq 0} a_j\partial^j, \qquad P_- = \sum_{j<0} a_j\partial^j.$$

Show that if $P(x,\partial), Q(x,\partial) \in \Psi$ are such that

$$\mathrm{Res}_{z=0}(P(x',z)e^{x'z}Q(x'',-z)e^{-x''z}) = 0,$$

then $(PQ^*)_- = 0$.

14.46. Consider now pseudodifferential operators of the form $P(x,\partial)$, where $x=(x_1,x_2,\dots)$, $\partial=\frac{\partial}{\partial x_1}$, and $x_2,x_3,\dots$ are some parameters, and denote $z\cdot x=\sum_{j\geq 1} z^j x_j$. Let $\Gamma(x,z)=e^{z\cdot x}\Gamma_+(z)^{-1}$, $\Gamma^*(x,z)=e^{-z\cdot x}\Gamma_+(z)$ (see (14.10.3)). Given a nonzero formal power series $\tau(x)$, introduce the associated wave function

$$w(x,z)=\Gamma(x,z)\tau(x)/\tau(x):=\hat{w}(x,z)e^{z\cdot x},$$

and the adjoint wave function

$$w^*(x,z)=\Gamma^*(x,z)\tau(x)/\tau(x):=\hat{w}^*(x,z)e^{-z\cdot x},$$

where $\hat{w}(x,z)=1+w_1(x)z^{-1}+w_2(x)z^{-2}+\cdots$, $\hat{w}^*(x,z)=1+w_1^*(x)z^{-1}+w_2^*(x)z^{-2}+\cdots$. We may write:

$$w(x,z)=Pe^{z\cdot x},\qquad w^*(x,z)=Qe^{-z\cdot x},$$

where $P=1+w_1(x)\partial^{-1}+\cdots$, $Q=1-w_1^*\partial^{-1}+w_2^*\partial^{-2}-\cdots$. Note that equation (14.11.5) of the KP hierarchy can be written as follows:

$$\text{(14.14.8)}\qquad \operatorname{Res}_{z=0} w(x',z)w^*(x'',z)=0.$$

Deduce from Exercise 14.45 that if τ is a solution of the KP hierarchy, then $Q=(P^*)^{-1}$.

14.47. Show that $\tau(x)$ is determined up to a constant function by its wave function, namely:

$$\frac{\partial}{\partial x_n}\log\tau=-\operatorname{Res}_{z=0} z^n\left(\sum_{j\geq 1} z^{-j-1}\frac{\partial}{\partial x_j}-\frac{\partial}{\partial z}\right)\log\hat{w}(x,z).$$

14.48. Let $L:=P\partial P^{-1}=\partial+u_1(x)\partial^{-1}+\cdots$, and let $B_n=(L^n)_+$. Show that: $B_0=1, B_1=\partial, B_2=\partial^2+2u_1, B_3=\partial^3+3u_1\partial+3u_2+3u_{1x_1}$, and that $u_1=(\log\tau)_{x_1x_1}$.

Prove that equation (14.14.8) implies

$$\text{(14.14.9)}\qquad \frac{\partial L}{\partial x_n}=[B_n,L],\quad n=1,2,\dots.$$

[Apply $\frac{\partial}{\partial x_n'}-B_n$ to (14.14.8) and use that $(\frac{\partial}{\partial x_n}-B_n)w(x,z)=(\frac{\partial P}{\partial x_n}+(L^n)_-P)e^{z\cdot x}$ to conclude from Exercise 14.45 that

$$\text{(14.14.10)}\qquad \frac{\partial P}{\partial x_n}=-(P\partial P^{-1})_-P\quad\text{(Sato equation)}.$$

Differentiating $LP=P\partial$ by x_n and using (14.14.10), obtain (14.14.9).]

14.49. Show that (14.14.9) implies
$(14.14.11)_{m,n}$

$$\left[\frac{\partial}{\partial x_n} - B_n, \frac{\partial}{\partial x_m} - B_m\right] = 0 \quad \text{(Zakharov–Shabat equation).}$$

Show that $(14.14.11)_{2,3}$ is the classical KP equation (14.11.13) on the function $u = 2u_1$.

14.50. Fix an integer $n \geq 2$. Show that the following conditions are equivalent:

a) $\tau(x)$ is independent of x_{jn}, $j \in \mathbb{N}$,

b) $\frac{\partial w}{\partial x_{jn}} = z^{nj} w$,

c) P is independent of x_{jn}, $j \in \mathbb{N}$,

d) L^n is a differential operator, i.e., $(L^n)_- = 0$. Deduce that if a solution τ of the KP hierarchy is independent of x_n, then it is independent of all x_{jn}, $j \in \mathbb{N}$.

14.51. Let $n = 2$. Then condition d) of Exercise 14.50 becomes $L^2 = \partial^2 + u(x) := S$, where $x = (x_1, x_3, \ldots)$, so that (14.14.9) reduces to

$$(14.14.12)_j \qquad \frac{\partial S^{1/2}}{\partial x_{2j+1}} = [(S^{j+1/2})_+, S^{1/2}].$$

Show that $(14.14.12)_1$ is the classical KdV equation (14.13.4).

14.52. Consider the following subsystem of the NLS hierarchy (cf. (14.13.6)):

$$D_1^2 \tau_n \cdot \tau_n + 2\tau_{n-1}\tau_{n+1} = 0, \quad n \in \mathbb{Z}.$$

Show that after letting $u_n(x) = (\log \frac{\tau_{n+1}}{\tau_n})(x, c_2, c_3, \ldots)$, this system becomes the classical Toda lattice:

$$(u_n)_{xx} = e^{u_n - u_{n-1}} - e^{u_{n+1} - u_n}.$$

14.53. Derive the following explicit form of the solution

$$\tau^{+\cdots+}_{a_1,\ldots,a_N; z_1,\ldots,z_N}(x, q) = \sum \tau_n(x) q^n$$

of the NLS hierarchy (see §14.13):

$$\tau_n = \sum_{1 \leq i_1 < \cdots < i_n \leq N} f_{i_1,\ldots,i_n}(x) \quad \text{if } 0 < n \leq N,$$
$$= 1 \text{ if } n = 0 \text{ and } = 0 \text{ otherwise, where}$$
$$f_{i_1,\ldots,i_n}(x) = \prod_{1 \leq p < q \leq n} (z_{i_p} - z_{i_q})^2 a_{i_1} \ldots a_{i_n} z_{i_1} \ldots z_{i_n}$$
$$\times \exp \sum_{k \geq 1} x_k (z_{i_1}^k + \cdots + z_{i_n}^k).$$

14.54. Derive the following formula for the nth derivative of the δ-function:

$$\delta_{z_1}^{(n)} = n[z_2^{-n-1}(1 - z_1/z_2)^{-n-1} + (-1)^n z_1^{-n-1}(1 - z_2/z_1)^{-n-1}].$$

In Exercises 14.55–14.63 we shall outline a more invariant approach to the affine algebra of type A_∞, the corresponding group and the infinite Grassmannian, and an application of this approach to the algebraic geometry of curves.

14.55. Let V be a vector space. Two subspaces A and B of V are called *commensurable*, write $A \sim B$, if $\dim A/(A \cap B) < \infty$ and $\dim B/(A \cap B) < \infty$. Show that this is an equivalence relation. Define the *relative dimension* of A and B by:

$$\dim A - \dim B = \dim A/(A \cap B) - \dim B/(A \cap B).$$

Show that this is well-defined.

14.56. Let $\mathrm{End}(V, A) = \{g \in \mathrm{End}\, V \mid gA + A \sim A\}$. Show that this is a subalgebra of the (associative) algebra $\mathrm{End}\, V$. Let $gl(V, A)$ be the associated Lie algebra and $GL(V, A) = \mathrm{End}(V, A)^\times$ the associated group. Show that

$$GL(V, A) = \{g \in (\mathrm{End}\, V)^\times \mid gA \sim A\}.$$

Show that the map $\deg : GL(V, A) \to \mathbf{Z}$ defined by $\deg(g) = \dim A - \dim gA$ is a homomorphism onto the additive group of $\mathbf{Z}$. Let $GL(V, A)^\circ$ denote the kernel of deg. Define the Grassmannian

$$\mathrm{Gr}(V, A) = \{U \subset V \mid U \sim A\},$$

and let $\mathrm{Gr}_n = \{U \in \mathrm{Gr}(V, A) \mid \dim U - \dim A = n\}$. Show that $GL(V, A)^\circ$ acts transitively on Gr_n. Show that taking $V = \mathbf{C}^\infty$ and $A = \mathbf{C}_0^\infty$, we recover the Grassmannian considered in Exercise 14.32, and that $\overline{a}_\infty$ is a subalgebra of the Lie algebra $gl(\mathbf{C}^\infty, \mathbf{C}_0^\infty)$.

14.57. Let $\mathrm{End}_{\mathrm{fin}}\, A = \{g \in \mathrm{End}\, A \mid \dim gA < \infty\}$, let $gl_{\mathrm{fin}} A$ be the associated Lie algebra and let $GL_{\mathrm{fin}} A = \{1 + g \mid g \in \mathrm{End}_{\mathrm{fin}}\, A,\ 1 + g \text{ is invertible}\}$ be the associated group. Choose a projection $\pi : V \to A$ and let

$$\widetilde{\mathrm{End}}(V, A) = \{(g, f) \in \mathrm{End}(V, A) \oplus \mathrm{End}\, A \mid \pi g \pi - f \in \mathrm{End}_{\mathrm{fin}}\, A\}.$$

Show that this is an algebra independent of the choice of π. Show that

$$0 \longrightarrow \mathrm{End}_{\mathrm{fin}}\, A \xrightarrow{i_2} \widetilde{\mathrm{End}}(V, A) \xrightarrow{p_1} \mathrm{End}(V, A) \longrightarrow 0, \tag{14.14.13}$$

where i_2 is the inclusion in the second summand and p_1 is the projection on the first summand, is an exact sequence of associative algebras. Considering the section $\mathrm{End}(V,A) \to \widetilde{\mathrm{End}}(V,A)$ defined by $g \mapsto (g, \pi g \pi)$, show that the associated Lie algebra $\widetilde{gl}(V,A)$ is isomorphic to the vector space $gl(V,A) \oplus gl_{\mathrm{fin}} A$ with the bracket:

$$[(g,f),(g',f')] = ([g,g'],[f,f'] + \Psi_A^\pi(g,g')),$$

where $\Psi_A^\pi(g,g') = [\pi g \pi, \pi g' \pi] - \pi[g,g']\pi$. Pushing out the subspace of traceless elements, we obtain from (14.14.13) the central extension of the Lie algebra $gl(V,A)$:

$$0 \to \mathbf{C} \to \widehat{gl}(V,A) \to gl(V,A) \to 0 \tag{14.14.14}$$

with the 2-cocycle $\psi_A^\pi(g,g') = \mathrm{tr}_V([\pi,g]g')$. Show that (14.14.14) gives the central extensions discussed in §7.12 and Exercise 7.28. Consider the group $\widetilde{GL}(V,A) = \widetilde{\mathrm{End}}(V,A)^\times$ and the embedding $GL_{\mathrm{fin}}(A) \to \widetilde{GL}(V,A)$ given by $f \mapsto (1,f)$. Show that we obtain thereby an exact sequence of groups:

$$1 \to GL_{\mathrm{fin}}(A) \to \widetilde{GL}(V,A) \to GL(V,A) \to 1.$$

Pushing out the subgroup of determinant 1 elements gives a central extension

$$1 \to \mathbf{C}^\times \to \widehat{GL}(V,A) \to GL(V,A) \to 1. \tag{14.14.15}$$

Let $\widetilde{GL}(V,A)^\circ = \{(g,f) \in \widetilde{GL}(V,A) \mid g \in GL(V,A)^\circ\}$.

14.58. Let $\overline{A}_\infty = \{a = (a_{ij})_{i,j\in\mathbf{Z}} \mid a \text{ is invertible and } a_{ij} - \delta_{ij} = 0 \text{ for } |i-j| \gg 0\} \subset GL(\mathbf{C}^\infty, \mathbf{C}_0^\infty)$ be the group corresponding to the Lie algebra $\overline{a}_\infty$, and let $\overline{A}_\infty^\circ = \overline{A}_\infty \cap GL(\mathbf{C}^\infty, \mathbf{C}_0^\infty)^\circ$. Formulas (14.9.4) extend to a representation of $\overline{a}_\infty$ (resp. $\overline{A}_\infty$) on the Clifford algebra Cl by derivations (resp. automorphisms), which we denote by p (resp. P). Show that the formula

$$\hat{r}(g,f)x|0\rangle = p(g)x|0\rangle + xr(g-f)|0\rangle$$

defines a representation of the Lie algebra $\widetilde{gl}(\mathbf{C}^\infty, \mathbf{C}_0^\infty)$ on F, which pushes down to $\widehat{gl}(\mathbf{C}^\infty, \mathbf{C}_0^\infty) \supset a_\infty$, producing thereby the representation $\hat{r}$ constructed in §14.9. Show that the formula

$$\widehat{R}_0(g,f)x|0\rangle = (P(g)x)R_0(gf^{-1})|0\rangle$$

defines a representation of the group $GL(\mathbf{C}^\infty, \mathbf{C}_0^\infty)^\circ$ on $F^{(0)}$, which pushes down to a projective representation of the group $\overline{A}_\infty^\circ$ on $F^{(0)}$, so that $\beta_0(\overline{A}_\infty^\circ|0\rangle) = \mathrm{Gr}_0$.

14.59. Let V be a vector space and let K be a subspace of V. Given a subspace D of V, let $H^0(D) = D \cap K$ and $H^1(D) = \frac{V}{D+K}$. Assuming that the latter two spaces are finite-dimensional, let $\chi(D) = \dim H^0(D) - \dim H^1(D)$. For two commensurable subspaces D_1 and D_2 of V prove the "abstract Riemann–Roch theorem":

$$\chi(D_1) - \chi(D_2) = \dim D_1 - \dim D_2.$$

14.60. Let X be a smooth algebraic curve. For a point $p \in X$, let A_p denote the formal completion of the local ring at p and let K_p be its field of fractions. Let $V_X = \prod_{p\in X}{}' K_p$ be the space of *adeles* on X (all but a finitely many components f_p of an adele are in A_p). Given a divisor $D = \sum_{p\in X} k_p p$, let $A(D) = \{(f_p)_p \in V_X \mid \nu_p(f_p) \geq -k_p\}$. Show that $\deg D = \dim A(D) - \dim A(\emptyset)$.

14.61. Let $\mathcal{K}$ be the field of rational functions on X and let $K = \{(f)_p \mid f \in \mathcal{K}\} \subset V_X$ be the subspace of *principal adeles*. Show that

$$H^i(X, \mathcal{O}_X(D)) = H^i(A(D)), i = 0, 1.$$

Deduce from Exercise 14.59 the "easy part" of the classical Riemann–Roch theorem.

14.62. Show that for commuting $g, g' \in gl(V, A)$, $\psi_A^\pi(g, g') = \psi_A(g, g')$ is independent of the choice of π, that $\psi_A(g, g') = \psi_{A'}(g, g')$ if $A \sim A'$, and that $\psi_A(g, g') = 0$ if $\dim A < \infty$ or $\dim V/A < \infty$. Show that if $g, g' \in gl(V, A) \cap gl(V, B)$ and $gg' = g'g$, then one has the "abstract residue theorem":

$$\psi_A(g, g') + \psi_B(g, g') = \psi_{A+B}(g, g') + \psi_{A\cap B}(g, g').$$

14.63. Considering f and $g \in \mathcal{K}$ as operators of multiplication in V_X, show that $\psi_{A_p}(f, g) = \mathrm{Res}_p\, g df$. Deduce from Exercise 14.62 the classical residue theorem:

$$\sum_{p\in X} \mathrm{Res}_p\, g df = 0, \quad g, f \in \mathcal{K}.$$

§14.15. Bibliographical notes and comments.

The exceptional role of the basic representation in the theory of affine algebras was pointed out in Kac [1978 A]. Quite a few explicit constructions of this representation are known. First, it is the principal vertex operator construction, obtained by Kac–Kazhdan–Lepowsky–Wilson [1981]. In the simplest case of $A_1^{(1)}$ this construction had been previously found by Lepowsky–Wilson [1978]. The second is the homogeneous vertex operator construction, obtained by Frenkel-Kac [1980] and, in a different form, by Segal [1981]. The third is the spin realization found independently by Frenkel [1981] and Kac–Peterson [1981]. The fourth is the infinite wedge representation found by Kac–Peterson [1981]. Kac–Peterson [1985 C] found a general approach, which attaches to every element w of the finite Weyl group $\mathring{W}$ (or, more generally, of $\operatorname{Aut}\mathring{Q}$) a vertex operator construction of the basic representation so that for $w = 1$ (resp. $w =$ Coxeter element) one recovers the homogeneous (resp. principal) construction; for $w = -1$ one recovers the construction of Frenkel–Lepowsky–Meurman [1984]. An exposition of this approach based on the vertex operator calculus was later given by Lepowsky [1985]. Further constructions of the basic representation are the path space realization by Date–Jimbo–Kuniba–Miwa–Okado [1989 A] given in the framework of statistical lattice models, and the standard monomial basis construction by Lakshmibai–Sheshadri [1989] done in the framework of geometric theory of infinite-dimensional flag varieties.

The construction of the integrable modules of higher levels in the framework of the vertex operator calculus is discussed by Lepowsky–Wilson [1984], Lepowsky–Primc [1985], Meurman–Primc [1987], Misra [1989 A], Frenkel–Lepowsky–Meurman [1989]. There are at least three explicit constructions of all integrable modules over $A_\ell^{(1)}$, the vertex operator calculus construction by Primc [1989], the path space construction by Date–Jimbo–Kuniba–Miwa–Okado [1989 B], and the standard monomial basis construction by Lakshmibai–Sheshadri [1989] and Lakshmibai [1989].

The vertex operator construction of basic representations of affine algebras of type different from A–D–E were studied by Goddard–Nahm–Olive–Schwimmer [1986], Bernard–Thierry–Mieg [1987 B], Misra [1989 A], and others.

The exposition of §§14.1, 14.3–14.7 closely follows Kac–Kazhdan–Lepowsky–Wilson [1981]. §14.2 is taken from Kac [1978 A]. The results of §14.2 in the case $r = 1$ are due to Kostant [1959]. His proof is longer, but does not use the case-by-case inspection of (14.2.4). Theorem 14.8 is due to Frenkel–Kac [1980], but the exposition in §14.8 is different and closer in spirit to the physicists' approach. Vertex operators $\Gamma_\alpha(z)$ and the Virasoro operators D_m play a crucial role in the dual string theory (see surveys by Mandelstam [1974], Schwartz [1973], [1982]). The work Frenkel–Kac [1980] linked this theory to the representation theory of affine algebras. Physicists were able to use this in the revival of the string theory started in 1984 (see Gross–Harvey–Martinec–Rohm [1985 B] and the book by Green–Schwartz–Witten [1987]). Note that using the vertex operators $\Gamma_\alpha(z)$, Frenkel [1985] and Goddard–Olive [1985 A] constructed certain (reducible) representations of (nonaffine) Kac–Moody algebras.

Since the pioneering work of Skyrme [1971] (see also Mandelstam [1975]), the boson–fermion correspondence has been playing an increasingly important role in 2-dimensional quantum field theory. More recently, it has become an important ingredient in the work of the Kyoto school on the KP hierarchy. The remarkable link of the theory of soliton equations to the infinite-dimensional Grassmannian was discovered by Sato [1981] and developed in the framework of the representation theory of affine algebras by Date–Jimbo–Kashiwara–Miwa [1981], [1982 A, B], [1983 A, B].

The infinite wedge representation was constructed in a more invariant form by Kac–Peterson [1981] and effectively used in the representation theory of the Virasoro algebra by Feigin–Fuchs [1982]. Theorem 14.10 is due to Date–Jimbo–Kashiwara–Miwa, who use the spinor formalism. Exposition of §§14.9 and 14.10 follows Kac–Peterson [1986] and Kac–Raina [1987]. The approach developed in §§14.11–14.13 is due to Kac–Wakimoto [1989]. The equation (14.11.9) of the KP hierarchy first appeared in Date–Jimbo–Kashiwara–Miwa [1981]; Proposition 14.11.2 is also due to them. Explicit formulas for the Hirota equations of the KP hierarchy were found by Lu [1989 A]. The fact that all Schur polynomials are solutions of the KP hierarchy (Proposition 14.11.1) was discovered by Sato [1981] and has become the starting point of this line of research.

The first part of Exercise 14.2 is due to Kostant [1959]; and its last part and Exercise 14.3 are due to Kac–Wakimoto [1989 A]. Exercises 14.5–14.8 are taken from Frenkel–Kac [1980].

Exercise 14.10 is taken from Kac–Peterson [1983]; the finite-dimensional case is due to Kostant. In the case of the $GL_n(\mathbf{C})$-module $\Lambda^k\mathbf{C}^n$, equation (14.13.3) is no other than the classical Plücker relations (see Exercise 14.27). This equation plays a key role in the theory of infinite-dimensional groups developed in Peterson–Kac [1983]. The main result of this paper is that the set of solutions of (14.13.3) actually coincides with $\mathcal{V}_\Lambda \cup \{0\}$ (cf. Lemma 14.11). This result has many important applications to the structure theory of Kac–Moody algebras; for instance, the conjugacy theorem of Cartan subalgebras is an easy consequence of it.

Exercises 14.13–14.16 are from Date–Jimbo–Kashiwara–Miwa [1982 A,B]. The general reduction procedure is discussed in their paper [1982 B]; and the BKP hierarchy, in their paper [1982 A]. The approach taken here to the BKP hierarchy follows that of Kac–Peterson [1986]; it was developed further by You [1989].

Note that the formula in the introduction of Date–Jimbo–Kashiwara–Miwa [1981] gives only the "big" cell of the "Birkhoff decomposition," and not the set of all polynomial solutions. The set of all solutions together with its Bruhat decomposition is described in Exercise 14.23. The background on the infinite flag varieties and Bruhat and Birkhoff decompositions (see Exercises 14.21 and 14.22) may be found in Garland–Raghunathan [1975], Pressley [1980], Lusztig [1983], Tits [1981], [1982], Kac–Peterson [1983], [1984 B], [1985 A], [1987], Peterson-Kac [1983], Atiyah–Pressley [1983], Goodman–Wallach [1984 A], Freed [1985], Pressley–Segal [1986], Kazhdan–Lusztig [1988], Kostant–Kumar [1986], [1987], Kumar [1984], [1985], [1987 A], [1988], Lakshmibai [1989], Lakshmibai–Seshadri [1989], Mathieu [1986 C], [1987], [1988], Arabia [1986], and others.

Exercises 14.30–14.34 are taken from Kac–Peterson [1986]. I also have benefited from lectures by and discussions with M. Kashiwara. Adler–Moser [1978] have studied some polynomials related to the KdV equation, found their degrees and a recurrent formula, but failed to find an explicit formula.

Exercise 14.38 is due to Kac, Ueno, and Yamada (see Yamada [1985]; the fact that Vir acts on Hir for KP was conjectured by Wakimoto–Yamada [1983]). The description of singular vectors of the Virasoro algebra given by Exercise 14.39 was conjectured by Goldstone and proved by Segal [1981] by a more complicated method. Exercise 14.42 seems to be new.

Given a representation of a group G on a vector space V and vectors $v \in V$, $v^* \in V^*$, the function $f_{v,v^*}(g) = \langle g \cdot v, v^* \rangle$ on G is called a *matrix coefficient*. It is well known (see, e.g., Vilenkin [1965]) that many special functions may be viewed as matrix coefficients restricted to some subset of G, and most of the properties of these functions may be derived from this fact. Exercise 14.26 shows that the skew Schur functions are special functions associated with the basic representation of GL_∞. As a result, many properties of these functions can be given a group-theoretical interpretation. A systematic study of matrix coefficients of the groups associated with Kac–Moody algebras was started by Kac–Peterson [1983].

Segal–Wilson [1985] gave a group-theoretical interpretation to the so-called quasi-periodic solutions of the KdV-type equations via the wedge representation.

All the representations constructed in this chapter are, in a certain sense, related to the projective line. A representation related to an elliptic curve is studied in Date–Jimbo–Kashiwara–Miwa [1983 B]. Representations related to algebraic curves are also considered by Cherednik [1983 A, B].

There are a number of papers on the applications of the theory of affine algebras to completely integrable Hamiltonian systems. Here are some of them: Adler–van Moerbeke [1980 A, B], [1982], Leznov–Saveliev–Smirnov [1981], Leznov–Saveliev [1983], Mikhailov–Olshanetsky–Perelomov [1981], Reiman–Semenov–Tjan–Shanskii [1979], [1981], Ueno–Takasaki [1984], Goodman–Wallach [1984 B].

Drinfeld–Sokolov [1981], [1984], were probably the first to notice a link between affine Lie algebras and the KdV-type equations. With each vertex of the Dynkin diagram of an affine Lie algebra they associate a KdV-type hierarchy of PDE, giving a uniform explanation for a large variety of scattered results in the area. Wilson [1984] gave a general approach to the construction of the solutions of these hierarchies of equations.

Exercises 14.44–14.51 are taken mainly from Date–Jimbo–Kashiwara–Miwa. Exercise 14.52 is taken from ten Kroode–Bergvelt [1986]; and Exercise 14.53, from Kac–Wakimoto [1989]. Exercises 14.55–14.58 use material from Kac–Peterson [1981], Arbarello–De Concini–Kac [1989], Pressley–Segal [1986]. Exercise 14.61 goes back to Chevalley [1951]. Exercises 14.62 and 14.63 are due to Tate [1968]. A similar approach was used by Arbarello–De Concini–Kac [1989] to prove an "abstract reciprocity law" and to derive from it the classical Weil reciprocity law.

Index of Notations and Definitions

Chapter 1.

§1.1.		generalized Cartan matrix
	$(\mathfrak{h}, \Pi, \Pi^\vee)$	realization of A
		indecomposable matrix
	Π	root basis
	$\Pi^\vee$	coroot basis
	$\alpha_1, \alpha_2, \ldots, \alpha_n$	simple roots
	$\alpha_1^\vee, \alpha_2^\vee, \ldots, \alpha_n^\vee$	simple coroots
	Q	root lattice
	Q_+	
	$\operatorname{ht} \alpha$	height of $\alpha \in Q$
	$\geq$	partial ordering on $\mathfrak{h}^*$
§1.2.	$\tilde{\mathfrak{g}}(A)$	
	$\tilde{\mathfrak{n}}_-$ and $\tilde{\mathfrak{n}}_+$	
	$\tilde{\mathfrak{r}}$	
§1.3.	$\mathfrak{g}(A)$	Lie algebra associated to A
	A	Cartan matrix of $\mathfrak{g}(A)$
	n	rank of $\mathfrak{g}(A)$
	$(\mathfrak{g}(A), \mathfrak{h}, \Pi, \Pi^\vee)$	quadruple associated to A
	$\mathfrak{g}(A)$	Kac–Moody algebra
	$\mathfrak{h}$	Cartan subalgebra
	$e_i, \ f_i \ (i = 1, \ldots, n)$	Chevalley generators
	$\mathfrak{g}'(A)$	derived algebra of $\mathfrak{g}(A)$
	$\mathfrak{h}' = \mathfrak{h} \cap \mathfrak{g}'(A)$	
	$\mathfrak{g}(A) = \bigoplus_{\alpha \in Q} \mathfrak{g}_\alpha$	root space decomposition
	$\mathfrak{g}_\alpha$	root space attached to $\alpha \in Q$
	$\operatorname{mult} \alpha$	multiplicity of $\alpha \in Q$
	Δ	set of roots
	Δ_+ and Δ_-	sets of positive and negative roots

	$\mathfrak{g}(A) = \mathfrak{n}_- \oplus \mathfrak{h} \oplus \mathfrak{n}_+$	triangular decomposition
	ω	Chevalley involution
§1.5.	$V = \bigoplus_{\alpha \in M} V_\alpha$	M-graded vector space
	$U = \oplus U \cap V_\alpha$	graded subspace
	$v \in V_\alpha$	homogeneous element
		formal topology on V
		formal completion of V
	$\mathfrak{g}(A) = \bigoplus_{j \in \mathbb{Z}} \mathfrak{g}_j(s)$	gradation of type s
	$\mathfrak{g}(A) = \bigoplus_{j \in \mathbb{Z}} \mathfrak{g}_j(1)$	principal gradation
	$\mathfrak{g}'(A)$	(direct definition)
§1.6.	$\mathfrak{c}$	center of $\mathfrak{g}(A)$ or $\mathfrak{g}'(A)$
§1.8.	$\mathfrak{g}_{-1} \oplus \mathfrak{g}_0 \oplus \mathfrak{g}_1$	local Lie algebra

Chapter 2.

§2.1.		symmetrizable matrix
	B	symmetrization of A
		symmetrizable algebra $\mathfrak{g}(A)$
	$\nu : \mathfrak{h} \to \mathfrak{h}*$	
§2.2.	$(.\|.)$	invariant symmetric bilinear form
§2.3.	$(.\|.)$	standard invariant form
§2.5.		restricted $\mathfrak{g}(A)$- or $\mathfrak{g}'(A)$-module
	Ω	generalized Casimir operator
	Ω_0	
	ρ	
§2.6.	$U(\mathfrak{g}(A)) = \bigoplus_{\beta \in Q} U_\beta$	
	$U(\mathfrak{g}'(A)) = \bigoplus_{\beta \in Q} U'_\beta$	
§2.7.	$\mathfrak{h}_{\mathbb{R}}$	
	$\mathfrak{k}(A)$	compact form of $\mathfrak{g}(A)$
	ω_0	compact involution
	$(x\|y)_0 = -(\omega_0(x)\|y)$	
§2.8.	Ω_2	
§2.9.		Heisenberg Lie algebra of order n

Chapter 3.

Section	Notation	Definition
§3.1.	$\mathfrak{g}({}^tA)$	dual Kac–Moody algebra
	$Q^\vee$	dual root lattice
	$\Delta^\vee$	dual root system
§3.4.		locally nilpotent element
		locally finite element
§3.6.	V	$\mathfrak{h}$- or $\mathfrak{h}'$-diagonalizable module
	V_λ	weight space
	λ	weight
	$\operatorname{mult}_V \lambda$	multiplicity of λ
		integrable $\mathfrak{g}(A)$- or $\mathfrak{g}'(A)$-module
§3.7.	r_i $(i = 1, \ldots, n)$	fundamental reflections
	W	Weyl group
§3.8.	G^π	group associated to an integrable $\mathfrak{g}(A)$-module π
	$\widetilde{W}^\pi$	
	D^π	
§3.11.	$w = r_{i_1} \ldots r_{i_s}$	reduced expression
	$\ell(w)$	length of w
§3.12.	C	fundamental chamber
	$w(C)$	chamber
	$X = \bigcup w(C)$	Tits cone
	$C^\vee$	dual fundamental chamber
	$X^\vee$	dual Tits cone
		Coxeter group
§3.14.		integrable Lie algebra

Chapter 4.

Section	Notation	Definition
§4.3.	A	of finite, affine or indefinite type
§4.7.	$S(A)$	Dynkin diagram of A
§4.8.	$a_0, \ldots, a_\ell$	labels of $S(A)$ in Table Aff
	$\delta = {}^t(a_0, a_1, \cdots, a_\ell)$	
§4.9.	θ	highest root

Chapter 5.

§5.1.	Δ^{re}	set of real roots
	Δ^{re}_{+}	set of positive real roots
	$\alpha^{\vee}$	dual real root
	r_α	reflection at $\alpha \in \Delta^{re}$
		long and short real roots
§5.2.	Δ^{im}	set of imaginary roots
	Δ^{im}_{+}	set of positive imaginary roots
§5.3.	$\operatorname{supp}\alpha$	support of $\alpha \in Q$
§5.6.	$\delta = \sum_{i=0}^{\ell} a_i\alpha_i$	
§5.7.		null root
§5.8.	Z	imaginary cone
§5.9.		a root basis
§5.10.	A	of hyperbolic type
§5.11.	Λ_n	Lorentzian lattice
§5.13.	Δ^{sim}	strictly imaginary roots

Chapter 6.

§6.0.		affine algebra
§6.1.	$a_0^{\vee}, \dots, a_\ell^{\vee}$	
	h and $h^{\vee}$	Coxeter number and dual Coxeter numb
	r	A belongs to Table Aff r
§6.2.	K	canonical central element
	d	scaling element
	$(.\vert.)$	normalized invariant form on affine algebra
	Λ_0	
	$\mathring{\mathfrak{h}}$	
	$\overline{\lambda}$ or $\overline{S}$	projection of $\lambda \in \mathfrak{h}*$ or $S \subset \mathfrak{h}*$ on $\mathring{\mathfrak{h}}*$
§6.3.	$\mathring{\mathfrak{g}}$	
	$\mathring{\Delta}$	
	$\mathring{Q}$	
	Δ^{re}_{s}, Δ^{re}_{l}	sets of short and long real roots
§6.4.	$\theta = \delta - a_0\alpha_0$	

	$(.\|.)$	normalized invariant form on finite-dimensional algebra
§6.5.	$\mathring{W}$	
	t_α	
	M	
	T	group of translations
		special vertex of $S(A)$
§6.6.	W_{af}	affine Weyl group
	C_{af}	fundamental alcove
§6.8	$\phi(x, y)$	Killing form
		integral lattice
Chapter 7.		
§7.1.	$\mathcal{L} = \mathbf{C}[t, t^{-1}]$	algebra of Laurent polynomials
	$\operatorname{Res} P(t)$	residue of $P(t) \in \mathcal{L}$ at $t = 0$
§7.2.		nontwisted affine algebra
	$\mathcal{L}(\mathring{\mathfrak{g}})$	loop algebra
	ψ	2-cocycle on $\mathcal{L}(\mathring{\mathfrak{g}})$
	$\tilde{\mathcal{L}}(\mathring{\mathfrak{g}})$	central extension of $\mathcal{L}(\mathring{\mathfrak{g}})$
	$\hat{\mathcal{L}}(\mathring{\mathfrak{g}})$	realization of a non-twisted affine algebra
§7.3.	$\mathfrak{d} = \bigoplus_{j \in \mathbf{Z}} \mathbf{C} d_j$	Lie algebra of regular vector fields on $\mathbf{C}^x$
	Vir	Virasoro algebra
	$Vir \ltimes \tilde{\mathcal{L}}(\mathring{\mathfrak{g}})$	semidirect product of Vir and affine algebra
§7.4	E_0 and F_0	
§7.7	$x^{(n)}$	$t^n \otimes x$
	$x(z)$	current
	$\delta(z_1 - z_2)$	formal δ-function
		primary field
		conformal weight
§7.8.	$\varepsilon(\alpha, \beta)$	asymmetry function
§7.9.	μ	diagram automorphism
§7.11.	A_∞, $A_{+\infty}$, B_∞, C_∞, D_∞	infinite affine matrices
	gl_∞	
§7.12.	$\overline{a}_\infty$, $\overline{b}_\infty$, $\overline{c}_\infty$, $\overline{d}_\infty$	

	$a_\infty, b_\infty, c_\infty, d_\infty$	completed infinite rank affine algebras
§7.13.	$\overset{\circ}{G}(\mathcal{L})$	loop group associated to $\overset{\circ}{G}$
	$T(z)$	energy-momentum tensor

Chapter 8.

§8.2.	$\mathcal{L}(\mathfrak{g},\sigma,m) = \bigoplus_{j\in\mathbb{Z}}(t^j \otimes \mathfrak{g}_{j \bmod m})$	
	$\tilde{\mathcal{L}}(\mathfrak{g},\sigma,m)$	
	$\hat{\mathcal{L}}(\mathfrak{g},\sigma,m)$	realization of a twisted affine algebra
§8.3	θ_0 and θ^0	
§8.4.	$\mathfrak{t}$	homogeneous Heisenberg subalgebra
§8.5.	$\alpha \xrightarrow{\gamma} \beta$	
§8.6.	$\sigma_{s;k}$	automorphisms of type $(s;k)$
	φ_σ	covering homomorphism
§8.7.		realization of type s of an affine algebra
§8.8.		quasirational automorphism
		rational automorphism

Chapter 9.

§9.1.	$P(V)$	set of weights of a $\mathfrak{g}(A)$-module V
	$v \in V_\lambda, v \neq 0$	weight vector of weight λ
	$D(\lambda) = \{\mu \in \mathfrak{h}* \mid \mu \leq \lambda\}$	
	$\mathcal{O}$	category of (certain) $\mathfrak{g}(A)$-modules
§9.2.		highest weight module
		highest weight
	v_Λ	highest weight vector
	$M(\Lambda)$	Verma module with highest weight Λ
	$M'(\Lambda)$	unique proper maximal submodule of $M($
	$U(\mathfrak{b}) \otimes_{U(\mathfrak{a})} V$	induced module
§9.3.	$L(\Lambda)$	irreducible module with highest weight Λ
		primitive vector
		primitive weight
§9.4.	$L^*(\Lambda)$	irreducible module with lowest weight-Λ
	B	contravariant bilinear form on $L(\Lambda)$
	$\langle v \rangle$	expectation value of v

	E_+	sequence of positive exponents of $\mathfrak{g}(A)$
	$\Gamma^\beta(z)$	vertex operators for $\mathfrak{g}(A)$
§14.8.		homogeneous vertex operator construction
	$\Gamma_\alpha(z)$	vertex operator
§14.9.	$\underline{i}_1 \wedge \underline{i}_2 \wedge \cdots$	semi-infinite monomial
	F	fermionic Fock space
		charge
	$\|m\rangle$	vacuum vector of charge m
	Par	set of all partitions
		energy
	ψ_j	wedging operator
	ψ_j^*	contracting operator
	Cl_V	Clifford algebra
		spin module
	r (resp. R or $\hat{r}$)	infinite wedge representation of gl_∞ (resp. GL_∞ or a_∞)
	$\overline{e} \in \overline{a}_\infty$	cyclic element
	$s \subset a_\infty$	principal subalgebra
§14.10.	α_n	free bosons
	B	bosonic Fock space
	$S_\lambda(x)$	Schur polynomial
	σ	boson-fermion correspondence
	$\Gamma_+(z)$ and $\Gamma_-(z)$	
	$\Gamma(z_1, z_2)$	vertex operator for a_∞
§14.11.	$\Omega_1 = \sum_{j\in\mathbb{Z}} \psi_j \otimes \psi_j^*$	
	$P(D)f \cdot g$	Hirota bilinear equation
		KP hierarchy
	$\frac{3}{4}u_{yy} = (u_t - \frac{3}{2}uu_x - \frac{1}{y}u_{xxx})_x$	Kadomtsev-Petviashvili equation
		N-solition solution of KP hierarchy
§14.12.	L_{high}	highest component of $L(\Lambda) \otimes L(\Lambda)$
	L_{low}	orthocomplement of L_{high}
	G	group generated by the $\exp te_i, \exp tf_i$
	$\mathcal{V}_\Lambda$	G-orbit of a highest weight vector v_Λ

References

Abe, E., Morita, J. [1988] Some Tits systems with affine Weyl groups in Chevalley groups over Dedekind domains, preprint.

Abe, E., Takeuchi, M. [1989] Groups associated with some types of infinite-dimensional Lie algebras, preprint.

Adler, M., Moser, J. [1978] On a class of polynomials connected with the Korteweg–de Vries equation, Comm. Math. Phys. 61 (1978), 1–30.

Adler, M., van Moerbeke, P. [1980 A] Completely integrable systems, Euclidean Lie algebras and curves, Advances in Math. 38 (1980), 267–317.

[1980 B] Linearization of Hamiltonian systems, Jacobi varieties and representation theory, Advances in Math. 38 (1980), 318–379.

[1982] Kowalevsky's asymptotic method, Kac–Moody Lie algebras and regularization, Comm. Math. Phys. 83 (1982), 83–106.

[1989] The Toda lattice, Dynkin diagrams, singularities and abelian varieties, preprint.

Alvares-Gaumé, L., Bost, J.-B., Moore, G., Nelson, P., Vafa, C. [1987] Bosonization on higher genus Riemann surfaces, Comm. Math. Phys. 112 (1987), 503–552.

Andersen, H. H. [1985] Schubert varieties and Demasure's character formula, Invent. Math. 79 (1985), 611–618.

Andrews, G. E. [1976] The theory of partitions, "Encyclopedia of Math.", vol. 2, 1976.

[1984] Hecke modular forms and the Kac–Peterson identities, Trans. Amer. Math. Soc. 283 (1984), 451–458.

[1986] q-series: their development and application in analysis, number theory, combinatorics, physics, and computer algebra, Regional Conf. Ser. in Math. 66, AMS, 1986.

Anick, D. [1983] The smallest Ω-irrational CW-complex, J. Pure Applied Algebra 28 (1983), 213–222.

Arabia, A. [1986] Cohomologie T-équivariante de G/B pour un groupe G de Kac–Moody, C. R. Acad. Sci. Paris, t. 302, Série I, No. 17 (1986).

Arbarello, E., De Concini, C. [1987] Another proof of a conjecture of S. P. Novikov on periods of abelian integrals on Riemann surfaces, Duke Math. J. 54 (1987), 163–178.

Arbarello, E., De Concini, C., Kac, V. G., Procesi C. [1988] Moduli spaces of curves and representation theory, Comm. Math. Phys. 117 (1988), 1–36.

Arbarello, E., De Concini, C., Kac, V. G. [1989], The infinite wedge representation and the reciprocity law for algebraic curves, Proceedings of the 1987 conference in Maine, Proc. Symposia in Pure Math. 49, 1989, 171–190.

Arcuri, R. C., Gomes, J. F., Olive, D. I. [1987] Conformal subalgebras and symmetric spaces, Nucl. Phys. B285 (1987), 327–339.

Atiyah, M. F., Pressley, A. N. [1983] Convexity and loop groups, in Arithmetic and Geometry, Progress in Math 36, 33–64, Birkhäuser, Boston, 1983.

Bais, F. A., Bouwknegt, P., Schoutens, K., Surridge, M. [1988 A] Extensions of the Virasoro algebra constructed from Kac–Moody algebras using higher order Casimir invariants, Nucl. Phys. B304 (1988) 348–370

[1988 B] Coset construction for extended Virasoro algebras, Nucl. Phys. B304 (1988), 371–391.

Bardaki, K., Halpern, M. [1971] New dual quark models, Phys. Rev. D3 (1971), 2493–2506.

Bausch, J., Rousseau, G. [1989] Algebres de Kac–Moody affine (automorphismes et formes reelles), Institut E. Cartan 11, 1989.

Beilinson, A. A. [1980] Residues and adeles, Funct. Anal. Appl. 14 (1980), 44–45.

Beilinson, A., Bernstein, I. N. [1981] Localization de $\mathfrak{g}$-modules, C. R. Acad. Sci. Paris 292 (1981), 15–18.

Beilinson, A. A., Manin Yu. I., Schechtman, V. V., [1987] Sheaves of Virasoro and Neveu–Schwarz algebras, Lecture Notes Math. 1289, Springer–Verlag, 1987, 52–66.

Beilinson, A. A., Schechtman, V. V. [1988] Determinant bundles and Virasoro algebra, preprint.

Belavin, A. A., Drinfeld, V. G. [1982] Equations of triangles and simple Lie algebras, Funkt. Analis i ego Prilozh. 16 (1982), No. 3, 1–29 (in Russian).

Belavin, A. A., Polyakov, A. M., Zamolodchikov, A. B. [1984 A] Infinite conformal symmetry of critical fluctuations in two dimensions, J. Stat. Phys. 34, (1984), 763–774.

[1984 B] Infinite conformal symmetry in two-dimensional quantum field theory, Nucl. Phys. B241 (1984), 333–380.

Benoist, Y. [1987] $\mathfrak{n}$-coinvariants des $\mathfrak{g}$-modules $\mathfrak{n}$-localement nilpotents, preprint.

Berman, S. [1976] On derivations of Lie algebras, Canad. J. Math. 27 (1976), 174–180.

[1989] On generators and relations for certain involutory subalgebras of Kac–Moody Lie algebras, Comm. Alg. 17(12) (1989), 3165–3185.

Berman, S., Lee, Y. S., Moody, R. V. [1989] The spectrum of a Coxeter transformation, affine Coxeter transformations, and the defect map, J. Algebra 121 (1989), 339–357.

Berman, S., Pianzola, A. [1987] Generators and relations for real forms of some Kac–Moody Lie algebras, Comm. Alg. 15(5) (1987), 935–959.

Berman, S., Moody, R. V. [1979] Multiplicities in Lie algebras, Proc. Amer. Math. Soc. 76 (1979), 223–228.

Bernard, D. [1987] String characters from Kac–Moody automorphisms, Nuclear Phys. B288 (1987), 628–648.

[1988] On the Wess–Zumino–Witten models on the torus, Nucl. Phys. B303 (1988), 77–93.

[1989] Towards generalized Macdonald's identities, in Infinite-dimensional Lie algebras and groups, Adv. Ser. in Math. Phys. 7, World. Sci., 1989, 467–482.

Bernard, D., Felder, G. [1989] Fock representation and BRST cohomology in $SL(2)$ current algebra, preprint.

Bernard, D., Thierry-Mieg, J. [1987 A] Bosonic Kac–Moody string theories, Phys. Letters B, 185 (1987), 65–72.

[1987 B] Level one representation of the simple affine Kac–Moody algebras in their homogeneous gradations, Comm. Math. Phys. 111 (1987), 181–246.

Bernstein, I. N., Gelfand, I. M., Gelfand, S. I. [1971] Structure of representations generated by highest weight vectors, Funkt. Analis i ego Prilozh. 5 (1971), No. 1, 1–9, English translation: Funct. Anal. Appl. 5 (1971), 1–8.

[1975] Differential operators on the basic affine space and a study of γ-modules, Lie groups and their representations. Summer school of the Bolyai János Math. Soc., Gelfand, I. M. (ed.), pp. 21–64. New York: Division of Wiley and Sons, Halsted Press, 1975.

[1976] On a category of $\mathfrak{g}$-modules, Funkt. Analis i ego Prilozh. 10 (1976) No. 2, 1–8. English translation: Funct. Anal. Appl. 10 (1976), 87–92.

Bernstein, I. N., Schvartzman, O. [1978] Chevalley theorem for complex crystallographic Coxeter groups, Funct. Anal. Appl. 12:4 (1978), 79–80.

Bershadsky, M., Ooguri, H. [1989] Hidden $SL(n)$ symmetry in conformal field theories, preprint.

Bien, F. [1989] Global representations of the diffeomorhism group of the circle, in: Infinite-dimensional Lie algebras and groups, Adv. Ser. in Math. Phys. 7, World Sci., 1989, 89–107.

Bilal, A., Gervais, J.-L. [1989] Conformal theories with nonlinearly extended Virasoro symmetries and Lie algebra classification, in: Infinite-dimensional Lie algebras and groups, Adv. Ser. in Math. Phys. 7, World Sci., 1989, 483–526.

Borcherds, R. [1986] Vertex algebras, Kac–Moody algebras, and the Monster, Proc. Natl. Acad. Sci. USA 83 (1986), 3068–3071.

[1988 A] Generalized Kac–Moody algebras, J. Algebra 115 (1988), 501–512.

[1988 B] Vertex algebras I, preprint.

[1989 A] Central extensions of generalized Kac–Moody algebras, J. Algebra.

[1989 B] The Monster Lie algebra, preprint.

Borcherds, R. E., Conway, J. H., Queen, L., Sloane, N. J. A. [1984] A Monster Lie algebra? Advances in Math. 53 (1984), 75–79.

Bott, R. [1977] On the characteristic classes of groups of diffeomorphisms, Enseign. Math. (1977), 23, 3–4, 209–220.

Bourbaki, N. [1968] Groupes et algèbres de Lie, Ch. 4–6, Hermann, Paris, 1968.

[1975] Groupes et algèbres de Lie, Ch. 7–8, Hermann, Paris, 1975.

Bouwknegt, P. [1989] Extended Conformal Algebras from Kac–Moody Algebras, in: Infinite-dimensional Lie algebras and groups, Adv. Ser. in Math. Phys. 7, World Sci. 1989, 527–555.

Bowick, M. J., Rajeer, S. R. [1987] String theory as the Kähler geometry of loop space, Phys. Rev. Lett. 58 (1987), 535–538.

Boyer, R. P. [1980] Representation theory of the Hilbert–Lie group $U(H)_2$, Duke Math. J. 47 (1980), 325–344.

[1988] Representation theory of $U_1(H)$, Proc. Amer. Math. Soc. 103 (1988), 97–104.

[1989] Representation theory of $U(\infty)$, Proc. of Symposia in Pure Math.

Brylinski, J.-L. [1988] Representations of loop groups, Dirac operators on loop space and modular forms, preprint.

[1989] Loop groups and noncommutative theta functions, preprint.

Brylinski, J.-L., Kashiwara, M. [1981] Kazhdan–Lusztig conjecture and holonomic systems, Invent. Math. 64 (1981), 387–410.

Cappelli, A., Itzykson, C., Zuber, J. B. [1987 A] Modular invariant partition functions in two dimensions, Nucl. Phys. B280 (1987), 445–460.

[1987 B] The *A-D-E* classification of minimal and $A_1^{(1)}$ conformal invariant theories, Comm. Math. Phys. 113 (1987), 1–26.

Cardy, J. [1986] Operator content of two-dimensional conformally invariant theories, Nucl. Phys. B270 (1986), 186–204.

Carey, A. L., Palmer, J. [1985] Infinite complex spin groups, preprint.

Carey, A. L., Rijsenaars, S. N. M. [1987] On fermion gauge groups, current algebras, and Kac–Moody algebras, Acta Applicandas Math. 10 (1987), 1–86.

Casian, L. [1989] Formules de multiplicité de Kazhdan–Lusztig dans le case de Kac–Moody, preprint.

Chari, V. [1985] Annihilators of the Verma modules of Kac–Moody Lie algebras, Invent. Math. 81 (1985), 47–58.

[1986] Integrable representations of affine Lie algebras, Invent. Math. 85 (1986), 317–335.

Chari, V., Ilangovan, S. [1984] On the Harish-Chandra homomorphism for infinite-dimensional Lie algebras, J. Algebra 90 (1984), 476-490.

Chari, V., Pressley, A. N. [1986] New unitary representations of loop groups, Math. Ann. 275 (1986), 87–104.

[1987] A new family of irreducible, integrable modules for affine Lie algebras, Math. Ann. 277 (1987), 543–562.

[1988 A] Integrable representations of twisted affine Lie algebras, J. Algebra 113 (1988), 438–464.

[1988 B] Unitary representations of the Virasoro algebra and a conjecture of Kac, preprint.

[1989] Integrable Representations of Kac–Moody Algebras: Results and Open Problems, in: Infinite-dimensional Lie algebras and groups, Adv. Ser. in Math. Phys. 7, World Sci. 1989, 3–24.

Chau, L.-L. [1983] Chiral fields, self-dual Yang–Mills fields as integrable systems and the role of Kac–Moody algebra, Brookhaven Lab., preprint.

Cherednik, I. B. [1983 A] On the definition of τ-functions for generalized affine Lie algebras, Funkt. Analis i ego Prilozh. 17 (1983) No. 3, 93–95 (in Russian).

[1983 B] On a group interpretation of Baker's function and τ-functions, Uspechi Mat. Nauk 38 (1983), No. 6, 133–134 (in Russian).

Chevalley, C. [1948] Sur la classification des algèbres de Lie simples et de leur représentations, C. R., 227 (1948), 1136–1138.

[1951] Introduction to the theory of algebraic functions of one variable, Math. Surveys 6, AMS, New York, 1951.

Chodos, A., Thorn, C. B. [1974] Making the massless string massive, Nucl. Phys. B72 (1974), 509–522.

Coleman, A. J. [1958] The Betti numbers of the simple Lie groups, Can. J. Math. 10 (1958), 349–356.

[1989] Killing and the Coxeter transformation of Kac–Moody algebras, Invent. Math. 95 (1989), 447–477.

Coleman, A. J., Howard, M. [1989] Root multiplication for general Kac–Moody algebras, C. R. Math. Rep. Acad. Sci. Canada, 11 (1989), 15–18.

Coley, R. [1983] On subalgebras of finite codimension in affine Lie algebras, MIT, preprint.

Conway, J. H. [1983] The automorphism group of the 26-dimensional even unimodular Lorentzian lattice, J. Algebra 80 (1983), 159–163.

Conway, J. H., Norton, S. P. [1979] Monstrous moonshine, Bull. London Math. Soc., 11 (1979), 308–339.

Dadok, J., Kac, V. G. [1985] Polar representations, J. Algebra, 92 (1985), 504–524.

Date, E., Jimbo, M., Kashiwara, M., Miwa, T. [1981] Operator approach to the Kadomtsev–Petviashvili equation. Transformation groups for soliton equations III, J. Phys. Soc. Japan, 50 (1981), 3806–3812.

[1982 A] A new hierarchy of soliton equations of KP-type. Transformation groups for soliton equations IV, Physics 4D (1982), 343–365.

[1982 B] Transformation groups for soliton equations. Euclidean Lie algebras and reduction of the KP hierarchy, Publ. RIMS, Kyoto University, 18 (1982), 1077–1110.

[1983 A] Transformation groups for soliton equations, in Proceedings of RIMS symposium (ed. M. Jimbo, T. Miwa), 39–120, World Sci., 1983.

[1983 B] Landau-Lifshitz equation: solitons, quasi-periodic solutions and infinite-dimensional Lie algebras, J. Phys. A.: Math. gen. 16 (1983), 221–236.

Date, E., Jimbo, M., Miwa, T. [1983] Method for generating discrete soliton equations III, J. Phys. Soc. Japan 52 (1983), 388–393.

Date, E., Jimbo, M., Kuniba, A., Miwa, T., Okada, M. [1987] Exactly solvable SOS models: Local height probabilities and theta function identities, Nucl. Phys. B290 [FS20] (1987), 231–273

[1988] Exactly solvable SOS models. II: Froof of the star-triangle relation an combinational identities. Adv. Stud. Pure Math. 16, Kinokuniya-Academic, 1988.

[1989 A] Path space realization of the basic representation of $A_n^{(1)}$, in: Infinite-dimensional Lie algebras and groups, Adv. Ser. in Math. Phys. 7, World Sci., 1989, 108–123.

[1989 B] Chemins, diagrammes de Maya et représentations de $\widehat{s\ell}(r, \mathbf{C})$, C. R. Acad. Sci. Paris, t. 308, (1989), 129–132.

[1989 C] Paths, Maya diagrams and representations of $\widehat{s\ell}(r, \mathbf{C})$, Adv. Stud. Pure Math. 19.

[1989 D] One-dimensional configuration sums in vertex models and affine Lie algebra characters, Lett. Math. Phys. 17 (1989), 69–77.

De Concini, C., Kac, V. G., Kazhdan, D. A. [1989] Boson-fermion correspondence over $\mathbb{Z}$, in: Infinite-dimensional Lie algebras and groups, Adv. Ser. in Math. Phys., World Sci., 1989, 124–137.

Demazure, M. [1974] Désingularisation des variétés de Schubert généralisées, Ann. Sci. Ec. Norm. Sup. 7 (1974), 53–88.

Deodhar, V. V. [1982] On the root system of a Coxeter group, Commun. in Algebra 10 (6) (1982), 611–630.

Deodhar, V. V., Gabber, O., Kac, V. G. [1982] Structure of some categories of representations of infinite-dimensional Lie algebras, Advances in Math. 45 (1982), 92–116.

Deodhar, V. V., Kumaresan, S. [1986] A finiteness theorem for affine Lie algebras, J. Algebra 103 (1986), 403–426.

Distler, J., Qiu, Z. [1989] BRS cohomology and Feigin–Fuchs representation of Kac–Moody and parafermionic theories, preprint.

Dixmier, J. [1974] Algèbres enveloppantes, Paris: Gauthier-Villars, 1974. English translation: Enveloping algebras, Minerva Translations, North Holland, 1977.

Dolan, L. [1984] Kac–Moody algebras and exact solvability in hadronic physics, Physics Reports 109 (1984), 1–94.

[1985] Why Kac–Moody algebras are interesting in physics, in Lectures in Applied Math. 21, 1985, 307–324.

Dolan, L., Duff, M. J. [1984] Kac–Moody symmetries in Kaluza–Klein theories, Phys. Rev. Lett. 52 (1984), 14–20.

Drinfeld, V. G. [1985] Hopf algebras and the quantum Yang–Baxter equation, Dokl. Akad. Nauk 10 (1985), 1060–1064. English translation: Soviet Math. Dokl. 32 (1985), 254–258.

[1986] Quantum groups, ICM proceedings, Berkeley, 1987, 798–820.

[1987] A new realization of Yangians and quantized affine algebras, Dokl. Akad. Nauk SSSR, 296 (1987). English translation: Soviet Math. Dokl. 36 (1988) 212–216.

[1989] On almost commutative Hopf algebras, Algebra and Analis 1 (1989), 30–46.

Drinfeld, V. G., Sokolov, V. V. [1981] Equations of Korteweg–de Vries type and simple Lie algebras, Doklady AN SSSR 258 (1981), No. 1, 11–16. English translation: Soviet Math. Dokl. 23 (1981), 457–462.

[1984] Lie algebras and equations of KdV type, Itogi Nauki i Techniki, ser. Sovrem. Problemy Matem 24, 1984, 81–180 (in Russian).

Dotsenko, V. S., Fateev, V. A. [1984] Conformal algebra and multipoint correlation function in 2D statistical models, Nucl. Phys. B240 (1984), 312–348.

Dyson, F. [1972] Missed opportunities, Bull. Amer. Math. Soc. 78 (1972), 635–652.

Eichler, M. [1966] Introduction to the theory of algebraic numbers and functions, Academic Press, New York, 1966.

Fateev, V. A., Lukyanov, S. L. [1988] The models of two-dimensional conformal quantum field theory with $\mathbb{Z}_n$ symmetry, Int. J. Mod. Phys. A3 (1988), 507–520.

Fateev, V. A., Zamolodchikov, A. B. [1987] Conformal quantum field theory models in two dimensions having $\mathbb{Z}_3$ symmetry, Nucl. Phys. B280 (1987), 644–660.

Feigin, B. L. [1984] Semi-infinite homology of Kac–Moody and Virasoro Lie algebras, Uspechi Mat. Nauk 39 (1984), No. 2, 195–196 (in Russian).

Feigin, B. L., Fuchs, D. B. [1982] Skew-symmetric invariant differential operators on a line and Verma modules over the Virasoro algebra, Funkt. Analis i ego Prilozh. 16 (1982), No. 2, 47–63 (in Russian).

[1983 A] Casimir operators in modules over the Virasoro algebra, Doklady AN SSSR 269 (1983), 1057–1060 (in Russian).

[1983 B] Verma modules over the Virasoro algebra, Funkt. Analis i ego Prilozh. 17 (1983), No. 3, 91–92 (in Russian).

[1984 A] Representations of the Virasoro algebra, preprint.

[1984 B] Verma modules over the Virasoro algebra, in Lecture Notes in Math. 1060, Springer–Verlag, 1984, 230–245.

[1989] Cohomology of some nilpotent subalgebras of the Virasoro and Kac–Moody Lie algebras, preprint.

Feigin, B. L., Frenkel, E. V. [1988] A family of representations of affine Lie algebras, Uspechi Mat. Nauk, 43:2 (1988), 227–228.

[1989 A] Representations of affine Kac–Moody algebras and bosonization, preprint.

[1989 B] Representations of affine Kac–Moody algebras, bosonisation and resolutions, preprint.

[1989 C] Affine Kac–Moody algebras and semi-infinite flag manifolds, preprint.

Feigin, B. L., Tsygan, B. L. [1983] Cohomology of Lie algebras of generalized Jacobi matrices, Funkt. Analis i ego Prilozh. 17:2 (1983), 86–87.

[1988] Riemann–Roch theorem and Lie algebra cohomology, preprint.

Feingold, A. J. [1980] A hyperbolic GCM Lie algebra and the Fibonacci numbers, Proc. Amer. Math. Soc. 80 (1980), 379–385.

Feingold, A. J., Frenkel, I. B. [1983] A hyperbolic Kac–Moody algebra and the theory of Siegel modular forms of genus 2, Math. Ann. 263 (1983), 87–144.

[1985] Classical affine algebras, Advances in Math. 56 (1985), 117–172.

Feingold, A. J., Lepowsky, J. [1978] The Weyl–Kac character formula and power series identities, Adv. Math. 29 (1978), 271–309.

Felder, G. [1989] BRST approach to minimal models, Nucl. Phys. B317 (1989), 215–236.

Felder, G., Fröhlich, J., Keller, G. [1989] On the structure of unitary conformal field theory. I. Existence of conformal blocks, ETH preprint.

Felder, G., Gawedzki, K., Kupiainen, A. [1988] Spectra of Wess-Zumino-Witten models with arbitrary simple groups, Commun. Math. Phys. 117 (1988), 127–158.

Fialowski, A. [1984] Deformations of nilpotent Kac–Moody algebras, Studia Sci. Math. Hung. 19 (1984).

Freed, D. S. [1985] Flag manifolds and Kaehler geometry, in Infinite-dimensional groups with applications, MSRI publ. 4, Springer–Verlag, 1985, 83–124.

Frenkel, I. B. [1981] Two constructions of affine Lie algebra representations and boson-fermion correspondence in quantum field theory, J. Funct. Analysis 44 (1981), 259–327.

[1982] Representations of affine Lie algebras, Hecke modular forms and Korteweg–de Vries type equations, in Lecture Notes in Math. 933, Springer–Verlag 1982, 71–110.

[1984] Orbital theory for affine Lie algebras, Invent. Math. 77 (1984), 301–352.

[1985] Representations of Kac–Moody algebras and dual resonance models, in Lectures in Applied Math. 21, 1985, 325–354.

[1986] Beyond affine Lie algebras, ICM Berkeley, 1986, 821–839.

Frenkel, I. B., Garland, H., Zuckerman, G. J. [1986] Semi-infinite cohomology and string theory, Proc. Natl. Acad. Sci. USA 83 (1986), 8442–8448.

Frenkel, I. B., Jing, N. [1988] Vertex representations of quantum affine algebras, Proc. Natl. Acad. Sci., USA 85 (1988), 9373–9377.

Frenkel, I. B., Kac, V. G. [1980] Basic representations of affine Lie algebras and dual resonance models, Invent. Math. 62 (1980), 23–66.

Frenkel, I. B., Lepowsky, J., Meurman, A. [1984] A natural representation of the Fischer–Griess Monster with the modular function J as character, Proc. Natl. Acad. Sci. USA 81 (1984), 3256–3260.

[1989] Vertex operator algebras and the Monster, Academic Press, 1989.

Friedan, D., Martinec, E., Shenker, S. [1986] Conformal invariance, supersymmetry and string theory, Nucl. Phys. B271 (1986), 93–130.

Friedan, D., Qiu, Z., Shenker, S. [1985] Conformal invariance, unitarity, and two-dimensional critical exponents, MSRI publ. 3, Springer–Verlag, 1985, 419–449.

[1986] Details of the non-unitarity proof, Comm. Math. Phys. 107 (1986), 535–542.

Friedan, D., Shenker, S. [1987] The analytic geiomety of two-dimensional conformal field theory, Nucl. Phys. B281, (1987), 509–545.

Fuchs, D. B. [1984] Cohomology of infinite-dimensional Lie algebras, Nauka, Moscow, 1984.

Furlan, F., Sotkov, G. M., Todorov, I. T. [1988] Two-dimensional conformal quantum field theory, Rivista del Nuovo Cimento.

Gabber, O., Kac, V. G. [1981] On defining relations of certain infinite-dimensional Lie algebras, Bull. Amer. Math. Soc. 5 (1981), 185–189.

Garland, H. [1975] Dedekind's η-function and the cohomology of infinite-dimensional Lie algebras, Proc. Natl. Acad. Sci. USA 72 (1975), 2493–2495.

[1978] The arithmetic theory of loop algebras, J. Algebra 53 (1978), 480–551.

[1980] The arithmetic theory of loop groups, Publ. Math. IHES 52 (1980), 5–136.

Garland, H., Lepowsky, J. [1976] Lie algebra homology and the Macdonald–Kac formulas, Invent. Math. 34 (1976), 37–76.

Garland, H., Raghunathan, M. S. [1975] A Bruhat decomposition for the loop space of a compact group: a new approach to results of Bott, Proc. Natl. Acad. Sci. USA 72 (1975), 4716–4717.

Gawedzki, K. [1987] Topological actions in two-dimensional quantum field theories, IHES preprint.

[1988] Conformal field theory, IHES preprint.

[1989] Wess-Zumino-Witten conformal field theory, IHES preprint.

Gawedzki, K., Kupianen, A. [1988] G/H conformal field theory from gauged WZW model, IHES preprint.

[1989] Coset construction from functional integrals, Helsinki preprint.

Gelfand, I. M., Fuchs, D. B. [1968] Cohomology of the Lie algebra of vector fields on a circle, Funct. Anal. Appl. 2:4 (1968) 92–93. English translation: Funct. Anal. Appl. 2 (1968) 342–343.

[1969] The cohomology of the Lie algebra of tangent vector fields on a smooth manifold, I, Funkt. Analis i ego Prilozh. 3 (1969), No. 3, 32–52. English translation: Funct. Anal. Appl. 3 (1969), 194–224.

[1970 A] The cohomology of the Lie algebra of tangent vector fields on a smooth manifold, II, Funkt. Analis i ego Prilozh. 4 (1970), No. 2, 23–32. English translation: Funct. Anal. Appl. 4 (1970), 110–119.

[1970 B] The cohomology of the Lie algebra of formal vector fields, Izvestija AN SSSR (ser. mat.), 34 (1970), 322–337. English translation: Math. USSR-Izvestija 34 (1970), 327–342.

Gepner, D. [1987 A] New conformal field theories associated with Lie algebras and their partition functions, Nucl. Phys. B290 (1987), 10–24.

[1987 B] On the spectrum of 2D-conformal field theories, Nucl. Phys. B287 (1987), 111–130.

Gepner, D., Witten, E. [1986] String theory on group manifold, Nucl. Phys. B278 (1986), 493–549.

Ginsparg, P. [1988] Applied conformal field theory, Les Houches 1988.

Goddard, P. [1989] Meromorphic conformal field theory, in: Infinite-dimensional Lie algebras and groups, Adv. Ser. in Math. Phys. 7, World Sci., 1989, 556–587.

Goddard, P., Kent, A., Olive, D. [1985] Virasoro algebras and coset space models, Phys. Lett. B 152 (1985), 88–93.

[1986] Unitary representations of the Virasoro and Super-Virasoro algebras, Comm. Math. Phys. 103 (1986), 105–119.

Goddard, P., Nahm, W., Olive, D. [1985] Symmetric spaces, Sugawara's energy momentum tensor in two dimensions and free fermions, Phys. Lett. B 160 (1985), 111–116.

Goddard, P., Nahm, W., Olive, D., Schwimmer, A. [1986] Vertex operators for non-simply-laced algebras, Comm. Math. Phys.107 (1986), 179–212.

Goddard, P., Olive, D. [1985 A] Algebras, lattices and strings, MSRI publ. 3, 1985, 51–96.

[1985 B] Kac–Moody algebras, conformal symmetry and critical exponents, Nucl. Phys. B257 (1985), 226–250.

[1986] Kac–Moody and Virasoro algebras in relation to quantum physics, Int. J. Mod. Phys. A1 (1986), 303–414.

Goddard, P., Olive, D., Waterson, G., Superalgebras, symplectic bosons and the Sugawara construction, Comm. Math. Phys. 112, (1987), 591–611.

Goodman, R., Wallach, N. [1984 A] Structure and unitary cocycle representations of loop groups and the group of diffeomorphisms of the circle, J. für reine und angewandte Math. 347 (1984), 69–133.

[1984 B] Classical and quantum mechanical systems of Toda-lattice type, Comm. Math Phys. 84 (1984), 177–217.

[1985] Projective unitary positive-energy representations of Diff(S^1), J. Funct. Analysis (1985).

Green, M. B., Schwarz, J. H. [1984] Anomaly cancellations in supersymmetric $D = 10$ gauge theory and superstring theory, Phys. Lett. 149B, 117–125.

Green, M. B., Schwarz, J. H., Witten, E. [1987] Superstring theory, Cambridge University Press, 1987.

Griess, R. [1982] The Friendly Giant, Invent. Math, 69 (1982), 1–102.

Gross, D. J., Harvey, J. A., Martinec, E., Rohm, R. [1985 A] The heterotic string, Phys. Rev. Lett. 54 (1985), 502–505.

[1985 B] Heterotic string theory I. The free heterotic string, Nucl. Phys. B256 (1985), 253–300

[1986] Heterotic string theory II. The interacting heterotic string, Nucl. Phys. B267 (1985), 75–100.

Guillemin, V. W. [1968] A Jordan–Hölder decomposition for a certain class of infinite-dimensional Lie algebras, J. Diff. Geom. 2 (1968), 313–345.

[1970] Infinite-dimensional primitive Lie algebras, J. Diff. Geom. 4 (1970), 257–282.

Guillemin, V. W., Kostant, B., Sternberg, S. [1988] Jesse Douglas's solution to the Plateau problem, Proc. Natl. Acad. Sci. USA

Guillemin, V. W., Quillen, D., Sternberg, S. [1966] The classification of the complex primitive infinite pseudogroups, Proc. Natl. Acad. Sci. USA, 55 (1966), 687–690.

Guillemin, V. W., Sternberg, S. [1964] An algebraic model of transitive differential geometry, Bull. Amer. Math. Soc. 70 (1964), 16–47.

Haddad Z. [1984] Infinite-dimensional flag varieties, Dissertation MIT, 1984.

Haefliger, A. [1976] Sur la cohomologie de l'algèbre de Lie de champs de vecteurs, Ann. Sci. Ec. Norm. Sup. 9 (1976), 503–532.

Harish-Chandra [1951] On some applications of the universal enveloping algebra of a semi-simple Lie algebra, Trans. Amer. Math. Soc., 70 (1951), 28–96.

de la Harpe, P. [1972] Classical Banach–Lie algebras and Banach–Lie groups of operators on Hilbert space, Lecture Notes in Math. 285, Springer–Verlag, 1972.

Hayashi, T. [1988] Sugawara operators and Kac–Kazhdan conjecture, Invent. Math. 94 (1988), 13–52.

Hawkins, T. [1982] Wilhelm Killing and the structure of Lie algebras, Archive for history of exact sciences, 26 (1982), 127–192.

Helgason, S. [1978] Differential geometry, Lie groups and symmetric spaces, Academic Press, 1978.

Hiller, H. [1982] Geometry of Coxeter groups, Pitman, Boston, 1982.

Hiramatsu, T. [1988] Theory of automorphic forms of weight 1, Adv. Ser. in Pure Math. 13 (1988), 503–584.

Hasegawa, K. [1988] Dual pairs of spinors, preprint.

Hoegh-Kron, R., Torresani, B [1986] Classification and construction of quasisimple Lie algebras, preprint.

Humphreys, J. E. [1972] Introduction to Lie algebras and representation theory, Berlin-Heidelberg-New York, Springer, 1972.

Imbens, H.-J. [1989] Drinfeld–Sokolov hierarchies and τ-functions, in: Infinite-dimensional Lie algebras and groups, Adv. Ser. in Math. Phys. 7, World Sci., 1989, 352–368.

Ismagilov, R. S. [1971] On unitary representations of the group of diffeomorphisms of the circle, Funkt. Analis i ego Prilozh. 5, (1971), No. 3, 45–53. English translation: Funct. Anal. Appl. 5 (1971), 209–216.

[1975] On unitary representations of the group of diffeomorphisms of $\mathbf{R}^n$, Mat. Sbornik 98 (140), (1975), 55–71. English translation: Math. USSR-Sbornik 27 (1975), 51–76.

[1976] On unitary representations of the group $C_0^\infty(X,G), G = SU_2$, Mat. Sbornik, 100 (142) (1976), 117–131. English translation: Math. USSR-Sbornik 29 (1976), 105–119.

[1980] Embedding of the group of measure preserving diffeomorphisms in a semidirect product and its unitary representations, Mat. Sbornik 113 (1980), 81–97 (in Russian).

Itzykson, C., Zuber, J.-B. [1986] Two-dimensional conformal invariant theories on a torus, Nucl. Phys. B275 (1986), 580–600.

Iwahori N., Matsumoto H. [1965] On some Bruhat decomposition and the structure of the Hecke rings of p-adic Chevalley groups, Publ. Math. IHES 25 (1965), 5–48.

Jacobson, N. [1962] Lie algebras, Interscience, New York, 1962.

Jakobsen, H. P., Kac, V. G. [1985] A new class of unitarizable highest weight representations of infinite-dimensional Lie algebras, in Non-linear equations in field theory, Lecture Notes in Physics 226, Springer–Verlag, 1985, 1–20. II. J. Funct. Anal. 82 (1989), 69–90.

Jantzen, J. C. [1979] Moduln mit einem höchsten Gewicht, Lecture Notes in Math. 750, Springer–Verlag, 1979.

[1983] Einhüllende Algebren halbeinfacher Lie-Algebren, Springer–Verlag, Berlin, Heidelberg, 1983.

Jimbo, M. [1985] A q-difference analogue of $U(\mathfrak{g})$ and the Yang–Baxter equation, Lett. Math. Phys. 10 (1985), 63–69.

Jimbo, M., Miwa, T. [1983] Solitons and infinite-dimensional Lie algebras, Publ. RIMS 19 (1983), 943–1001.

[1984] Irreducible decomposition of fundamental modules for $A_\ell^{(1)}$ and $C_\ell^{(1)}$, and Hecke modular forms, Adv. Stud. Pure Math. 4 (1984), 97–119.

[1985] On a duality of branching rules for affine Lie algebras, Adv. Stud. Pure Math. 6 (1985), 17–65.

Jimbo, M., Miwa, T., Okado, M. [1986] Solvable lattice models with broken $\mathbf{Z}_N$-symmetry and Hecke's indefinite modular forms, Nucl. Phys. B 275 (1986), 517–545.

[1987 A] Solvable lattice models whose states are dominant integral weights of $A_{n-1}^{(1)}$. Lett. Math. Phys. 14 (1987), 123–131.

[1987 B] An $A_{n-1}^{(1)}$ family of solvable lattice models. Mod. Phys. Lett. B1 (1987), 73–79.

[1988 A] Local state probabilities of solvable lattice models. An $A_{n-1}^{(1)}$ family, preprint RIMS 594, Kyoto Univ., Nucl. Phys. B300 [FS22] (1988), 74–108.

[1988 B] Symmetric tensors of the $A_{n-1}^{(1)}$ family, in Algebraic analysis, Academic Press, 1988.

Julia, B. [1982] Gravity, supergravity and integrable systems, ENS preprint.

[1983] Infinite-dimensional groups acting on (super)-gravity phase spaces, ENS preprint.

[1985] Kac–Moody symmetry of gravitation and supergravity theories, in Lectures in Applied Math. 21, 1985, 355–374.

Kac, V. G. [1967] Simple graded Lie algebras of finite growth, Funkt . Analis i ego Prilozh. 1 (1967), No. 4, 82–83. English translation: Funct. Anal. Appl. 1 (1967), 328–329.

[1968 A] Graded Lie algebras and symmetric spaces, Funkt. Analis. i ego Prilozh. 2 (1968), No. 2, 93–94. English translation: Funct. Anal. Appl. 2 (1968), 183–184.

[1968 B] Simple irreducible graded Lie algebras of finite growth, Izvestija AN USSR (ser. mat.) 32 (1968), 1923–1967. English translation: Math. USSR-Izvestija 2 (1968), 1271–1311.

[1969 A] Automorphisms of finite order of semi-simple Lie algebras, Funkt. Analis i ego Prilozh. 3 (1969), No. 3, 94–96. English translation: Funct. Anal. Appl. 3 (1969), 252–254.

[1969 B] An algebraic definition of compact Lie groups, Trudy MIEM, No. 5, (1969), 36–47 (in Russian).

[1974] Infinite-dimensional Lie algebras and Dedekind's η-function, Funkt. Analis i ego Prilozh. 8 (1974), No. 1, 77–78. English translation: Funct. Anal. Appl. 8 (1974), 68–70.

[1975] On the question of the classification of orbits of linear algebraic groups, Uspechi Math. Nauk. 30 (1975), No. 6, 173–174 (in Russian).

[1977] Lie superalgebras, Advances in Math. 26, No. 1 (1977), 8–96

[1978 A] Infinite-dimensional algebras, Dedekind's η-finction, classical Möbious function and the very strange formula, Advances in Math. 30 (1978), 85–136.

[1978 B] Highest weight representations of infinite-dimensional Lie algebras, in Proceeding of ICM, 299–304, Helsinki, 1978.

[1979] Contravariant form for infinite-dimensional Lie algebras and superalgebras, in Lecture Notes in Physics 94, Springer–Verlag, 1979, 441–445.

[1980 A] Infinite root systems, representations of graphs and invariant theory, Invent. Math. 56 (1980), 57–92.

[1980 B] An elucidation of "Infinite-dimensional algebras ... and the very strange formula." $E_8^{(1)}$ and the cube root of the modular invariant j, Advances in Math. 35 (1980), 264–273.

[1980 C] On simplicity of certain infinite-dimensional Lie algebras, Bull. Amer. Math. Soc. 2 (1980), 311–314.

[1980 D] Some remarks on nilpotent orbits, J. Algebra 64 (1980), 190–213.

[1980 E] A remark on the Conway–Norton conjecture about the "Monster" simple group, Proc. Natl. Acad. Sci. USA, 77 (1980), 5048–5049.

[1981] Simple Lie groups and the Legendre symbol, in Lecture Notes in Math. 848, Springer–Verlag 1981, 110–124.

[1982 A] Infinite root systems, representations of graphs and invariant theory II, J. Algebra 78 (1982), 141–162.

[1982 B] Some problems on infinite-dimensional Lie algebras and their representations, in Lecture Notes in Math. 933, Springer–Verlag, 1982, 117–126.

[1983] Montecatini lectures on invariant theory, in Lecture Notes in Math. 996, Springer–Verlag, 1983, 74–108.

[1984] Laplace operators of infinite-dimensional Lie algebras and theta functions, Proc. Natl. Acad. Sci. USA 81 (1984), 645–647.

[1985 A] Torsion in cohomology of compact Lie groups and Chow rings of algebraic groups, Invent. Math. 80 (1985), 69–79.

[1985 B] Constructing groups associated to infinite-dimensional Lie algebras, in Infinite-dimensional groups with applications, MSRI publ. 4, Springer–Verlag, 1985, 167–216.

[1986] Highest weight representations of conformal current algebras, Symposium on Topological and Geometrical methods in Field theory, Espoo, Finland, 1986. World Sci., 1986, 3–16.

[1988] Modular invariance in mathematics and physics, Address at the centennial of the AMS, 1988.

Kac, V. G., Kazhdan, D. A. [1979] Structure of representations with highest weight of infinite-dimensional Lie algebras, Advances in Math. 34 (1979), 97–108.

Kac, V. G., Kazhdan, D. A., Lepowsky, J., Wilson, R. L. [1981] Realization of the basic representation of the Euclidean Lie algebras, Advances in Math. 42 (1981), 83–112.

Kac, V. G., van de Leur, J. [1987] Super boson-fermion correspondence, Ann. de L'Institure Fourier, 37 (1987), 99–137.

[1989 A] Super boson-fermion correspondence of type B, in: Infinite-dimensional Lie algebras and groups, Adv. Ser. in Math. Phys. 7, World Sci., 1989, 369–406.

[1989 B] On classification of superconformal algebras, in: Strings 88, World Sci., 1989, 77–106

Kac, V. G., Moody, R. V., Wakimoto, M. [1988], On E_{10}, Proceedings of the 1987 conference on differential-geometrical methods in physics, Kluwer, 1988, 109–128.

Kac, V. G., Peterson, D. H. [1980] Affine Lie algebras and Hecke modular forms, Bull. Amer. Math. Soc. 3 (1980), 1057–1061.

[1981] Spin and wedge representations of infinite-dimensional Lie algebras and groups, Proc. Natl. Acad. Sci. USA 78 (1981), 3308–3312.

[1983] Regular functions on certain infinite-dimensional groups, in Arithmetic and Geometry, 141–166, Progress in Math. 36, Birkhäuser, Boston, 1983.

[1984 A] Infinite-dimensional Lie algebras, theta functions and modular forms, Advances in Math. 53 (1984), 125–264.

[1984 B] Unitary structure in representations of infinite-dimensional groups and a convexity theorem, Invent. Math. 76 (1984), 1–14.

[1985 A] Defining relations of certain infinite-dimensional groups, in Proceedings of the Cartan conference, Lyon 1984, Astérisque, 1985, Numero hors serie, 165–208.

[1985 B] Generalized invariants of groups generated by reflections, in

Proceedings of the Conference Giornate di Geometria, Rome 1984. Progress in Math. 60, Birkhäuser, 1985, 231–250.

[1985 C] 112 constructions of the basic representation of the loop group of E_8, Proceedings of the conference "Anomalies, geometry, topology" Argonne, 1985. World Sci., 1985, 276–298.

[1986] Lectures on the infinite wedge representation and the MKP hierarchy, seminaire de Math. Superieures, Les Presses de L'Université de Montréal, 102 (1986), 141–186.

[1987] On geometric invariant theory for infinite-dimensional groups, in: Lecture Notes in Math. 1271, 1987, 109–142.

Kac, V. G., Popov, V. L., Vinberg, E. B. [1976] Sur les groupes linear algebriques dont l'algèbre des invariants est libre, C. R. Acad. Sci. Paris, 283 (1976), 875–978.

Kac, V. G., Raina, A. K. [1987] Bombay lectures on highest weight representations, World Sci., Singapore, 1987.

Kac, V. G., Todorov, I. T. [1985] Superconformal current algebras and their unitary representations, Comm. Math. Phys., 102 (1985), 337–347.

Kac, V. G., Wakimoto, M. [1986] Unitarizable highest weight representations of the Virasoro, Neveu–Schwartz and Ramond algebras, in Proceedings of the Symposium on conformal groups and structures, Claustal, 1985. Lecture Notes in Physics 261 (1986), 345–372.

[1988 A] Modular and conformal invariance constraints in representation theory of affine algebras, Advances in Math. 70 (1988), 156–234.

[1988 B] Modular invariant representations of infinite-dimensional Lie algebras and superalgebras, Proc. Natl. Acad. Sci. USA, 85 (1988), 4956–4960.

[1989 A] Exceptional hierarchies of soliton equations, Proceedings of the 1987 conference on theta functions in Maine, Proc. Symposia in Pure Math. 49, 1989, 191–237.

[1989 B] Classification of modular invariant representations of affine algebras, in: Infinite-dimensional Lie algebras and groups, Adv. Ser. in Math. Phys. 7, World Sci., 1989, 138–177.

Kantor, I. L. [1968] Infinite-dimensional simple graded Lie algebras, Doklady AN SSR 179 (1968), 534–537. English translation: Soviet Math. Dokl. 9 (1968), 409–412.

[1970] Graded Lie algebras, Trudy sem. Vect. Tens. Anal. 15 (1970), 227–266 (in Russian).

Kaplansky, I., Santharoubane, L.-J. [1985] Harish-Chandra modules over the Virasoro algebra, in Infinite-dimensional groups with applications, MSRI series, Springer–Verlag, 1985, 217–232.

Kashiwara, M. [1989] Kazhdan–Lusztig conjecture for symmetrizable Kac–Moody Lie algebras, preprint.

Kashiwara, M., Tanisaki, T. [1990] Kazhdan–Lusztig conjecture for symmetrizable Kac–Moody algebras II. Intersection cohomology of Schubert varieties, Progress in Math. 92, Birkhäuser, Boston, 1990, 159–196.

Kassel, C. [1984] Kaehler differentials and coverings of complex simple Lie algebras extended over a commutative algebra, J. Pure Applied Algebra 34 (1984), 265–275.

Kawamoto, N., Namikawa, Y., Tsuchiya, A., Yamada, Y. [1988] Geometric realization of conformal field theory on Riemann surfaces, Comm. Math. Phys. 116 (1988), 247–308.

Kazhdan, D. A., Lusztig, G. [1979] Representations of Coxeter groups and Hecke algebras, Invent. Math. 53 (1979), 165–184.

[1980] Schubert varieties and Poincaré duality, Proc. Symp. in Pure Math. of AMS 36 (1980), 185–203.

[1988] Fixed point varieties on affine flag manifolds, Israel J. Math. 62 (1988), 129–168.

Kerov, S., Vershik, A. [1982] Characters and factor representations of the infinite unitary group, Soviet Math. Dokl. 26 (1982), 570–574.

Kirillov, A. A. [1973] Representations of an infinite-dimensional unitary group, Doklady AN SSSR 212 (1973), 288–290. English translation: Soviet Math. Doklady 14 (1973), 1355–1358.

[1974 A] Unitary representations of the group of diffeomorphisms and some of its subgroups, IPM, preprint (in Russian).

[1974 B] Representations of certain infinite-dimensional Lie groups, Vestnik MGU 29 (1974), No. 1, 75–83 (in Russian).

[1982] Infinite-dimensional Lie groups, their orbits, invariants and representations. The geometry of moments, in: Lecture Notes in Math. 970, Springer–Verlag, 1982, 101–123.

[1987] A Kähler structure on K-orbits of the group of diffeomorphisms of a circle, Funct. Anal. Appl. 21:2 (1987), 42–45 (in Russian).

Kirillov, A. A., Yurjev, D. V. [1987] Kähler geometry of the infinite-dimensional homogeneous space $M = \mathrm{Diff}_+(S^1)/\mathrm{Rot}(S^1)$, Funct. Anal. Appl. 21:4 (1987), 34–46.

Kirillov, A. N., Reshetikhin, N. Yu. [1989] Representations of the algebra $U_q(s\ell(2))$, q-orthogonal polynomials and invariants of links, in: Infinite-dimensional Lie algebras and groups, Adv. Ser. in Math. Phys. 7, World Sci., 1989, 285–339.

Knizhnik, V. G., Zamolodchikov, A. B. [1984] Current algebra and Wess–Zumino model in two dimensions, Nucl. Phys. B247 (1984), 83–103.

Knizhnik, V. G., Polyakov, A. M., Zamolodchikov, A. B. [1988] Fractal structure in 2D quantum gravity, Mod. Phys. Lett. A3 (1988), 819–826.

Knopp, M. [1970] Modular functions in analytic number theory, Markham Publishing Co., Chicago, 1970.

Kontsevich, M. L. [1987] Virasoro algebra and Teichmüller spaces, Funkt. Analis i ego Prilozh. 21:2 (1987), 78–79.

Kostant, B. [1959] The principal three-dimensional subgroup and the Betti numbers of a complex simple Lie group, Amer. J. Math. 81 (1959), 973–1032.

[1976] On Macdonald's η-function formula, the Laplacian and generalized exponents, Advances in Math. 20 (1976), 179–212.

[1985] The McKay correspondence, the Coxeter element and representation theory, Asterisque, hors series, 1985, 209–255.

Kostant, B., Kumar, S. [1986] The nil Hecke ring and cohomology of G/B for a Kac–Moody group G, Advances in Math. 62 (1986), 187–237.

[1987] T-equivariant K-theory of generalized flag varieties, Proc. Natl. Acad. Sci. USA 84 (1987), 4351–4354.

Kostant, B., Sternberg, S. [1987] Symplectic reduction, BRS cohomology and infinite-dimensional Clifford algebras, Ann. of Physics 176 (1987), 49–113.

[1988] The Schwartzian derivative and hyperbolic geometry, Lett. Math. Phys.

Krichever, I. M., Novikov, S. P. [1987 A] Virasoro type algebras, Riemann surfaces and structures of soliton theory, Funkt. Analis i ego Prilozh. 21:2 (1987), 46–63.

[1987 B] Virasoro type algebras, Riemann surfaces and strings in Minkowski space, Funkt. Analis i ego Prilozh. 214 (1987), 47–60.

[1989] Virasoro type algebras, stress energy tensor and operator decompositions on Riemann surfaces, Funkt. Analis i ego Prilozh. 23:1 (1989).

Ku, J.-M. [1987 A] The Jantzen filtration of a certain class of verma modules, Proc. Amer. Math. Soc. 99 (1987), 35–40.

[1987 B] Local submodules and the multiplicities of irreducible subquotients i category $\mathcal{O}$, J. Algebra 106 (1987), 403–412.

[1988 A] On the uniqueness of embeddings of Verma modules defined by the Shapovalov form, J. Algebra 118 (1988), 85–101.

[1988 B] Structure of the Verma module $M(-\rho)$ over Euclidean Lie algebras, J. Algebra.

[1988 C] Relative version of Weyl–Kac character formula, J. Algebra.

[1989 A] The structure of Verma modules over the affine Lie algebra $A_1^{(1)}$, Trans. Amer. Math. Soc.

[1989 B] Projection principle in Verma modules, Trans. Amer. Math. Soc.

Kumar, S. [1984] Geometry of Schubert cells and cohomology of Kac–Moody Lie algebras, J. Diff. Geometry 20 (1984), 389–431.

[1985] Rational homotopy theory of flag varieties associated to Kac–Moody groups, in Infinite-dimensional groups with applications, MSRI publ. 4, Springer–Verlag, 1985.

[1986 A] A homology vanishing theorem for Kac–Moody algebras with coefficients in the category $\mathcal{O}$, J. Algebra 102 (1986), 444–462.

[1986 B] Non-representability of cohomology classes by bi-invariant forms (gauge and Kac–Moody groups), Comm. Math. Phys. 106 (1986), 177–181.

[1987 A] Demazure character formula in arbitrary Kac–Moody setting, Invent. Math. 89 (1987), 395–423.

[1987 B] Extension of the category $\mathcal{O}^g$ and a vanishing theorem for the Ext function for Kac–Moody algebras, J. Algebra 108 (1987), 472–491.

[1988] A connection of equivariant K-theory with the singularity of Schubert varieties, preprint.

[1989] Existence of certain components in the tensor product of two integrable highest weight modules for Kac–Moody algebras, in: Infinite-dimensional Lie algebras and groups, Adv. Ser. in Math. Phys. 7, World Sci., 1989, 25–38.

Kupershmidt, B. A. [1984] Isotopic Lie algebras and matrix Lax equations, Algebras, Groups and Geometries 1 (1984), 144–153.

Lakshmibai, V., [1989] Schubert varieties in $\widehat{SL}_n$, preprint.

Lakshmibai, V., Seshadri, C. S. [1989] Standard monomial theory for $\widehat{SL}_2$, in: Infinite-dimensional Lie algebras and groups, Adv. Ser. in Math. Phys. 7, World Sci., 1989, 178–234.

Langlands [1988] On unitary representations of the Virasoro algebra, in Infinite-dimensional Lie algebras and their applications, World Sci., 1988, 141–159.

Lepowsky, J. [1978] Lie algebras and combinatorics, in Proceedings of ICM, Helsinki 1978.

[1979] Generalized Verma modules, loop space cohomology and Macdonald-type identities, Ann. Sci. École Norm. Sup. 12 (1979), 169–234.

[1985] Calculus of twisted vertex operators, Proc. Natl. Acad. Sci. USA.

Lepowsky, J., Milne, S. [1978] Lie algebraic approaches to classical partition identities, Advances in Math. 29 (1978), 15–59.

Lepowsky, J., Moody, R. V. [1979] Hyperbolic Lie algebras and quasi-regular cusps on Hilbert modular surfaces, Math. Ann. 245 (1979), 63–88.

Lepowsky, J., Primc, M. [1985] Structure of the standard modules for the affine Lie algebra $A_1^{(1)}$, Contemporary Math. 46, AMS, Providence, 1985.

Lepowsky, J., Wilson, R. L. [1978] Construction of the affine Lie algebra $A_1^{(1)}$, Comm. Math. Phys. 62 (1978), 43–53.

[1981] A new family of algebras underlying the Rogers–Ramanujan identities and generalizations, Proc. Natl. Acad. Sci. USA (1981), 7254–7258.

[1982] A Lie theoretic interpretation and proof of the Rogers–Ramanujan identities, Advances in Math. 45 (1982), 21–72.

[1984] The structure of standard modules, I: Universal algebras and the Rogers–Ramanujan identities, Invent. Math. 77 (1984), 199–290. II. The case $A_1^{(1)}$, principal gradation, Invent. Math. 79 (1985), 417–442.

Lerche, W., Schellerens, A. N. Warner, N. P. [1989] Lattices and strings, Phys. Reports 177 (1989), 1–140.

Levstein, F. [1988] A classification of involutive automorphisms of an affine Kac–Moody Lie algebra, J. Algebra 114 (1988), 489–518.

Leznov, A. N., Saveliev, M. V., Smirnov [1981] Representation theory and integration of non-linear dynamical systems, Theor. Math. Phys. 48 (1981), 3–12 (in Russian).

Leznov, A. N., Saveliev, M. V. [1983] Two-dimensional exactly and completely integrable dynamical systems, Comm. Math. Phys. 89 (1983), 59–75.

Li, S.-P., Moody, R. V., Nicolescu, M., Patera, J. [1986] Verma bases for representations of classical simple Lie algebras, J. Math. Phys. 27 (1986), 668–677.

Looijenga, E. [1976] Root systems and elliptic curves, Invent. Math. 38 (1976), 17–32.

[1980] Invariant theory for generalized root systems, Invent. Math. 61 (1980), 1–32.

Lusztig, G. [1983] Singularities, character formulas and a q-analog of weight multiplicities, Astérisque, 101–102 (1983), 208–239.

[1984] Cells in affine Weyl groups, Proc. of International Symposium on algebraic groups, Katata, Japan, 1984.

[1988] Quantum deformations of certain simple modules over enveloping algebras, Adv. Math. 70 (1988) 237–249

[1989 A] Modular representations and quantum groups, Contemporary Math. 82 (1989), 59–77.

[1989 B] On quantum groups, preprint.

[1989 C] Quantum groups at roots of 1, preprint.

Lu, S.-R. [1989 A] Explicit equations of the KP and MKP hierarchies, Proc. Symposia in Pure Math. 49, 1989, 101–105.

[1989 B] On modular invariant parition functions in non-unitary theories, Phys. Lett. B 218 (1989) 46–50.

[1989 C] Some results on modular invariant representations, in: Infinite-dimensional Lie algebras and groups, Adv. Ser. in Math. Phys. 7, World Sci., 1989, 235–253.

Macdonald, I. G. [1972] Affine root systems and Dedekind's η-function, Invent. Math. 15 (1972), 91–143.

[1979] Symmetric functions and Hall polynomials, Oxford University Press, 1979.

Malikov, F. G. [1989] Singular vectors in Verma modules over affine algebras, Funkt. Analis i ego Prilozh. 23:1 (1989), 76–77.

Malikov, F. G., Feigin, B. L., Fuchs, D. B. [1986] Singular vectors in Verma modules over Kac–Moody algebras, Funkt. Analis i ego Prilozh. 20:2 (1986), 25–37.

Mandelstam, S. [1974] Dual resonance models, Physics Rep. 13 (1974), 259–353.

[1975] Soliton operators for the quantized Sine–Gordon equation, Phys. Rev. D 11 (1975), 3026–3030.

Marcuson, R. [1975] Tits' system in generalized non-adjoint Chevalley groups, J. Algebra 34 (1975), 84–96.

Mathieu, O. [1986 A] Sur un problème de V. G. Kac: La classification de certain algèbres de Lie graduées simples, J. Algebra 102 (1986), 505–536.

[1986 B] Classification des algèbres de Lie graduées simples de croissance ≤ 1, Invent. Math. 86 (1986), 371–426.

[1986 C] Formule de Demazure–Weil et theoreme de Borel–Weil–Bott pour les algèbres de Kac–Moody générales, C. R. Acad. Sci. Paris 303 (1986) 391–394.

[1987] Classes canonique de variétés de Schubert et algèbres affine, C. R. Acad. Sci. Paris 305 (1987), 105–108.

[1988] Formules de charactères pour les algèbres de Kac–Moody générales, Astérisque 159–160, 1988.

[1989 A] Construction d'un groupe de Kac–Moody et applications, Compositio Math. 6 9 (1989), 37–60.

[1989 B] Frobenius action on the B-cohomology, in: Infinite-dimensional Lie algebras and groups, Adv. Ser. in Math. Phys. 7, World Sci., 1989, 39–51.

[19–] Classification of Harish-Chandra modules over the Virasoro algebra,

[19–] Classification of simple graded Lie algebras of finite growth

Mathur, S. D., Mukhi, S., Sen, A. [1988] On the classification of rational conformal field theories, Phys. Lett. B 213 (1988), 303–308.

Meurman, A. [1982] Characters of rank two hyperbolic Lie algebras as functions at quasiregular cusps, J. Algebra 76 (1982), 494–504.

Meurman, A., Primc, M. [1987] Annihilating ideals of standard modules of $s\ell(2,\mathbb{C})^{\sim}$ and combinational identities, Advances in Math., 64 (1987), 177–240.

Mickelsson, J. [1985] Representations of Kac–Moody algebras by step algebras, J. Math. Phys. 26 (1985), 377–382.

[1987] Kac–Moody groups, topology of the Dirac determinant bundle and fermionization, Comm. Math. Phys. 110 (1983), 173–183.

[1989] Current algebra in 3 + 1 space-time dimensions, in: Infinite-dimensional Lie algebras and groups, Adv. Ser. in Math. Phys. 7, World Sci., 1989, 254–272.

Mickelsson, J., Rajeev, S. [1988] Current algebras in $d+1$ dimensions and determinant bundles over infinite-dimensional Grassmannians, Comm. Math. Phys. 116 (1988), 365–390.

Mikhailov, A. V., Olshanetsky, M. A., Perelomov, A. M. [1981] Two-dimensional generalized Toda lattice, Comm. Math. Phys. 79 (1981), 473–488.

Misra, K. C. [1984 A] Structure of certain modules for $A_n^{(1)}$ and the Rogers–Ramanujan identities, J. Algebra 88 (1984), 196–227.

[1984 B] Structure of some standard modules for $C_n^{(1)}$, J. Algebra 90 (1984), 385–409.

[1987] Basic representations of some affine Lie algebras and generalized Euler identities, J. Austral. Math. Soc. (Ser. A) 42 (1987), 296–311.

[1989 A] Realization of the level two standard $s\ell(2k+1,\mathbb{C})^{\sim}$-modules, preprint.

[1989 B] Realization of the level one standard $\widetilde{C}_{2k+1}$-modules, preprint.

Mitchell, S. A. [1987] A Bott filtration on a loop group, Seattle preprint.

[1988] Quillen's theorem on buildings and the loops on a symmetric space, L'Enseignement Math. 34 (1988), 123–166.

Mitzman, D. [1985] Integral bases for affine Lie algebras and their universal enveloping algebras, Contemporary Math. 40, 1985.

Moody, R. V. [1967] Lie algebras associated with generalized Cartan matrices, Bull. Amer. Math. Soc., 73 (1967), 217–221.

[1968] A new class of Lie algebras, J. Algebra 10 (1968), 211–230.

[1969] Euclidean Lie algebras, Canad. J. Math. 21 (1969), 1432–1454.

[1975] Macdonald identities and Euclidean Lie Algebras, Proc. Amer. Math. Soc., 48 (1975), 43–52.

[1979] Root systems of hyperbolic type, Advances in Math. 33 (1979), 144–160.

[1982] A simplicity theorem for Chevalley groups defined by generalized Cartan matrices, preprint.

Moody, R. V., Pianzola, A. [1988] On infinite root systems, preprint.

Moody, R. V., Teo, K. L. [1972] Tits systems with crystallographic Weyl groups, J. Algebra 21 (1972), 178–190.

Moody, R. V., Yokonuma, T. [1982] Root systems and Cartan matrices, Canad. J. Math. 34 (1982), 63-79.

Moore, G., Seiberg, N. [1988] Polynomial equations for conformal field theories, Phys. Lett. 212B (1988), 451–460

[1989] Classical and quantum conformal field theory, Princeton preprint.

Moreno, C. J., Rocha-Caridi, A. [1987] Rademacher-type formulas for the multiplicities of irreducible highest weight representations of affine Lie algebras, Bull. Amer. Math. Soc. 16 (1987), 292–296

Morita, J. [1979] Tits' systems in Chevalley groups over Laurent polynomial rings, Tsukuba Math. J. 3 (1979), 41–51.

[1984] Conjugacy classes of three-dimensional simple Lie subalgebras of the affine Lie algebra $A_l^{(1)}$, University of Tsukuba, preprint.

[1987] Commutator relations in Kac–Moody groups, Proc. Japan Acad. 63A (1987), 21–22.

Morita, J., Rehman, U. [1989 A] Symplectic K_2 of Laurent polynomials, associated Kac–Moody groups and Witt rings, preprint.

[1989 B] A Matsumoto type theorem for Kac–Moody groups, preprint.

Morita, J., Wakimoto, M. [1987] A Lie algebraic approach to the Diophantine equation $x_1^2+x_2^2+\cdots+x_n^2=y^2$ $(n\leq 9)$, Jap. J. Math. 13 (1987), 163–167.

Mulase, M. [1984] Cohomolagical structure in soliton equations and Jacobian varieties, J. Diff. Geometry 19 (1984), 403–430.

[1989] Infinite-dimensional Grassmannians, vector bundles on curves and commuting differential operators, preprint.

Mumford, D. [1983] Tata lectures on theta I, Progress in Math. 28, Birkhäuser, Boston, 1983.

[1984] Tata lectures on theta II, Progress in Math. 43, Birkhäuser, Boston, 1984.

Neretin, Ju. A. [1983] Unitary representation with highest weight of the group of diffeomorphism of the circle, Funct. Analis i ego Prilozh. 17:2 (1983), 85–86 (in Russian).

[1987] On the complex semigroup containing the group of diffeomorphisms of the circle, Funkt. Analis i ego Prilozh. 21:2 (1987), 82–83.

Okamoto, K., Sakurai, T. [1982 A] On a certain class of irreducible unitary representations of the infinite-dimensional rotation group II, Hiroshima Math. J. 13 (1982).

[1982 B] An analogue of Peter–Weyl theorem for the infinite-dimensional unitary group, Hiroshima Math. J. 12 (1982), 529–542.

Olshanskii, G. I. [1978] Unitary representations of the infinite-dimensional classical groups $U(p,\infty), SO_0(p,\infty)$, $Sp(p,\infty)$ and the corresponding motion groups, Funkt. Analis i ego Prilozh. 12 (1978), No. 3, 32–44. English translation: Funct. Anal. Appl. 12 (1978), 185–195.

[1980] Construction of unitary representations of infinite-dimensional classical groups, Doklady AN SSSR 250 (1980), 284–288. English translation: Soviet Math. Dokl. 21 (1980), 66–70.

[1982] Spherical functions and characters on the group $U(\infty)^X$, Uspechi Mat. Nauk. 37, No. 2 (1982), 217–218 (in Russian).

[1983] Unitary representations of infinite pairs (G, K) and R. Howe formalism, Doklady AN SSSR 269 (1983), 33–36 (in Russian).

Paunov, R. R., Todorov, I. T. [1989] Local extensions of the $U(1)$ current algebra and their positive energy representations, in: Infinite-dimensional Lie algebras and groups, Adv. Ser. in Math. Phys. 7, World Sci., 1989, 588–604.

Peterson, D. H. [1982] Affine Lie algebras and theta functions, in Lecture Notes in Math. 933, Springer–Verlag 1982, 166–175.

[1983] Freudental-type formulas for root and weight multiplicities, preprint.

Peterson, D. H., Kac, V. G. [1983] Infinite flag varieties and conjugacy theorems, Proc. Natl. Acad. Sci. USA, 80 (1983), 1778–1782.

Piatetsky-Shapiro, I. I., Shafarevich, I. R. [1971] A Torelli theorem for algebraic surfaces of type $K3$, Izvestija AN USSR (ser. mat.) 35 (1971), 530–572.

Pickrell, D. [1987] Measures on infinite-dimensional Grassmann manifolds, J. Funct. Anal. 70 (1987), 323–356.

Popov, V. L. [1976] Representations with a free module of covariant, Funkt. Analis i ego Prilozh. 10:3 (1976), 91–92.

Post, G. F. [1986 A] Pure Lie algegraic approach to the modified KdV equation, J. Math. Phys. 27 (1986), 678–681.

[1986 B] Lie algebraic approach to τ-functions and its equations, Lett. Math. Phys. 11 (1986), 253–257.

[1987] On the τ-functions of $A_2^{(2)}$, J. Math. Phys. 28 (1987), 1691–1696.

Pressley, A. N. [1980] Decompositions of the space of loops on a Lie group, Topology 19 (1980), 65–79.

Pressley, A. N., Segal, G. B. [1985] Loop groups and their representations, Oxford University Press, 1985.

Previato, E., Wilson, G. [1989] Vector bundles over curves and solutions of the KP equations, preprint.

Primc, M. [1989] Standard representations of $A_n^{(1)}$, in: Infinite- dimensional Lie algebras and groups, Adv. Ser. in Math. Phys. 7, World Sci., Singapore, 1989, pp. 273–284.

Reiman, A. G., Semenov-Tjan-Shanskii, M. A. [1979] Reduction of Hamiltonian systems, affine Lie algebras and Lax equations, Invent. Math. 54 (1979), 81–100.

[1981] Reduction of Hamiltonian systems, affine Lie algebras and Lax equations II, Invent. Math. 63 (1981), 423–432.

Reshetikhin, N. Yu. [1989] Quasitriangular Hopf algebras, solutions of the Yang–Baxter equation and invariants of links, Algebra and Analysis 1 (1989), 169–194.

Ringel, C. M. [1989] Hall algebras and quantum groups, preprint.

Rocha, A., Wallach, N. R. [1982] Projective modules over graded Lie algebras, Math. Z. 180 (1982), 151–177.

[1983 A] Highest weight modules over graded Lie algebras: Resolutions, filtrations and character formulas, Trans. Amer. Math. Soc., 277 (1983), 133–162.

[1983 B] Characters of irreducible representations of the Lie algebra of vector fields on the circle, Invent. Math. 72 (1983), 57–75.

Roussseau, G. [1989] Almost split K-forms of Kac–Moody algebras, in: Infinite-dimensional Lie algebras and groups, Adv. Ser. in Math. Phys. 7, World Sci., 1989, 70–88.

Rudakov, A. N. [1969] Automorphism groups of infinite-dimensional Lie algebras, Izvestija AN SSSR (ser. mat.) 33 (1969), 748–764.

[1974] Irreducible representations of infinite-dimensional Lie algebras of Cartan type, Izvestija AN SSSR (ser. mat.) 38 (1974), 835–866. English translation: Math. USSR-Izvestija 8 (1974), 836–866.

[1975] ditto II, 39 (1975), 496–511. English translation: Math. USSR-Izvestija 9 (1974), 465–480.

Saito, K. [1985] Extended affine root systems I, (Coxeter transformations), Publ. RIMS, 21 (1985), 75–179.

[1988] Extended affine root systems II (Flat invariants), RIMS-633, preprint.

Sato, M. [1981] Soliton equations as dynamical systems on infinite dimensional Grassmann manifolds, RIMS Kokyuroku 439 (1981) 30–46.

Schellekens, A. N., Warner, N. P. [1986] Conformal subalgebras of Kac-Moody algebras, Phys. Rev. D 34 (1986), 3092–4010.

[1987] Anomalies, characters and strings, Nucl. Phys. B 287 (1987), 317–340.

Schoenberg, B [1973] Elliptic modular forms, Springer–Verlag, 1973.

Schoefield, A. H., [1988 A] Subspaces in general position of a vector space, preprint.

[1988 B] The internal structure of real Schur representations, preprint.

[1988 C] Generic representations of quivers, preprint.

Schwarz, J. H. [1973] Dual resonance theory, Physics Rep. 8 (1973), 269–335.

[1982] Superstring theory, Physics Rep. 83 (1982), 223–322.

[1989] Superconformal symmetry and superstring compactification, preprint.

Segal, G. [1981] Unitary representations of some infinite-dimensional groups, Comm. Math. Phys. 80 (1981), 301–342.

[1988] The definition of conformal field theory, in: Differential Geometrical methods in theoretical physics, Kluwer Acad. Publ., 1988, 165–172.

Segal, G., Wilson, G. [1985] Loop groups and equations of KdV type, Publ. IHES 61 (1985), 5–65.

Seligman, G. B. [1986] Kac–Moody modules and generalized Clifford algebras, Proc. Canad. Math. Soc. Summer Meeting, 1986.

Sen, C. [1984] The homology of Kac–Moody Lie algebras with coefficients in a generalized Verma module, J. Algebra 90 (1984), 10–17.

Serre, J.-P. [1966] Algèbres de Lie semi-simple complexes, Benjamin, New York and Amsterdam, 1966.

[1970] Cours d'arithmétique, Presses Universitaires de France, 1970.

Shafarevich, I. R. [1981] On some infinite-dimensional groups II, Izvestija

AN SSR (ser. mat.) 45 (1981), 214–226. English translation: Math. USSR Izvestija 18 (1982), 185–191.

Shapovalov, N. N. [1972] On a bilinear form on the universal enveloping algebra of a complex semisimple Lie algebra, Funkt. Analis i ego Prilozh. 6 (1972), No. 4, 65–70. English translation: Funct. Anal. Appl. 6 (1972), 307–312.

Shiota T. [1986] Characterization of Jacobian varieties in terms of soliton equations, Invent. Math. 83 (1986), 332–382.

[1989] Prym varieties and soliton equations, in: Infinite-dimensional Lie algebras and groups, Adv. Ser. in Math. Phys. 7, World Sci., 1989, 407–448.

Shue, J. R. [1960] Hilbert space methods in the theory of Lie algebras, Trans. Amer. Math. Soc. 95 (1960), 69–80.

[1961] Cartan decomposition for L* algebras, Trans. Amer. Math. Soc. 98 (1961), 333–349.

Singer, I. M., Sternberg, S. [1965] On infinite groups of Lie and Cartan, J. Analyse Math. 15 (1965), 1–114.

Skyrme, T. H. R. [1971] Kinks and the Dirac equation, J. Math. Phys. 12 (1971), 1735–1743.

Slodowy, P. [1981] Chevalley groups over $\mathbb{C}((t))$ and deformations of simply elliptic singularities, RIMS Kokyuroku 415, 19–38, Kyoto University, 1981.

[1985 A] A character approach to Looijenga's invariant theory for generalized root systems, Compositio Math. 55 (1985), 3–32.

[1985 B] An adjoint quotient for certain groups attached to Kac–Moody algebras, in Infinite-dimensional groups with applications, MSRI publ. 4, Springer–Verlag, 1985. 307–334.

[1986 A] Beyond Kac–Moody algebras, and inside, Canadian Math. Soc. Conference Proc. 5 (1986), 361–371.

[1986 B] On the geometry of Schubert varieties attached to Kac–Moody Lie algebras, Canadian Math. Soc. Conference Proc. 6 (1986), 405–442.

Stanley, R. [1980] Unimodal sequences arising from Lie algebras, in Young day proceedings, Marcel Dekker, 1980, 127–136.

[1965] Regular elements of semisimple algebraic groups, Publ. IHES 25 (1965), 281–312.

Steinberg, R. [1967] Lectures on Chevalley groups, Yale University, 1967.

Stratila, S. [1982] Some representations of $U(\infty)$, Proc. Symp. in Pure Math. 38 (1982) 515–520.

Sugawara, H. [1968] A field theory of currents, Phys. Rev. 176 (1968), 2019–2025.

Takasaki, K. [1983] A new approach to self-dual Yang–Mills equations, RIMS-459, preprint.

Tate, J. [1968] Residues of differentials on curves, Ann. Sci. Ecole Norm. Sup. 1 (1968), 149–159.

Takhtadjan, L. A., Faddeev, L. D. [1986] Hamiltonian approach in soliton theory, Nauka, Moscow, 1986.

Taubes, C. H. [1988] S^1 actions and elliptic genera, Harvard preprint.

ten Kroode, A. D. E., Bergvelt, M. F. [1986] The homogeneous realization of the basic representation of $A_1^{(1)}$ and the Toda lattice, Lett. Math. Phys. 12 (1986), 139–147.

Thierry-Mieg J. T. [1985] Remarks concerning the $E_8 \times E_8$ and D_{16} string theories, Phys. Lett. 156B (1985) 199–202.

[1986] Anomaly cancellation and fermionization in 10, 18 and 26 dimension, preprint.

[1987] BRS analysis of Zamolodchikov's spin 2 and 3 current algebra, preprint.

Tits, J. [1981] Resumé de cours, Annuaire du Collège de France 1980–81, Collège de France, Paris, 75–87.

[1982] Resumé de cours, Annuaire du Collège de France 1981–82, Collège de France, Paris, 91–106.

[1985] Groups and group functors attached to Kac–Moody data, in Lecture Notes in Math 1111, Springer–Verlag, 1985.

[1987] Uniqueness and presentation of Kac–Moody groups over fields, J. Algebra 105 (1987), 542–573.

Thorn, C. B. [1984] Computing the Kac determinant using dual model techniques and more about the no-ghost theorem, Nucl. Phys. B248 (1984), 551–569.

Todorov, I. T. [1985] Current algebra approach to conformal invariant two-dimensional models, Phys. Lett. 153B (1985), 77–81.

[1988] Finite temperature 2-dimensional QFT models of conformal current algebra, in Infinite-dimensional Lie algebras and quantum field theory, World Sci., 1988, 97–123.

Tsuchiya, A., Kanie, Y. [1986 A] Fock representations of the Virasoro algebra. Intertwining operators, Nagoya University, Publ. RIMS, Kyoto University, 22 (1986) 259–327.

[1986 B] Unitary representations of the Virasoro algebra, Duke Math. J. 53 (1986), 1013–1046.

[1987] Unitary representations of the Virasoro algebra and branching law of representations of affine Lie algebras, preprint.

[1988] Vertex operators in conformal field theory on $\mathbb{P}^1$ and monodromy representations of braid group, Adv. Stud. Pure Math. 16 (1988), 297–372.

Tsuchiya, A., Ueno, K., Yamada, Y. [1989] Conformal field theory on universal family of stable curves with gauge symmetries, Adv. Stud. Pure Math. 19 (1989), 495–595.

Tsygan, B. L. [1983] Homology of the matrix algebras over rings and the Hochschild homology, Uspechi Mat. Nauk 38 (1983), No. 2, 217–218 (in Russian).

Ueno, K. [1983] Infinite-dimensional Lie algebras acting on chiral fields and the Riemann–Hilbert problem, Publ. RIMS, Kyoto University 19 (1983), 59–82.

Ueno, K., Nakamura, Y. [1983] Infinite-dimensional Lie algebras and transformation theories for non-linear field equations, in Proceedings on non-linear integrable systems (ed. Jimbo, M. and Miwa, T.), World Sci., 1983, 241–272.

Ueno, K., Takasaki, K. [1984] Toda lattice hierarchy, Adv. Studies in Pure Math. (1984).

Ueno, K., Yamada, H. [1986] A supersymmetric extension of non-linear integrable systems, in Topological and geometrical methods in field theory, World Sci., 1986, 59–72.

Vafa, C., Warner, N. P. [1989] Catastrophe theory and the classification of conformal theories, Phys. Lett. 218 B (1989), 51–60.

Van Asch, A. [1976] Modular forms and root systems, Math. Ann. 222 (1976), 145–170.

Varadarajan, V. S. [1984] Lie groups, Lie algebras and their representations, Springer–Verlag, 1984.

Vergne, M. [1977] Seconde quantification et groupe symplectique, C. R. Acad. Sci. (Paris) 285A (1977), 191–194.

Verlinde, E. [1988] Fusion rules and modular transformations in 2D conformal field theory, Nucl. Phys. B 300 (1988), 360–375.

Verma, D.-N. [1968] Structure of certain induced representations of complex semisimple Lie algebras, Bull. Amer. Math. Soc. 74 (1968), 160–166.

Vershik, A. M., Gelfand, I. M., Graev, M. I. [1973] Representations of the group $SL(2, R)$, where R is a ring of functions, Uspechi Mat. Nauk 28 (1973), No. 5, 87–132.

[1975] Representations of the group of diffeomorphisms, Uspechi Mat. Nauk 30 (1975) No. 6, 3–50. English translation: Russian Math. Surveys 30 (1975), No. 6, 1–50.

[1980] Representations of the group of functions taking values in a compact Lie group, Compositio Math. 42 (1980), 217–243.

[1983] A commutative model of representations of the current group $SL(2, R)^X$ related to a unipotent subgroup, Funkt. Analis i ego Prilozh. 17 (1983), No. 2, 70–72 (in Russian).

Vilenkin, N. Ja. [1965] Special functions and the theory of group representations, Nauka, Moscow, 1965. English translation: Providence, AMS, 1968.

Vinberg, E. B. [1971] Discrete linear groups generated by reflections, Izvestija AN USSR (ser. mat.) 35 (1971), 1072–1112. English translation: Math. USSR-Izvestija 5 (1971), 1083–1119.

[1972] On the group of units of some quadratic forms, Matem. Sbornik 87 (1972), 18–36.

[1975] Some arithmetical discrete groups in Lobachevskii spaces, in "Discrete subgroups of Lie groups," Oxford University Press, 1975, pp. 323–348.

[1976] The Weyl group of a graded Lie algebra, English translation: Math. USSR-Izvestija 10 (1976), 463–495.

[1985] Hyperbolic reflection groups, Uspekhi Mat. Nauk 40 (1985), 29–66.

Vinberg, E. B., Kac, V. G. [1967] Quasi-homogeneous cones, Mat. Zametki 1 (1967), 347–354. English translation: Math. Notes 1 (1967).

Virasoro, M. A. [1970] Subsidiary conditions and ghosts in dual resonance models, Phys. Rev. D1 (1970), 2933–2936.

Wakimoto, M. [1983] Two formulae for specialized characters of Kac–Moody Lie algebras, Hiroshima University, preprint.

[1984 A] Basic representations of extended affine Lie algebras, Hiroshima University, preprint.

[1984 B] Virasoro algebras and highest weight modules of affine Lie algebras, Hiroshima University, preprint.

[1986] Fock representations of affine Lie algebra $A_1^{(1)}$, Comm. Math. Phys. 104 (1986), 605–609.

Wakimoto, M., Yamada, H. [1983] Irreducible decompositions of Fock representations of the Virasoro algebra, Letters Math. Phys. 7 (1983), 513–516.

Warner, N. P. [1989] Lectures on N=2 superconformal theories and singularity theory, MIT preprint.

Weisfeiler, B. Ju. [1968] Infinite-dimensional filtered Lie algebras and their connection with graded Lie algebras, Funkt. Analis i ego Prilozh. 2 (1968), No. 1, 94–95. English translation: Funct. Anal. Appl. 2 (1968), 88–89.

Weisfeiler, B. Ju., Kac, V. G. [1971] Exponentials in Lie algebras of characteristic *p*. Izvestija AN USSR (ser. mat.) 35 (1971), 762–788. English translation: Math. USSR-Izvestija 5 (1971), 777–803.

Wilson, G. [1982] The affine Lie algebra $C_2^{(1)}$ and an equation of Hirota and Satsuma, Phys. Lett. A.

[1984] Habillage et fonctions τ, C. R. Acad. Sci. Paris 299 (1984), 587–590.

[1985 A] Infinite-dimensional Lie groups and algebraic geometry in soliton theory, Phil. Trans. R. Soc. Lond. A 315 (1985), 393–404

[1985 B] Algebraic curves and soliton equations, in Geometry to-day, Progress in Math. 60, 1985, 303–329.

Witten, E. [1984] Non-abelian bosonization in two dimensions, Comm. Math. Phys. 92 (1984), 455–472.

[1986] Physics and geometry, Proc. of ICM, Berkeley, 1986.

[1988 A] Quantum field theory, Grassmannians, and algebraic curves, Commun. Math. Phys. 113 (1988), 529–600.

[1988 B] Coadjoint orbits of the Virasoro group, Comm. Math. Phys. 114 (1988), 1–53.

[1988 C] Free fermions on an algebraic curve, in Proc. Symposia in Pure Math. 48, 1988.

[1989 A] Quantum field theory and the Jones polynomial, in Braid group, knot theory and statistical mechanics, Adv. Ser. in Math. Phys. 9, World Sci., 1989, 239–329.

[1989 B] Geometry and quantum field theory, Address at the centennial of the AMS, 1989.

[1989 C] The central charge in three dimensions, preprint.

Yamada, H. [1985] The Virasoro algebra and the KP hierarchy, in Infinite-dimensional groups with applications, MSRI series 4, Springer–Verlag, 1985, 371–380.

You, Y.-C. [1989] Polynomial solutions of the BKP hierarchy and projective representations of symmetric groups, in: Infinite-dimensional Lie algebras and groups, Adv. Ser. in Math. Phys. 7, World Sci., 1989, 449–466.

Zamolodchikov, A. B. [1984] Conformal bootstrap in two dimensions, preprint.

[1986] Infinite additional symmetries in two-dimensional conformal field theory, Theor. Mat. Phys. 65 (1986), 1205–1213.

Zel'manov, E. [1984] Lie algebras with a finite gradation, Mat. Sbornik 124 (1984), 353-392 (in Russian).

Conference Proceedings and Collections of Papers

Jimbo, M., Miwa, T., eds., Non-linear integrable systems — classical theory and quantum theory (Proceedings of RIMS symposium in Kyoto, May 1981), World Sci. 1983.

Winter, D., ed., Lie algebras and related topics (Proceedings of the conference at Rutgers University, May 1981), Lecture Notes in Math 933, Springer-Verlag, 1982.

Lepowsky, J., Mandelstam, S., Singer, I. M., eds., Vertex operators in mathematics and physics (Proceedings of the conference at the MSRI, Berkeley CA, November 1983), MSRI publ. 3, Springer-Verlag, 1985.

Kac, V. G., ed., Infinite-dimensional groups with applications (Proceedings of the conference at the MSRI, Berkeley CA, May 1984), MSRI publ. 4, Springer-Verlag, 1985.

Britten, D. J., Lemire, F. M., Moody, R. V., eds., Lie algebras and related topics (Proceedings of a conference in Windsor, July 1984), CMS conference prodeedings 5, Amer. Math. Soc., 1986.

Bardeen, W. A., White, A. R., eds., Symposium on anomalies, geometry, topology (Proceedings of the conference at Argonne National Laboratory, University of Chicago, March 1985), World Sci., 1985.

Hietarinta, J., Westerholm, J., eds., Topological and geometrical methods in field theory (Symposium in Espoo, Finland, June 1986), World Sci., 1986.

Kass, S. N., ed., Infinite-dimensional Lie algebras and their applications (Proceedings of the conference at Montreal University, May 1986), World Sci., 1988.

Yau, S. T., ed., Mathematical aspects of string theory (Proceedings of the conference at University of California, San Diego, July 1986), Adv. Ser. in Math. Phys. 1, World Sci., 1987.

Doebner, H. D., Hennig, J. D., Palev, T. D., eds., Infinite-dimensional Lie algebras and quantum field theory (Proceedings of the Varna summer school, 1987), World Sci., 1988.

Bleuler, K., Werner, M., eds., Differential geometrical methods in physics (Proceedings of a conference in Como, Italy, August 1987) Series C: 250, Kluwer Acad. Publ., 1988.

Goddard, P., Olive, D., eds., Kac-Moody and Virasoro algebras (a reprint volume for physicists) Adv. Ser. in Math. Phys. 3, World Sci., 1988.

Itzykson, C., Saleur, H., Zuber, J.-B., eds., Conformal invariance and applications to statistical mechanics, World Sci., 1988.

Gates, S. J., Preitschopf, C. R., Siegel, W., eds., Strings 88 (Proceedings of a conference at University of Maryland, May 1988), World Sci., 1989.

Kac, V. G., ed., Infinite-dimensional Lie algebras and groups (Proceedings of a conference in Lumini, Marseille, July 1988), Adv. Ser. in Math. Physics 7, World Sci., 1989.

Yang, C. N., Ge, M. L. eds., Braid group, knot theory and statistical mechanics, Adv. Ser. in Math. Physics 9, World Sci., 1989.

Gunning, R. C., Ehrenpries, L., eds., Theta functions, Bowdoin, 1987, Proceedings of Symposia in Pure Math. 49, Amer. Math. Soc., 1989.

Brezin, E., Zinn-Justin, J., eds., Fields, strings and critical phenomena, Les Houches XLIX, Elsevier Sci. Publ., 1989.

Printed in the United States
By Bookmasters